Analyzing Environmental Data

Analyzing Environmental Data

Walter W. Piegorsch
University of South Carolina
Columbia, South Carolina

A. John Bailer
Miami University
Oxford, Ohio

John Wiley & Sons, Ltd

Other Wiley Editorial Offices

John Wiley & Sons, Inc. 111 River Street, Hoboken, NJ 07030, USA

Jossey-Bass, 989 Market Street, San Francisco, CA 94103-1741, USA

Wiley-VCH Verlag GmbH, Boschstr. 12, D-69469 Weinheim, Germany

John Wiley & Sons Australia Ltd, 33 Park Road, Milton, Queensland, 4064, Australia

John Wiley & Sons (Asia) Pte Ltd, 2 Clementi Loop #02-01, Jin Xing Distripark, Singapore 129809

John Wiley & Sons Canada Ltd, 22 Worcester Road, Etobicoke, Ontario, Canada, M9W 1L1

Wiley also publishes its books in a variety of electronic formats. Some content that appears in print
may not be available in electronic books.

Library of Congress Cataloging in Publication Data

Piegorsch, Walter W.
    Analyzing environmental data / Walter W. Piegorsch, A. John Bailer.
      p.   cm.
    Includes bibliographical references and indexes.
    ISBN 0-470-84836-7 (acid-free paper)
    1. Environmental sampling. 2. Regression analysis. 3. Correlation (Statistics) I. Bailer,
A. John. II. Title.

    GE45.S25P54 2005
    363.7'0072'7—dc22

                                                                                    2004057769

British Library Cataloguing in Publication Data

A catalogue record for this book is available from the British Library

ISBN 0-470-84836-7 (HB)

Typeset in 10/12pt Times by Integra Software Services Pvt. Ltd, Pondicherry, India
Printed and bound in Great Britain by TJ International Ltd, Padstow, Cornwall
This book is printed on acid-free paper responsibly manufactured from sustainable forestry
in which at least two trees are planted for each one used for paper production.

To Karen and Jenny for their patience and
encouragement; to Sara, Jacob, Chris, and
Emily for sharing time with research;
and to our readers with thanks for their interest.

Contents

Preface

Data collected by environmental scientists cover a highly diverse set of application areas, ranging from public health studies of toxic environmental exposures to meteorological investigations of chaotic atmospheric phenomena. As a result, analysis of environmental data has itself become a highly diverse effort. In this text we provide a selection of methods for undertaking such analyses, keying on the motivating environmetric features of the observations. We emphasize primarily regression settings where some form of predictor variable is used to make inferences on an outcome variable of environmental interest. (Where possible, however, we also include allied topics, such as uncertainty/sensitivity analysis in Chapter 4 and environmental sampling in Chapter 8.)

This effort proved challenging: the broader field of *environmetrics* has experienced rapid growth in the past few decades (El-Shaarawi and Hunter, 2002; Guttorp, 2003), and it became clear to us that no single text could possibly survey all the modern, intricate statistical methods available for analyzing environmental data. We do not attempt to do so here. In fact, the environmetric paradigm under study in some chapters often leads to a basic, introductory-level presentation, while in other chapters it forces us to describe rather advanced data-analytic approaches. For the latter cases we try where possible to emphasize the simpler models and methods, and also guide readers to more advanced material via citations to the literature. Indeed, to keep the final product manageable, some advanced environmetric topics have been given little or no mention. These include issues in survival analysis (Kalbfleisch and Prentice, 2002), extreme-value analysis (Coles, 2001), experimental design (Mason *et al.*, 2003), Bayesian methods (Carlin and Louis, 2000), geographic information systems (Longley *et al.*, 2001), and applications such as ordination or other multivariate methods popular in quantitative ecology (McGarigal *et al.*, 2000). Readers interested in these important topic areas may benefit from the many books that discuss them in detail, including those cited above.

For an even larger perspective, we recommend the collection of articles given in Wiley's *Encyclopedia of Environmetrics* (El-Shaarawi and Piegorsch, 2002), a project in which we had the pleasure of participating. The *Encyclopedia* was envisioned and produced to give the sort of broad coverage to this diverse field that a single book cannot; in the text below we often refer readers to more in-depth material from the *Encyclopedia* when the limits of our own scope and intent are reached. Alongside and in addition to these references, we also give sourcebook references to the many fine texts that delve into greater detail on topics allied with our own presentation. (As one reviewer quite perceptively remarked, for essentially every topic we present there exists a recent, single sourcebook devoted to that material, although perhaps not with an environmental motivation attached. Our goal was to bring these various topics together under a single cover and with purposeful environmental focus; however, we also try where appropriate to make the reader aware of these other, dedicated products.) We hope the result will be a coherent collection of topics that we have found fundamental for the analysis of environmental data.

Individuals who will benefit most from our presentation are students and researchers who have a sound grounding in statistical methods; we recommend a minimum of two semesters of graduate study in statistical methodology. Even with this background, however, many portions of the book will require more advanced quantitative skills; typically a familiarity with integral and differential calculus, vector and matrix notation/manipulation, and often also knowledge of a few advanced concepts in statistical theory such as probability models and likelihood analysis. We give brief reminders throughout the text on some of these topics; for readers who require a 'refresher' in the more advanced statistical concepts, however, we recommend a detailed study of the review of probability and statistical inference in Appendix A and the references therein. We have also tried to separate and sequester the calculus/linear algebra-based material to the best extent possible, so that adept instructors who wish to use the text for students without a background in calculus and linear algebra may do so with only marginal additional effort.

An integral component of our presentation is appeal to computer implementation for the more intricate analyses. A wealth of computer packages and programming languages are available for this purpose and we give in selected instances Internet URLs that guide users to potentially useful computer applications. (All URLs listed herein are current as of the time of this writing.) For 'hands-on' use, we highlight the SAS® system (SAS Institute Inc., 2000). SAS's ubiquity and extent make it a natural choice, and we assume a majority of readers will already be familiar with at least basic SAS mechanics or can acquire such skills separately. (Users versed in the S-Plus® computer package will find the text by Millard and Neerchal, 2001, to be of complementary use.) Figures containing sample SAS computer code and output are displayed throughout the text. Although these are not intended to be the most efficient way to program the desired operations, they will help illustrate use of the system and (perhaps more importantly) interpretation of the outputs. Outputs from SAS procedures (versions 6.12 and 8.2) are copyright ©2002–2003, SAS Institute Inc., Cary, NC, USA. All Rights Reserved. Reproduced with permission of SAS Institute Inc., Cary, NC. We also appreciate the kind permission of Chapman & Hall/CRC Press to adapt selected material from our earlier text on *Statistics for Environmental Biology and Toxicology* (Piegorsch and Bailer, 1997).

All examples end with the symbol ✿. Large data sets used in any examples and exercises in Chapters 5 and 6 have been archived online at the publisher's website, http://www.wiley.com/go/environmental. In the text, these are presented in reduced tabular form to show only a few representative observations. We indicate this wherever it occurs.

By way of acknowledgments, our warmest gratitude goes to our colleague Don Edwards, who reviewed a number of chapters for us and also gave extensive input into the material in Chapters 5 and 6. Extremely helpful suggestions and input came also from Timothy G. Gregoire, Andrew B. Lawson, Mary C. Christman, James Oris, R. Webster West, Philip M. Dixon, Dwayne E. Porter, Oliver Schabenberger, Jay M. Ver Hoef, Rebecca R. Sharitz, John M. Grego, Kerrie P. Nelson, Maureen O. Petkewich, and three anonymous reviewers. We are also indebted to the Wiley editorial group headed by Siân Jones, along with her colleague Helen Ramsey, for their professionalism, support, and encouragement throughout the preparation of the manuscript. Of course, despite the fine efforts of all these individuals, some errors may have slipped into the text, and we recognize these are wholly our own responsibility. We would appreciate hearing from readers who identify any inconsistencies that they may come across. Finally, we hope this book will help our readers gain insights into and develop strategies for analyzing environmental data.

WALTER W. PIEGORSCH AND A. JOHN BAILER

Columbia, SC and Oxford, OH

May 2004

1

Linear regression

Considerable effort in the environmental sciences is directed at predicting an environmental or ecological response from a collection of other variables. That is, an
observed *response variable*, Y, is recorded alongside one or more *predictor variables*,
and these latter quantities are used to describe the deterministic aspects of Y. If we
denote the predictor variables as $x_1, x_2, \ldots, x_p$, it is natural to model the deterministic aspects of the response via some function, say, $g(x_1, x_2, \ldots, x_p; \boldsymbol{\beta})$, where
$\boldsymbol{\beta} = [\beta_0 \; \beta_1 \ldots \beta_p]^T$ is a column vector of $p + 1$ unknown parameters. (A *vector* is an
array of numbers arranged as a row or column. The superscript ^{T} indicates transpos-
ition of the vector, so that, for example, $[a_1 a_2]^T = \begin{bmatrix} a_1 \\ a_2 \end{bmatrix}$. More generally, a *matrix* is
an array of numbers arranged in a square or rectangular fashion; one can view a
matrix as a collection of vectors, all of equal length. Background material on
matrices and vectors appears in Appendix A. For a more general introduction to
the use of matrix algebra in regression, see Neter *et al.*, 1996, Ch. 5.) We use the
function $g(\cdot)$ to describe how Y changes as a function of the x_js.

As part of the model, we often include an additive error term to account for any
random, or *stochastic*, aspects of the response. Formally, then, an observation Y_i is
assumed to take the form

$$Y_i = g(x_{i1}, x_{i2}, \ldots, x_{ip}; \boldsymbol{\beta}) + \varepsilon_i, \tag{1.1}$$

$i = 1, \ldots, n$, where the additive error terms ε_i are assigned some form of probability
distribution and the *sample size* n is the number of recorded observations. Unless
otherwise specified, we assume the Y_is constitute a random sample of statistically
independent observations. If Y represents a continuous measurement, it is common
to take $\varepsilon_i \sim$ i.i.d. $N(0, \sigma^2)$, 'i.i.d.' being a shorthand notation for *i*ndependent and
*i*dentically *d*istributed (see Appendix A). Coupled with the additivity assumption in
(1.1), this is known as a *regression* of Y on the x_js.

Analyzing Environmental Data   W. W. Piegorsch and A. J. Bailer
© 2005 John Wiley & Sons, Ltd   ISBN: 0-470-84836-7 (HB)

Note also that we require the x_j predictor variables to be fixed values to which no stochastic variability may be ascribed (or, at least, that the analysis be conditioned on the observed pattern of the predictor variables).

We will devote a large portion of this text to environmetric analysis for a variety of regression problems. In this chapter, we give a short review of some elementary regression models, and then move on to a selection of more complex forms. We start with the most basic case: simple linear regression.

1.1 Simple linear regression

The simple linear case involves only one predictor variable ($p = 1$), and sets $g(x_{i1}; \boldsymbol{\beta})$ equal to a linear function of x_{i1}. For simplicity, when $p = 1$ we write x_{i1} as x_i. Equation (1.1) becomes

$$Y_i = \beta_0 + \beta_1 x_i + \varepsilon_i,$$

$i = 1, \ldots, n$, and we call $\beta_0 + \beta_1 x_i$ the *linear predictor*. The linear predictor is the deterministic component of the regression model. Since this also models the population mean of Y_i, we often write $\mu(x_i) = \beta_0 + \beta_1 x_i$, and refer to $\mu(x)$ as the *mean response function*.

The simple linear regression model can also be expressed as the matrix equation $\mathbf{Y} = \mathbf{X}\boldsymbol{\beta} + \boldsymbol{\varepsilon}$ where $\mathbf{Y} = [Y_1 \ldots Y_n]^{\mathrm{T}}$, $\boldsymbol{\varepsilon} = [\varepsilon_0 \ldots \varepsilon_n]^{\mathrm{T}}$ and $\mathbf{X}$ is a matrix whose columns are the two vectors $\mathbf{J} = [1 \ldots 1]^{\mathrm{T}}$ – i.e., a column vector of ones – and $[x_1 \ldots x_n]^{\mathrm{T}}$.

As a first step in any regression analysis, we recommend that a graphical display of the data pairs (x_i, Y_i) be produced. Plotted, this is called a *scatterplot*; see Fig. 1.1 in Example 1.1, below. The scatterplot is used to visualize the data and begin the process of assessing the model fit: straight-line relationships suggest a simple linear model, while curvilinear relationships suggest a more complex model. We discuss nonlinear regression modeling in Chapter 2.

Under the common assumptions that $\mathrm{E}[\varepsilon_i] = 0$ and $\mathrm{Var}[\varepsilon_i] = \sigma^2$ for all $i = 1, \ldots, n$, the model parameters in $\boldsymbol{\beta} = [\beta_0\ \beta_1]^{\mathrm{T}}$ have interpretations as the Y-intercept (β_0) and slope (β_1) of $\mu(x_i)$. In particular, for any unit increase in x_i, $\mu(x_i)$ increases by β_1 units. To estimate the unknown parameters we appeal to the least squares (LS) method, where the sum of squared errors $\sum_{i=1}^{n}\{Y_i - \mu(x_i)\}^2$ is minimized (LS estimation is reviewed in §A.4.1). The LS estimators of β_0 and β_1 here are

$$b_0 = \overline{Y} - b_1 \overline{x}$$

and

$$b_1 = \frac{\sum_{i=1}^{n}(x_i - \overline{x})(Y_i - \overline{Y})}{\sum_{i=1}^{n}(x_i - \overline{x})^2} = \frac{\sum_{i=1}^{n} x_i Y_i - \frac{1}{n}\sum_{i=1}^{n} x_i \sum_{i=1}^{n} Y_i}{\sum_{i=1}^{n} x_i^2 - \frac{1}{n}\left(\sum_{i=1}^{n} x_i\right)^2}, \tag{1.2}$$

where $\overline{Y} = \sum_{i=1}^{n} Y_i/n$ and $\overline{x} = \sum_{i=1}^{n} x_i/n$. The algebra here can be simplified using matrix notation: if $\mathbf{b} = [b_0 \ b_1]^T$ is the vector of LS estimators, then $\mathbf{b} = (\mathbf{X}^T\mathbf{X})^{-1}\mathbf{X}^T\mathbf{Y}$, $(\mathbf{X}^T\mathbf{X})^{-1}$ being the *inverse* of the matrix $\mathbf{X}^T\mathbf{X}$ (see §A.4.3).

If we further assume that $\varepsilon_i \sim$ i.i.d. $N(0, \sigma^2)$, then the LS estimates will correspond to maximum likelihood (ML) estimates for β_0 and β_1. (ML estimation is reviewed in §A.4.3.) The LS/ML estimate of the mean response, $\mu(x) = \beta_0 + \beta_1 x$, for any x, is simply $\hat{\mu}(x) = b_0 + b_1 x$.

We should warn that calculation of b_1 can be adversely affected by a number of factors. For example, if the x_is are spaced unevenly, highly separated values of x_i can exert strong *leverage* on b_1 by pulling the estimated regression line too far up or down. (See the web applet at http://www.stat.sc.edu/~west/javahtml/Regression. html for a visual demonstration. Also see the discussion on regression diagnostics, below.) To avoid this, the predictor variables should be spaced as evenly as possible, or some transformation of the x_is should be applied before performing the regression calculations. The natural logarithm is a typical choice here, since it tends to compress very disparate values. If when applying the logarithm, one of the x_i values is zero, say $x_1 = 0$, one can average the other log-transformed x_is to approximate an equally spaced value associated with $x_1 = 0$. This is *consecutive-dose average spacing* (Margolin *et al.*, 1986): denote the transformed predictor by $u_i = \log(x_i), i = 2, \ldots, n$. Then at $x_1 = 0$, use

$$u_1 = u_2 - \frac{u_n - u_2}{n - 1}. \tag{1.3}$$

A useful tabular device for collecting important statistical information from a linear regression analysis is known as the *analysis of variance (ANOVA) table*. The table lays out *sums of squares* that measure variation in the data attributable to various components of the model. It also gives the *degrees of freedom* (df) for each component. The df represent the amount of information in the data available to estimate that particular source of variation. The ratio of a sum of squares to its corresponding df is called a *mean square*.

For example, to identify the amount of variability explained by the linear regression of Y on x, the sum of squares for regression is SSR $= \sum_{i=1}^{n}(\hat{Y}_i - \overline{Y})^2$, where $\hat{Y}_i = b_0 + b_1 x_i$ is the ith *predicted value* (also called a *fitted value*). SSR has degrees of freedom equal to the number of regression parameters estimated minus one; here, $df_r = 1$. Thus the mean square for regression when $p = 1$ is MSR $=$ SSR/1.

We can also estimate the unknown variance parameter, σ^2, via ANOVA computations. Find the sum of squared errors SSE $= \sum_{i=1}^{n}(Y_i - \hat{Y}_i)^2$ and divide this by the error df (the number of observations minus the number of regression parameters estimated), $df_e = n - 2$. The resulting *mean squared error* is

$$\text{MSE} = \frac{\sum_{i=1}^{n}(Y_i - \hat{Y}_i)^2}{n - 2},$$

and this is an unbiased estimator of σ^2. We often call $\sqrt{\text{MSE}}$ the *root mean squared error*.

We do not go into further detail here on the construction of sums of squares and ANOVA tables, although we will mention other aspects of linear modeling and ANOVA below. Readers unfamiliar with ANOVA computations can find useful expositions in texts on linear regression analysis, such as Neter *et al.* (1996) or Christensen (1996).

We use the MSE to calculate the *standard errors* of the LS/ML estimators. (A standard error is the square root or estimated square root of an estimator's variance; see §A.4.3.) Here, these are

$$se[b_0] = \sqrt{\mathrm{MSE}\left\{\frac{1}{n} + \frac{\bar{x}^2}{\sum_{i=1}^{n}(x_i - \bar{x})^2}\right\}}$$

and

$$se[b_1] = \sqrt{\frac{\mathrm{MSE}}{\sum_{i=1}^{n}(x_i - \bar{x})^2}}. \tag{1.4}$$

Standard errors (and variances) quantify the variability of the point estimator, helping to gauge how meaningful the magnitude of a given estimate is. They also give insight into the impact of different experimental designs on estimating regression coefficients. For example, notice that $se[b_0]$ is smallest for x_is chosen so that $\bar{x} = 0$, while $se[b_1]$ is minimized when $\sum_{i=1}^{n}(x_i - \bar{x})^2$ is taken to be as large as possible.

Similarly, the standard error of $\hat{\mu}(x)$ is

$$se[\hat{\mu}(x)] = \sqrt{\mathrm{MSE}\left\{\frac{1}{n} + \frac{(x - \bar{x})^2}{\sum_{i=1}^{n}(x_i - \bar{x})^2}\right\}}.$$

Notice that, as with $\hat{\mu}(x)$, $se[\hat{\mu}(x)]$ varies with x. It attains its minimum at $x = \bar{x}$ and then increases as x departs from $\bar{x}$ in either direction. One may say, therefore, that precision in $\hat{\mu}(x)$ is greatest near the center of the predictor range – i.e., at $\bar{x}$ – and diminishes as x moves away from it. Indeed, if one drives x too far away from the predictor range, $se[\hat{\mu}(x)]$ can grow so large as to make $\hat{\mu}(x)$ essentially useless. This illustrates the oft-cited concern that *extrapolation* away from the range of the data leads to imprecise, inaccurate, and in some cases even senseless statistical predictions.

The standard errors are used in constructing statistical inferences on the β_js or on $\mu(x)$. For example, notice that if $\beta_1 = 0$ then the predictor variable has no effect on the response and the simple linear model collapses to $Y_i = \beta_0 + \varepsilon_i$, a 'constant + error' model for Y. To assess this, assume that the $N(0, \sigma^2)$ assumption on the ε_is is valid. Then, a $1 - \alpha$ *confidence interval* for β_1 is

$$b_1 \pm t_{\alpha/2}(n - 2)se[b_1].$$

(The theory of confidence intervals is reviewed in §A.5.1.) An alternative inference is available by conducting a *hypothesis test* of the null hypothesis $H_0: \beta_1 = 0$ vs. the alternative hypothesis $H_a: \beta_1 \neq 0$. (The theory of hypothesis tests is reviewed in §A.5.3.) Here, we find the test statistic

$$|t_{calc}| = \frac{|b_1|}{se[b_1]}$$

based on Student's t-distribution (§A.2.11), and reject H_0 when $|t_{calc}| \geq t_{\alpha/2}(n-2)$. (We use the subscript 'calc' to indicate a statistic that is wholly calculable from the data.) Equivalently, we can reject H_0 when the corresponding *P-value*, here

$$P = 2P\left[t(n-2) \geq \frac{|b_1|}{se(b_1)}\right],$$

drops below the preset *significance level* α (see §A.5.3).

For testing against a one-sided alternative such as $H_a: \beta_1 > 0$, we reject H_0 when $t_{calc} = b_1/se[b_1] \geq t_\alpha(n-2)$. The *P*-value is then $P[t(n-2) \geq b_1/se(b_1)]$. Similar constructions are available for β_0; for example, a $1-\alpha$ confidence interval is $b_0 \pm t_{\alpha/2}(n-2)se[b_0]$.

All these operations can be conducted by computer, and indeed, many statistical computing packages perform simple linear regression. Herein, we highlight the SAS® system (SAS Institute Inc., 2000), which provides LS/ML estimates $\mathbf{b} = [b_0 \ b_1]^T$, their standard errors $se[b_j]$, an ANOVA table that includes an unbiased estimator of σ^2 via the MSE, and other summary statistics, via its PROC GLM or PROC REG procedures.

Example 1.1 (Motor vehicle CO₂) To illustrate use of the simple linear regression model, consider the following example. In the United Kingdom (and in most other industrialized nations) it has been noted that as motor vehicle use increases, so do emissions of various byproducts of hydrocarbon combustion. Public awareness of this potential polluting effect has bolstered industry aspirations to 'uncouple' detrimental emissions from vehicle use. In many cases, emission controls and other efforts have reduced the levels of hazardous pollutants such as small particulate matter (PM) and nitrogen oxides. One crucial counter-example to this trend, however, is the ongoing increases in the greenhouse gas carbon dioxide (CO_2). For example, Redfern *et al.* (2003) discuss data on $x = $ UK motor vehicle use (in kilometers per year) vs. $Y = CO_2$ emissions (as a relative index; 1970 = 100). Table 1.1 presents the data.

A plot of the data in Table 1.1 shows a clear, increasing, linear trend (Fig. 1.1). Assuming that the simple linear model with normal errors is appropriate for these data, we find the LS/ML estimates to be $b_0 = 28.3603$ and $b_1 = 0.7442$. The corresponding standard errors are $se[b_0] = 2.1349$ and $se[b_1] = 0.0127$. Since $n = 28$, a 95% confidence interval for β_1 is $0.7742 \pm t_{0.025}(26) \times 0.0127 = 0.7742 \pm 2.056 \times 0.0127 = 0.7742 \pm 0.0261$. (We find $t_{0.025}(26)$ from Table B.2 or via the SAS function tinv; see Fig. A.4.) Based on this 95% interval, the CO_2 index increases approximately 0.75 to 0.80 units (relative to 1970 levels) with each additional kilometer.

Table 1.1   Yearly CO_2 emissions (rel. index; 1970 = 100) vs. motor vehicle use (rel. km/yr; 1970 = 100) in the United Kingdom, 1971–1998

Year	1971	1972	1973	1974	1975	1976	1977
x = vehicle use	105.742	110.995	116.742	114.592	115.605	121.467	123.123
$Y = CO_2$	104.619	109.785	117.197	114.404	111.994	116.898	119.915
Year	1978	1979	1980	1981	1982	1983	1984
x = vehicle use	127.953	127.648	135.660	138.139	141.911	143.707	151.205
$Y = CO_2$	126.070	128.759	130.196	126.409	130.136	134.212	140.721
Year	1985	1986	1987	1988	1989	1990	1991
x = vehicle use	154.487	162.285	174.837	187.403	202.985	204.959	205.325
$Y = CO_2$	143.462	153.074	159.999	170.312	177.810	182.686	181.348
Year	1992	1993	1994	1995	1996	1997	1998
x = vehicle use	205.598	205.641	210.826	214.947	220.753	225.742	229.027
$Y = CO_2$	183.757	185.869	186.872	185.100	192.249	194.667	193.438

Source: Redfern *et al.* (2003).

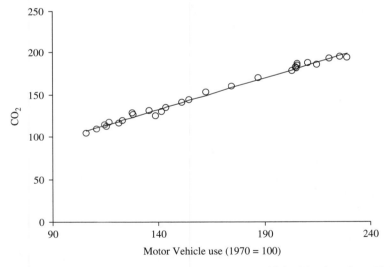

Figure 1.1   Scatterplot and estimated LS line for motor vehicle CO_2 data from Table 1.1

Alternatively, we can test the significance of the slope with these data. Specifically, since one would expect *a priori* that increased motor vehicle use would increase CO_2 emissions, the hypotheses $H_0: \beta_1 = 0$ vs. $H_a: \beta_1 > 0$ are a natural choice. Suppose we set our significance level to $\alpha = 0.01$. For these data, the test statistic is $t_{calc} = b_1/se[b_1] = 0.7742/0.0127 = 60.96$, with corresponding P-value $P[t(26) \geq 60.96] < 0.0001$. This is well below α, hence we conclude that a significant, increasing effect exists on CO_2 emissions associated with the observed pattern of motor vehicle use in the UK between 1971 and 1998.

The sample size in Example 1.1, $n = 28$, is not atypical for a simple linear regression data set, but of course analysts can encounter much larger sample sizes in environmental practice. We will study selected examples of this in the chapters on nonlinear regression (Chapter 2), temporal data (Chapter 5), and spatially correlated data (Chapter 6), below.

Once a model has been fitted to data, it is important to assess the quality of the fit in order to gauge the validity of the consequent inferences and predictions. In practice, any statistical analysis of environmental data should include a critical examination of the assumptions made about the statistical model, in order to identify if any unsupported assumptions are being made and to alert the user to possible unanticipated or undesired consequences. At the simplest level, a numerical summary for the quality of a regression fit is the *coefficient of determination* $\{\sum_{i=1}^{n}(x_i - \bar{x})(Y_i - \bar{Y})\}^2 / \{\sum_{i=1}^{n}(x_i - \bar{x})^2 \sum_{i=1}^{n}(Y_i - \bar{Y})^2\}$, denoted as R^2. This may also be computed from the ANOVA table as $R^2 = \mathrm{SSR}/\{\mathrm{SSR} + \mathrm{SSE}\}$. Under a linear model, R^2 has interpretation as the proportion of variation in Y_i that can be attributed to the variation in x_i. If the predictor variable explains Y precisely (i.e., the x_i, Y_i pairs all coincide on a straight line), R^2 attains its maximum value of 1.0. Alternatively, if there is *no* linear relationship between x_i and Y_i (so $\beta_1 = 0$), $R^2 = 0.0$. As such, higher values of R^2 indicate higher-quality explanatory value in x_i.

More intricate *regression diagnostics* can include a broad variety of procedures for assessing model fit (Davison and Tsai, 1992; Neter *et al.*, 1996, Ch. 3). Most basic among these is study of the *residuals* $r_i = Y_i - \hat{Y}_i$. Almost every analysis of a regression relationship should include a graph of the residuals, r_i, against the predicted values, $\hat{Y}_i$ (or, if $p = 1$, against x_i). Such a *residual plot* can provide information on a number of features. For instance, if there is an underlying curvilinear trend in the data that was not picked up by the original scatterplot, the residual plot may highlight the curvilinear aspects not explained by the simple linear terms. Or, if the assumption of variance homogeneity is inappropriate – i.e., if $\mathrm{Var}[\varepsilon_i]$ is not constant over changing x_i – the residual plot may show a fan-shaped pattern of increasing or decreasing residuals (or both) as $\hat{Y}_i$ increases. Figure 1.2 illustrates both these sorts of patterns. Notice in Fig. 1.2(b) that variability increases with increasing mean response; this sort of pattern is not uncommon with environmental data.

If the residual plot shows a generally uniform or random pattern, then evidence exists for a reasonable model fit.

Example 1.2 (Motor vehicle CO_2, cont'd) Returning to the data on motor vehicle use in the UK, we find $\mathrm{SSR} = 26\,045.2953$ and $\mathrm{SSE} = 196.0457$. This gives $R^2 = 0.9925$, from which it appears that variation in CO_2 emissions is strongly explained by variation in motor vehicle use.

Figure 1.3 shows the residual plot from the simple linear model fit. The residual points appear randomly dispersed, with no obvious structure or pattern. This suggests that the variability in CO_2 levels about the regression line is constant and so the homogeneous variance assumption is supported. One could also graph a histogram or normal probability plot of the residuals to assess the adequacy of the normality assumption. If the histogram appears roughly bell-shaped, or if the normal plot produces a roughly straight line, then the assumption of normal errors may be reasonable. For the residuals in Fig. 1.3, a normal probability plot constructed using **PROC UNIVARIATE** in SAS (via its `plot` option; output suppressed) does plot as

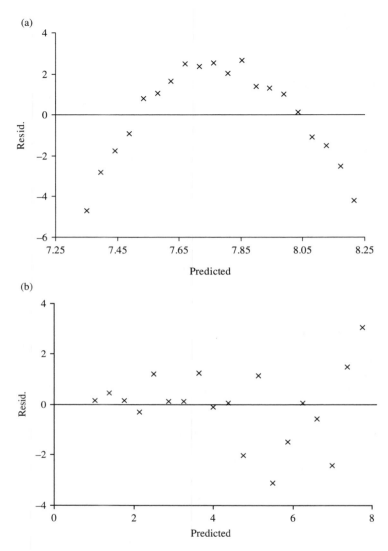

Figure 1.2   Typical residual plots in the presence of model misspecification. (a) Curvilinear residual trend indicates curvilinearity not fit by the model. (b) Widening residual spread indicates possible variance heterogeneity. Horizontal reference lines indicate residual $= 0$

roughly linear. Or one can call for normal probability plots directly in PROC REG, using the statement

```
plot nqq.*r. npp.*r.;
```

The plot statement in PROC REG can also be used to generate a residual plot, via

```
plot r.*p.;
```

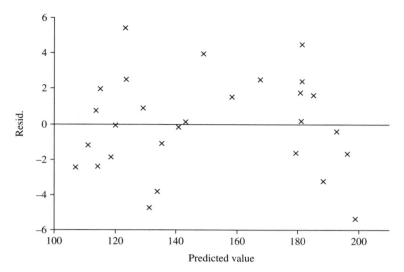

Figure 1.3 Residual plot for motor vehicle CO_2 data from Table 1.1. Horizontal bar indicates residual $= 0$

or an overlay of the data and the predicted regression line, via

```
plot Y*x p.*x/overlay;
```

When the residual plot identifies a departure from variance homogeneity, inferences on the unknown parameters based on the simple linear fit can be incorrect, and some adjustment is required. If the heterogeneous variation can be modeled or otherwise quantified, it is common to weight each observation in inverse proportion to its variance and apply weighted least squares (WLS; see §A.4.1). For example, suppose it is known or anticipated that the variance changes as a function of x_i, say $\mathrm{Var}[Y_i] \propto h(x_i)$. Then, a common weighting scheme employs $w_i = 1/h(x_i)$.

For weights given as w_i, $i = 1, \ldots, n$, the WLS estimators become

$$\tilde{b}_0 = \left(\sum_{i=1}^{n} w_i \right)^{-1} \left(\sum_{i=1}^{n} w_i Y_i - b_1 \sum_{i=1}^{n} w_i x_i \right) \tag{1.5}$$

and

$$\tilde{b}_1 = \frac{\sum_{i=1}^{n} w_i x_i Y_i - \left(\sum_{i=1}^{n} w_i \right)^{-1} \left(\sum_{i=1}^{n} w_i x_i \sum_{i=1}^{n} w_i Y_i \right)}{\sum_{i=1}^{n} w_i x_i^2 - \left(\sum_{i=1}^{n} w_i \right)^{-1} \left(\sum_{i=1}^{n} w_i x_i \right)^2}. \tag{1.6}$$

The standard errors require similar modification; for example,

$$se[\tilde{b}_1] = \frac{\sqrt{\widetilde{MSE}}}{\sqrt{\sum_{i=1}^{n} w_i x_i^2 - \left(\sum_{i=1}^{n} w_i\right)^{-1} \left(\sum_{i=1}^{n} w_i x_i\right)^2}},$$

where $\widetilde{MSE}$ is the weighted mean square $\sum_{i=1}^{n} w_i(Y_i - \tilde{b}_0 - \tilde{b}_1 x_i)^2/(n-2)$. Inferences on β_1 then mimic those described above for the simple linear case. In SAS, both PROC GLM and PROC REG can incorporate these (or any other) weighting schemes, using the `weight` statement. Neter *et al.* (1996, §10.1) give further details on the use of WLS methods.

If appropriate weights cannot be identified, it is often possible to stabilize the variances by transforming the original observations. A common transformation in many environmental applications is the (natural) logarithm: $V_i = \log(Y_i)$. This is part of a larger class of transformations, known as the Box–Cox power transformations (Box and Cox, 1964). The general form is $V_i = (Y_i^\lambda - 1)/\lambda$, for some specified transformation parameter λ. The natural logarithm is the limiting case at $\lambda = 0$. Other popular transformations include the square root ($\lambda = 1/2$), the quadratic ($\lambda = 2$), and the reciprocal ($\lambda = -1$). One can also estimate λ from the data, although this can lead to loss of independence among the V_is. Users should proceed with caution when estimating a power transformation parameter; see Carroll and Ruppert (1988) for more on this and other issues regarding data transformation in regression. Another useful transformation, often employed with percentage data, is the *logit transform*: if Y_i is a percentage between 0 and 100, take $V_i = \log\{Y_i/(100 - Y_i)\}$. We employ this in Example 1.5, below.

Many other procedures are available for diagnosing and assessing model fit, correcting for various model perturbations and inadequacies, and analyzing linear relationships. A full description of all these methods for the simple linear model is beyond the scope of this chapter, however. Details can be found in the targeted textbook by Belsley *et al.* (1980), or in general texts on statistics such as Samuels and Witmer (2003, Ch. 12) and Neter *et al.* (1996, Chs. 1–5).

1.2 Multiple linear regression

The simplest statistical model for the case of $p > 1$ predictor variables in (1.1) employs a linear term for each predictor: set $g(x_{i1}, x_{i2}, \ldots, x_{ip}; \beta) = \beta_0 + \beta_1 x_{i1} + \cdots + \beta_p x_{ip}$. This is a *multiple linear regression* model. The parameter β_j may be interpreted as the change in $E[Y_i]$ that occurs for a unit increase in x_{ij} – the 'slope' of the jth predictor – assuming all the other x-variables are held fixed. (When it is not possible to vary one predictor while holding all others constant, then of course this interpretation may not make sense. An example of such occurs with polynomial regression models; see §1.5.) We require $n > p + 1$.

Assuming, as above, that the errors satisfy $E[\varepsilon_i] = 0$ and $Var[\varepsilon_i] = \sigma^2$ for all $i = 1, \ldots, n$, the LS estimators for $\beta = [\beta_0\ \beta_1 \ldots \beta_p]^T$ can be derived using multivariable calculus. When the additional assumption is made that $\varepsilon_i \sim$ i.i.d. $N(0, \sigma^2)$, these LS estimates will correspond to ML estimates.

Unfortunately, the LS/ML estimators for $\boldsymbol{\beta}$ are not easily written in closed form. The effort can be accomplished using vector and matrix notation in similar fashion to that mentioned in §1.1, although actual calculation of the estimates is most efficiently performed by computer. Almost any statistical computing package can fit a multiple linear regression via LS or WLS methods; in Example 1.3, below, we illustrate use of SAS.

Similar to the simple linear case, we can test whether any particular predictor variable, x_{ij}, is important in modeling $E[Y_i]$ via appeal to a t-test: find $t_{\text{calc}} = b_j/se[b_j]$ and reject $H_0: \beta_j = 0$ in favor of $H_a: \beta_j \neq 0$ when $|t_{\text{calc}}| = |b_j|/se[b_j] \geq t_{\alpha/2}(n - p - 1)$. Note that this tests the significance of the jth predictor variable given that all the other predictor variables are present in the model. In this sense, we call it an *adjusted test* or a *partial test* of significance. Confidence intervals are similar; for example, a pointwise $1 - \alpha$ confidence interval for β_j is $b_j \pm t_{\alpha/2}(n - p - 1)se[b_j]$; $j = 1, \ldots, p$. Notice the change in df$_e$ from the simple linear case where $p = 1$: estimation of each additional β_j results in a loss of 1 additional df for error, so we have gone from df$_e = n - 2$ to df$_e = n - (p + 1)$.

We can also make statements on subsets or groupings of the β-parameters. For example, consider a test of the null hypothesis that a group of $k > 1$ of the β_js is equal to zero, say, $H_0: \beta_{j+1} = \cdots = \beta_{j+k} = 0$. Rejection of H_0 suggests that the corresponding group of k predictor variables has a significant impact on the regression relationship. A general approach for such a test involves construction of *discrepancy measures* that quantify the fit of the general (or *full*) model with all $p + 1$ of the β-parameters, and the *reduced model* with $p - k + 1$ (non-zero) β-parameters. For the multiple regression model with normally distributed errors, a useful discrepancy measure is the sum of squared errors $\text{SSE} = \sum_{i=1}^{n} (Y_i - \hat{Y}_i)^2$, where $\hat{Y}_i = b_0 + b_1 x_{i1} + \cdots + b_p x_{ip}$ is the ith predicted value under the full model. For clarity, we augment the SSE notation by indicating if it is calculated under the full model (FM) or under the reduced model (RM): SSE(FM) or SSE(RM). The SSEs are used to quantify the relative quality of each model's fit to the data: if H_0 is false, we expect SSE(RM) to be larger than SSE(FM), since the model under which it is fitted fails to include important predictor variables. Corresponding to these terms, we also write the degrees of freedom associated with each error terms as df$_e$(FM) and df$_e$(RM), respectively. The difference between the two is $\Delta_e = \text{df}_e(\text{RM}) - \text{df}_e(\text{FM})$. Here, df$_e$(FM) $= n - p - 1$, while df$_e$(RM) $= n + k - p - 1$, so that $\Delta_e = k$ is the number of parameters constrained by the null hypothesis.

To use this discrepancy approach for testing H_0, calculate the test statistic

$$F_{\text{calc}} = \frac{\{\text{SSE(RM)} - \text{SSE(FM)}\}/\Delta_e}{\text{SSE(FM)}/\text{df}_e(\text{FM})}, \tag{1.7}$$

which under H_0 is distributed as per an F-distribution with Δ_e and df$_e$(FM) degrees of freedom (§A.2.11). We denote this as $F_{\text{calc}} \sim F[\Delta_e, \text{df}_e(\text{FM})]$. Reject H_0 in favor of an alternative that allows at least one of the β_js in H_0 to be non-zero when F_{calc} exceeds the appropriate upper-α F-critical point, $F_\alpha(\Delta_e, \text{df}_e[\text{FM}])$. For the multiple regression setting, this is $F_{\text{calc}} \geq F_\alpha(k, n - p - 1)$. The P-value is $P = P[F(k, n - p - 1) \geq F_{\text{calc}}]$. This testing strategy corresponds to a form of generalized likelihood ratio test (§A.5).

In many cases, the various measures in (1.7) can be read directly from an ANOVA table for the full model (Neter *et al.*, 1996, Ch. 16). For example, if SSR(FM) is the full model's sum of squares for regression and the reduced model contains only the intercept β_0 (so $k = p$), $F_{calc} = \{SSR(FM)/p\}/MSE$. Also, an extension of the coefficient of determination from the simple linear setting is the *coefficient of multiple determination*: $R^2 = SSR(FM)/\{SSR(FM) + SSE(FM)\}$. As in the simple linear case, R^2 measures the proportion of variation in Y_i that can be accounted for by variation in the collection of x_{ij}s.

For this approach to be valid, the parameters represented under the RM must be a true subset of those under the FM. We say then that the models are *nested*. If the relationship between the RM and FM does not satisfy a nested hierarchy, F_{calc} under H_0 may not follow (or even approximate) an F-distribution. The family of models are then said to be *separate* (Cox, 1961, 1962); inferences for testing separate families are still an area of developing environmetric research (Hinde, 1992; Schork, 1993).

Example 1.3 (Soil pH) Edenharder *et al.* (2000) report on soil acidity in west-central Germany, as a function of various soil composition measures. For $Y = $ soil pH, three predictor variables (all percentages) were employed: $x_{i1} = $ soil texture (as clay), $x_{i2} = $ organic matter, and $x_{i3} = $ carbonate composition ($CaCO_3$ by weight). The $n = 17$ data points are given in Table 1.2.

Table 1.2 Soil pH vs. soil composition variables in west-central Germany

$x_1 = \%$ Clay	$x_2 = \%$ Organics	$x_3 = $ Carbonate	$Y = $ pH
51.1	4.3	6.1	7.1
22.0	2.6	0.0	5.4
17.0	3.0	2.0	7.0
16.8	3.0	0.0	6.1
5.5	4.0	0.0	3.7
21.2	3.3	0.1	7.0
14.1	3.7	16.8	7.4
16.6	0.7	17.3	7.4
35.9	3.7	15.6	7.3
29.9	3.3	11.9	7.5
2.4	3.1	2.8	7.4
1.6	2.8	6.2	7.4
17.0	1.8	0.3	7.5
32.6	2.3	9.1	7.3
10.5	4.0	0.0	4.0
33.0	5.1	26.0	7.1
26.0	1.9	0.0	5.6

Source: Edenharder *et al.* (2000).

To fit the multiple linear regression model $Y_i = \beta_0 + \beta_1 x_{i1} + \beta_2 x_{i2} + \beta_3 x_{i3} + \varepsilon_i$ to the data in Table 1.2, we employ the SAS procedure PROC GLM, although one could also apply PROC REG profitably here, especially with its regression diagnostic plots as noted in Example 1.2. Sample SAS code is given in Fig. 1.4. (In some installations of SAS, the statement `run;` may be required at the end of the input code; however, we do not use it in the examples herein.)

The SAS output (edited for presentation purposes) is given in Fig. 1.5. (We will not go into particulars of the SAS code and output here, since we expect readers to be familiar with this level of detail for multiple linear regression. Where appropriate in future chapters, however, we will supply more background on selected SAS code and/or output.)

```
*  SAS code to fit mult. lin. regr.;
   data soil;
   input Yph x1clay x2org x3carb @@;
   datalines;
         7.1    51.1   4.3    6.1      5.4    22.0   2.6    0.0
         7.0    17.0   3.0    2.0      6.1    16.8   3.0    0.0
         3.7     5.5   4.0    0.0      7.0    21.2   3.3    0.1
         7.4    14.1   3.7   16.8      7.4    16.6   0.7   17.3
         7.3    35.9   3.7   15.6      7.5    29.9   3.3   11.9
         7.4     2.4   3.1    2.8      7.4     1.6   2.8    6.2
         7.5    17.0   1.8    0.3      7.3    32.6   2.3    9.1
         4.0    10.5   4.0    0.0      7.1    33.0   5.1   26.0
         5.6    26.0   1.9    0.0

proc glm;
   model Yph = x1clay x2org x3carb;
```

Figure 1.4   Sample SAS program to fit multiple linear regression model to soil pH data

```
                         The SAS System
                 General Linear Models Procedure

     Dependent Variable: Yph
                             Sum of
     Source            DF    Squares    Mean Square  F Value    Pr > F
     Model              3  9.30703059   3.10234353     2.78    0.0834
     Error             13 14.53296941   1.11792072
     Corrected Total   16 23.84000000

              R-Square   Coeff Var     Root MSE     Yph Mean
              0.390396    16.01997     1.057318     6.600000

                                    Standard
     Parameter       Estimate        Error     t Value    Pr > |t|
     Intercept      7.030707780    0.85651288     8.21     <.0001
     x1clay         0.016908806    0.02203381     0.77     0.4566
     x2org         -0.424258440    0.26104819    -1.63     0.1281
     x3carb         0.078999750    0.03537659     2.23     0.0437
```

Figure 1.5   SAS output (edited) from multiple linear regression fit for soil pH data

Environmental questions of interest with these data include: (i) do the soil texture and organic matter predictors (x_{i1} and x_{i2}, respectively) affect soil pH; and (ii) does carbonate composition (as quantified via x_{i3}) affect soil pH above and beyond any effects of soil texture and organic matter? To address issue (i), we can apply the discrepancy-measure approach from (1.7). Set the null hypothesis to H_0: $\beta_1 = \beta_2 = 0$. Figure 1.5 represents information on the full model, so we find SSE(FM) = 14.533 with df_e(FM) = 13. To fit the reduced model, we can use code similar to that in Fig. 1.4, replacing only the call to PROC GLM:

```
proc glm; model Yph=x3carb;
```

This produces SSE(RM) = 17.775 with df_e(RM) = 15 (output not shown; notice that this reduced model corresponds to a simple linear regression fit of Y on x_{i3}). Thus (1.7) becomes $F_{calc} = (3.242/2)/(14.533/13) = 1.621/1.118 = 1.450$. At $\alpha = 0.05$, we refer this to $F_{0.05}(2, 13) = 3.806$. (Critical points for F-distributions are best acquired by computer; see Fig. A.8.) Since 1.450 fails to exceed 3.806, we cannot reject H_0, leading us to conclude that addition of x_{i1} and x_{i2} does not significantly improve the model that already contains x_{i3} as a predictor variable. Note that in SAS, it is possible to acquire the test statistic F_{calc} directly, using PROC REG and the command

```
test x1clay=0, x2org=0;
```

To address issue (ii) we can use the SAS output to construct a 95% confidence interval for β_3. This is $b_3 \pm t_{0.025}(df_e)se[b_3]$, where for these data $df_e = 17 - 3 - 1 = 13$. From Table B.2 we find $t_{0.025}(13) = 2.1604$. Thus, using the information supplied in the last block of output in Fig. 1.5, we find the 95% interval to be $0.079 \pm 2.1604 \times 0.035 = 0.079 \pm 0.076$, or $0.003 < \beta_3 < 0.155$. That is, a unit (i.e. 1%) increase in $CaCO_3$ by weight appears to increase soil pH by between 0.003 and 0.155 units when both other predictors are held fixed. In terms of x_{i3}, we conclude that the carbonate predictor does have a significant effect on soil pH since this interval does not contain $\beta_3 = 0$. Notice that the corresponding P-value from Fig. 1.5 is $P = 0.0437$.

Exercise 1.4 studies the residuals under the full and reduced models in detail. ✪

An important consideration when employing multiple linear regression models is the differing contributions of the various predictor variables. It can be the case that information in one of the predictors closely duplicates or overlaps with information in another predictor. The effect is known as *multicollinearity* among the predictor variables, and it can lead to computational instabilities, large statistical correlations, and inflated standard errors in the LS/ML estimators for β. A number of strategies exist to identify multicollinearity in a set of posited predictor variables. Most simply, examining pairwise scatterplots among all the predictor variables is an easy way to see if any two predictor variables are providing redundant information. More complicated diagnostics include the so-called *variance inflation factor (VIF)* for each of the p predictors. This is defined as $VIF_j = 1/(1 - R_j^2)$, where R_j^2 is the coefficient of multiple determination found from regressing the jth predictor, x_{ij}, on the other $p - 1$ predictor variables ($j = 1, \ldots, p$). In SAS, PROC REG computes VIFs via the `vif` option in the `model` statement. A set of predictors whose maximum VIF exceeds 10 is felt to be highly collinear. For the three predictor variables in Example 1.3, above, none of the VIFs exceeds 1.2, implying no serious concern regarding multicollinearity (Exercise 1.4).

Remedies for multicollinearity are primarily design-related: (i) to the best extent possible, identify predictors that provide separate information; (ii) avoid inclusion of predictors that are essentially the same variable (e.g., species abundance and species density in an ecological field study); (iii) always try to keep the number of predictor variables manageable (see below). One can also try transforming highly collinear predictor variables – via, say, a logarithmic or reciprocal transform – to decrease their VIFs. In some cases, as simple an operation as centering each variable about its mean by replacing x_{ij} by $u_{ij} = x_{ij} - \bar{x}_i$ can help alleviate multicollinearity. We employ this strategy with polynomial regression models in §1.5, below. One can also apply different forms of multivariate data reduction, such as principal components analysis, to the set of predictors to make them more amenable to regression modeling. We will not go into detail on such approaches here, however. For more on multivariate statistical methods see, for example, Manly (1994).

Notice that VIFs are determined from only the x_{ij}s, and so they may be computed prior to acquisition of the Y_is. Thus, where possible, the VIFs can be calculated and a check for multicollinearity can precede sampling. Highly collinear predictors can be reassessed as to their anticipated value to the analysis, and removed if felt to be non-essential or marginal.

The construction and specification of predictor variables for a multiple linear regression can be a complex process. A natural approach, seen in many environmental studies, is to identify as many possible predictors as the investigator feels might affect the response and include all of these in the regression equation. Combinations of the original predictors are also possible, such as higher-order polynomial and cross-product terms; see §1.5. For example, a multiple linear regression with three predictor variables, x_{i1}, x_{i2}, and x_{i3}, may be expanded to second order by including the quadratic terms $x_{i4} = x_{i1}^2, x_{i5} = x_{i2}^2$, and $x_{i6} = x_{i3}^2$ and the second-order cross products $x_{i7} = x_{i1}x_{i2}, x_{i8} = x_{i1}x_{i3}$, and $x_{i9} = x_{i2}x_{i3}$. (The cross products are called 'interaction' terms. Note that it often does not make sense to include a higher-order term in a model without also including all its lower-order siblings. Thus, for example, if we were to include $x_{i9} = x_{i2}x_{i3}$ we would typically also include x_{i2} and x_{i3}. Similarly, if we include $x_{i6} = x_{i3}^2$ we would also include x_{i3}, etc.) Thus from three original x-variables we can construct a full second-order model with $p = 9$ predictors. Not all of these nine variables will necessarily be important in the model, but they can nonetheless be included and assessed for significance (as long as $n > p + 1 = 10$ observations are recorded). Exercise 1.5 explores this sort of modeling approach.

It is also possible to automate the variable selection effort. This involves studying in a stepwise, systematic fashion all possible models (or some prespecified subset, such as all possible second-order models) available from a given set of original x-variables. From these, an operating model is chosen that meets or exceeds some optimality criterion. For example, one might wish to minimize the MSE or maximize R^2 after adjusting for the total number of predictor variables employed in the model. From these exploratory efforts, conclusions can be drawn on the selected model, using the sorts of inferential and diagnostic methods described above. (The inferences should be viewed as preliminary or tentative, however, since they are typically not corrected for the repeated *a priori* testing required to identify the selected model. Potscher and Novak, 1998, and Zhang, 1992, discuss some theoretical complexities associated with making formal inferences after variable selection; see also Olden and Jackson, 2000.)

The systematic search can be performed backwards from a maximal model (*backward*, or *backstep, elimination*), or forwards from some very simple progenitor model (*forward selection*); see Neter *et al.* (1996, Ch. 8). The effort is by necessity computationally intensive, and therefore is best performed by computer; for example, in SAS use PROC STEPWISE, or apply PROC REG with the `selection=` option in the `model` statement. (Indeed, all 2^p possible regression models for a set of p predictor variables can be constructed using PROC RSQUARE.) We should warn the user, however, that blindly or inattentively ceding to the computer the final decision(s) on which predictors to include in a regression model is often foolhardy, since the computer cannot incorporate important interpretations from the underlying subject-matter. No automated variable selection procedure should ever replace informed scientific judgment.

1.3   Qualitative predictors: ANOVA and ANCOVA models

1.3.1   ANOVA models

In many environmental regression problems, variables that are thought to affect the mean response may be qualitative rather than quantitative; for example, sex or ethnic status in an environmental health investigation, or habitat type in an ecological study. In such settings, it is still desirable to try and relate the mean response to some sort of linear predictor. A natural approach is to assign a set of codes or scores to the different levels of the qualitative variable(s) and build from these a series of quantitative predictor variables. For instance, suppose we remove the intercept term from the model to avoid complications in interpretation. Then, we could set $x_{i1} = 1$ if the ith observation corresponds to the first level of the (first) qualitative predictor (zero otherwise), $x_{i2} = 1$ if the ith observation corresponds to the second level of the (first) qualitative predictor (zero otherwise), etc. Other coding schemes are also possible (Neter *et al.*, 1996, §16.11).

In this manner, a multiple linear regression model can be constructed to account for qualitative as well as quantitative predictor variables. For the qualitative setting, however, interpretation of the regression coefficients as slopes, or as changes in effect, becomes questionable when using simple coding schemes such as those suggested above. As a result, we often write the model in a more traditional parameterization, based on a so-called ANOVA structure. (The name comes from use of an analysis of variance to assess the effects of the qualitative factor; see Neter *et al.*, 1996, Ch. 16.) To each qualitative factor the model assigns certain effect parameters, say, α_i for factor A, β_j for factor B, γ_k for factor C, etc. The factor indices vary over all levels of each factor. For instance, if a single factor, A, has a levels, then $i = 1, \ldots, a$.

Assuming n_i observations are recorded at each combination of this single factor, we write

$$Y_{ij} = \theta + \alpha_i + \varepsilon_{ij}, \qquad (1.8)$$

$i = 1, \ldots, a$, $j = 1, \ldots, n_i$. As above, we assume that the random error terms satisfy $\varepsilon_{ij} \sim$ i.i.d. $N(0, \sigma^2)$. The standalone parameter θ may be interpreted as the *grand mean* of the model when the α_i parameters are viewed as deviations from θ due to the effects of factor A. This is a *one-factor ANOVA model*. The total sample size is $N = \sum_{i=1}^{a} n_i$. If n_i is constant at each level of i, say $n_i = n$, then we say the ANOVA is *balanced*, otherwise it is *unbalanced*.

Unfortunately, under this factor-effect parameterization there is not enough information in the data to estimate every parameter. As currently stated, the model in (1.8) has $a + 1$ regression parameters, $\theta, \alpha_1, \ldots, \alpha_a$, but only a different groups from which to estimate these values. Thus we can estimate all a values of α_i (and σ^2, using the usual MSE term), but not θ. To accommodate estimation of θ, we impose an *estimability constraint* on the α_is. Many possible constraints exist; two of the more common are the *zero-sum constraint* $\sum_{i=1}^{a} \alpha_i = 0$ and the *corner-point constraint* $\alpha_a = 0$. If applied properly, the constraints do not affect tests or estimates of the differences between groups or estimates of other linear combinations of the group means; however, the interpretation of the parameter estimates is directly impacted by the constraint used. For example, under the zero-sum constraint, θ is the mean of all the a group means and α_i is the difference between the ith factor-level mean and the overall mean. By contrast, under the corner-point constraint θ is the mean of the last factor level while α_i is the difference between the ith factor-level mean and this last factor level's mean.

If a second factor, B, were considered in the study design such that n_{ij} observations are recorded at each combination of the two factors, we write $Y_{ijk} = \theta + \alpha_i + \beta_j + \varepsilon_{ijk}$, $i = 1, \ldots, a$, $j = 1, \ldots, b$, $k = 1, \ldots, n_{ij}$. This is a *two-factor, main-effects ANOVA model*, so named since it contains effects due to only the two main factors. Allowing for the additional possibility that the two factors may interact calls for expansion of the model into

$$Y_{ijk} = \theta + \alpha_i + \beta_j + \gamma_{ij} + \varepsilon_{ijk}, \qquad (1.9)$$

$i = 1, \ldots, a$, $j = 1, \ldots, b$, $k = 1, \ldots, n_{ij}$, with total sample size $N = \sum_{i=1}^{a} \sum_{j=1}^{b} n_{ij}$. The γ_{ij}s are the *interaction parameters*. As in (1.8), estimability constraints are required under this factor-effects parameterization. The zero-sum constraints are $\sum_{i=1}^{a} \alpha_i = 0$, $\sum_{j=1}^{b} \beta_j = 0$, $\sum_{i=1}^{a} \gamma_{ij} = 0$ for all j, and $\sum_{j=1}^{b} \gamma_{ij} = 0$ for all i. The corner-point constraints are $\alpha_a = 0$, $\beta_b = 0$, $\gamma_{aj} = 0$ for all j, and $\gamma_{ib} = 0$ for all i. Extensions to higher-factor models are also possible (Neter *et al.*, 1996, Ch. 23).

To analyze data under an ANOVA model, closed-form equations exist only in select cases. These provide point estimates of each mean level's effect and sums of squares for the ANOVA table; see Samuels and Witmer (2003, Ch. 11) or Neter *et al.* (1996, Ch. 16). Because of their correspondence to multiple linear regression models, however, ANOVA-type models may also be studied using regression methods, via computer. In SAS, for example, PROC GLM can analyze ANOVA-type models by invoking the `class` statement; this is used to identify which predictors are qualitative 'classification' variables. PROC GLM can produce sequential sums of squares (SS) for testing the significance of model components in the sequential order in which they are fitted, partial SS for testing the significance of model components when they are fitted last in sequential order, the various corresponding mean squares, and from these the ANOVA table. Ratios of the factor-specific mean squares to the MSE produce F-statistics to assess the significance of any factor or any multi-factor interaction, using the general approach embodied in equation (1.7); see Neter *et al.* (1996, Chs. 16–22). For example, the null hypothesis of no A × B interaction under a full two-factor model is $H_0: \gamma_{11} = \cdots = \gamma_{ab} = 0$. Assess this via the F-statistic

$$F_{\text{calc}} = \frac{\text{MS}[\text{A} \times \text{B}|\text{A}, \text{B}]}{\text{MSE}}$$

and reject H_0 at significance level δ if $F_{\text{calc}} \geq F_\delta([a-1][b-1], \text{df}_e)$, where df_e are the df associated with the error term. (The symbol '|' is used to indicate sequencing or conditioning. Read it as 'fitted after' or 'conditional on'.) For the full two-factor model, $\text{df}_e = N - ab + 1$.

As with our earlier comment on model construction with higher-order quantitative predictors, it does not make sense to include a higher-order qualitative term in a model without also including all associated lower-order terms. Thus, for example, if we were to include a two-factor interaction using the γ_{ij}s, we would typically also include both main effects terms via the α_is and β_js. This suggests a natural ordering for hypothesis testing: test interaction first via $H_0: \gamma_{ij} = 0$ for all i,j. If the interaction is insignificant, follow with separate tests of each main effect via $H_0: \alpha_i = 0$ for all i and $H_0: \beta_j = 0$ for all j. In some settings this strategy has a mathematical interpretation: if the interaction terms represent a departure from the additive effects of the two factors, testing interaction first is equivalent to evaluating whether a departure from additive effects is present before examining the separate additive effects of each main effect.

Example 1.4 (Macroinvertebrate ratios) Gonzalez and Manly (1998) give data on the change in abundance of two freshwater benthic macroinvertebrates (ephemeroptera and oligochaetes) in three New Zealand streams over the four different annual seasons. Ephemeroptera are pollution-sensitive, so by measuring the ratio Y_{ij} of ephemeroptera to oligochaetes we can assess potential stream pollution. We view the $a = 3$ streams as one factor, and the $b = 4$ seasons as a second factor; $n = 3$ independent random samples were taken at each of the 12 stream × season combinations. (Following Gonzalez and Manly, we assume that the samples were separated enough in time so that no temporal correlation exists within streams across the different seasons for these data.) The data are given in Table 1.3. For them, we assume the full two-factor model (1.9). A SAS analysis via PROC GLM is given via the sample code in Fig. 1.6.

The output (edited) from PROC GLM appears in Fig. 1.7. In it, we begin by assessing the stream × season interaction for these data. That is, compare the full model with linear predictor $\theta + \alpha_i + \beta_j + \gamma_{ij}$ to a reduced model whose linear predictor contains only the main-effect additive terms $\theta + \alpha_i + \beta_j$. Using either sequential

Table 1.3   Ratios of ephemeroptera to oligochaetes in three New Zealand streams

Stream	Season			
	Summer	Autumn	Winter	Spring
A	0.7, 8.5, 7.1	0.5, 1.1, 1.4	0.1, 0.1, 0.2	0.2, 0.1, 0.4
B	1.2, 0.7, 0.8	10.7, 19.9, 9.4	2.0, 6.3, 4.8	2.0, 1.6, 1.9
C	7.3, 10.4, 9.5	46.6, 20.3, 24.0	1.2, 0.8, 6.1	0.2, 0.1, 0.1

Source: Gonzalez and Manly (1998).

```
* SAS code to fit 2-factor ANOVA;
  data macroinv;
  input stream $ season $ Y  @@;
  datalines;
    A sum  0.7    A sum  8.5    A sum  7.1
       ⋮             ⋮             ⋮
    C spr  0.2    C spr  0.1    C spr  0.1

proc glm;
  class stream season;
  model Y = stream season stream*season;
```

Figure 1.6   Sample SAS program to fit two-factor ANOVA model to macroinvertebrate ratio data

```
                        The SAS System
                General Linear Models Procedure

                    Class Level Information
                 Class      Levels     Values
                 stream        3       A B C
                 season        4       fal spr sum win

Dependent Variable: Y
                             Sum of
Source              DF       Squares    Mean Square  F Value  Pr > F
Model               11    2512.389722   228.399066    10.18   <.0001
Error               24     538.233333    22.426389
Corrected Total     35    3050.623056

              R-Square       Coeff Var    Root MSE    Y Mean
              0.823566       81.84514     4.735651    5.786111

Source              DF    Type I SS    Mean Square  F Value  Pr > F
stream               2    478.203889   239.101944    10.66   0.0005
season               3   1080.727500   360.242500    16.06   <.0001
stream*season        6    953.458333   158.909722     7.09   0.0002

Source              DF    Type III SS  Mean Square  F Value  Pr > F
stream               2    478.203889   239.101944    10.66   0.0005
season               3   1080.727500   360.242500    16.06   <.0001
stream*season        6    953.458333   158.909722     7.09   0.0002
```

Figure 1.7   SAS output (edited) from two-factor ANOVA fit for macroinvertebrate ratio data

sums of squares (also called `Type I SS` in **PROC GLM**) or partial sums of squares (called `Type III SS` in **PROC GLM**), the F-statistic is $F_{calc} = (953.4583/6)/(538.2333/24) = 7.09$. (Since the design here is balanced, the Type I and Type III SS are identical.) Referring this to $F(6,24)$, **PROC GLM** gives the interaction P-value as $P = 0.0002$, implying that a significant stream $\times$ season interaction exists. Standard practice requires therefore that further analysis of the stream or season main effects is inappropriate, since the main effects cannot be disentangled from each other (Neter *et al.*, 1996, §20.3). Further analysis of the stream and/or season effect must be stratified across levels of the other factor; for example, to assess if there is an effect due to season, perform an analysis of the season effect at each level of the stream factor. (That is, perform three separate one-factor ANOVAs.) A similar stratification is required to assess if there is a stream effect. In Exercise 1.8, such an analysis identifies selected significant effects for both seasons and streams. ✪

We should warn that complications can occur in two-factor (and higher) ANOVA if the design is unbalanced, that is, if $n_i \neq n$ for any i. Due to the lack of balance, point estimates of a factor's effects when averaged over the levels of the other factor(s) may not correspond to the marginal quantities of interest, and this can carry over into tests of hypotheses and other inferences on that factor. Estimation by what are called *least squares means* is often employed in these instances. (For example, in SAS one can apply the `lsmeans` command.) The details extend beyond the scope of our presentation here, however, and we note only that investigators must be careful when analyzing unbalanced multi-factor designs. For more on least squares means and analyses for unbalanced data, see Neter *et al.* (1996, Ch. 22) or Littell *et al.* (2002, §6.3).

1.3.2 ANCOVA models

It is also possible to work with a mix of both quantitative and qualitative predictor variables. The simplest example of this occurs when there is a single qualitative factor under study, but where it is known that the data vary according to some other, quantitative variable. Interest exists in assessing the effects of the qualitative factor, adjusted for any variation due to the quantitative factor. In this case, it is common to call the quantitative factor a 'covariate,' and apply an *analysis of covariance* (*ANCOVA*). The model may be written as

$$Y_{ij} = \theta + \alpha_i + \beta(x_{ij} - \bar{x}) + \varepsilon_{ij}, \qquad (1.10)$$

$i = 1, \ldots, a$, $j = 1, \ldots, n_i$. The total sample size is $N = \sum_{i=1}^{a} n_i$. In (1.10), α_i represents the differential factor-level effect and β is the slope parameter associated with the covariate. (The slope parameter is held constant under this model; this *equal-slopes assumption* requires that the covariate have the same effect across all levels of the qualitative factor.) An estimability constraint is again required on the α_is, such as the zero-sum constraint $\sum_{i=1}^{a} \alpha_i = 0$. The error terms take the usual assumption: $\varepsilon_{ij} \sim$ i.i.d. $N(0, \sigma^2)$.

Notice the correction in (1.10) of the covariate by its mean $\bar{x} = \sum_{i=1}^{a} \sum_{j=1}^{n_i} x_{ij} / \sum_{i=1}^{a} n_i$. Many authors view this primarily as a computational aid, used to adjust the different estimates of the factor-level effects for any differential covariate effects. It yields a pertinent interpretation, however, as an evaluation of the differences among levels of the qualitative factor at a fixed level of $x = \bar{x}$ (Neter *et al.*, 1996, §25.2).

Inferences in an ANCOVA are most often directed at the qualitative variable, after 'adjustment' for the covariate effect. This is performed by constructing an ANOVA table with the covariate fitted before the qualitative factor, and assessing significance via the sequential sum of squares for the qualitative factor: SS[factor| covariate]. Applied in equation (1.7), this compares the full model in (1.10) to the reduced model $Y_{ij} = \theta + \beta(x_{ij} - \bar{x}) + \varepsilon_{ij}$. The resulting test statistic is $F_{calc} = $ MS[factor|covariate]/MSE; this reflects the added contribution of the qualitative factor to the model. Reject the null hypothesis of no factor effect, $H_0: \alpha_1 = \cdots = \alpha_a$, at significance level δ if $F_{calc} \geq F_\delta(a - 1, N - a - 1)$.

Example 1.5 (Faba bean genotoxicity) Menke *et al.* (2000) report on studies of the faba bean plant (*Vicia faba*) for use as an ecotoxicological indicator species. One specific endpoint investigated was DNA damage (percentage of DNA separating out during single-cell gel electrophoresis) in nuclei from *V. faba* root tip cells. Greater damage is indicated by higher percentages of separated DNA. Of interest was whether changing the treatment of the cells with different endonucleases produced differences in DNA damage. The study protocol also called for a variety of treatment times, since increasing treatment time was thought to increase genotoxic response and hence enhance the assay's potential as a ecotoxicological screen.

Listed in Table 1.4 are the mean percentage responses of separated DNA as a function of treatment time ($x = 0, 10, 20,$ or 30 min) and endonuclease treatment (*FokI*, *EcoRI*, DNaseI + $MgCl_2$, or DNaseI + $MnCl_2$). Notice the highly unbalanced structure of the experimental design. We view exposure time as the quantitative covariate, and endonuclease treatment as the qualitative predictor to be tested.

One additional concern is the necessary assumption in (1.10) of variance homogeneity. Since percentage data are notorious for their variance heterogeneity, we apply a variance–stabilizing transform here: the logit of the observed percentages. That is, if P_{ij} is the mean percentage response for treatment i at time level j given in Table 1.4, we take $Y_{ij} = \log\{P_{ij}/(100 - P_{ij})\}$ as the response variable for analysis. (An alternative transformation is the arcsine square root: $V_{ij} = \sin^{-1} \sqrt{P_{ij}/100}$.) A plot of the data (Fig. 1.8) shows that the logits are roughly linear across the time span under study.

Figure 1.9 presents sample SAS code for fitting the equal-slopes ANCOVA model from (1.10) to the data in Table 1.4. Notice our continuing use of PROC GLM for the fit. (The `solution` option in the `model` statement calls for LS estimates of the α_is and of β. The estimates of α_i depend on the choice of estimability constraint and so require careful interpretation. SAS indicates this in its output with a warning message that the estimators are 'not unique'; see Fig. 1.10. The LS estimate of β is, however, a valid, unique, LS estimate of the common slope. The corresponding standard error can be used to construct confidence limits or test hypotheses about β; see the comment below.)

Table 1.4 Mean percentage DNA separation in *V. faba* root tip cell nuclei

Endonuclease treatment	Treatment time (min)			
	0	10	20	30
*Fok*I	20.5	65.4	68.4	78.1
	13.4	56.0	54.7	72.8
	17.2	57.3	62.8	81.9
	19.6	59.2	61.8	79.4
*Eco*RI	6.8	10.5	49.0	62.1
	13.0	12.6	40.6	50.2
DNaseI + MgCl$_2$	18.2	24.5	28.0	48.2
	21.9	28.0	31.6	47.8
	2.6	–	56.2	–
	2.4	–	56.0	–
	9.9	–	52.7	–
	14.4	–	42.6	–
DNaseI + MnCl$_2$	58.9	91.5	96.6	–
	65.2	89.9	96.6	–

Dashes indicate no observation at that time–treatment combination. Source: Menke *et al.* (2000).

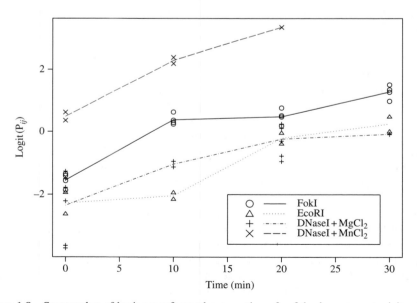

Figure 1.8 Scatterplot of logit-transformed proportions for faba bean genotoxicity data

```
* SAS code to fit equal-slopes ANCOVA;
 data faba;
 input etrt $ time   percent   @@;
    y = log(percent / (100 - percent));
    timebar = 13.47826;
    covar = time - timebar;
 datalines;
 FokI    0 20.5   FokI    0 13.4   FokI    0 17.2   FokI    0 19.6
          :                :                :                :
 MnC12 10 91.5   MnC12 10 89.9   MnC12 20 96.6   MnC12 20 96.6

 proc glm;
    class etrt;
    model Y = covar etrt / solution;
```

Figure 1.9   Sample SAS program to fit equal-slopes ANCOVA for faba bean genotoxicity data

```
                          The SAS System
                  General Linear Models Procedure

      Dependent Variable: Y
                                 Sum of
      Source               DF    Squares    Mean Square  F Value  Pr > F
      Model                 4  96.5575940    24.1393985    70.53  <.0001
      Error                41  14.0331928     0.3422730
      Corrected Total      45 110.5907868

                R-Square      Coeff Var      Root MSE       Y Mean
                0.873107     -232.1263      0.585041     -0.252036

      Source       DF     Type I SS    Mean Square  F Value  Pr > F
      covar         1   38.19447717   38.19447717   111.59  <.0001
      etrt          3   58.36311682   19.45437227    56.84  <.0001

      Source       DF    Type III SS   Mean Square  F Value  Pr > F
      covar         1   45.39120872   45.39120872   132.62  <.0001
      etrt          3   58.36311682   19.45437227    56.84  <.0001

      Parameter           Estimate      Std. Error  t Value  Pr > |t|
      Intercept         2.362511350 B   0.24047205     9.82  <.0001
      covar             0.092543930     0.00803616    11.52  <.0001
      etrt     EcoRI   -3.567986935 B   0.31850293   -11.20  <.0001
      etrt     FokI    -2.352084490 B   0.28293472    -8.31  <.0001
      etrt     MgC12   -3.380745038 B   0.28078673   -12.04  <.0001
      etrt     MnC12    0.000000000 B        .           .      .
      NOTE: The X'X matrix has been found to be singular, and a
            generalized inverse was used to solve the normal
            equations. Terms whose estimates are followed by the
            letter 'B' are not uniquely estimable.
```

Figure 1.10   SAS output (edited) from equal-slopes ANCOVA fit for faba bean genotoxicity data

We find from Fig. 1.10 that the F-test of $H_0: \alpha_1 = \alpha_2 = \alpha_3 = \alpha_4$, that is, whether endonuclease treatment makes a difference in the generation of DNA damage, produces a test statistic of $F_{calc} = 56.84$, on 3 and 41 df (using either the Type I or Type III SS term to build the F-statistic). The corresponding P-value is $P < 0.0001$, and is highly significant. We conclude that there is a clear difference among endonuclease treatments, after adjusting for possible differences in percent DNA damage due to treatment time.   ✦

Some environmental phenomena yield data that are best studied using an ANCOVA-type model where the quantitative predictor is itself of interest, making the null hypothesis $H_0: \beta = 0$ a primary concern. If so, then it is natural to reverse the order of the fit, and first test $H_0: \beta = 0$ via the F-statistic $F_{\text{calc}, \beta} = \text{MS}$ [quantitative predictor|qualitative factor]/MSE. Here, we reject H_0 at significance level δ if $F_{\text{calc}, \beta} \geq F_\delta(1, N - a - 1)$. Failure to reject H_0 then allows for testing of the qualitative factor; that is, if $F_{\text{calc}, \beta} < F_\delta(1, N - a - 1)$, we can test $H_0: \alpha_1 = \cdots = \alpha_a$ via $F_{\text{calc}, \alpha} = \text{MS}$[qualitative factor]/MSE. Reject the new H_0 at significance level δ if $F_{\text{calc}, \alpha} \geq F_\delta(a - 1, N - a - 1)$. If appropriate, (1.10) also can include polynomial terms for the covariate, such as the quadratic ANCOVA model

$$Y_{ij} = \theta + \alpha_i + \beta_1(x_{ij} - \overline{x}) + \beta_2(x_{ij} - \overline{x})^2 + \varepsilon_{ij}, \tag{1.11}$$

$i = 1, \ldots, a, \ j = 1, \ldots, n_i$.

As noted previously, the ANCOVA analysis as we have presented it makes the strong assumption that the covariate effect is the same across all levels of the qualitative factor. We can extend the model in (1.10) to assess this assumption: simply allow for differential slopes. That is, permit β to vary with the qualitative level index, producing

$$Y_{ij} = \theta + \alpha_i + \beta_i(x_{ij} - \overline{x}) + \varepsilon_{ij}, \tag{1.12}$$

$i = 1, \ldots, a, \ j = 1, \ldots, n_i$. Computationally, this is equivalent to including an interaction term; for example, in Fig. 1.9, use the alternative `model` statements

```
model y=covar etrt covar*etrt / solution;
```

or equivalently

```
model y=covar | etrt / solution;
```

Exercise 1.9 explores this with the faba bean genotoxicity data from Example 1.5.

1.4   Random-effects models

The ANOVA and ANCOVA models (1.8)–(1.12) employ a single, additive error term to describe any random variation (the *stochastic* components) in the data. All the other terms are assumed to be constant or fixed (the *deterministic* components of the model), and we often say that this combination represents a *fixed-effects model*. In some settings, however, random variation may be associated with one or more of the factors under study, and hence the fixed-effects assumption may be untenable. For instance, the factor levels under study may themselves be a random sample of all possible levels for that particular factor. This is called a *random effect*.

The consequences of a random-effect assumption on model equations such as (1.8) or (1.9) are relatively straightforward. Consider (1.8): if the treatment's effect is itself random, make the assumption that $\alpha_i \sim$ i.i.d. $N(0, \tau^2)$. (We also assume that the α_is are statistically independent of the ε_{ij}s.) Notice that since the α_is are now viewed as normal random variables, it makes no sense to place any estimability conditions on them.

The consequences of a random-effect assumption on the observations and on the analysis are somewhat more subtle. Equation (1.8) now leads to $E[Y_{ij}] = \theta$ and $\mathrm{Var}[Y_{ij}] = \sigma^2 + \tau^2$. For simplicity, assume the design is balanced, so $n_i = n$. In this simple case with only a single qualitative factor, estimation of the factor effect parameters α_i is typically no longer relevant, since they are now random. Instead, we center attention on the contribution of each model component to the total variance $\mathrm{Var}[Y_{ij}]$. These *variance components* (Searle *et al.*, 1992) may be estimated as $\hat{\sigma}^2 = \mathrm{MSE}$ (again), and $\hat{\tau}^2 = (\mathrm{MSA} - \mathrm{MSE})/n$, where MSA is the mean square associated with the single (A) factor. Notice, however, that there is no guarantee that $\hat{\tau}^2$ will be non-negative, even though we assume $\tau^2 \geq 0$. A simple correction truncates $\hat{\tau}^2$ at zero if it falls below that value. More complex estimation schemes to accommodate this non-negativity, along with adaptations for designs when the experimental design is not balanced, are also possible; see Hocking (1996, Ch. 17), Kelly and Mathew (1994), or the dedicated text by Searle *et al.* (1992). To test for an effect due to factor A, we now write $H_0: \tau^2 = 0$. Rejection of H_0 occurs at significance level δ when $F_{\mathrm{calc}} = \mathrm{MSA}/\mathrm{MSE} \geq F_\delta(a - 1, a[n - 1])$.

Another useful quantity associated with the variance components in a random-effects model is the *intraclass correlation coefficient*: $\tau^2/(\tau^2 + \sigma^2)$. This is the correlation between Y_{ij} and Y_{ih} ($j \neq h$) for any fixed level of i, and it measures the proportion of variation in Y accounted for by variation in the random effect.

Confidence intervals for the variance components or for certain functions of them, such as the intraclass correlation, are often straightforward to construct. For example, a $1 - \delta$ confidence interval for $\tau^2/(\tau^2 + \sigma^2)$ has the form $L/(1 + L) < \tau^2/(\tau^2 + \sigma^2) < U/(1 + U)$, for

$$L = \frac{1}{n}\left\{\frac{F_{\mathrm{calc}}}{F_{(1-\delta)/2}(a - 1, a[n - 1])} - 1\right\}$$

$$U = \frac{1}{n}\left\{\frac{F_{\mathrm{calc}}}{F_{\delta/2}(a - 1, a[n - 1])} - 1\right\},$$

and where $F_{\mathrm{calc}} = \mathrm{MSA}/\mathrm{MSE}$ from above.

Extensions to the two-factor case in (1.9) follow in similar form. For example, if α_i, β_j, and γ_{ij} all represent random effects in (1.9), then we make the assumptions that $\alpha_i \sim$ i.i.d. $N(0, \tau_\alpha^2)$, $\beta_j \sim$ i.i.d. $N(0, \tau_\beta^2)$, and $\gamma_{ij} \sim$ i.i.d. $N(0, \tau_\gamma^2)$, where the random variables are all assumed mutually independent of each other and of ε_{ijk}. Under such a multi-factor random-effects model, estimation and inference proceed in a manner similar to that for the single-factor case; see Neter *et al.* (1996, §24.2).

When some of the factors in a multi-factor model are random but others remain fixed we have a *mixed-effects model*, or simply a *mixed model*. In a mixed model, the inferences must change slightly to accommodate the mixed-effect features. We will not go into additional detail here on the subtleties of estimation and inference with mixed-effects models, although see §5.6.2 for a special case with temporally correlated data. Readers unfamiliar with mixed models may find Neter *et al.* (1996,

Ch. 24) a useful source; also see the review by Drum (2002). For fitting mixed models in SAS, the workbook by Littell *et al.* (1996) is particularly helpful.

1.5 Polynomial regression

In many environmental applications, the mean response is affected in a much more complex fashion than can be described by simple linear relationships. A natural extension of the simple linear relationship is the addition of higher-order polynomial terms. We hint at this above with the quadratic ANCOVA model of equation (1.11). Made formal, a pth-order polynomial regression model is

$$Y_i = \beta_0 + \beta_1(x_i - \bar{x}) + \beta_2(x_i - \bar{x})^2 + \cdots + \beta_p(x_i - \bar{x})^p + \varepsilon_i, \qquad (1.13)$$

$i = 1, \ldots, n$. We make the usual homogeneous-variance, normal-error assumption: $\varepsilon_i \sim$ i.i.d. $N(0, \sigma^2)$. Also, we assume $p < n - 1$, and require that there be at least $p + 1$ distinguishable values among the x_is.

  This model is used to represent simple curvilinear relationships between $E[Y_i]$ and x_i. Notice that for any $p \geq 1$, (1.13) is in fact a special case of the multiple regression model from §1.2, where $x_{ij} = (x_i - \bar{x})^j$. (Centering the x_is accomplishes two objectives: (i) it provides an interpretation for β_0 as the mean response at $x = \bar{x}$; and (ii) it helps avoid possible multicollinearity in the predictors (Bradley and Srivastava, 1979). Where possible, one can also *design* the study to avoid multicollinearity and still operate with a polynomial response model. That is, if the values of x_i can be chosen in advance, special forms called *orthogonal polynomials* exist that guarantee no multicollinearity among the polynomial regressors. See, for example, the review by Narula, 1979.) Thus all the statistical machinery available for fitting and analyzing a multiple linear regression is also available for fitting and analyzing polynomial regression models. This fact underlies much of the justification for employing these models: curvilinear regression relationships in the environmental sciences are often more complex than can be described by a simple polynomial function. Equation (1.13) becomes useful, however, when it can provide an approximation to the true nonlinear relationship over the range of x under study. In this case, the simplicity of use made possible by connection to multiple linear regression methodology becomes an attractive motivation. For example, to test the effect of the x-variable, assess $H_0: \beta_1 = \beta_2 = \cdots = \beta_p = 0$ via an F-statistic as in (1.7). Or simple hypothesis tests on the individual regression coefficients follow from standard t-distribution theory. For instance, if $p = 2$ a test of $H_0: \beta_2 = 0$ using $t_{calc} = |b_2|/se[b_2]$ assesses whether the quadratic curvature is required in a model that already has a linear term present by comparing the full model with linear predictor $\beta_0 + \beta_1(x_i - \bar{x}) + \beta_2(x_i - \bar{x})^2$ to a reduced model with the simple linear predictor $\beta_0 + \beta_1(x_i - \bar{x})$. Note, however, that in the polynomial context β_j loses its interpretation as change in response to the jth predictor when holding the other predictors fixed. Clearly, one cannot vary $(x_i - \bar{x})^j$ while holding $(x_i - \bar{x}), \ldots, (x_i - \bar{x})^{j-1}, (x_i - \bar{x})^{j+1}, \ldots, (x_i - \bar{x})^p$ fixed.

The best-known and most useful polynomials are the quadratic (or parabolic) model at $p = 2$ and the cubic model at $p = 3$. The former is useful for curvilinear response that changes once and only once from strictly increasing (decreasing) to strictly decreasing (increasing). The latter is useful if the curvilinear response inflects such that its *rate of change* shifts from strictly decreasing to strictly increasing or vice versa; for example, the response might move from increasing to decreasing, and back to increasing. Of course, any order of polynomial may be fitted if the data are rich enough. We do not recommend use of (1.13) past $p = 3$ or perhaps $p = 4$, however, unless there is strong motivation from the underlying environmental subject-matter. Also, as noted in §1.2, it generally does not make sense to include a higher-order term without also including all the lower orders. Thus, whenever we include a certain order of polynomial, p, in a regression model we generally also include all the lower orders, $p - 1, p - 2, \ldots, 2, 1$.

Example 1.6 (Rice yield)   Seshu and Cady (1984) present data on yield of irrigated rice (*Oryza sativa* L.) as a function of minimum temperature, with a goal of understanding the environmental conditions that affect yield. (The temperatures were averages of daily minima over a 30-day period immediately after the rice plants flowered.) The data appear in Table 1.5. A plot (Fig. 1.11) shows a clear curvilinear shape to the temperature–yield curve, for which a quadratic polynomial may provide a good approximation. Towards this end, we consider the polynomial regression model (1.13) with $p = 2$.

Table 1.5   Rice yield (t/ha) vs. minimum temperature (°C)

x = min. temp.	Y = yield	x = min. temp.	Y = yield
29.2	2.3	23.4	3.0
28.1	3.1	23.4	4.4
27.2	2.8	23.2	3.1
26.4	2.4	23.1	2.6
26.3	3.6	23.1	4.8
26.2	2.3	23.0	3.2
26.2	3.8	22.9	3.3
26.0	3.1	22.5	3.1
25.9	2.4	22.5	3.4
25.7	2.1	22.4	3.2
24.5	3.5	21.7	4.2
24.4	3.8	21.2	4.5
24.0	3.1	20.0	4.7
23.9	2.9	19.2	5.0
23.9	3.2	19.0	6.2
23.7	3.0	19.0	6.0
23.7	4.5	18.8	6.1
23.7	3.5	18.0	7.3
23.6	3.3	18.0	6.6
23.5	3.7	17.4	6.2

Source: Seshu and Cady (1984).

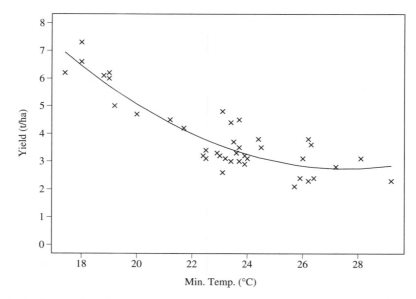

Figure 1.11 Scatterplot of observations ($\times$) and estimated LS parabola (——) for rice yield data from Table 1.5

A number of issues are of interest with these data. In particular, to assess if temperature significantly affects yield we set the null hypothesis to $H_0: \beta_1 = \beta_2 = 0$ and use an *F*-statistic based on the discrepancy measure in equation (1.7). Figure 1.12 gives sample SAS code to perform the LS fit, with associated output (edited) in Fig. 1.13. Therein, the *F*-statistic for testing whether all of the non intercept terms are zero is found under F Value in the main Analysis of Variance display. This is precisely the test of interest here, so for testing H_0 we take $F_{\text{calc}} = 78.09$. Referred to an $F(2,37)$ distribution, we see from the **PROC REG** output under Pr > F that $P < 0.0001$. Clearly, there is a significant effect of temperature on yield.

```
*  SAS code to fit quadratic regression;
data rice;
   input yield   mint @@;
      mintbar = 23.1975;
      x = mint - mintbar;
      x2 = x*x;
   datalines;
   2.8    27.2       3.1    28.1       2.3    29.2       3.0    23.4
    :      :          :      :          :      :          :      :
   6.0    19.0       6.1    18.8       7.3    18.0       6.6    18.0

proc reg;
   model yield = x x2 ;
```

Figure 1.12 Sample SAS program to fit quadratic regression for rice yield data

```
                          The SAS System
                          The REG Procedure

                       Analysis of Variance
                               Sum of       Mean
         Source          DF     Squares     Square    F Value   Pr > F
         Model            2    53.93171    26.96586     78.09   <.0001
         Error           37    12.77604     0.34530
         Corrected Total 39    66.70775

                        Parameter Estimates
                         Parameter     Standard
         Variable    DF   Estimate       Error     t Value   Pr > |t|
         Intercept    1    3.50874      0.11915      29.45    <.0001
         x            1   -0.35563      0.03372     -10.55    <.0001
         x2           1    0.04076      0.00939       4.34     0.0001
```

Figure 1.13   SAS output (edited) from quadratic regression fit for rice yield data

To study this effect further, we construct confidence intervals on the β_1, β_2 pair of regression coefficients. Pointwise intervals on each parameter are $b_j \pm t_{\alpha/2}(37)se[b_j]$, $j = 1, 2$, where the parameter estimates and standard errors can be read directly from Fig. 1.13 (under `Parameter Estimate`). Since the resulting intervals are based on the same set of data, however, some adjustment for multiplicity is required. The simplest is a Bonferroni adjustment (§A.5.4). Here, this affects only the critical point: the adjustment reduces α by a factor equal to the number of intervals being computed. Thus, the (two) Bonferroni-adjusted joint confidence intervals have the form $b_j \pm t_{\alpha/4}(37)se[b_j]$, $j = 1, 2$. At $\alpha = 0.05$, the necessary critical point is $t_{0.0125}(37)$. Using the SAS function `tinv` (see Fig. A.4), we find this as $t_{0.0125}(37) = 2.3363$. For β_1, this produces the Bonferroni-adjusted interval $-0.3556 \pm 2.3363 \times 0.0337 = -0.3556 \pm 0.0788$. For β_2, we find $0.0408 \pm 2.3363 \times 0.0094 = 0.0408 \pm 0.0219$. Both intervals fail to contain zero, suggesting that each term contributes significantly to the model (and, again, that the temperature–yield relationship is significant).

We can study the residuals from this LS/ML by plotting them against the predicted values; that is, plot $r_i = Y_i - \hat{Y}_i$ vs. $\hat{Y}_i = b_0 + b_1 x_i + b_2 x_i^2$. Sample SAS code to perform this is

```
proc reg;
   model yield = x x2 ;
   plot r. *p.;
```

The resulting residual plot (not shown) indicates no substantial diagnostic problems.   ✿

Applications of polynomial models in the environmental sciences are far-reaching and diverse. The combination of curvilinear features and simple regression structure often allows for a number of interesting quantitative questions to be addressed. For example, the quadratic form ($p = 2$) of (1.13) using the mean-centered predictor $x - \bar{x}$ reaches a single optimum point (a minimum if the parabola is convex, a maximum if the parabola is concave) at $x_{\text{opt}} - \bar{x} = -\beta_1/2\beta_2$, or $x_{\text{opt}} = \bar{x} - \beta_1/2\beta_2$. At the core of this quantity is a ratio of parameters for which, under a constant-variance, normal-error model, the ML estimator is simply the ratio of the point estimators. This leads to

$$\hat{x}_{\text{opt}} = \bar{x} - \frac{b_1}{2b_2}.$$

Unfortunately, although this point estimator is straightforward, statistical inferences on the ratio of parameters in x_{opt} can be difficult to construct. One possibility in this setting is to find confidence limits using a result due to Fieller (1940). *Fieller's theorem*, as it is known, gives a $1 - \alpha$ confidence interval on the ratio using the point estimates b_1 and b_2, the standard errors $se[b_1]$ and $se[b_2]$, and an estimate of the covariance between b_1 and b_2: $\hat{\sigma}_{12} = \text{Cov}[b_1, b_2]$. (Not to be confused with analysis of covariance from §1.3, the *covariance* here is a summary measure of the joint variability between b_1 and b_2; see §A.1.) These quantities are typically provided on standard computer outputs; for example, to display the estimated covariance in PROC REG use the `covb` option in the `model` statement. With these values, the Fieller confidence limits for x_{opt} can be calculated as

$$\hat{x}_{\text{opt}} + \frac{\gamma}{1 - \gamma} \left\{ (\hat{x}_{\text{opt}} - \bar{x}) + \frac{\hat{\sigma}_{12}}{2se^2[b_2]} \right\}$$

$$\pm \frac{t_{\alpha/2}(\nu)}{2(1-\gamma)|b_2|} \left\{ se^2[b_1] + 4(\hat{x}_{\text{opt}} - \bar{x}) \left[\hat{\sigma}_{12} + se^2[b_2](\hat{x}_{\text{opt}} - \bar{x}) \right] \right. \tag{1.14}$$

$$\left. - \gamma \left(se^2[b_1] - \frac{\hat{\sigma}_{12}^2}{se^2[b_2]} \right) \right\}^{1/2},$$

where $\gamma = [t_{\alpha/2}(\nu)]^2 se^2(b_2)/b_2^2$ measures departure from symmetry in the distribution of $\hat{x}_{\text{opt}}$ ($\gamma \to 0$ indicates greater symmetry), and $\nu = \text{df}_e$ from the ANOVA table for the quadratic regression fit. (See Exercise 1.16 for a derivation of Fieller's result.) In some applications of Fieller's theorem appeal is made to large-sample approximations, and in those cases we often take $\nu = \infty$, that is, we use the $N(0,1)$ limiting distribution for $t(\nu)$. In this case, set $t_{\alpha/2}(\infty) = z_{\alpha/2}$.

One might also consider use of the delta method (§A.6) to build approximate $1 - \alpha$ limits for x_{opt}. That is, from §A.5.1 we know that a $1 - \alpha$ Wald confidence interval is $\hat{x}_{\text{opt}} \pm z_{\alpha/2}se[\hat{x}_{\text{opt}}]$, and to complete the interval all we need is the standard error $se[\hat{x}_{\text{opt}}]$. This is the square root of the variance, $\text{Var}[\hat{x}_{\text{opt}}]$, which since $\bar{x}$ is assumed fixed simplifies to $\text{Var}[\hat{x}_{\text{opt}}] = \text{Var}[\bar{x} - b_1/2b_2] = \text{Var}[-b_1/2b_2] = \frac{1}{4}\text{Var}[b_1/b_2]$. We use the delta method to approximate the variance of the ratio (see Example A.3) and use this in $\hat{x}_{\text{opt}} \pm z_{\alpha/2}\sqrt{\text{Var}[\hat{x}_{\text{opt}}]}$. This approximation will often give comparable results to that provided by Fieller's theorem, but can also suffer from selected instabilities in its true confidence level, relative to the Fieller result (Buonaccorsi, 2002). In general, we recommend use of Fieller's theorem (1.14) over the delta method in situations such as this.

Example 1.7 (Rice yield, cont'd) Returning to the rice yield data in Table 1.5, suppose there is interest in determining the temperature at which the minimum yield is attained, and, in particular, in providing a confidence interval for this value. (Once the daily minimum temperature enters this range, agricultural managers could be alerted that poor yield conditions are imminent.) From the SAS analysis in Fig. 1.13, we find $\hat{x}_{opt} = \bar{x} - b_1/2b_2 = 23.197 + 4.362 = 27.559\,°C$.

To calculate 95% Fieller limits from (1.14), we require $se[b_1] = 0.0337$, $se[b_2] = 0.0094$, and $t_{0.025}(37) = 2.0262$. For the estimated covariance, $\hat{\sigma}_{12}$, return to PROC REG and update the model statement to

```
model yield = x x2 / covb;
```

in Fig. 1.12. This produces the output

```
                  Covariance of Estimates
COVB              INTERCEP              X                   X2
INTERCEPT    0.0141959035      -0.000528327       -0.000700488
X           -0.000528327        0.0011370764        0.0000665211
X2          -0.000700488        0.0000665211        0.0000881977
```

from which we read $\hat{\sigma}_{12} = 6.65211 \times 10^{-5}$. Appeal to (1.14) then yields the 95% limits $25.903 < x_{opt} < 31.856\,°C$. Over this range of minimum temperatures, we are 95% confident that the poorest conditions for average rice production will occur. ✪

There are many other aspects to proper use and application of polynomial regression models, including proper regression diagnostics, variable selection, optimal design, higher-order polynomial models, and response-surface models. The exercises below study some of these issues to a limited extent. Interested readers can find more on these topics in regression textbooks, including Neter *et al.* (1996. §7.7) and Christensen (1996, §§7.11–12).

Exercises

1.1. Dalgård *et al.* (1994) reported on a study of mercury (Hg) toxicity in pregnant Faroe islanders, where potentially high mercury body burdens occur through the islanders' large consumption of pilot whale meat. Interest included the relation between $x =$ daily Hg ingestion (calculated in µg) and $Y = $ Hg concentration in the woman's umbilical cord blood (in µmol/l, recorded immediately after giving birth). A sample of $n = 12$ women produced the following data:

$x =$ Hg ingestion	1.4	49	90	96	108	125
$Y =$ cord blood Hg	0.007	0.23	0.43	0.46	0.52	0.60
$x =$ Hg ingestion	146	153	233	324	354	671
$Y =$ cord blood Hg	0.70	0.73	1.12	1.56	1.70	3.22

(a) Plot the data. Does a simple linear model seem reasonable for representing the relationship between x and Y?

(b) Assume the simple linear model $Y_i = \beta_0 + \beta_1 x_i + \varepsilon_i$, and find the LS/ML estimators b_0 and b_1. Also find the associated standard errors $se[b_0]$ and $se[b_1]$. Use these to find a 95% confidence interval for β_1.

(c) A natural question to ask is whether increasing Hg intake increases Hg cord blood concentrations. This translates into a test of $H_0: \beta_1 = 0$ vs. $H_a: \beta_1 > 0$. Test these hypotheses at significance level $\alpha = 0.01$. What do you conclude?

(d) Calculate the predicted values and residuals from this LS/ML fit and construct a residual plot. What does the plot suggest?

1.2. Todd *et al.* (2001) report data on lead (Pb) concentrations in human bone tissue. For example, in human tibias they note that core Pb concentrations (in $\mu g/g$) can often be predicted from surface concentrations.

(a) At the proximal end of the tibia (closest to the knee), data are available from $n = 9$ independent cadavers:

x = surface Pb ($\mu g/g$)	22.5	5.2	11.8	16.5	12.5
Y = core Pb ($\mu g/g$)	12.0	3.8	6.7	9.0	8.1
x = surface Pb ($\mu g/g$)	21.5	22.4	13.6	5.8	
Y = core Pb ($\mu g/g$)	13.9	14.2	7.8	3.1	

Analyze these data by fitting a simple linear regression model. Test if there is a significant linear relationship between surface and core Pb concentrations at significance level $\alpha = 0.05$. What do you conclude?

(b) Calculate the predicted values and residuals from the fit in part (a) and construct a residual plot. What does the plot suggest?

(c) Predict the mean value of core Pb concentration at the new surface Pb level $x = 25\,\mu g/g$, that is, find $\hat{\mu}(25)$. Also find the standard error of your estimate and use it to calculate a 95% confidence interval on the true value of $\mu(25)$.

(d) At the distal end of the tibia (farthest from the knee), $n = 9$ similar observations are taken:

x = surface Pb ($\mu g/g$)	25.8	6.4	16.7	20.2	15.2
Y = core Pb ($\mu g/g$)	14.4	5.4	8.8	8.2	11.0
x = surface Pb ($\mu g/g$)	24.2	25.6	11.0	10.0	
Y = core Pb ($\mu g/g$)	15.7	16.3	7.6	4.7	

Analyze these data by fitting a simple linear regression model. Test if there is a significant linear relationship between surface and core Pb concentrations at significance level $\alpha = 0.05$. What do you conclude?

(e) Calculate the predicted values and residuals from the fit in part (d) and construct a residual plot. What does the plot suggest?

1.3. In a study of agricultural nutrient management and environmental quality, data were collected on the response of a feed crop to increasing nitrogen (N) fertilizer. (The goal was to use the minimum nitrogen possible to achieve target yields, consistent with good environmental practice.) The response variable was $Y =$ yield (bu/acre), vs. $x =$ applied nitrogen (lb/acre). The data are:

$x =$ Nitrogen (lb/acre)	20	60	120	180
$Y =$ yield (bu/acre)	22	17	39	61
	15	28	46	36
	43	34	42	71

(a) Calculate the sample mean and sample variance at each level of x_i. Is the variance stable?

(b) If your calculations in part (a) indicate that there is variance heterogeneity, fit a weighted LS regression, using x_i as the single predictor variable. (Use for your weights the reciprocals of the observed sample variances at each level of x_i. That is, calculate the sample variances, S_i^2, for the three observations at each x_i, and use $w_i = 1/S_i^2$ for the weights. Notice that for all observations within a triplet, the weights will be equal.) Test the hypothesis that mean yield does not vary across nitrogen applications. Set $\alpha = 0.01$.

1.4. Return to the multiple linear regression analysis for the soil pH data from Example 1.3.

(a) Calculate the VIFs for each of the predictor variables. Is multicollinearity a concern here?

(b) Acquire the residuals and the predicted values from the full three-predictor fit, and construct a residual plot. Also construct individual plots of the residuals vs. each of the predictor variables. (In PROC REG, this can be accomplished by invoking the command

```
plot r.*(p. x1clay x2org x3carb);
```

after the `model` command.) Do you see anything unusual in the plots?

1.5. As part of a study on water resource management, Lu *et al.* (1999) present data on seven-day streamflow rates (in ft^3/s) over a series of streams in west-central Florida. To model two-year recurrence flow, four possible hydrological predictor variables were collected: $x_{i1} =$ drainage area (m^2), $x_{i2} =$ basin slope (ft/m), $x_{i3} =$ soil-infiltration index (inches), and $x_{i4} =$ rainfall index (inches). The response variable, Y_i, was taken to be the logarithm of the two-year recurrence flow rates. After removing any zero/censored flow rates, the $n = 45$ data values were:

Y_i	x_{i1}	x_{i2}	x_{i3}	x_{i4}	Y_i	x_{i1}	x_{i2}	x_{i3}	x_{i4}
13	180	0.93	2.66	5.27	1.7	24	3.66	2.05	10.06
170	640	1.09	4.14	3.15	1.2	65.3	5.83	2.05	10.06
3.1	41.2	14.96	12.2	6.19	4.0	80.0	4.93	2.05	8.56
13	469	2.00	4.12	8.54	9.1	149	5.03	2.55	9.56
1.1	83.6	2.04	3.08	5.34	40	135	4.96	2.70	8.98
4.8	89.2	1.78	2.28	5.34	12	107	3.60	2.05	8.98
4.7	75.0	3.57	5.38	5.34	43	335	3.45	2.14	4.94
0.9	23.0	4.17	5.38	5.34	20.0	28	2.07	5.38	4.94
16	58.9	1.66	4.88	5.34	0.1	9.5	6.70	2.05	8.98
8.9	60.9	3.12	4.21	8.04	4.8	110	3.52	2.72	8.98
9.1	38.8	4.9	4.38	8.04	70	220	3.87	2.77	4.94
36	379	1.4	2.88	8.04	0.8	375	2.41	2.73	4.94
8.1	109	3.81	5.14	8.04	1.4	35	4.12	2.23	4.94
7.7	160	1.26	5.20	8.98	2.4	72.5	3.54	2.42	7.23
26.0	390	1.25	4.05	8.98	0.9	182	1.97	4.11	7.23
4.8	121	3.88	2.05	8.98	30.0	570	1.35	2.70	11.42
100	826	1.38	4.05	8.04	2.3	145	1.75	2.71	11.42
3.2	330	1.68	2.08	6.93	73.0	810	1.24	5.37	11.42
110	367	1.3	3.04	6.93	13.0	40	2.10	5.38	9.27
1.7	132	4.06	2.03	6.93	330	1825	0.86	3.89	9.27
1.1	218	2.79	2.05	6.93	15.0	300	1.58	2.00	3.33
3	223	2.54	2.00	6.93	0.1	26	5.9	2.00	3.33
0.6	87.5	2.00	2.00	10.06					

(a) Calculate the VIFs for each predictor variable. Is multicollinearity a concern?

(b) Fit a multiple linear regression model to these data, using all four predictor variables. Perform a single (2 df) test to see if x_{i2} and x_{i4} are required in the model. Set $\alpha = 0.01$.

(c) Lu *et al.* studied models with polynomial and cross-product terms to see if these could provide a better fit for the data. To explore this, define the new predictor variables $x_{i5} = (x_{i3} - \bar{x}_3)^2$ and $x_{i6} = (x_{i3} - \bar{x}_3)^3$, where $\bar{x}_3 = \sum_{i=1}^{45} x_{i3}/45$. Fit the four predictors $x_{i1}, x_{i3} - \bar{x}_3, x_{i5}$, and x_{i6} to the data, and calculate the new MSE. Is this smaller than the MSE from the fit in part (b)? (If so, it would indicate a more precise fit, since both models employ the same number of predictor variables.)

(d) To your fit in part (c) also add the following cross-product terms: $x_{i7} = x_{i1}(x_{i3} - \bar{x}_3)$ and $x_{i8} = x_{i1}(x_{i3} - \bar{x}_3)^2$. Test to see if either or both these terms contribute significantly to the model fit. Operate at $\alpha = 0.01$ throughout.

1.6. Under the one-factor ANOVA model (1.8), the null hypothesis of no effect due to the qualitative factor is $H_0: \alpha_1 = \cdots = \alpha_a$. Show that under either the zero-sum constraint or the corner-point constraint, this is equivalent to $H_0: \alpha_1 = \cdots = \alpha_a = 0$.

1.7. As part of a study on control of hazardous waste, Stelling and Sjerps (1999) report data on total polycyclic aromatic hydrocarbons (PAHs) observed in lots of fractionated demolition waste (in mg/kg). The original data are as follows:

Lot 1:	44.5, 86, 115, 120, 127, 136, 142, 147, 154, 240
Lot 2:	74.9, 85.4, 85.9, 97.5, 102, 130, 139, 151, 169, 245
Lot 3:	3.4, 11.2, 11.6, 21.9, 24.0, 29.8, 29.9, 29.9, 41.3, 51.3
Lot 4:	76.3, 97.7, 117, 120, 120, 121, 127, 132, 145, 249
Lot 5:	55, 74.2, 86.2, 114, 122, 125, 130, 137, 226, 258
Lot 6:	53.3, 56.3, 73.1, 85.2, 90.7, 91.8, 96.5, 99.4, 102, 126

Concentration data can often depart from normality. To adjust for this, apply a log-transformation and analyze the transformed data to assess if there is any difference in log-PAH concentrations across lots. Set your significance level to 10%. Include a residual analysis to check for any model irregularities.

1.8. Return to the macroinvertebrate ratio data in Table 1.3 and complete the two-factor analysis as follows.

(a) Since a significant interaction was observed, study the season effect by performing three separate one-factor ANOVAs, stratified over levels of stream. Is there a significant effect due to season at any of the three streams? Set your significance level to 1%.

(b) Since, in effect, you performed three different tests in part (a), adjust your inferences via a Bonferroni correction (see §A.5.4). Do the results change at familywise significance level $\alpha = 0.01$?

(c) Mimic the analysis in part (a) and study the stream effect by performing four separate one-factor ANOVAs, stratified over levels of season. Is there a significant effect due to streams in any of the four seasons? Set your significance level to 1%.

(d) Since you performed four different tests in part (c), adjust your inferences via a Bonferroni correction (see §A.5.4). Do the results change at familywise significance level $\alpha = 0.01$?

1.9. Return to the faba bean genotoxicity data from Example 1.5.

(a) Test to see if the standard ANCOVA assumption of common slopes is valid for the logit-transformed data; that is, under (1.12) test $H_0: \beta_1 = \cdots = \beta_a$. Operate at the 5% significance level.

(b) Reanalyze these data using a one-factor ANOVA model, ignoring the covariate. Comment on whether it was beneficial to adjust for potential covariate differences in this analysis.

1.10. Malling and Delongchamp (2001) reported data on organ weight changes in laboratory mice exposed to the chemical mutagen ethylnitrosourea (ENU). Of interest was whether ENU exposure affects organ weights in these mice, compared to a group of control animals. The study protocol allowed for different-aged mice to be used, hence some adjustment for murine age should be included in any analysis of the ENU effect. This is essentially an ANCOVA problem: we view age as the quantitative covariate, and exposure condition (ENU or control) as the qualitative variable of interest. For spleen weights, the data are:

Controls: Age (d)	434	434	427	422	408
Spleen weight (mg)	71.4	85.9	72.9	75.3	77.2
ENU: Age (d)	392	435	434	427	427
Spleen weight (mg)	187.6	123.0	137.2	125.3	111.2

(a) Plot the spleen weights as a function of age, overlaying the control and exposure groups on the same graph. Does the assumption of equal slopes – as used in (1.10) – appear valid for these data?

(b) Based on your results from part (a), analyze these data using an appropriate ANCOVA model. Assess if there is an effect on spleen weights due to ENU exposure, after adjusting for any age effects. What do you conclude at the 5% significance level?

(c) Perform a residual analysis to check for any model irregularities.

1.11. Return to the rice yield data in Example 1.6.

(a) Seshu and Cady (1984) included a number of factors beyond minimum temperature in their study of environmental effects on rice yields. For example, they also measured x_{i3} = solar radiation (mWh/cm^2) at each location. The full data are:

Y_i	x_{i1}	x_{i3}	Y_i	x_{i1}	x_{i3}
2.3	29.2	421	3.0	23.4	483
3.1	28.1	362	4.4	23.4	537
2.8	27.2	404	3.1	23.2	418
2.4	26.4	340	2.6	23.1	451
3.6	26.3	431	4.8	23.1	618
2.3	26.2	427	3.2	23.0	419
3.8	26.2	595	3.3	22.9	472
3.1	26.0	365	3.1	22.5	490
2.4	25.9	372	3.4	22.5	484
2.1	25.7	435	3.2	22.4	423
3.5	24.5	457	4.2	21.7	428
3.8	24.4	609	4.5	21.2	373

3.1	24.0	469	4.7	20.0	438
2.9	23.9	459	5.0	19.2	326
3.2	23.9	481	6.2	19.0	521
3.0	23.7	379	6.0	19.0	609
4.5	23.7	637	6.1	18.8	528
3.5	23.7	529	7.3	18.0	566
3.3	23.6	455	6.6	18.0	315
3.7	23.5	512	6.2	17.4	370

Study the impact of this additional predictor by fitting it along with $x_{i1} =$ minimum temperature and $x_{i2} = x_{i1}^2$ (from Example 1.6). Also include a term for possible quadratic effects of $x_{i3} =$ solar radiation; that is, include $x_{i4} = x_{i3}^2$ in the fit. (Remember to center each predictor variable about its mean.) Test to determine if the quadratic solar radiation term is significant at $\alpha = 0.01$. If the quadratic term is insignificant, test to determine if solar radiation contributes significantly to the model, after allowing for a quadratic effect due to minimum temperature. Here also, operate at $\alpha = 0.01$.

(b) To your fit in part (a), add an additional term for interaction between temperature and radiation, $x_{i5} = x_{i1}x_{i3}$. Test to determine if the interaction is significant at $\alpha = 0.01$.

(c) If either the solar radiation terms or the radiation $\times$ temperature interaction (or all) were significant in your fit from parts (a) and (b), acquire the predicted values, $\hat{Y}_i$, and the residuals, $Y_i - \hat{Y}_i$, from the fit. Graph (i) a residual plot to assess the quality of the fit, and (ii) $\hat{Y}_i$ vs. x_{i1} and x_{i3} in three dimensions. The latter plot will visualize the *response surface* (Khuri and Cornell, 1996) created by the joint effects of the two predictor variables. What do you conclude from the two plots?

1.12. Bates and Watts (1988, §A1.1) present data on the concentration of polychlorinated biphenyl (PCB, in parts per million) residues in trout from a northeastern US lake, as a function of age of the fish (in years). (PCBs are mixtures of organochlorines and other chemicals used in a wide variety of industrial and commercial applications prior to 1977. Their ubiquitous use led to ecosystem contamination in many developed nations. Study of their accumulation throughout the food web is an important concern.) The data are as follows:

Age	1	1	1	1	2	2	2
PCB conc.	0.6	1.6	0.5	1.2	2.0	1.3	2.5
Age	3	3	3	4	4	4	5
PCB conc.	2.2	2.4	1.2	3.5	4.1	5.1	5.7
Age	6	6	6	7	7	7	8
PCB conc.	3.4	9.7	8.6	4.0	5.5	10.5	17.5
Age	8	8	9	11	12	12	12
PCB conc.	13.4	4.5	30.4	12.4	13.4	26.2	7.4

(a) Concentration data often depart from normality. To adjust for this, apply a log transformation to the PCB concentrations. Plot the data. What pattern do you see?

(b) Bates and Watts (1988) suggest that a simple linear fit to these data can be improved by considering a transformation of the age variable. They suggest $x = (\text{age})^{1/3}$. Make this transformation and plot the transformed data. Comment on the result.

(c) Fit a simple linear model to the transformed data; that is, fit $\log(\text{concentration}) = \beta_0 + \beta_1(\text{age})^{1/3} + \varepsilon$. Test if there is a significant effect due to (transformed) age at $\alpha = 0.01$.

(d) Calculate residuals from your fit in part (c) and plot them against age. Comment on the pattern. Based on your results, suggest a revised model and fit it to the data. Identify whether this adds significantly to the model fit. Continue to operate at $\alpha = 0.01$.

1.13. In a study of the long-term effects of radiation exposure in humans, data were taken on the frequencies of mutations (as average chromosome aberrations in peripheral blood cells) seen in survivors of atomic radiation exposure in Hiroshima, Japan. Subjects were also assessed for their estimated radiation exposures (in rems). The data are as follows:

Radiation exposure	1.980	2.540	4.670	5.000	6.890	8.120
Mutation frequency	40.0057	41.2645	45.9816	45.9828	52.2148	54.7815
Radiation exposure	9.995	11.578	14.400	15.001	16.870	19.560
Mutation frequency	59.6586	63.0588	70.0128	71.3794	75.5131	80.6950

Since mutation frequencies often exhibit a right skew (to higher values), take as your response variable the natural logarithms: $Y_i = \log(\text{mutation frequency})$. Assume that these transformed variates are normally distributed with constant variance.

(a) For $x_i = $ estimated radiation exposure, plot Y_i vs. x_i. What does the plot show?

(b) Fit a simple linear model and assess if radiation exposure has a significant effect on log-mutant frequencies. Operate at $\alpha = 0.01$.

(c) Find the predicted values and from these calculate residuals from your fit in part (b). Construct a residual plot and comment on the fit.

(d) Based on your results in part (c), suggest an additional term for the model. Fit this term and identify whether it adds significantly to the model. Continue to operate at $\alpha = 0.01$.

1.14. Ribakov et al. (2001) report on engineering efforts to study building response to earthquake stress after reinforcement with active viscous damping. Observations were taken on $Y_i = $ peak displacements (cm) as a function of height above ground level (represented here as x_i = story number). Data from the 1989 Loma-Prieta, CA, earthquake were given as follows:

Story	1	2	3	4	5	6	7
Displacement	0.2	0.41	0.55	0.71	0.85	0.94	0.95

(a) Plot Y vs. x. What does the plot show?

(b) Fit a quadratic regression model and assess if there is an effect of height on displacement. Use story number as a surrogate for height. (Remember to center the predictor variable about its mean.) Operate at $\alpha = 0.01$.

(c) The effect of centering the predictor variable is especially striking here. What is the sample correlation between $x_i - \bar{x}$ and $(x_i - \bar{x})^2$? Examine the estimated covariance between b_1 and b_2: fit the model with and without centering and study $\mathrm{Cov}[b_1, b_2]$.

(d) Estimate the story height at which peak displacement occurs. Include a 99% confidence interval (using Fieller's theorem) for this point. What cautions might you give regarding these estimates?

1.15. Similar to the study in Exercise 1.14, Ribakov *et al.* (2001) also report data on Y_i = peak displacements (cm) as a function of height above ground level (as x_i = story number) from the 1995 Eilat, Israel, earthquake, this time for buildings equipped with electrorheological dampers. The data are:

Story	1	2	3	4	5	6	7
Displacement	0.37	0.73	0.99	1.27	1.47	1.61	1.68

(a) Plot Y vs. x. What does the plot show?

(b) Fit a quadratic regression model and assess if there is an effect of height (as story number) on displacement. Remember to center the predictor variable about its mean. Operate at $\alpha = 0.05$.

(c) Estimate the story height at which peak displacement occurs. Include a 95% confidence interval (using Fieller's theorem) for this point. What cautions might you give regarding these estimates?

1.16. Derive the general version of Fieller's theorem (Buonaccorsi, 2002) via the following steps:

(a) Assume we have two parameters, θ_1 and θ_2, for which we have unbiased estimators $\hat{\theta}_1$ and $\hat{\theta}_2$; that is, $E[\hat{\theta}_j] = \theta_j$, $j = 1, 2$. We desire confidence limits for the ratio $\varphi = \theta_1/\theta_2$. Start with $D = \hat{\theta}_1 - \varphi\hat{\theta}_2$. Show that $E[D] = 0$.

(b) Suppose the standard error of $\hat{\theta}_j$ in part (a) is $se[\hat{\theta}_j]$, $j = 1, 2$. Also let $\hat{\sigma}_{12} = \mathrm{Cov}[\hat{\theta}_1, \hat{\theta}_2]$. Show that the standard error of D, $se[D]$, is the square root of $se^2[\hat{\theta}_1] + \varphi^2 se^2[\hat{\theta}_2] - 2\varphi\hat{\sigma}_{12}$.

(c) Assume that when standardized, D is approximately normal, $(D - E[D])/se[D] \dot\sim N(0, 1)$, in large samples. (The dot notation above the $\sim$ indicates that the distributional relationship is only approximate. The approximation improves as $n \to \infty$.) Then a set of confidence limits could be derived from the relationship $P\{|D - E[D]|/se[D] \leq z_{\alpha/2}\} = 1 - \alpha$. Show that this equivalent to

$$P\left\{(D - E[D])^2/se^2[D] \leq z_{\alpha/2}^2\right\} = 1 - \alpha.$$

(d) Express the event from the probability statement in part (c) as $\{D - E[D]\}^2 \leq z_{\alpha/2}^2 se^2[D]$. (Why does the inequality's direction remain the same?) Substitute the appropriate quantities from parts (a) and (b) into this expression for each quantity involving D. Show that this leads to a quadratic inequality in φ.

(e) Take the inequality in part (d) and operate with the points at equality. Solve for the two roots of this quadratic equation. For simplicity, let $\gamma = z_{\alpha/2}^2 se^2[\hat\theta_2]/\hat\theta_2^2$.

(f) Manipulate the two roots to produce

$$\hat\varphi + \frac{\gamma}{1 - \gamma}\left\{\hat\varphi - \frac{\hat\sigma_{12}}{se^2[\hat\theta_2]}\right\}$$

$$\pm \frac{z_{\alpha/2}}{(1 - \gamma)|\hat\theta_2|}\left\{se^2[\hat\theta_1] + \hat\varphi\left(se^2[\hat\theta_2]\hat\varphi - 2\hat\sigma_{12}\right) - \gamma\left(se^2[\hat\theta_1] - \frac{\hat\sigma_{12}^2}{se^2[\hat\theta_2]}\right)\right\}^{1/2},$$

where $\hat\varphi = \hat\theta_1/\hat\theta_2$. Note that if the large-sample approximation in part (c) is in fact exact in small samples, then we would write $(D - E[D])/se[D] \sim t(\nu)$, where ν are the df associated with estimating the standard error of D. If so, we would replace $z_{\alpha/2}^2$ with $t_{\alpha/2}^2(\nu)$.

2

Nonlinear regression

When a curvilinear relationship exists between the mean response and a posited predictor variable, the basic linear forms described in Chapter 1 may be inadequate to model the relationship. Even the low-order polynomial models described in §1.5 may not suffice, since the strict polynomial forms are not always flexible enough to model great curvature or asymptotic patterns in the response. Fortunately, many other functions can serve as viable regression models for environmental data. These functions cannot typically be transformed or linearized into a multiple linear regression model, and are therefore truly nonlinear. In this chapter, we fit regression models to normal response variables where the mean is a nonlinear function of the model parameters, expanding on the paradigm in Chapter 1 where the models could be nonlinear in the predictor variables but were linear in the parameters. (In Chapter 3, we will consider regression models for non-normal responses that also may be nonlinear in the model parameters.)

To begin, we work with a single predictor variable, x_i, and reapply the regression relationship from (1.1):

$$Y_i = g(x_i; \boldsymbol{\beta}) + \varepsilon_i, \tag{2.1}$$

$i = 1, \ldots, n$, where now $g(x; \boldsymbol{\beta})$ is some prespecified function, nonlinear in one or more of x and $\boldsymbol{\beta} = [\beta_0 \ \beta_1 \ldots \beta_p]^T$, and where $\varepsilon_i \sim$ i.i.d. $N(0, \sigma^2)$. Examples of such nonlinear models include: (i) a simple exponential model $g(x; \boldsymbol{\beta}) = \beta_1 e^{\beta_2 x}$; (ii) a biexponential model $g(x; \boldsymbol{\beta}) = \beta_0 + \beta_1 e^{\beta_2 x} + \beta_3 e^{\beta_4 x}$ (and its multiexponential extensions); or (iii) rational polynomials such as $g(x; \boldsymbol{\beta}) = (\beta_0 + \beta_1 x)/(\beta_2 + \beta_3 x)$. (We discuss a broad range of nonlinear models in the sections that follow.)

Analyzing Environmental Data W. W. Piegorsch and A. J. Bailer
© 2005 John Wiley & Sons, Ltd   ISBN: 0-470-84836-7 (HB)

2.1 Estimation and testing

Estimation of the parameters in $\boldsymbol{\beta}$ again proceeds via least squares, using the objective quantity

$$D_g = \sum_{i=1}^{n}\{Y_i - g(x_i; \boldsymbol{\beta})\}^2$$

as in equation (A.19). Minimizing D_g with respect to $\boldsymbol{\beta}$ produces the nonlinear LS estimates $\mathbf{b} = [b_0\ b_1 \ldots b_p]^{\mathrm{T}}$. Under a normal distribution assumption on the additive errors, these nonlinear LS estimates coincide with the ML estimator for $\boldsymbol{\beta}$. If prior evidence or preliminary analysis suggests that the errors have heterogeneous variances, the objective quantity can be extended to include weights, w_i, that are inversely proportional to some measure of the changing variance, as in equation (A.20). Similar to the construction for the simpler linear model in equations (1.5) and (1.6), we use the w_is to produce nonlinear weighted LS (WLS) estimates for $\boldsymbol{\beta}$.

Unfortunately, for most choices of $g(x; \boldsymbol{\beta})$ it is impossible to express the nonlinear LS/ML estimators in closed form. Thus, a numerical solution for the $\boldsymbol{\beta}$ estimator is usually obtained. One starts with a set of initial estimates for $\boldsymbol{\beta}$ and then iterates towards a stationary point of D_g, typically by employing information about the partial derivatives (with respect to the β_js) of D_g. Known as *gradients*, these derivatives reflect the change of the objective function with respect to each of the parameters. They provide information on two relevant aspects of the nonlinear regression fit: they give a direction for the fitting algorithm to move so that the iterative value of D_g improves; and they suggest when the fitting exercise has converged (e.g., when each gradient element is sufficiently close to zero).

Parameter estimation often requires knowledge of these gradients and hence of $g(x; \boldsymbol{\beta})$. As a result, to fit nonlinear regression models one must make a number of important decisions regarding details of the computational process. These include:

(a) *Choice of iterative algorithm to minimize D_g*. Traditionally, nonlinear LS is implemented via appeal to the Gauss–Newton method (Bates and Watts, 1988, §2.2.1) or modifications thereof, where the nonlinear LS surface is approximated by a linearized expression. Using the gradients, iterative steps towards the LS solution are taken based on the linear approximation, until no further improvement can be made in D_g. At that point, we say the steps have *converged* to a nonlinear LS solution. Unfortunately, one cannot always guarantee that such a LS solution will be achieved; in some cases the iteration can converge to a local minimum or 'pocket', and not the global LS solution.

An important modification to Gauss–Newton is known as the Levenberg–Marquardt algorithm (Levenberg, 1944; Marquardt, 1963), which is used to increase stability in the iterative process. (In most cases, we recommend the Levenberg–Marquardt method when performing nonlinear LS fits.) Both Gauss–Newton and Levenberg–Marquardt employ information in the gradients of D_g, and hence require specification of the partial derivatives of $g(x; \boldsymbol{\beta})$.

See Press *et al.* (1992, §15.5) or Smythe (2002) for more on these methods of nonlinear optimization.

(b) *Use of derivative-free methods.* Some implementations of nonlinear LS allow for derivative-free optimization; that is, they approximate or estimate information in the gradient from previous or current iterations, and hence do not require formal specification of the gradient functions. Although these tend to be slower than their derivative-based counterparts, the algorithms can be useful when it is difficult to make the derivative/gradient specification.

(c) *Good initial values.* Most nonlinear LS algorithms will perform poorly if the initial estimates are far from the optimum point. The analyst must therefore provide intelligent, accurate starting values whenever employing iterative optimization algorithms. In some cases, it may be necessary to provide a grid of multiple initial values to explore the stability of a numerical solution. We will highlight considerations along these lines throughout the examples in this chapter.

(d) *Software.* This effort is by nature computer-intensive. Many packages and programs exist for performing nonlinear optimizations, and in particular nonlinear LS regression. Below, we will illustrate use of PROC NLIN in SAS.

(e) *Creativity.* Solutions to some nonlinear functions may be difficult to obtain. If an initial calculation using a certain method fails to converge, the analyst must reassess any or all of the decisions made above and try other options (different algorithms, updated starting values, different parameterizations of the model, combinations of approaches, etc.). Indeed, for highly nonlinear functions and with small sample sizes two different analysts may very well arrive at two (slightly) different LS solutions after fitting the same model to the same set of data! Usually, however, inferences derived from the two LS fits will agree within practical bounds.

When these various choices lead to a nonlinear LS solution, the point estimates can be used to construct predicted values, $\hat{Y}_i = g(x_i; \mathbf{b})$, find standard errors, $se[b_j]$, and compute $\text{SSE} = \sum_{i=1}^{n} (Y_i - \hat{Y}_i)^2$. From these, statistical inferences may be conducted. For example, to test the significance of some subset of $\boldsymbol{\beta}$, say, $\{\beta_1, \beta_2, \ldots, \beta_k\}$, fit the full model (FM) with all $p+1$ parameters and compare it to the reduced model (RM) with those k β_js set equal to zero. This is just the discrepancy-measure approach from (1.7), where we compute SSE(RM), SSE(FM), and the resulting F-statistic, F_{calc}. In large samples $F_{\text{calc}} \overset{\cdot}{\sim} F(k, n - p - 1)$, so we view the collection of k parameters as significant when $F_{\text{calc}} \geq F_\alpha(k, n - p - 1)$. (The dot notation above the $\sim$ indicates that the distributional relationship is only approximate. The approximation improves as $n \to \infty$.) When the errors are normally distributed, the F-based discrepancy approach also corresponds to a likelihood ratio (LR) test. (LR tests are reviewed in §A.5.3.)

For individual inferences on any specific β_j, we can construct confidence intervals or perform hypothesis tests using the t-distribution. For example, an approximate $1 - \alpha$ confidence interval on β_j is $b_j \pm t_{\alpha/2}(n - p - 1) \, se[b_j]$. For small n, and depending on the nonlinear function being modeled, the t-approximation here can be unreliable, so users should proceed with caution. For simultaneous $1 - \alpha$ confidence *bands* on the mean response function $g(x_i; \boldsymbol{\beta})$, see Khorasani and Milliken (1982) and Cox and Ma (1995).

In the following sections, we describe a series of different nonlinear regression models, and illustrate their use in selected environmental applications. Our presentation is not intended to be complete; readers wishing a broader discussion of the issues and details of nonlinear regression modeling should consult dedicated sources such as Bates and Watts (1988), Seber and Wild (1989), or the handbook by Ratkowsky (1990).

2.2 Piecewise regression models

Perhaps one of the most intriguing nonlinear models seen in environmental applications is the piecewise model, also known as a *segmented regression model*. An elementary example is a piecewise linear model – sometimes called a *bilinear model* – made up of two straight lines separated at a *change point* somewhere along the range of the predictor variable. When the change point is unknown the model is highly nonlinear, hence its intriguing features. Segmented models can be continuous (if the segments join at the change point), or discontinuous (if at the change point a sudden jump or drop occurs in the mean response). If continuous, the change point is also called a *join point* or *knot*. For continuous models, the segmentation can be smooth or abrupt, depending on whether the function's derivative with respect to x exists at the change point. All of these features can be modeled via manipulation of the unknown parameters of the segmented model.

In the simplest case, one of the straight lines is a horizontal plateau. This is a *plateau model*, also called a *threshold model* when the horizontal response corresponds to some threshold phenomenon. For a left-plateau model, we write

$$g(x_i; \boldsymbol{\beta}) = \begin{cases} \beta_0 & \text{if } x_i \leq \tau, \\ \beta_1 + \beta_2(x_i - \tau) & \text{if } x_i > \tau, \end{cases} \qquad (2.2)$$

for use in (2.1). Notice the inclusion of the change point, $x = \tau$, directly in the model expression. This facilitates parameter estimation and inference. We assume that τ is unknown, so that technically it is part of the unknown parameter vector $\boldsymbol{\beta}$: under (2.2), $p = 4$. In the special case of $\beta_0 = 0$, we say (2.2) is a *truncated linear regression model*. More generally, if the horizontal level of β_0 is a known constant, the analysis can simplify somewhat; cf. Exercise 2.2.

The construction in (2.2) assumes that the mean response at the change point is discontinuous. Thus $g(x_i; \boldsymbol{\beta})$ jumps (or drops) a distance of $\beta_1 - \beta_0$ at $x = \tau$. If continuity is imposed, (2.2) simplifies to

$$g(x_i; \boldsymbol{\beta}) = \begin{cases} \beta_0 & \text{if } x_i \leq \tau, \\ \beta_0 + \beta_2(x_i - \tau) & \text{if } x_i > \tau \end{cases} \qquad (2.3)$$

(notice that the restriction $\beta_0 = \beta_1$ has led to loss of the β_1 parameter); see Exercise 2.1. Both (2.2) and (2.3) can be easily modified to incorporate a right rather than a left plateau.

For unknown τ, any plateau model is inherently nonlinear. (For known τ, the model can be fitted via multiple linear regression; see Neter *et al.*, 1996, §11.5.) To find point estimates for the β-parameters and τ, we turn to nonlinear LS. Inferences on these parameters follow using the large-sample methods outlined in §2.1. The next example illustrates the approach using PROC NLIN in SAS.

Example 2.1 (Lizard development)   Braña and Ji (2000) report data on embryogenic response in wall lizards (*Podarcis muralis*) by manipulating selected environmental conditions associated with their embryonic development. For example, after incubating $n = 32$ *P. muralis* eggs at temperatures near the upper extreme of their usual range (32 °C), Braña and Ji studied $Y =$ hatchling mass (g) as a function of $x =$ snout-vent length (mm). Their research suggested that hatchling mass responded in a possibly segmented fashion: at small snout-vent lengths hatchling masses appeared roughly constant, but after a certain point the length-to-mass relationship appeared to increase in a linear fashion. This could be viewed as a threshold response, calling for use of a model such as (2.2) or (2.3). We will choose the latter, since the continuity assumption seems reasonable here. The data, as read from Braña and Ji's Fig. 1, appear in Table 2.1. The data are displayed as part of Fig. 2.1.

To fit the continuous (left) threshold model we employ PROC NLIN with the Levenberg–Marquardt fitting algorithm. This requires the partial derivatives of (2.3) with respect to the unknown parameters. To facilitate determination of these expressions, we employ indicator functions, $I_{\mathbb{A}}(x)$. (An indicator function has the

Table 2.1   Hatchling mass (g) vs. snout-vent length (mm) in wall lizards (*Podarcis muralis*) after eggs were incubated at 32 °C

$x =$ length	$Y =$ mass	$x =$ length	$Y =$ mass
22.87	0.294	25.26	0.323
23.45	0.302	25.36	0.353
23.49	0.265	25.47	0.354
23.65	0.297	25.52	0.350
23.76	0.294	25.61	0.361
24.36	0.338	25.76	0.362
24.44	0.295	25.82	0.327
24.44	0.347	25.86	0.354
24.51	0.338	25.91	0.309
24.61	0.333	25.96	0.361
24.91	0.358	25.96	0.366
24.95	0.350	26.15	0.344
24.95	0.331	26.20	0.358
25.00	0.327	26.27	0.348
25.16	0.345	27.12	0.371
25.26	0.334	27.28	0.421

Source: Braña and Ji (2000).

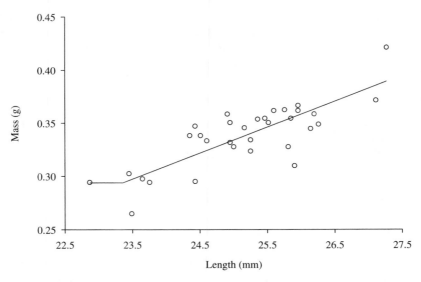

Figure 2.1 Scatterplot for lizard development data. Also included are the segmented regression predicted lines from a nonlinear LS fit of model (2.3)

form $I_\mathbb{A}(t) = 1$ if t is contained in the set $\mathbb{A}$, and $I_\mathbb{A}(t) = 0$ otherwise; cf. §A.2.1.) The segmented linear function can be written as $g(x; \boldsymbol{\beta}) = \beta_0 + \beta_2(x - \tau)I_{(\tau, \infty)}(x)$, where $I_{(\tau, \infty)}(x)$ equals 1 when $x > \tau$, and 0 when $x \leq \tau$. From this, the first derivatives with respect to the β-parameters are simply $\partial g/\partial \beta_0 = 1$ and $\partial g/\partial \beta_2 = (x - \tau)I_{(\tau, \infty)}(x)$. For $\partial g/\partial \tau$, notice that the indicator $I_{(\tau, \infty)}(x)$ is identical to the indicator $I_{(-\infty, x)}(\tau)$. Hence, we can write $g(\cdot)$ directly as a function of τ: $g(x; \boldsymbol{\beta}) = \beta_0 + \beta_2(x - \tau)I_{(-\infty, x)}(\tau)$. From this we see that the slope of this function with respect to τ is $-\beta_2$ over $x > \tau$, and 0 over $x \leq \tau$. Since the slope is also the first derivative, this can be expressed compactly as $\partial g/\partial \tau = -\beta_2 I_{(-\infty, x)}(\tau) = -\beta_2 I_{(\tau, \infty)}(x)$.

For the initial estimates, Fig. 2.1 suggests that the segmentation may occur between the fifth and sixth data points. Hence, we initialize the join point as the mean of these two snout-vent lengths: $\tau_0 = \frac{1}{2}(23.76 + 24.36) = 24.06$. For the horizontal plateau, β_0, take the mean of the first five hatching masses: $\beta_{00} = (0.294 + 0.302 + \cdots + 0.294)/5 = 0.2904$. For the slope of the right linear segment, regress the remaining 28 data pairs using the simple linear predictor $\beta_{00} + \beta_{20}(x_i - 24.06)$. We find $\beta_{00} = 0.3195$ and $\beta_{20} = 0.0189$. Notice that the two initial estimates of β_0 differ slightly, which is not unexpected. We could use either initial estimate, or even take their average. In the code that follows, we use the mean of the first five egg masses as our initial value for β_0. (But, readers should experiment with other starting values such as $\beta_{00} = 0.3195$ or $\beta_{00} = \frac{1}{2}(0.2904 + 0.3195) = 0.3050$.) Our PROC NLIN code appears in Fig. 2.2. Critical features of the code include: (i) invocation of the Levenberg–Marquardt algorithm via the `method= marquardt` option in the procedure call; (ii) the `parameters` statement, which is required to set the initial estimates; (iii) the `if-else` construction to segment the model as per (2.3); and (iv) `der.` statements that give the partial derivatives. The consequent output (edited) appears in Fig. 2.3. (Note that in Fig. 2.3 and throughout

```
* SAS code to fit left plateau model;
data lizard;
  input mass length @@;
datalines;
22.87  0.294      23.45  0.302      23.49  0.265      23.65  0.297
  :       :                 :                 :
26.20  0.358      26.27  0.348      27.12  0.371      27.28  0.421

proc nlin method=marquardt;
  parameters  b0 = 0.2904    b2 = 0.0189    tau = 24.06 ;
  if length <= tau then do;               *left plateau;
    model mass = b0 ;
    der.b0 = 1;
    der.b2 = 0;
    der.tau = 0;
  end;
  else do;                                *post-threshold;
    model mass = b0 + b2*(length-tau);
    der.b0 = 0;
    der.b2 = length-tau;
    der.tau = -b2;
  end;
```

Figure 2.2 Sample SAS program to fit left-plateau model to lizard development data

```
                        The SAS System
                        The NLIN Procedure

                    Method:  Marquardt
                    Iterative Phase
         Iter        b0           b2          tau      Sum of Squ.
          0        0.2904       0.0190      24.0600       0.0321
          1        0.2904       0.0190      22.5244       0.0109
          2        0.2904       0.0232      23.2609       0.0102
          3        0.2934       0.0239      23.2068       0.0102
          4        0.2940       0.0239      23.3216       0.00998
          5        0.2940       0.0241      23.3597       0.00997
          6        0.2940       0.0241      23.3605       0.00997
NOTE: Convergence criterion met.

                          Sum of        Mean                Approx
    Source          DF   Squares       Square    F Value    Pr > F
    Regression       3    3.6697       1.2232     26.10     <.0001
    Residual        29    0.00997      0.000344
    Uncorrected Total 32  3.6797
    Corrected Total 31    0.0279

                          Approx          Approximate 95%
    Parameter    Estimate  Std Error     Confidence Limits
    b0            0.2940     0.0185      0.2561      0.3319
    b2           0.0241     0.00353     0.0168      0.0313
    tau         23.3605     0.3086     22.7293     23.9917

                   Approximate Correlation Matrix
                       b0            b2            tau
        b0        1.0000000     0.0000000     0.0000000
        b2        0.0000000     1.0000000     0.8938052
        tau       0.0000000     0.8938052     1.0000000
```

Figure 2.3 SAS output (edited) from fit of left-plateau model for lizard development data

the text, the information in our SAS outputs may differ from that generated by different versions of the software, or by the same version of the software on different platforms. Analysts should be ready to experiment with convergence issues and other disparities in such cases.)

From the output in Fig. 2.3, we see that convergence was attained in 6 steps from the initial estimates. The minimized SSE is listed under Sum of Squares: SSE = 0.009 97, with $df_e = 29$. The final LS estimates are given under Estimate: $b_0 = 0.294$, $b_2 = 0.024$, and $\hat{\tau} = 23.36$. A residual plot (Fig. 2.4) suggests no gross anomalies with this model fit.

Large-sample pointwise 95% confidence intervals are provided for each parameter under Approximate 95% Confidence Limits. For instance, we can infer with approximate 95% confidence that under this model the join point lies between 22.73 and 23.99 mm.

Modern versions (SAS ver. 6.12 or later) of PROC NLIN are equipped with automatic derivatives, so that the Gauss–Newton and Levenberg–Marquardt algorithms may be applied without the need to specify the der. statements. In this case, the listsder option can be used to list the derivatives actually employed by the procedure. Readers may verify that without the der. statements in Fig. 2.2, the output in Fig. 2.3 would not be materially affected.

Example 2.2 (Simple change-point model) Environmental chemists often study the response of polymers to changes in environmental conditions (temperature, humidity, atmospheric pressure, etc.). For example, Gray and Bergbreiter (1997) gave data on the hydrogenation rate of a nitroarene polymer as it related to temperature variation. Previous research suggested that the thermoresponsive character of such polymers can change abruptly with temperature; Gray and Bergbreiter's experiment was designed to study this phenomenon. The response variable was $Y = \log$-'velocity' of the hydrogenation rate, taken against the predictor variable

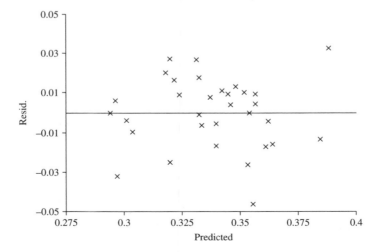

Figure 2.4 Residual plot from fit of left-plateau model for lizard development data

$x = 10^3$/temperature (where temperature is taken in kelvin). The data, as read from Gray and Bergbreiter's Fig. 6, appear in Table 2.2; they are plotted in Fig. 2.5.

We see that the response is essentially flat for all x, except that it changes abruptly between $x = 3.21$ and $x = 3.27$. Thus we consider for these data a special case of (2.2) where the two segments are simple plateaus:

$$g(x_i; \beta) = \begin{cases} \beta_1 & \text{if } x_i \leq \tau, \\ \beta_2 & \text{if } x_i > \tau. \end{cases}$$

Table 2.2 Log-hydrogenation rate ('velocity') as a function of reciprocal temperature ($\times 10^3$) for a poly(N-isopropylacrylamide)-bound nitroarene

$x = \frac{1000}{\text{temp.}}$	$Y = \log(\text{velocity})$
3.04	-8.48
3.14	-8.70
3.21	-8.58
3.27	-3.03
3.39	-3.03
3.52	-2.60
3.66	-2.93

Source: Gray and Bergbreiter (1997).

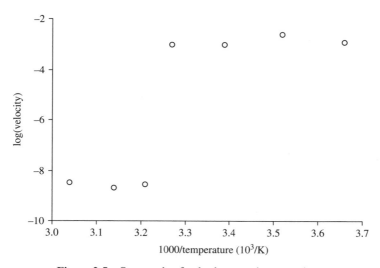

Figure 2.5 Scatterplot for hydrogenation rate data

For unknown τ this is a nonlinear model, but we can fit it using multiple regression software such as **PROC GLM** through some slight manipulation. To begin, suppose that τ is known, say $3.21 \leq \tau < 3.27$, and restrict the analysis to this assumption. (Note: we are not specifying any particular value for τ, only that it appears to be some value in this interval. One could say, of course, that this is clear from Table 2.2; however, such an indication is a result of observing a stochastic phenomenon and hence is not known *a priori*.) Then, fit the model by building a simple two-variable linear predictor: $Y_i = \beta_1 u_{1i} + \beta_2 u_{2i} + \varepsilon_i$, where $u_{1i} = 1$ if $x_i \leq \tau$ (zero otherwise) and $u_{2i} = 1$ if $x_i > \tau$ (zero otherwise). Notice the lack of an intercept term (β_0) in the linear predictor. This fits the left plateau via u_{1i} and the right plateau via u_{2i}. Figure 2.6 gives sample code using **PROC GLM**.

The **PROC GLM** output (edited) appears in Fig. 2.7. Notice that the LS point estimates for the β-parameters are just the sample means of the left segment's and right segment's observations, respectively: $b_1 = -8.5867$ and $b_2 = -2.9875$, which

```
*  SAS code to fit simple change point model;
data polymer;
input   x   left right   Ylogv ;
datalines;
        3.04    1    0     -8.48
        3.14    1    0     -8.70
        3.21    1    0     -8.58
        3.27    0    1     -3.03
        3.39    0    1     -3.03
        3.52    0    1     -2.60
        3.66    0    1     -2.93

proc glm;
model Ylogv = left right / noint;
```

Figure 2.6 Sample SAS program to fit simple two-plateau model to polymer data

```
                            The SAS System
                          The GLM Procedure

        Dependent Variable: Ylogv
                                    Sum of
        Source              DF      Squares      Mean Square   F Value   Pr > F
        Model                2    254.7745583    127.3872792   4276.42   <.0001
        Error                5      0.1489417      0.0297883
        Uncorrected Total    7    254.9235000

        Source      DF       Type I SS      Mean Square    F Value      Pr > F
        left         1     221.1925333     221.1925333     7425.48      <.0001
        right        1      33.5820250      33.5820250     1127.35      <.0001

                                    Standard
        Parameter        Estimate      Error      t Value     Pr > |t|
        left          -8.586666667   0.09964660    -86.17      <.0001
        right         -2.897500000   0.08629649    -33.58      <.0001
```

Figure 2.7 SAS output (edited) from **PROC GLM** fit of simple two-plateau model for polymer data

seems reasonable here. To be precise, we should write these as functions of τ, since they are found by assuming that $\tau = 3.24$, or more precisely, that $3.21 \leq \tau < 3.27$: $b_1(\tau) = b_1(3.24) = -8.5867$ and $b_2(\tau) = b_2(3.24) = -2.9875$. As seen below, similar notation is appropriate for all other statistics derived from Fig. 2.7.

A necessary adjustment concerns MSE(τ): df_e is listed as $5 = 7 - 2$, which PROC GLM found as (#observations) − (#parameters fitted); that is, since in effect we did not 'tell' PROC GLM about τ, no correction was made for it in df_e. Since in reality τ is unknown, the correct MSE(τ) should have divisor $7 - 3 = 4$: MSE(3.24) = $0.1489/4 = 0.0372$. (A corresponding adjustment should also be made to MSR(τ).) Note that SSE(τ), which is the unscaled objective quantity for the LS fit here, is correct as it stands.

We should note that this model structure limits identification of τ: without some additional, external, or prior information about the unknown parameter(s), all the nonlinear LS algorithm can do is form an interval for the estimated change point. The final LS statement on τ will be that the SSE is minimized for any τ between two consecutive predictor values, such as $3.21 \leq \tau < 3.27$.

Of course, the restricted LS fit in Fig. 2.7 corresponds to only one possible specification for τ. To find the unrestricted nonlinear LS estimates for β_1, β_2, and τ, we repeat this analysis for each possible change-point interval. That is, assume now that $3.14 \leq \tau < 3.21$ and repeat the approach in Fig. 2.6. In particular, capture SSE(τ). Since our goal is to minimize the overall SSE, we repeat this effort over all eight possible intervals for τ, and report the final LS fit as that corresponding to the smallest SSE(τ). (Remember to adjust df_r and df_e for estimation of τ.)

Applying this approach to the polymer data (Exercise 2.3) leads to the summary statistics presented in Table 2.3. (At the two extremes, $\tau < 3.04$ and $\tau \geq 3.66$, we are forcing $\beta_1 = 0$ and $\beta_2 = 0$, respectively, so that the restricted model is one of the simplest possible: 'constant + error.' The point estimate of the non-zero β-parameter is just $\overline{Y}$, while SSE(τ) = $\sum_{i=1}^{n} (Y_i - \overline{Y})^2$.) From the table, we see that our initial indications were correct: the minimum SSE(τ) occurs for $3.21 \leq \tau < 3.27$. From this, we conclude that the unrestricted LS estimates are $b_1 = -8.5867$ and $b_2 = -2.9875$, while the unbiased estimate of σ^2 is MSE = $0.1489/4 = 0.0372$. Confidence intervals

Table 2.3 Iterative SSE calculations for simple two-plateau change-point model with polymer data

Assumption on τ	$b_1(\tau)$	$b_2(\tau)$	SSE(τ)
$\tau < 3.04$	−	−5.3357	55.6346
$3.04 \leq \tau < 3.14$	−8.4800	−4.8117	44.1003
$3.14 \leq \tau < 3.21$	−8.5900	−4.0340	25.9815
$3.21 \leq \tau < 3.27$	−8.5867	−2.9875	0.1489
$3.27 \leq \tau < 3.39$	−7.1975	−2.8533	23.2829
$3.39 \leq \tau < 3.52$	−6.3640	−2.7650	37.1306
$3.52 \leq \tau < 3.66$	−5.7367	−2.9300	48.8825
$\tau \geq 3.66$	−5.3357	−	55.6346

and other inferences on the model parameters should be based on these final LS quantities. Tracing an objective function over a range of plausible values while estimating other model traits is sometimes referred to as a *profile estimation strategy*.

This example has touched on the problem of detecting a change in a series of environmental data (Esterby, 2002), often called *change-point analysis*. Segmented regression models are special cases of the general change-point problem, but the issue extends well past these specific models. More details on this area of statistical theory and practice are available in, for example, Carlstein *et al.* (1995), Jandhyala *et al.* (2002), and the references therein.   ☙

The two-segment model in (2.2) can be extended to involve multiple plateaus. For instance, a discontinuous model with both a right and a left plateau is

$$g(x_i; \boldsymbol{\beta}) = \begin{cases} \beta_0 & \text{if } x_i \leq \tau_1, \\ \beta_1 + \beta_2 x_i & \text{if } \tau_1 < x_i \leq \tau_2, \\ \beta_3 & \text{if } x_i > \tau_2, \end{cases} \qquad (2.4)$$

where now τ_1 is the left change point and τ_2 is the right change point. Continuity constraints may also be imposed on (2.4) at either or both join points.

As suggested earlier, the simple linear plateau model is a special case from a larger class of segmented regression models. We write the general two-segment (or *two-phase*) model as

$$g(x_i; \boldsymbol{\beta}) = \begin{cases} h_1(x_i - \tau; \boldsymbol{\beta}) & \text{if } x_i \leq \tau, \\ h_2(x_i - \tau; \boldsymbol{\beta}) & \text{if } x_i > \tau, \end{cases} \qquad (2.5)$$

where $h_1(\cdot; \cdot)$ and $h_2(\cdot; \cdot)$ are functions of $x_i - \tau$ and $\boldsymbol{\beta}$ specified by the investigator, with or without a continuity constraint. (Under continuity, we set $h_1(0; \boldsymbol{\beta}) = h_2(0; \boldsymbol{\beta})$ and determine how this affects the regression parameters in $\boldsymbol{\beta}$.) From this, the general, discontinuous, piecewise linear regression model becomes

$$g(x_i; \boldsymbol{\beta}) = \begin{cases} \beta_0 + \beta_1(x_i - \tau) & \text{if } x_i \leq \tau, \\ \beta_2 + \beta_3(x_i - \tau) & \text{if } x_i > \tau, \end{cases} \qquad (2.6)$$

with $\boldsymbol{\beta}$ made up of $p = 5$ unknown parameters, including the change point at $x = \tau$. Under continuity at τ, (2.6) becomes

$$g(x_i; \boldsymbol{\beta}) = \begin{cases} \beta_0 + \beta_1(x_i - \tau) & \text{if } x_i \leq \tau, \\ \beta_0 + \beta_3(x_i - \tau) & \text{if } x_i > \tau. \end{cases} \qquad (2.7)$$

In this parameterization, β_0 corresponds to the expected response at $x = \tau$. Partial derivatives for either (2.6) or (2.7) can be derived using methods similar to those seen in Example 2.1.

An alternative segmented-model parameterization is $g(x_i; \boldsymbol{\beta}) = \beta_0 + \beta_1 x_i + \beta_2(x_i - \tau)_+$, where $(x_i - \tau)_+ = \max\{0, x_i - \tau\}$. Here, β_0 is the intercept for the first segment, while $\beta_0 - \beta_2\tau$ is the intercept for the second segment. β_1 retains its interpretation as the slope of the first segment, but now $\beta_1 + \beta_2$ is the slope of the second segment.

Example 2.3 (Pond water levels)   Small, isolated bodies of water within larger wetland ecosystems often exhibit substantial variation in their water levels, due ostensibly to factors associated with their supply sources. For example, in the US state of Georgia levels of many small ponds are affected by the nearby deep Floridian aquifer. To study this phenomenon, Blood *et al.* (1997) collected data for one such pond as $Y =$ pond water level vs. $x =$ aquifer level; see Table 2.4.

A plot of these data (Exercise 2.5) indicates a clear nonlinear pattern, with what appears to be a change in the slope of the response at aquifer levels past about $\tau = 41$.

Table 2.4   Pond water level vs. deep aquifer level in a wetland pond

$x =$ aquifer level	$Y =$ pond level	$x =$ aquifer level	$Y =$ pond level
36.56	53.487	41.61	54.237
36.58	53.487	41.64	54.177
36.58	53.487	41.87	54.207
36.58	53.527	41.92	54.227
36.71	53.487	41.95	54.207
36.98	53.487	42.21	54.257
37.18	53.487	42.25	54.197
38.06	53.487	42.37	54.237
38.15	53.607	42.37	54.237
38.51	53.627	42.45	54.217
39.28	53.807	42.61	54.257
39.41	53.827	42.63	54.247
39.71	53.897	42.75	54.247
39.74	53.957	42.85	54.227
39.93	53.987	42.92	54.207
39.99	54.007	43.01	54.347
40.20	53.957	43.13	54.227
40.38	54.087	43.17	54.247
40.63	54.197	43.31	54.217
40.64	54.257	43.45	54.217
40.79	53.987	43.58	54.267
40.80	54.237	43.60	54.227
40.90	54.157	43.60	54.267
41.09	54.227	43.78	54.247
41.29	54.217	43.88	54.267
41.38	54.227	43.99	54.267
41.49	54.187	44.03	54.227
41.52	54.237	44.22	54.247
41.54	54.247	44.53	54.257
41.54	54.267		

Source: Blood *et al.* (1997).

To assess this, and to study these data in more detail, we fit model (2.7) via nonlinear LS. For initial estimates, we set the initial join point to $\tau_0 = 41$, and then carry out two simple linear regressions: the first over all (x_i, Y_i) pairs whose aquifer level is less than 41, and the second over all pairs whose aquifer level is greater than 41. For the first segment, we fit $\beta_0 + \beta_1(x_i - 41)$; this produces $\beta_{00} = 54.1532$ and $\beta_{10} = 0.1641$. For the second segment, we fit the model $\beta_0 + \beta_3(x_i - 41)$; this produces $\beta_{00} = 54.2178$ and $\beta_{30} = 0.0114$. As in Example 2.1, the two initial estimates of β_0 differ slightly. In order to use both pieces of information, we update our initial estimate to be the average of the two: $\beta_{00} = \frac{1}{2}(54.1532 + 54.2178) = 54.1855$.

To find the partial derivatives, we again write the model using indicator functions: here, (2.7) becomes $g(x; \boldsymbol{\beta}) = \beta_0 + \beta_1(x - \tau)I_{(-\infty, \tau]}(x) + \beta_3(x - \tau)I_{(\tau, \infty)}(x)$. From this, we find $\partial g/\partial \beta_0 = 1$, $\partial g/\partial \beta_1 = (x - \tau)I_{(-\infty, \tau]}(x)$, and $\partial g/\partial \beta_3 = (x - \tau)I_{(\tau, \infty)}(x)$. Also, $\partial g/\partial \tau$ is $-\beta_1 I_{[x, \infty)}(\tau) - \beta_3 I_{-\infty, x)}(\tau) = -\beta_1 I_{(-\infty, \tau]}(x) - \beta_3 I_{(\tau, \infty)}(x)$. The corresponding PROC NLIN code appears in Fig. 2.8. The consequent output (edited) appears in Fig. 2.9.

The output in Fig. 2.9 indicates rapid convergence, with a stable fit. In particular, the join point is estimated as $\hat{\tau} = 41.3671$, with approximate 95% confidence interval given as $41.0124 < \tau < 41.7217$. Notice also that the approximate 95% confidence interval for β_3 contains zero. This may be an indication that the right linear segment does not deviate significantly from a horizontal plateau. Along with some other aspects of the fit (including a residual plot), these issues are studied in more detail as part of Exercises 2.5 and 2.6. ✪

```
* SAS code to fit segmented regr. model;
data pond;
  input  aquifer  pond @@;
  datalines;
  41.54   54.267        43.60   54.227        42.92   54.207
    :                     :                     :
  36.98   53.487        37.18   53.487        38.15   53.607

proc nlin method=marquardt;
  parameters   b0 = 54.1855
               b1 =  0.1641
               b3 =  0.0114
               tau = 41.0    ;
  if aquifer <= tau then do;              *left segment;
               model pond = b0 + b1*(aquifer-tau) ;
               der.b0 = 1;
               der.b1 = aquifer-tau;
               der.b3 = 0;
               der.tau= -b1;
  end;
  else do;                                *right segment;
               model pond = b0 + b3*(aquifer-tau);
               der.b0 = 1;
               der.b1 = 0;
               der.b3 = aquifer-tau;
               der.tau= -b3;
  end;
```

Figure 2.8 Sample SAS program to fit continuous two-segment model to pond water level data

```
                          The SAS System
                          The NLIN Procedure

                          Method:  Marquardt
                          Iterative Phase
      Iter        b0          b1          b3          tau      Sum of Squ.
       0       54.1855     0.1641      0.0114      41.0000      0.2512
       1       54.2227     0.1641      0.0113      41.4239      0.1938
       2       54.2209     0.1664      0.0118      41.3604      0.1931
       3       54.2216     0.1663      0.0115      41.3671      0.1931
       4       54.2215     0.1663      0.0115      41.3671      0.1931
      NOTE:  Convergence criterion met.

                                  Sum of        Mean                    Approx
      Source               DF     Squares       Square     F Value      Pr > F
      Regression            4     172485       43121.4     399.64      <.0001
      Residual             55     0.1931       0.00351
      Uncorrected Total    59     172486
      Corrected Total      58     4.4026

                                   Approx         Approximate 95%
      Parameter    Estimate      Std Error       Confidence Limits
      b0           54.2215        0.0199        54.1816      54.2615
      b1            0.1663        0.00711        0.1521       0.1806
      b3            0.0115        0.0113        -0.0110       0.0341
      tau          41.3671        0.1770        41.0124      41.7217

                          Approximate Correlation Matrix
                   b0            b1            b3            tau
      b0      1.0000000    -0.0611817    -0.8331591     0.7290529
      b1     -0.0611817     1.0000000     0.0000000    -0.5973490
      b3     -0.8331591     0.0000000     1.0000000    -0.5642136
      tau     0.7290529    -0.5973490    -0.5642136     1.0000000
```

Figure 2.9 SAS output (edited) from fit of continuous two-segment model for pond water level data

2.3 Exponential regression models

Among the functional forms available for modeling nonlinear regression relationships, those based on the base, e, of the natural logarithm are perhaps the most flexible. A wide variety of functions can be formed by taking e to some power of x, and then by adding/subtracting and multiplying/dividing the individual results. In some cases, the final model may be linearized; that is, it may be possible to convert the nonlinear form into a model that has a underlying simple (or multiple) linear structure. Perhaps the best example of this is the exponential form $g(x_i; \beta) = \theta_0 e^{-\theta_1 x_i}$: clearly $\log\{g(x_i; \beta)\} = \log(\theta_0) - \theta_1 x_i$, or, for $\beta_0 = \log(\theta_0)$ and $\beta_1 = -\theta_1, \beta_0 + \beta_1 x_i$. (If $\theta_1 \geq 0$, this is a simple *exponential decay model*. If $\theta_1 \leq 0$, it is a simple *exponential growth model*.) If the errors are multiplicative on the original scale, then by taking the natural logarithm of the observations we recover the simple linear model, and the methods of §1.1 are appropriate. If, however, the errors are additive (and the variances are constant) on the original scale – that is, Y_i satisfies (2.1) with $g(x_i; \beta) = \theta_0 e^{-\theta_1 x_i}$ – then linearizing is contraindicated. Indeed, an LS fit to the log-transformed data is only optimal (in the sense of minimizing the residual sum of squares) on the log scale: values such as 10 and 0.1 are viewed as equally distant from a fitted value of 1.0. As a result, responses less than 1.0 will have greater influence on the fit than those on the original scale of measurement. See Bailer and Portier (1990) for an illustration of this effect.

In many cases, the exponential form cannot be linearized, and a fit via nonlinear LS is the primary option. The two basic forms of general exponential model are the multiexponential, $g(x_i; \boldsymbol{\beta}) = \beta_0 + \beta_1 e^{-\beta_2 x_i} + \cdots + \beta_{p-1} e^{-\beta_p x_i}$, and the reciprocal multiexponential,

$$g(x_i; \boldsymbol{\beta}) = \frac{1}{\beta_0 + \beta_1 e^{-\beta_2 x_i} + \cdots + \beta_{p-1} e^{-\beta_p x_i}}.$$

In both cases, the order of the model, p, is typically taken to be no higher than about $p = 6$. (Similar to the case for polynomial regression in §1.5, large values of p may fit the data well, but interpretation and generalizations to other environmental contexts may be unclear.)

The basic multiexponential models are the *monoexponential model*

$$g(x_i; \boldsymbol{\beta}) = \beta_0 + \beta_1 e^{-\beta_2 x_i}, \tag{2.8}$$

the *biexponential model*

$$g(x_i; \boldsymbol{\beta}) = \beta_0 + \beta_1 e^{-\beta_2 x_i} + \beta_3 e^{-\beta_4 x_i}, \tag{2.9}$$

and the *triexponential model*

$$g(x_i; \boldsymbol{\beta}) = \beta_0 + \beta_1 e^{-\beta_2 x_i} + \beta_3 e^{-\beta_4 x_i} + \beta_5 e^{-\beta_6 x_i}. \tag{2.10}$$

(We caution that the triexponential can sometimes be difficult to fit; Ratkowsky, 1990, §4.5, gives some suggestions for alternatives.) These forms are especially useful in pharmacological studies where $Y =$ concentration of a chemical toxin, pharmaceutical product or metabolite, radioactive tracer, etc., is studied as a function of $x =$ time. Called *concentration–time models*, the functions are derived from basic suppositions about the pharmacokinetic aspects (absorption, excretion, elimination, etc.) of the chemical, often in concert with differential equations that characterize their rate(s) of change over time. For example, the simplest concentration–time model is associated with a single bolus injection or exposure, assumed to impact a single 'compartment' in the organism under study; this is a form of *compartmental modeling* (Matis and Wehrly, 1994; Becka, 2002). The underlying kinetics often are described via differential equations. For instance, exponential decay can arise from a simple one-compartment kinetic model when flux in the model is described by the differential equation $\frac{dg}{dx} = -\beta_2 g(x; \boldsymbol{\beta})$, where the initial condition $g(0; \boldsymbol{\beta}) = \beta_1$ is imposed. The solution to this differential equation is the exponential decay model $g(x; \boldsymbol{\beta}) = \beta_1 e^{-\beta_2 x}$. Bates and Watts (1988, Ch. 5) provide a detailed discussion on many of the statistical issues associated with fitting compartmental or other differential equation-based models.

Of course, the use of compartmental models is not restricted to pharmacological studies, and such models can prove useful in many environmental problems. These include environmental pharmacology, as above, but also environmental toxicokinetics, enzyme kinetics and other chemometric applications, quantitative risk assessment, ecological population dynamics, etc. In all such cases, the investigator typically has previous subject-matter information or experimental indications as to which construction/model/analysis will be appropriate for the data at hand. (In some cases, the effort can be enhanced by advanced computer software; see, for example, Nash and Quon, 1996.) Given a model specification, the issue becomes one of determining the nonlinear LS fit and making inferences on pertinent model parameters. This is accomplished via the general methods described above. The following examples provide some illustrations.

Example 2.4 (Pine tree abundance) The US National Forest Service (NFS) records extensive data on the nature, inventory, abundance, status, health, and other features of NFS stands in its Forest Inventory and Analysis Data Base Retrieval System; see http://srsfia.usfs.msstate.edu/scripts/ewdbrs.htm. Among these data, a rough measure of abundance is the number of trees (in millions) of a particular species in the forests, taken as a function of tree diameter (in inches). For example, data on loblolly pines (*Pinus taeda*) in national forests of the US Southeast (defined here as the states of Alabama, Florida, Georgia, Mississippi, North Carolina, South Carolina, and Tennessee) appear in Table 2.5; Fig. 2.10 plots the data, indicating a clear curvilinear relationship.

To model these data we use the simple monoexponential model from (2.8), which can be fitted via PROC NLIN. Here again, we prefer the Levenberg–Marquardt

Table 2.5 Loblolly pine abundance vs. minimum diameter (inches) for southeastern US national forests between 1990 and 1999

$x =$ min. diameter	$Y =$ trees ($\times 10^6$)
1.0	120.905
3.0	72.846
5.0	41.289
7.0	25.103
9.0	15.020
11.0	10.489
13.0	7.630
15.0	7.090
17.0	4.302
19.0	2.563
21.0	2.838
29.0	0.107

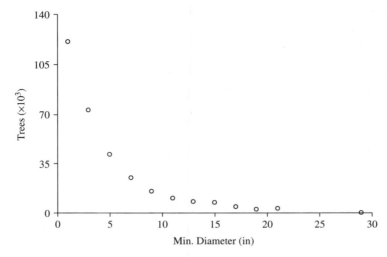

Figure 2.10 Scatterplot for pine tree abundance data

iterative fitting algorithm. For the partial derivatives we find $\partial g/\partial \beta_0 = 1, \partial g/\partial \beta_1 = e^{-\beta_2 x}$, and $\partial g/\partial \beta_2 = -x\beta_1 e^{-\beta_2 x}$.

To set the initial estimates, start by recognizing in (2.8) that if the response decays over time such that $\beta_2 > 0, E[Y_i] \to \beta_0$ as $x \to \infty$. Thus we can set β_{00} equal to the observed response at the largest value of x_i. (For most data sets this will be $Y_{(1)} = \min_{i \in \{1,\dots,n\}}\{Y_i\}$, although we do not require that $\beta_{00} = Y_{(1)}$. The symbol '∈' is read as 'is an element of.') Here, this is $\beta_{00} = 0.107$. Next, notice that $g(x_i; \boldsymbol{\beta}) - \beta_0 = \beta_1 e^{-\beta_2 x_i}$, so we expect $\log\{Y_i - \beta_0\} \approx \log\{\beta_1\} - \beta_2 x_i$. Use this fact to regress the pseudo-variable $U_i = \log\{Y_i - \beta_{00}\}$ on $z_i = -x_i$ (ignore any values where $Y_i - \beta_{00} \leq 0$). From this, take β_{10} as e raised to the estimated intercept and β_{20} as the estimated slope. For the data in Table 2.5, this produces $\beta_{10} = 110.7678$ and $\beta_{20} = 0.1936$. The corresponding **PROC NLIN** code appears in Fig. 2.11. The consequent output (edited) appears in Fig. 2.12.

```
* SAS code to fit monoexponential model;
data pine;
  input diam trees @@;
  datalines;
  1.0 120.905    3.0 72.846    5.0 41.289    7.0 25.103
  9.0  15.020   11.0 10.489   13.0  7.630   15.0  7.090
  17.0  4.302   19.0  2.563   21.0  2.838   29.0  0.107

proc nlin method=marquardt;
  parameters  b0 = 0.107   b1 = 110.767782   b2 = 0.19358046;
  model trees = b0 + b1*exp(-b2*diam) ;
  der.b0 = 1;
  der.b1 = exp(-b2*diam);
  der.b2 = -diam*b1*exp(-b2*diam);
```

Figure 2.11 Sample SAS program to fit monoexponential model to pine tree abundance data

```
                            The SAS System
                            The NLIN Procedure

                          Method:  Marquardt
                          Iterative Phase
          Iter          b0              b1              b2      Sum of Squ.
           0          0.1070          110.8          0.1936        1032.9
           1          3.6662          150.8          0.2770       42.1585
           2          2.1878          156.1          0.2705       15.3500
           3          2.1774          156.1          0.2707       15.3347
           4          2.1776          156.1          0.2707       15.3347
       NOTE: Convergence criterion met.

                                     Sum of       Mean                  Approx
        Source               DF      Squares      Square    F Value     Pr > F
        Regression            3      22721.4      7573.8    4314.83     <.0001
        Residual              9      15.3347      1.7039
        Uncorrected Total    12      22736.7
        Corrected Total      11      14719.0

                                       Approx        Approximate 95%
        Parameter    Estimate        Std Error       Confidence Limits
        b0              2.178          0.6032        0.813        3.542
        b1            156.1            2.1267      151.3        160.9
        b2              0.271          0.0067        0.256        0.286
```

Figure 2.12 SAS output (edited) from fit of monoexponential model for pine tree abundance data

The output in Fig. 2.12 indicates rapid convergence, with a stable fit. A residual plot (Fig. 2.13) appears generally reasonable, although there is a slight hint of a decrease in variance for large predicted values. From Fig. 2.12, we see that an approximate 95% confidence interval for β_2 is $0.256 < \beta_2 < 0.286$.

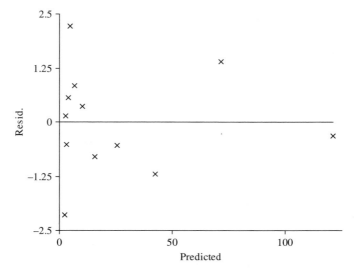

Figure 2.13 Residual plot from fit of monoexponential model for pine tree abundance data. Horizontal reference line indicates residual $= 0$

One might also be interested in making inferences on β_0, which for these data can be viewed as a limiting asymptote for the minimum abundance with large-diameter trees. To test H_0: $\beta_0 = 0$, we appeal to the discrepancy-measure approach in (1.7) and fit a monoexponential model under H_0. Our goal is to acquire the SSE(RM) statistic under the reduced model that $\beta_0 = 0$. Doing so here (results not shown) leads to SSE(RM) = 36.5796. With SSE(FM) = 15.3347 from Fig. 2.12, we use (1.7) to construct the F-statistic:

$$F_{calc} = \frac{(36.5796 - 15.3347)/1}{15.3347/9} = \frac{21.2449}{1.7039} = 12.4687.$$

At $\alpha = 0.05$, we compare F_{calc} to $F_{0.05}(1,9) = t_{0.025}^2(9) = 2.2622^2 = 5.1175$ (from Table B.2). Since F_{calc} clearly exceeds this critical point, we can conclude that β_0 does appear to deviate significantly from zero. (This is confirmed by the 95% confidence interval for β_0 in Fig. 2.12: $0.813 < \beta_0 < 3.542$ does not contain zero.)   ⊕

Example 2.5 (Nitrite utilization)   Bates and Watts (1988, §A1.12) present data on nitrite usage by bush beans (*Phaseolus vulgaris* L.) after exposure to varying intensities of natural light (in $\mu E/m^2 s$). Measured is the utilization of nitrite (as nmol/g-hr) in the plants' primary leaves. The data appear in Table 2.6. Notice that three replicate observations are taken at each level of light exposure.

A plot of the data (Fig. 2.14) shows a clear increase in nitrite utilization, followed by a possible drop-off as light intensity grows. Bates and Watts note that one possible model to account for this curvilinear effect is a special case of the biexponential in (2.9), $g(x_i; \boldsymbol{\beta}) = \beta_1(e^{-\beta_2 x_i} - e^{-\beta_4 x_i})$. One important aspect of this model is its 'regression

Table 2.6   Nitrite utilization in bush beans (*Phaseolus vulgaris* L.) after exposure to light

x = Light intensity ($\mu E/m^2 s$)	Y = Nitrite util. (nmol/g-hr)	x = Light intensity ($\mu E/m^2 s$)	Y = Nitrite util. (nmol/g-hr)
2.2	256	27.0	9884
2.2	685	27.0	11597
2.2	1537	27.0	10221
5.5	2148	46.0	17319
5.5	2583	46.0	16539
5.5	3376	46.0	15047
9.6	3634	94.0	19250
9.6	4960	94.0	20282
9.6	3814	94.0	18357
17.5	6986	170.0	19638
17.5	6903	170.0	19043
17.5	7636	170.0	17475

Source: Bates and Watts (1988, §A1.12).

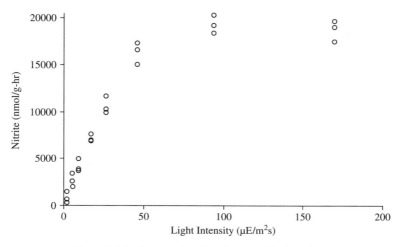

Figure 2.14   Scatterplot for nitrite utilization data

through the origin' feature: $g(0; \boldsymbol{\beta}) = 0$. Nitrite utilization is thought be to be essentially zero when the plants are in darkness.

We fit this model to the nitrite utilization data using the Levenberg–Marquardt
nonlinear LS algorithm in PROC NLIN. For the partial derivatives, we find
$\partial g/\partial \beta_1 = e^{-\beta_2 x} - e^{-\beta_4 x}$, $\partial g/\partial \beta_2 = -x\beta_1 e^{-\beta_2 x}$, and $\partial g/\partial \beta_4 = x\beta_1 e^{-\beta_4 x}$.

To set the initial estimates, consider the following. Denote the value where $g(x; \boldsymbol{\beta})$
attains its maximum as x^*. From differential calculus, we know the derivative at this
point, $\frac{\partial g}{\partial x}\big|_{x=x^*} = g'(x^*; \boldsymbol{\beta})$, is 0. We find that $g'(x; \boldsymbol{\beta}) = \beta_1\beta_4 e^{-\beta_4 x} - \beta_1\beta_2 e^{-\beta_2 x}$, so solv-
ing $g'(x^*; \boldsymbol{\beta}) = 0$ for x^* yields

$$x^* = \frac{\log(\beta_2) - \log(\beta_4)}{\beta_2 - \beta_4}$$

(Exercise 2.8). We denote the corresponding value of $g(x^*; \boldsymbol{\beta})$ by g^*:

$$g^* = \beta_1 \left(\frac{\beta_2}{\beta_4}\right)^{-\beta_2/(\beta_2-\beta_4)} \left(1 - \frac{\beta_2}{\beta_4}\right).$$

Similarly, the point at which $g(x; \boldsymbol{\beta})$ inflects from concave to convex occurs when
$g''(x; \boldsymbol{\beta}) = 0$. Denote this by x^+. Since $g''(x; \boldsymbol{\beta}) = \beta_1(\beta_2^2 e^{-\beta_2 x} - \beta_4^2 e^{-\beta_4 x})$, we can solve
$g''(x^+; \boldsymbol{\beta}) = 0$ for x^+ to find $x^+ = 2x^*$ (Exercise 2.8). The corresponding value of
$g(x^+; \boldsymbol{\beta})$ is

$$g^+ = \beta_1 \left(\frac{\beta_2}{\beta_4}\right)^{-2\beta_2/(\beta_2-\beta_4)} \left\{1 - \left(\frac{\beta_2}{\beta_4}\right)^2\right\}.$$

These quantities are manipulated to find initial values for the nonlinear LS fit. First,
since $g'(x^*; \boldsymbol{\beta}) = 0$, we know $\beta_1 e^{-\beta_4 x^*} = \beta_1\beta_2\beta_4^{-1}e^{-\beta_2 x^*}$ and hence $g^* = \beta_1$ $(e^{-\beta_2 x^*} -$
$e^{-\beta_4 x^*}) = \beta_1 e^{-\beta_2 x^*}(1 - \beta_2/\beta_4)$. Now, if we assume $\beta_4 > \beta_2$, then $1 - \beta_2/\beta_4 \approx 1$ and

hence $g^* \approx \beta_1 e^{-\beta_2 x^*}$. For $Y_{(n)} = \max_{i \in \{1, \ldots, n\}} \{Y_i\}$, let $x^{(*)}$ be the value of x_i at $Y_{(n)}$. Use these as approximations to g^* and x^*: $g^* \approx Y_{(n)}$ and $x^* \approx x^{(*)}$. Thus, we have

$$\beta_2 \approx -\frac{1}{x^{(*)}} \log \left\{ \frac{Y_{(n)}}{\beta_1} \right\}. \tag{2.11}$$

Next, recognize that since $g''(x^+; \boldsymbol{\beta}) = 0$, we have $\beta_1 e^{-\beta_4 x^+} = \beta_1 \beta_2^2 \beta_4^{-2} e^{-\beta_2 x^+}$. From this, we see $g^+ = \beta_1 e^{-\beta_2 x^+} [1 - (\beta_2/\beta_4)^2]$, and, if $\beta_4 > \beta_2, g^+ \approx \beta_1 e^{-\beta_2 x^+}$. Suppose we identify approximate values $Y^{(+)}$ and $x^{(+)}$ for g^+ and x^+, respectively, from a plot of the data. Then we have

$$\beta_2 \approx -\frac{1}{x^{(+)}} \log \left\{ \frac{Y^{(+)}}{\beta_1} \right\}. \tag{2.12}$$

Combining (2.11) and (2.12) and solving for β_1 produces

$$\beta_1 = \exp \left\{ \frac{x^{(+)} \log[Y_{(n)}] - x^{(*)} \log[Y^{(+)}]}{x^{(+)} - x^{(*)}} \right\} \tag{2.13}$$

(Exercise 2.8). We use (2.13) as our initial estimate, β_{10}.

With (2.13) in hand, notice then that using it in (2.12) leads to $\beta_{20} = -(1/x^{(+)}) \log\{Y^{(+)}/\beta_{10}\}$. Lastly, consider β_{40}. Exercise 2.10 shows that $g(x; \boldsymbol{\beta})$ is approximately linear near $x = 0$. Thus we could perform a simple linear regression of Y_i on x_i for, say, $x_i < 10$, and set the estimated slope, m_{init}, from that regression equal to the function's slope at zero, $g'(0; \boldsymbol{\beta}) = \beta_1(\beta_4 - \beta_2)$. Given initial values β_{10} and β_{20}, we can then solve for β_{40}: $\beta_{40} = (m_{init}/\beta_{10}) - \beta_{20}$.

To apply these initial calculations to the data in Table 2.6, we need to specify $x^{(*)}$ and $x^{(+)}$. The former is clearly $x^{(*)} = 94$, but the latter is unclear. The inflection appears to be outside the range of the data, so to avoid any wildly subjective estimates we simply set $x^{(+)}$ equal to $x_{(n)} = \max_{i \in \{1, \ldots, n\}} \{x_i\}$; here, $x^{(+)} = 170$. With this, $Y^{(+)}$ is the observed mean response at that point: $Y^{(+)} = 18\,718.6667$. This leads to $\beta_{10} = 20\,035.5359$ and $\beta_{20} = 3.9992 \times 10^{-4}$. With these, regressing Y_i on x_i over $x_i < 10$ yields an initial slope estimate of $m_{init} = 443.414$, so that $\beta_{40} = 0.0217$. The corresponding PROC NLIN code appears in Fig. 2.15. The consequent output (edited) appears in Fig. 2.16.

The output in Fig. 2.16 indicates rapid convergence, in spite of the surprising fact that the LS estimate for β_2 converged to a value over ten times larger than the initial value of β_{20}. The parameter estimates are consistent with our expectations; for example, $b_4 > b_2$, and the estimated inflection point, $\hat{x}^+ = 2\{\log(b_2) - \log(b_4)\}/(b_2 - b_4) = 226.078\,\mu\text{E/m}^2\text{s}$, is well past $x = 170$. Also, the estimated point at which the response reaches a maximum (which is of environmental interest here) is $\hat{x}^* = \{\log(b_2) - \log(b_4)\}/(b_2 - b_4) = 113.039\,\mu\text{E/m}^2\text{s}$.

The asymptotic confidence intervals are a bit unusual, however. For instance, the 95% interval on β_1 is $-3277.2 < \beta_1 < 87204.5$. Since this contains $\beta_1 = 0$, one could conclude that the mean response does not deviate significantly from zero

```
* SAS code to fit biexponential model;
data nitrite;
  input light nitrite @@;
  datalines;
  2.2    256         2.2    685         2.2   1537
  :                  :                  :
  170.0 19638        170.0 19043        170.0 17475

proc nlin method=marquardt;
  parameters  b1  = 20035.53587
              b2  = 0.000399919
              b4  = 0.021731467    ;
      model nitrite = b1*exp(-b2*light) - b1*exp(-b4*light) ;
      der.b1 = exp(-b2*light) - exp(-b4*light);
      der.b2 = -light*b1*exp(-b2*light);
      der.b4 = light *b1*exp(-b4*light);
```

Figure 2.15 Sample SAS program to fit specialized biexponential model to nitrite utilization data

```
                          The SAS System
                          The NLIN Procedure

                        Method:   Marquardt
                        Iterative Phase
     Iter        b1          b2          b4       Sum of Squ.
       0      20035.5      0.00040     0.0217      95647197
       1      31477.2      0.00342     0.0176      95341701
       2      41067.2      0.00408     0.0164      21472106
       3      41894.2      0.00403     0.0165      17701562
       4      41971.7      0.00403     0.0165      17701320
       5      41963.7      0.00403     0.0165      17701320
    NOTE: Convergence criterion met.
```

Source	DF	Sum of Squares	Mean Square	F Value	Approx Pr > F
Regression	3	3.5241E9	1.1747E9	1393.61	<.0001
Residual	21	17701320	842920		
Uncorrected Total	24	3.5418E9			
Corrected Total	23	1.1584E9			

Parameter	Estimate	Approx Std Error	Approximate 95% Confidence Limits	
b1	41963.7	21754.6	-3277.2	87204.5
b2	0.00403	0.00217	-0.00048	0.00855
b4	0.0165	0.00488	0.00636	0.0267

```
               Approximate Correlation Matrix
                    b1             b2              b4
    b1       1.0000000      0.9965057      -0.9937344
    b2       0.9965057      1.0000000      -0.9827334
    b4      -0.9937344     -0.9827334       1.0000000
```

Figure 2.16 SAS output (edited) from fit of specialized biexponential model for nitrite utilization data

(which is clearly not the case; cf. Fig. 2.14). To study this more carefully, consider a formal test of $H_0: \beta_1 = 0$. Appealing to the discrepancy-measure approach from (1.7), we fit a reduced model under H_0 to find SSE(RM). Since H_0 stipulates that the mean response is zero, SSE here will correspond simply to the 'uncorrected'

SSE in Fig. 2.16. (Try it: fit $E[Y_i] = 0$ to these data using PROC REG with the `model` statement

```
model Y =/ noint;
```

and recover the consequent SSE.) This is $SSE(RM) = 3\,541\,813\,740.0$ on 24 df. Applying equation (1.7), we find $F_{calc} = 1393.61$. Notice that this is just the ratio of mean squares from Fig. 2.16. Compared to $F_{0.01}(3, 21) = 4.87$, the result is highly significant (indeed, $P < 0.0001$). Hence we can reject H_0 with some certainty. This reinforces the recommendation we gave above: where possible, inferences on the unknown parameters in a nonlinear regression should be based on F-statistics such as (1.7), since operations with measures such as $b_1/se[b_1]$ can be unstable with small data sets in some nonlinear models. (The effect also extends to the parameter estimates: notice the very strong correlations between the b_js under `Asymptotic Correl-ation Matrix` in Fig. 2.16. The values off the diagonal are $Corr[b_j, b_k], j \neq k$, and they all sit very close to ± 1. With such highly correlated estimators, inferences on the unknown parameters must be performed with caution.)

For model adequacy assessment, Fig. 2.17 plots the residuals from this biexponential fit. We see a higher fraction of negative residuals, especially for lower light intensity. This could indicate a problem of overprediction at lower light levels. There is also a pattern of possible increasing variability as light intensity increases; that is, $Var[Y_i]$ may not be constant with respect to x_i. To remedy this, Exercise 2.9 suggests a weighted LS approach for the data analysis; see also Example 2.8.

Example 2.5 illustrates an important concern with many nonlinear regression analyses: the model parameterization can leave the estimators highly correlated. This can disrupt inferences on the unknown parameters. To remedy this, and often to improve convergence, one can reparameterize the model. For instance,

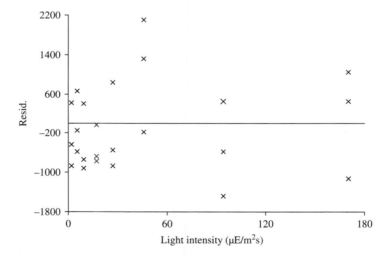

Figure 2.17 Residual plot from fit of biexponential model for nitrite utilization data Horizontal reference line indicates residual $= 0$

$g(x; \boldsymbol{\beta}) = \beta_0 + \beta_1 e^{-\beta_2 x}$ may also be written as $\beta_0 + \beta_1 \beta_3^x, \beta_0 + (\beta_4 - \beta_0)\beta_3^x$, or $\beta_0\{1 - e^{-\beta_2(x-\beta_5)}\}$, depending on the application and interpretation under study. Indeed, this simple model goes by a variety of names, again depending on the motivating application and/or subject area; examples include *Mitscherlich's law* or the *von Bertalanffy model* (Ratkowsky, 1990, §4.3).

For some data sets, it is appropriate to combine the features of an exponential model and a polynomial model to describe the mean response. A common example is also one of the simplest: $g(x_i; \boldsymbol{\beta}) = (\beta_0 + \beta_1 x_i)e^{-\beta_2 x_i}$. This model rises to a maximum before dropping back down (for $\beta_2 > 0$) and inflecting towards a horizontal asymptote at zero. Its behavior can in some cases appear similar to a biexponential function.

2.4 Growth curves

Exponential and other parametric functions can also be useful when modeling the growth of an organism or system over time or over some other pertinent predictor variable. Two of the simplest growth curves have already been described in other contexts: the simple linear model from §1.1, $g(x_i; \boldsymbol{\beta}) = \beta_0 + \beta_1 x_i$, and the simple exponential curve from §2.3, $g(x_i; \boldsymbol{\beta}) = \beta_1 e^{-\beta_2 x_i}$, where 'growth' corresponds to $\beta_2 < 0$. These two models make strong assumptions about the nature of the growth phenomenon. The linear model assumes a constant rate of growth (β_1), while the exponential model assumes growth is proportional to population size. Indeed, the structure of the two models becomes clear when the growth pattern is specified by a differential equation: in the first, linear growth occurs when $\frac{dg}{dx} = \beta_1$; while in the second, exponential growth occurs when $\frac{dg}{dx} = -\beta_2 g(x; \boldsymbol{\beta})$. Further, both models can allow for unlimited growth in the sense that, with proper choice of the β-parameters, $g(x; \boldsymbol{\beta}) \to \infty$ as $x \to \infty$. In most environmental scenarios, this is unlikely (although the functions may still provide reasonable approximations for the effect under study over the observed range of x_i).

Often, more complex growth phenomena can be modeled using sigmoidal (S-shaped) functions, or portions thereof, as depicted in Fig. 2.18. A number of these

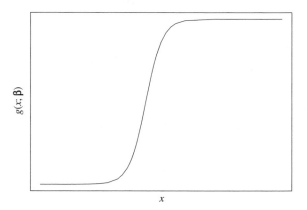

Figure 2.18 A typical sigmoidal curve

sigmoidal curves are popular enough to have formal names, including the Gompertz model, the logistic model, and the Weibull model. We study each of these in turn, below.

We should note that when modeling growth, an important consideration is whether or not the data being collected are statistically independent. It is not uncommon and often perfectly natural to record data on the same organism or subject when studying growth over time. If so, these are *repeated measurements* (Kenward, 2002), and correlations induced among repeated observations on the same subject may violate the independence assumption. In most cases, however, the formal model for $g(x_i; \beta)$ is unaffected. Repeated observations over time are also known as *longitudinal data*, although both these terms can refer to experimental settings beyond the growth curve paradigm, for example, patient monitoring in a clinical trial. For purposes of illustration in this chapter, and unless otherwise specified, we will continue to assume the data satisfy (2.1). (For cases where the independence assumption is invalid, of course, the analysis should take the inter-observational correlations into account. We provide an introduction to this, and to the larger issue of analyzing data that are temporally correlated, in Chapter 5.) Readers interested specifically in the analysis of repeated measures should consult specialized sources on growth curve modeling, such as Meredith and Stehman (1991), von Ende (1993), or the text by Kshirsagar and Smith (1995). More general texts on longitudinal analysis may also prove useful; these include Diggle *et al.* (1994) and Vonesh and Chinchilli (1996).

2.4.1 Gompertz model

The *Gompertz regression model* is based on the nonlinear function

$$g(x_i; \beta) = \beta_0 \exp\{-e^{-\beta_1 - \beta_2 x_i}\}. \tag{2.14}$$

While the double exponentials in the model expression make the function appear rather complex, it takes on a fairly simple form: for $\beta_2 > 0$ the function is a strictly increasing sigmoid, with vertical intercept at $\beta_0 \exp\{-e^{-\beta_1}\}$, a left horizontal asymptote (as $x \to -\infty$) at the horizontal axis, and a right horizontal asymptote (as $x \to \infty$) at β_0. For $\beta_2 < 0$ the sigmoid is strictly decreasing, and the asymptotes are reversed. Inferences often concern β_2, since it controls the time-response function. Notice, however, that by setting $\beta_2 = 0$ we are in effect forcing $g(x_i; \beta) = $ constant. Thus, assessing $\beta_2 = 0$ is tantamount to assessing $H_0: \beta_1 = \beta_2 = 0$.

The function – actually, the equivalent form $\beta_0 \psi^{\delta^x}$ – was suggested by Gompertz (1825) to describe mortality rates in humans, and it is a mainstay component for mortality and financial contingency modeling. Starting in the early 1900s, however, the function was also used to describe monotone growth in a number of environmental and biological applications; Winsor (1932) gave an early and lucid review. This popularity is due in part to the flexible, asymmetric nature by which it models growth: the curve's height at inflection is β_0/e, which lies below (i.e., not at) the center of its vertical range, and the rate at which $g(x; \beta)$ in (2.14) approaches its upper horizontal asymptote at β_0 differs from the rate at which it approaches its lower horizontal asymptote at zero.

The inflection point of the Gompertz growth curve occurs at $x^+ = -\beta_1/\beta_2$ (Exercise 2.13). This is often a quantity of interest, since it is the point at which the rate of growth is steepest. From the nonlinear LS estimates, b_j $(j = 0, 1, 2)$, of the individual regression parameters we estimate the infection point via $\hat{x}^+ = -b_1/b_2$. Corresponding $1 - \alpha$ confidence limits can be computed using Fieller's theorem from Exercise 1.16: given b_1, b_2, their standard errors $se[b_1]$ and $se[b_2]$, and an estimated covariance between b_1 and b_2, $\hat{\sigma}_{12} = \text{Cov}[b_1, b_2]$, $1 - \alpha$ Fieller limits for x^+ can be calculated here as

$$\hat{x}^+ + \frac{\gamma}{1-\gamma}\left\{\hat{x}^+ + \frac{\hat{\sigma}_{12}}{se^2[b_2]}\right\}$$

$$\pm \frac{t_{\alpha/2}(\nu)}{(1-\gamma)|b_2|}\left\{se^2[b_1] + \hat{x}^+(2\hat{\sigma}_{12} + se^2[b_2]\hat{x}^+) - \gamma\left(se^2[b_1] - \frac{\hat{\sigma}_{12}^2}{se^2[b_2]}\right)\right\}^{1/2},$$

(2.15)

where $\gamma = [t_{\alpha/2}(\nu)]^2 se^2(b_2)/b_2^2$ measures departure from symmetry in the distribution of $\hat{x}^+$ ($\gamma \to 0$ indicates greater symmetry), and $\nu = df_e$ from the nonlinear LS fit. Inferences on the height at inflection are somewhat simpler: estimate $g^+ = \beta_0/e$ using $\hat{g}^+ = b_0/e$, with approximate $1 - \alpha$ confidence limits given by $\hat{g}^+ \pm t_{\alpha/2}(df_e)\frac{se[b_0]}{e}$.

We fit (2.14) to data via a nonlinear LS algorithm such as Levenberg–Marquardt. Partial derivatives useful for implementing the algorithm are $\partial g/\partial\beta_0 = \exp\{-e^{-\beta_1-\beta_2 x}\}$, $\partial g/\partial\beta_1 = \beta_0\exp\{-\beta_1 - \beta_2 x - e^{-\beta_1-\beta_2 x}\}$, and $\partial g/\partial\beta_2 = x\beta_0\exp\{-\beta_1 - \beta_2 x - e^{-\beta_1-\beta_2 x}\}$. For the initial estimates, recognize that since β_0 is an upper asymptote, a simple estimate is the largest observed response $Y_{(n)} = \max_{i\in\{1,...,n\}}\{Y_i\}$. Thus we set $\beta_{00} = Y_{(n)}$. For β_1 and β_2, we know $g(x_i; \boldsymbol{\beta})/\beta_0 = \exp\{-e^{-\beta_1-\beta_2 x_i}\}$, so $-\log\{-\log[g(x_i; \boldsymbol{\beta})/\beta_0]\} = \beta_1 + \beta_2 x_i$. Thus, if we construct the pseudo-variate

$$U_i = -\log\left(-\log\left\{\frac{Y_i}{\beta_{00}}\right\}\right)$$

and regress this on x_i (excluding the value of U_i corresponding to $Y_{(n)}$ and any values of $Y_i \le 0$), the resulting intercept and slope estimates can serve as β_{10} and β_{20}, respectively. The next example illustrates the approach.

Example 2.6 (Fruit diameters) Reporting on a study of growth in fruit under different microclimatic conditions, De Silva *et al.* (1997) display data on apple growth over $x = $ time (days after pollination). For $Y = $ fruit diameters (in mm), representative data (read from De Silva *et al.*'s Fig. 2) appear in Table 2.7.

A plot of the data (Exercise 2.14) shows a clear sigmoidal response. We use PROC NLIN to fit the Gompertz curve (2.14) to these data; sample code appears in Fig. 2.19, with corresponding output (edited) in Fig. 2.20.

Table 2.7   Apple diameters (mm) over time (days after pollination)

$x =$ Days after pollination	$Y =$ Diameter (mm)
14	6.393
28	15.082
42	20.984
60	31.803
74	40.492
88	50.164
102	58.033
116	64.098
130	68.787
146	75.262

```
* SAS code to fit Gompertz curve;
data apple;
  input days diam @@;
  datalines;
  14   6.393    28 15.082    42 20.984    60 31.803    74 40.492
  88 50.164   102 58.033   116 64.098   130 68.787   146 75.262

proc nlin method=marquardt;
  parameters   b0  =  75.262
               b1  = -1.3829114
               b2  =  0.02742912 ;
       model diam = b0*exp(-exp(-b1-b2*days));
       der.b0 = exp(-exp(-b1-b2*days)) ;
       der.b1 = b0*exp(-b1-b2*days-exp(-b1-b2*days));
       der.b2 = days*b0*exp(-b1-b2*days-exp(-b1-b2*days));
```

Figure 2.19   Sample SAS program to fit Gompertz curve to fruit diameter data

The output in Fig. 2.20 indicates rapid convergence and a relatively stable fit. A residual plot from the fit shows a generally random spread (Exercise 2.14). To test for a growth effect over time, the null effect is H_0: $\beta_1 = \beta_2 = 0$. Under H_0, we fit the reduced model that $g(x_i;\ \beta)$ is constant over time. (Use the SAS commands

```
proc reg; model Y = ;
```

to fit an intercept-only model, $Y_i = \beta_0 + \varepsilon_i$, and find SSE(RM) from the ANOVA table.) Here SSE(RM) = 5163.9793 on 9 df (output not shown, although here SSE(RM) corresponds to the Corrected Total Sum of Squares in Fig 2.20). Appeal to equation (1.7) yields $F_{calc} = \frac{1}{2}(5157.8053/0.8820) = 2923.926$. When compared to an $F(2,7)$ reference distribution, the result is highly significant ($P < 0.0001$), and we conclude that the data exhibit a significant growth response.

```
                              The SAS System
                              The NLIN Procedure

                          Method:   Marquardt
                          Iterative Phase
        Iter          b0             b1            b2       Sum of Squ.
         0        75.2620        -1.3829       0.0274        76.5159
         1        81.6094        -1.2712       0.0224        32.3592
         2        89.6951        -1.1793       0.0192        11.6368
         3        91.9526        -1.1825       0.0189         6.1766
         4        91.9810        -1.1824       0.0189         6.1740
    NOTE: Convergence criterion met.

                              Sum of        Mean                    Approx
    Source              DF    Squares       Square     F Value      Pr > F
    Regression           3    23742.4       7914.1     8972.88     <.0001
    Residual             7     6.1740       0.8820
    Uncorrected Total   10    23748.5
    Corrected Total      9     5164.0

                                   Approx           Approximate 95%
    Parameter     Estimate       Std Error       Confidence Limits
    b0            91.9810          2.8739        85.1854      98.7767
    b1            -1.1824          0.0328        -1.2600      -1.1047
    b2             0.0189          0.000963       0.0167       0.0212

                    Approximate Correlation Matrix
                            b0               b1               b2
         b0         1.0000000        0.5793100       -0.9388058
         b1         0.5793100        1.0000000       -0.7989324
         b2        -0.9388058       -0.7989324        1.0000000
```

Figure 2.20   SAS output (edited) from fit of Gompertz model for fruit diameter data

If interest includes the time at which the growth rate attains its maximum, we calculate $\hat{x}^{+} = -b_1/b_2 = 1.1824/0.0189 = 62.561$ days. A 95% confidence interval on x^{+} is based on the Fieller result in (2.15). Figure 2.20 provides the information we need. Begin with $se[b_1] = 0.0328$, $se[b_2] = 9.63 \times 10^{-4}$, and $\text{Corr}[b_1, b_2] = -0.7989$. Then, the covariance between b_1 and b_2 is the product of these three values: $\hat{\sigma}_{12} = se[b_1]se[b_2]\text{Corr}[b_1, b_2] = 0.0328 \times 9.63 \times 10^{-4} \times (-0.7989) = -2.5235 \times 10^{-5}$. With these, and using $t_{0.025}(7) = 2.365$, we find from (2.15) that $58.1 < x^{+} < 68.1$ days, or from about 8 to $9\frac{1}{2}$ weeks. ✪

2.4.2 Logistic growth curves

Another important model used to describe growth over time is the *logistic growth curve*

$$g(x_i; \boldsymbol{\beta}) = \frac{\beta_0}{1 + e^{-\beta_1 - \beta_2 x_i}}. \tag{2.16}$$

Notice that this can also be written as $g(x_i; \boldsymbol{\beta}) = \beta_0(1 + e^{\beta_1 + \beta_2 x_i})^{-1}e^{\beta_1 + \beta_2 x_i}$.

Verhulst (1845) derived a form of (2.16) to describe dynamic population growth in the presence of restricted resources. (He assumed that the population's growth rate was directly proportional to its size and to the unused carrying capacity of the conjoining environment.) Like the Gompertz curve, (2.16) has three unknown parameters; however, it differs from the Gompertz in many of its operational characteristics. Most important among these is that (2.16) is a *symmetric* sigmoid: its rate of change is symmetric about its inflection point. Thus the two curves complement each other, (2.14) serving to represent asymmetric growth patterns and (2.16) serving to represent symmetric patterns.

For $\beta_2 > 0$ the logistic function is strictly increasing, with vertical intercept at $\beta_0/(1 + e^{-\beta_1})$, a left horizontal asymptote (as $x \to -\infty$) at the horizontal axis, and a right horizontal asymptote (as $x \to \infty$) at β_0. When modeling growth, the upper horizontal asymptote is often viewed as the theoretical maximum size of the population and/or the carrying capacity of the environmental system under study.

For $\beta_2 < 0$ the sigmoid is strictly decreasing, and the asymptotes are reversed (Exercise 2.16). As with the Gompertz function, inferences on (2.16) often concern β_2, since it controls the time-response function: rejection of the null hypothesis $H_0: \beta_2 = 0$ indicates a significant sigmoid growth effect over x.

We should warn readers that a number of other, related expressions take the name 'logistic' for modeling and analyzing environmental data. One is a probability function, the *logistic distribution* (Johnson et al., 1995, Ch. 23), used as an alternative to the normal distribution for continuous data. Another is called the *logistic regression model*, a form of generalized linear model for data that are distributed as binomial rather than normal. (We describe this regression model in §3.3.2.) We will reserve the term 'logistic regression model' for this latter case, and hence refer to (2.16) as a 'logistic growth curve' or a 'logistic growth model.' Readers should be careful not to confuse (2.16) with either the logistic distribution for continuous data or with the logistic regression model for binomial data.

Similar to (2.14), the logistic curve in (2.16) attains its maximum rate of growth (if $\beta_2 > 0$) or decay (if $\beta_2 < 0$) at its inflection point $x^+ = -\beta_1/\beta_2$, with height $g^+ = \beta_0/2$. Estimation of and inferences for these values mimic those in the Gompertz case: given point estimates of the individual regression parameters, b_j ($j = 0, 1, 2$), estimate x^+ via $\hat{x}^+ = -b_1/b_2$. Corresponding $1 - \alpha$ confidence limits can be computed using Fieller's theorem; the form in equation (2.15) remains valid. For the inflection height, use $\hat{g}^+ = b_0/2$, with approximate $1 - \alpha$ confidence limits given by $\hat{g}^+ \pm \frac{1}{2} t_{\alpha/2}(\mathrm{df_e})se[b_0]$.

To fit the logistic growth curve, we appeal to a nonlinear LS algorithm such as Levenberg–Marquardt. Partial derivatives useful for implementing the algorithm are $\partial g/\partial \beta_0 = (1 + e^{-\beta_1 - \beta_2 x})^{-1}$, $\partial g/\partial \beta_1 = \beta_0 e^{-\beta_1 - \beta_2 x}(1 + e^{-\beta_1 - \beta_2 x})^{-2}$, and $\partial g/\partial \beta_2 = x\beta_0 e^{-\beta_1 - \beta_2 x}(1 + e^{-\beta_1 - \beta_2 x})^{-2}$. For the initial estimates, since β_0 is an upper asymptote, take β_{00} as the largest observed response, $\beta_{00} = Y_{(n)}$. For β_1 and β_2, notice that $g(x_i; \boldsymbol{\beta})/\beta_0 = (1 + e^{-\beta_1 - \beta_2 x_i})^{-1}$, so $\log\{g(x_i; \boldsymbol{\beta})/[\beta_0 - g(x_i; \boldsymbol{\beta})]\} = \beta_1 + \beta_2 x_i$. Thus, if we construct the pseudo-variate

$$V_i = \log\left\{\frac{Y_i}{\beta_{00} - Y_i}\right\}$$

and regress this on x_i (excluding the value of V_i corresponding to $Y_{(n)}$ and any values of $Y_i \leq 0$), the resulting intercept and slope estimates can serve as β_{10} and β_{20}, respectively.

Example 2.7 (Mold growth rate) Larralde-Corona *et al.* (1997) report on a study of environmental/nutrient dependencies in the growth kinetics of the mold *Aspergilus niger*. Growth was measured in terms of the organism's ability to extend/elongate its hyphae (tubular filaments) over time, when exposed to a certain initial concentration of glucose. Data were taken as $Y =$ mean hyphal length (in μm) from independent groups of 30 hyphae over $x =$ time (hr). At an initial glucose concentration of 10 g/L, the resulting observations, as read from Larralde-Corona *et al.*'s Fig. 1, appear in Table 2.8.

The authors felt that for these data a logistic growth model was reasonably well motivated by the underlying growth kinetics, and a scatterplot (Fig. 2.21) supports this. To fit (2.16) to the data, we find initial estimates $\beta_{00} = 456.33$, $\beta_{10} = -13.913$, and $\beta_{20} = 1.298$. Sample PROC NLIN code for performing the fit appears in Fig. 2.22.

Table 2.8 Elongation kinetics of *Aspergilus niger* over time at an initial glucose concentration of 10 g/L

$x =$ Time (hr)	$Y =$ Hyphal length (μm)
6	0.00
7	3.25
8	12.99
9	58.44
10	128.25
11	224.03
12	394.58
14	456.33

Source: Larralde-Corona *et al.* (1997).

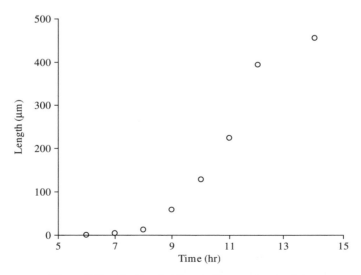

Figure 2.21 Scatterplot for mold growth rate data

```
*  SAS code to fit logistic growth curve;
data mold;
   input time length @@;
   datalines;
 6     0.00    7     3.25    8    12.99    9    58.44
10   128.25   11   224.03   12   394.58   14   456.33

proc nlin method=marquardt;
   parameters  b0  =  456.33
               b1  =  -13.9130705492334
               b2  =  1.29775227563122   ;
   model length = b0/(1+exp(-b1-b2*time)) ;
   der.b0 = 1/(1+exp(-b-b2*time));
   der.b1 = b0*exp(-b1-b2*time)/((1+exp(-b1-b2*time))**2);
   der.b2 = time*b0*exp(-b1-b2*time)/((1+exp(-b1-b2*time))**2);
```

Figure 2.22 Sample SAS program to fit logistic growth curve to mold growth rate data

The results from performing the fit in Fig. 2.22 (not shown) indicate good convergence. A plot of the $n = 8$ residuals from the fit (Fig. 2.23) suggests that variability increases as time increases, although with such a small number of observations strong conclusions about the residual pattern are difficult to defend. Still, the heterogeneous variance pattern may be reasonable: at earlier times the mean response of an increasing logistic curve is more limited and may not be able to range as widely. (A similar argument holds for very late times where the response approaches β_0. For these data, however, observations were not taken at large enough values of x for this latter effect to manifest itself.) To correct for possible variance heterogeneity, we weight each observation inversely to variance and apply a nonlinear weighted least squares (WLS) fit.

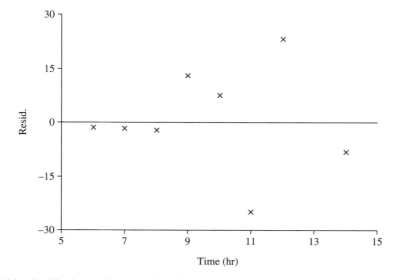

Figure 2.23 Residual plot from LS fit of logistic growth model for mold growth rate data. Horizontal reference line indicates residual $= 0$

We discuss a similar WLS fit in Exercise 2.9. In that exercise, estimates of the differential variance are available from observed sample variances, S_i^2, at each x_i. For the data in Table 2.8, however, this is not possible, so instead we will attempt to model the form of the variance heterogeneity. Since we accept the possibility that variation increases as time increases, we model the variance as strictly proportional to time, that is, $\sigma_i^2 \propto x_i$. Then, we weight inversely to variance via $w_i = 1/x_i$. With these weights, we invoke PROC NLIN as in Fig. 2.22 but also include a _weight_=*name* statement, where *name* is the name of the SAS variable containing the w_is. This produces the output (edited) in Fig. 2.24.

The results from the WLS fit show good convergence. To test for a growth effect over time, the null effect is $H_0: \beta_1 = \beta_2 = 0$. Under H_0, we fit the reduced model that $g(x_i; \boldsymbol{\beta})$ is constant over time (output not shown; cf. Example 2.6). Here the weighted SSE under H_0 is SSE(RM) $= 21\,552.9322$ on 7 df (this is also the Corrected Total Sum of Squares in Fig. 2.24). Appeal to equation (1.7) yields $F_{calc} = \frac{1}{2}(21\,422.2377/26.1389) = 409.777$. When compared to an $F(2,5)$ reference distribution, the result is highly significant ($P < 0.0001$), and we conclude that the data exhibit a significant growth response.

Interest in this experiment was targeted at understanding the growth rates of the organism, hence an estimate of the maximum growth rate was considered important. Using the results in Fig. 2.24, we estimate that the maximum growth rate occurs at $\hat{x}^+ = -b_1/b_2 = 12.5541/1.1477 = 10.94\,\text{hr}$. A 95% confidence interval on x^+ is based on Fieller's theorem in (2.15). From Fig. 2.24 we see $se[b_1] = 1.314$, $se[b_2] = 0.128$,

```
                        The SAS System
                        The NLIN Procedure

                      Method:  Marquardt
                       Iterative Phase
    Iter          b0           b1          b2     Weighted SS
     0          456.3      -13.9131      1.2978        220.2
     1          481.6      -11.9445      1.0864        145.5
     :            :            :           :             :
     6          480.0      -12.5541      1.1477        130.7
     7          480.0      -12.5541      1.1477        130.7
    NOTE: Convergence criterion met.

                          Sum of      Mean                 Approx
    Source           DF   Squares     Square    F Value    Pr > F
    Regression        3   34327.4    11442.5     437.76   <.0001
    Residual          5     130.7    26.1389
    Uncorrected Total 8   34458.1
    Corrected Total   7   21552.9

                              Approx          Approximate 95%
    Parameter   Estimate    Std Error       Confidence Limits
    b0            480.0      23.4339         419.8       540.2
    b1          -12.5541      1.3140       -15.9316     -9.1765
    b2            1.1477      0.1280         0.8187      1.4766

                Approximate Correlation Matrix
                      b0             b1             b2
    b0         1.0000000      0.5433790     -0.6031152
    b1         0.5433790      1.0000000     -0.9949179
    b2        -0.6031152     -0.9949179      1.0000000
```

Figure 2.24 SAS output (edited) from WLS fit of logistic growth curve for mold growth rate data

and $\mathrm{Corr}[b_1, b_2] = -0.9949$. The covariance between b_1 and b_2 is the product of these three values: $\hat{\sigma}_{12} = se[b_1]se[b_2]\mathrm{Corr}[b_1, b_2] = 1.314 \times 0.128 \times (-0.9949) = -0.1673$. With these, and using $t_{0.025}(5) = 2.5706$, we find from (2.15) that $10.61 < x^+ < 11.39$ hr.

Of additional interest here is an estimate of the growth rate *at* its steepest point. This is $g^+ = g'(x^+; \boldsymbol{\beta}) = \beta_0\beta_2/4$. Given LS/ML estimates of b_0 and b_2, we estimate g^+ with $\hat{g}^+ = b_0b_2/4 = 137.72\,\mu\text{m/hr}$. To find an approximate $1 - \alpha$ confidence interval for g^+, we appeal to the delta method (§A.6). In Example A.2 we show that a delta-method approximation to the variance of a product such as b_0b_2 is

$$\mathrm{Var}[b_0b_2] \approx b_2^2 se^2[b_0] + 2b_0b_2\hat{\sigma}_{02} + b_0^2 se^2[b_2],$$

where $\hat{\sigma}_{02} = \mathrm{Cov}[b_0, b_2]$. Then an approximate $1 - \alpha$ Wald interval for g^+ is

$$\hat{g}^+ \pm t_{\alpha/2}(\mathrm{df_e})\frac{\sqrt{b_2^2 se^2[b_0] + 2b_0b_2\hat{\sigma}_{02} + b_0^2 se^2[b_2]}}{4}. \tag{2.17}$$

From Fig. 2.24, we calculate $\hat{\sigma}_{02}$ as $se[b_0]se[b_2]\mathrm{Corr}[b_0, b_2] = 23.434 \times 0.128 \times (-0.603) = -1.809$, so that $se[\hat{g}^+] = \frac{1}{4}\{1.148^2 23.434^2 - (2 \times 1.148 \times 480.0 \times 1.809) + 480.0^2 0.128^2\}^{1/2} = 12.51$. Using this, an approximate 95% confidence interval for g^+ is $137.719 \pm 2.5706 \times 12.51 = 137.72 \pm 32.16\,\mu\text{m/hr}$.

Use of the simple Wald construction in (2.17) relies on the critical assumption that the (large-sample) distribution of $\hat{g}^+$ is approximately $t(\mathrm{df_e})$. This can be questionable when dealing with products or ratios derived from other parameters. (Estimated products or ratios are likely to possess skewed sampling distributions that do not approximate well the symmetric features of a t-distribution.) To address this concern, we can operate instead on the natural logarithm of the product or ratio when building point and interval estimates, and then exponentiate the results to return to the original scale of measurement. For example, an estimate of $\log\{g^+\}$ is $\log\{\hat{g}^+\} = \log(b_0) + \log(b_2) - \log(4)$, so that an alternate $1 - \alpha$ Wald confidence interval for g^+ is

$$\exp\{\log[\hat{g}^+] \pm t_{\alpha/2}(\mathrm{df_e})se[\log(\hat{g}^+)]\}.$$

To find the standard error of $\log\{\hat{g}^+\}$, use the fact that $se^2[\log(\hat{g}^+)] = se^2[\log(b_0) + \log(b_2)] = se^2[\log(b_0)] + se^2[\log(b_2)] + 2\mathrm{Cov}[\log(b_0), \log(b_2)]$. For the individual standard errors, application of the delta method (Example A.1) shows $se[\log(b_j)] \approx se[b_j]/|b_j|$, $j = 1, 2$. Also, for the covariance a multivariate version of the delta method leads to $\mathrm{Cov}[\log(b_0), \log(b_2)] \approx \hat{\sigma}_{02}/b_0b_2$; cf. equation (A.26). Thus we find

$$se[\log\{\hat{g}^+\}] \approx \sqrt{\frac{se^2[b_0]}{b_0^2} + \frac{2\hat{\sigma}_{02}}{b_0b_2} + \frac{se^2[b_2]}{b_2^2}}.$$

Applying this in our transformed Wald interval produces

$$\exp\left\{\log\left[\frac{b_0 b_2}{4}\right] \pm t_{\alpha/2}(\mathrm{df_e})\sqrt{\frac{se^2[b_0]}{b_0^2} + \frac{2\hat{\sigma}_{02}}{b_0 b_2} + \frac{se^2[b_2]}{b_2^2}}\right\}$$

as an approximate $1 - \alpha$ confidence interval on g^+. For the data in Table 2.8 this yields the 95% interval $109.04 < g^+ < 173.95\,\mu\mathrm{m/hr}$, which has roughly the same length but is shifted slightly higher than the simpler interval in (2.17). ✪

In some settings, the logistic model can be reparameterized to represent certain effects in a more explicit fashion. For example, suppose x_i quantifies an environmental stimulus whose effects lead to decreases in an organism's body weight or biomass. A common measure used to summarize the detrimental effects of the stimulus is the *median effective dose* (Trevan, 1927), which is the dose, concentration, exposure, etc., that leads to a 50% reduction from the control/non-exposed response. (We present other uses and interpretations of the median effective dose in §4.1.1, when discussing environmental risk assessment.) Denote this as $x = \mathrm{ED}_{50}$, so that $g(\mathrm{ED}_{50}; \boldsymbol{\beta}) = \frac{1}{2}g(0; \boldsymbol{\beta})$. A variation of (2.16) that builds the ED_{50} explicitly into the parameterization is

$$g(x_i; \boldsymbol{\beta}) = \frac{\beta_0}{1 + (x_i/\mathrm{ED}_{50})^{\beta_1}} \tag{2.18}$$

(Van Ewijk and Hoekstra, 1993). Clearly, $g(\mathrm{ED}_{50}; \boldsymbol{\beta}) = \beta_0/2 = \frac{1}{2}g(0; \boldsymbol{\beta})$, as desired. β_0 takes the role as the spontaneous population mean response, and β_1 is now the dose-response parameter: when $\beta_1 = 0$, the mean response is independent of x.

Notice in (2.18) that for ED_{50} to make sense, the mean response must be non-increasing; that is, the environmental stimulus cannot cause any increase in $g(x; \boldsymbol{\beta})$ as x grows. In terms of the unknown parameters, this is $\beta_1 \geq 0$. To incorporate this restriction into the nonlinear LS algorithm, one never allows the point estimate, b_1, to drop below zero. In PROC NLIN, for example, this is instituted by use of the `bounds` statement: following the `der.` statements, include the command

```
bounds b1 >= 0;
```

(If this bounding operation were not available, an alternative would be to write β_1 as e^{λ_1} and estimate λ_1 directly. This guarantees $\beta_1 > 0$, since e^{λ_1} is always positive for any λ_1.) Note that if this model is applied to data that are in fact increasing in x, it is likely that the algorithm will not converge. Van Ewijk and Hoekstra (1993) describe additional extensions of (2.16) and (2.18).

For some settings, the three-parameter logistic model is too restrictive (as is the Gompertz), since it forces the lower horizontal asymptote to be zero. We overcome this limitation by extending the model with an additional parameter: the *four-parameter logistic growth model* is

$$g(x_i; \boldsymbol{\beta}) = \beta_0 + \frac{\beta_1 - \beta_0}{1 + e^{-\beta_2 - \beta_3 x_i}}, \tag{2.19}$$

where now β_0 is the lower horizontal asymptote and β_1 is the upper horizontal asymptote. (One can similarly extend the Gompertz function for cases where an asymmetric sigmoid is desired. The four-parameter Gompertz curve is $g(x_i; \boldsymbol{\beta}) = \beta_0 + (\beta_1 - \beta_0) \exp\{-e^{-\beta_2 - \beta_3 x_i}\}$.)

Inferences on (2.19) often begin with β_3, since it controls the time-response function. Notice, however, that by setting $\beta_3 = 0$ we are in effect forcing $g(x_i; \boldsymbol{\beta}) = $ constant. Thus assessing $\beta_3 = 0$ is tantamount to assessing $H_0: \beta_1 = \beta_2 = \beta_3 = 0$. One can also test if the simpler three-parameter logistic curve from (2.16) is sufficient by studying $H_0: \beta_0 = 0$. In both cases, we recommend use of the discrepancy-measure approach based on (1.7).

The four-parameter logistic curve attains its maximum rate of growth (if $\beta_3 > 0$) or decay (if $\beta_3 < 0$) at the inflection point $x^+ = -\beta_2/\beta_3$, with height $g^+ = (\beta_0 + \beta_1)/2$. Estimation of and inferences for these values mimic those for the three-parameter model: given point estimates of the individual regression parameters, b_j $(j = 0, 1, 2, 3)$, estimate x^+ via $\hat{x}^+ = -b_2/b_3$. Corresponding $1 - \alpha$ confidence limits can be computed using Fieller's theorem; equation (2.15) remains valid for x^+ with appropriate modification for the shift in subscripts. For the inflection height, use $\hat{g}^+ = (b_0 + b_1)/2$, with approximate $1 - \alpha$ confidence limits given by

$$\hat{g}^+ \pm \frac{t_{\alpha/2}(\mathrm{df}_e)}{2} \sqrt{se^2[b_0] + 2\hat{\sigma}_{01} + se^2[b_1]},$$

where $\hat{\sigma}_{01} = \mathrm{Cov}[b_0, b_1]$.

One often sees (2.19) employed in concert with a heterogeneous variance assumption, requiring either a nonlinear WLS fit (see Example 2.8, below), or a more complex nonlinear regression model that allows σ^2 to vary as a function of x. Readers interested in the latter construction may consult, for example, Davidian and Carroll (1987), Bellio *et al.* (2000), and the references therein.

Example 2.8 (Herbicide bioassay) The four-parameter logistic growth model in (2.19) can be applied to more than just time-response data, of course. Consider the data presented by Bellio *et al.* (2000) on the toxic potential to rutabagas (*Brassica napus*) of the sulfonylurea herbicide chlorsulfuron. Chlorsulfuron is a general-use pesticide employed for weed control. Improper or inattentive application may produce collateral damage to desired species, and study of the chemical's environmental toxicity was of interest.

Over a series of increasing doses (in nmol/L) of the herbicide, the callus area (in mm^2) of replicate tissue cultures from exposed plants was measured; decreasing area indicates increasing toxicity of the chemical. For $Y = \log\{$callus area$\}$ and $x = \log_{10}\{$dose$\}$, the data appear in Table 2.9, and are plotted in Fig. 2.25. (To define $x = \log_{10}\{$dose$\}$ when dose $= 0$, we applied consecutive-dose average spacing from (1.3).) Bellio *et al.* note that although the response appears to drop close to the horizontal axis as log-dose increases, it is possible that some response is still present at very high exposures to the herbicide. Thus they, and we, employ a four-parameter logistic model (2.19) to allow for a non-zero lower asymptote.

Table 2.9 Log-callus areas in *Brassica napus* after exposure to chlorsulfuron (nmol/L)

$\log_{10}$(dose)	Replicate observations	Sample variance
−3.60902	7.244, 7.268, 7.308, 7.553, 7.577, 7.588, 7.845, 8.184	0.1028
−3.09691	6.593, 6.914, 7.417, 7.873	0.3168
−2.60206	7.31, 7.65, 7.69, 7.71	0.0361
−1.60206	6.928, 7.004, 7.411, 7.517, 7.702	0.1116
−1.50864	6.707, 7.504, 7.584, 7.714, 7.819	0.1944
−1.09691	6.797, 6.879, 7.020, 7.457, 7.899	0.2130
0.0	2.015, 2.989, 3.077, 4.825, 5.144	1.7608
0.49136	0.513, 1.051, 1.515, 2.062, 2.550	0.6468
1.0	0.068, 0.718, 0.880, 0.916, 1.356	0.2177

Source: Bellio *et al.* (2000).

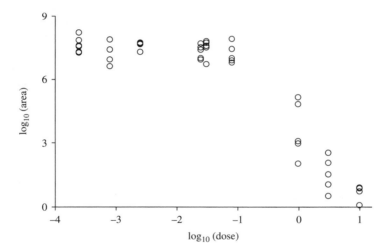

Figure 2.25 Scatterplot for herbicide bioassay data

Notice also in the scatterplot that variability appears to increase at the medium-to-high log-doses, even after application of the logarithmic transform to the data. This is corroborated in Table 2.9, where the sample variances within each dose group exhibit a heterogeneous pattern. As a result, we consider use of nonlinear WLS to estimate and make inferences on the unknown parameters. By contrast to the approach taken in Example 2.7, with the data in Table 2.9 we do have concomitant estimates of variability at each log-dose, in the sample variances S_i^2. Thus we use as our weights the reciprocals of these sample variances, $w_i = 1/S_i^2$.

To fit the model, we appeal again to the Levenberg–Marquardt method. Partial derivatives useful for implementing the algorithm are $\partial g/\partial \beta_0 = (1 + e^{\beta_2 + \beta_3 x})^{-1}$,

$\partial g / \partial \beta_1 = (1 + e^{-\beta_2 - \beta_3 x})^{-1}$, $\partial g / \partial \beta_2 = (\beta_1 - \beta_0) e^{-\beta_2 - \beta_3 x} (1 + e^{-\beta_2 - \beta_3 x})^{-2}$, and $\partial g / \partial \beta_3 = x(\beta_1 - \beta_0) e^{-\beta_2 - \beta_3 x} (1 + e^{-\beta_2 - \beta_3 x})^{-2}$. For the initial estimates, since β_0 is a lower asymptote, take β_{00} as the smallest observed response, $\beta_{00} = Y_{(1)} = \min_{i \in \{1, \ldots, n\}} \{Y_i\}$. Similarly, since β_1 is an upper asymptote, take β_{10} as the largest observed response, $\beta_{10} = Y_{(n)} = \max_{i \in \{1, \ldots, n\}} \{Y_i\}$. For β_2 and β_3, notice that $\log\{[g(x_i; \boldsymbol{\beta}) - \beta_0]/[\beta_1 - g(x_i; \boldsymbol{\beta})]\} = \beta_2 + \beta_3 x_i$. Thus, if we construct the pseudo-variate

$$W_i = \log\left\{ \frac{Y_i - \beta_{00}}{\beta_{10} - Y_i} \right\}$$

and regress this on x_i (excluding the undefined values of W_i at $Y_{(n)}$ and $Y_{(1)}$), the resulting intercept and slope estimates can serve as β_{20} and β_{30}, respectively.

For the data in Table 2.9, we find $\beta_{00} = 0.0680$, $\beta_{10} = 8.184$, $\beta_{20} = -0.2324$, and $\beta_{30} = -0.9867$. Sample PROC NLIN code for performing the four-parameter fit appears in Fig. 2.26. Notice use of the `_weight_=` statement to define the weights for the WLS fit. Output (edited) from the code appears in Fig. 2.27, indicating good convergence and a stable fit.

A natural question of interest is whether chlorsulfuron has a dose-response effect, that is, whether $g(x_i; \boldsymbol{\beta})$ is constant over log-dose. Here the reduced model sets $H_0: \beta_1 = \beta_2 = \beta_3 = 0$, producing the weighted sum of squares $\tilde{S}SE(RM) = 1254.528$ on 45 df. In Fig. 2.27 we see $\tilde{S}SE(FM) = 39.863$ on 42 df, so the corresponding, weighted version of equation (1.7) yields $F_{calc} = \frac{1}{45-42}(1214.665/0.9491) = 426.6$. Compared to an $F(3,42)$ reference distribution, the result is highly significant ($P < 0.0001$), and we conclude that the data do exhibit a significant dose response.   ☉

```
*  SAS code to fit 4-param. logistic growth curve;
data herbicid;
   input logc logarea s2 @@;
   w = 1/s2;
   datalines;
   -3.60902   7.244   0.102771          -3.60902   7.268   0.102771
       :                                    :
    1.00000   0.916   0.217693           1.00000   1.356   0.217693

proc nlin method=marquardt;
   parameters   b0  =  0.068
                b1  =  8.184
                b2  =  -0.232407070200315
                b3  =  -0.986708359946539  ;
   model logarea = b0 + ((b1-b0)/(1+exp(-b2-b3*logc)))  ;
   der.b0 = 1/(1+exp(b2+b3*logc));
   der.b1 = 1/(1+exp(-b2-b3*logc));
   der.b2 = (b1-b0)*exp(-b2-b3*logc)/((1+exp(-b2-b3*logc))**2);
   der.b3 = logc*(b1-b0)*exp(-b2-b3*logc)
                               /((1+exp(-b2-b3*logc))**2);
_weight_ = w;
```

Figure 2.26 Sample SAS program to fit four-parameter logistic growth curve to herbicide bioassay data

```
                            The SAS System
                            The NLIN Procedure

                         Method:  Marquardt
                           Iterative Phase
      Iter       b0         b1          b2          b3     Weighted SS
       0      0.0680     8.1840     -0.2324     -0.9867         199.9
       1     -2.8111     7.1249      0.9551     -1.3548         162.4
       2     -1.1160     7.5780      0.5950     -1.7677       54.0089
       :        :          :           :           :             :
       7      0.4697     7.5446     -0.2637     -2.8382       39.8632
       8      0.4698     7.5446     -0.2637     -2.8383       39.8632
NOTE: Convergence criterion met.

                                     Sum of       Mean               Approx
        Source                DF     Squares      Square    F Value   Pr > F
        Regression             4     16621.1      4155.3     426.59   <.0001
        Residual              42     39.8632      0.9491
        Uncorrected Total     46     16661.0
        Corrected Total       45      1254.5

                                        Approx       Approximate 95%
        Parameter    Estimate       Std Error    Confidence Limits
        b0             0.4698          0.3454     -0.2273      1.1668
        b1             7.5446          0.0671      7.4092      7.6800
        b2            -0.2637          0.3200     -0.9095      0.3821
        b3            -2.8383          0.4903     -3.8277     -1.8489
```

Figure 2.27 SAS output (edited) from fit of four-parameter logistic growth curve for herbicide bioassay data

Another way to extend the three-parameter logistic model is to increase the complexity of the factor involving the exponential term. *Richards's* (1959) *model* (also called the *Chapman–Richards model*) is given by

$$g(x_i; \boldsymbol{\beta}) = \frac{\beta_0}{\left(1 + e^{-\beta_1 - \beta_2 x_i}\right)^{1/\beta_3}},$$

where for $\beta_3 = 1$ the model simplifies to (2.16). The function is derived from differential equations that model organism growth via the difference between its anabolic (synthesis/construction) and catabolic (reduction/destruction) activities. Although useful in a number of applications with forestry and other plant growth (Gregorczyk, 1998), the basic statistical features of this model make it highly nonlinear. Convergence of the nonlinear LS algorithm is not always achieved, and inferences using the resulting estimators can be unstable (Ratkowsky, 1990, §2.5.4). Some simplicity is attained by restricting $\beta_1 = 0$, but not enough to overcome the model's other statistical disadvantages. We recommend application of this logistic extension only when no other sigmoidal alternative is available and the subject-matter demands its use.

2.4.3 Weibull growth curves

Another sigmoid curve based on the exponential operator is the *Weibull growth model*. The model takes a slightly different approach to relate the unknown parameters with the predictor variable and, as a result, it can allow for a variety of curvilinear shapes. The curve is

$$g(x_i; \boldsymbol{\beta}) = \beta_0 + \beta_1 \exp\{-\beta_2 x_i^{\beta_3}\}. \tag{2.20}$$

As with the logistic model, one must be careful not to confuse this with the Weibull distribution for continuous data from §A.2.9, or with another form of regression model for binomial data called a Weibull dose-response model. (We describe the Weibull dose-response model in §4.2.1.) These various entities have similar mathematical forms, but are applied in different statistical settings. We will reserve the term 'Weibull dose-response model' for its use in §4.2, and hence refer to (2.20) as a 'Weibull growth curve' or a 'Weibull growth model.' Readers must be careful not to confuse (2.20) with either the Weibull distribution for continuous data or with the Weibull dose-response model for binomial data.

Calculation of (2.20) is typically restricted to $x_i > 0$ unless β_3 is a positive integer. The curve can fit a variety of complex nonlinear patterns. For instance, if $\beta_2 > 0$ and $\beta_3 > 0$, then at $x = 0$ the curve has a vertical intercept of $g(0; \boldsymbol{\beta}) = \beta_0 + \beta_1$, and as $x \to \infty$, the curve approaches a horizontal asymptote of β_0. If $\beta_2 < 0$, however, then $g(0; \boldsymbol{\beta}) = \beta_0 + \beta_1$, and for $\beta_3 > 0, g(x; \boldsymbol{\beta})$ diverges to ∞ (if $\beta_1 > 0$) or $-\infty$ (if $\beta_1 < 0$) as $x \to \infty$. For $\beta_3 < 0, g(0; \boldsymbol{\beta})$ diverges to ∞, while as $x \to \infty, g(x; \boldsymbol{\beta})$ approaches a horizontal asymptote of $\beta_0 + \beta_1$. In this latter sense, it approximates a simple hyperbola.

Inferences often concern β_2 or β_3, since if either is zero, $g(x; \boldsymbol{\beta})$ is constant with respect to x. The β_3 parameter tends to produce slightly more stable point estimates and hence more reliable inferences than β_2, however. To evaluate the null effect $g(x; \boldsymbol{\beta}) = $ constant, we prefer the discrepancy measure from (1.7) which, in effect, assesses $H_0: \beta_1 = \beta_2 = \beta_3 = 0$.

As above, we fit (2.20) to data via a nonlinear LS algorithm such as Levenberg–Marquardt. Partial derivatives useful for implementing the algorithm are $\partial g/\partial \beta_0 = 1, \partial g/\partial \beta_1 = \exp\{-\beta_2 x^{\beta_3}\}, \partial g/\partial \beta_2 = -x^{\beta_3}\beta_1 \exp\{-\beta_2 x^{\beta_3}\}$, and $\partial g/\partial \beta_3 = -x^{\beta_3} \log(x)\beta_1 \beta_2 \exp\{-\beta_2 x^{\beta_3}\}$. For initial estimates, the variety of shapes the Weibull curve takes on makes it difficult to recommend one omnibus strategy. Analysts must graph the data (always!) and identify from the scatterplot which particular shape (sigmoid, monotone decreasing, etc.) the fitted curve might resemble. From this, one calculates initial estimates by mimicking the sorts of approaches seen with previous nonlinear curves (above). The next example gives an illustration.

Example 2.9 (Pesticide ecotoxicity)   Strodl Andersen *et al.* (1998) report on ecotoxicity (as growth inhibition) in the freshwater algae *Selenastrum capricornutum* after exposure to the pesticide 2-methyl-4,6-dinitro-phenol (DNOC). Samples of the algae were exposed to the pesticide at a variety of concentrations (in cL/L), and their growth rates were measured. From these, observed growth inhibition rates were calculated. The data appear in Table 2.10. (The original DNOC concentration levels were widely dispersed, so we use a logarithmic transform to space them more evenly.

Table 2.10 Growth inhibition rates in *Selenastrum capricornutum* after exposure to varying concentrations (cL/L) of 2-methyl-4,6-dinitro-phenol

$x = \log(\text{conc.})$	$Y = $ Inhibition rate	$x = \log(\text{conc.})$	$Y = $ Inhibition rate
0.3751	0.00503	6.1972	0.32243
0.3751	−0.04372	6.1972	0.35901
0.3751	−0.02340	6.1972	0.37592
0.3751	0.00792	7.8066	0.93736
0.3751	0.03938	7.8066	0.96860
0.3751	0.01478	7.8066	1.00491
2.0000	−0.02918	8.4998	0.97165
2.0000	0.00327	8.4998	1.03250
2.0000	0.00631	8.4998	1.03262
4.5878	0.10319		
4.5878	0.07838		
4.5878	0.18259		

Data based on growth rates reported in Strodl Andersen *et al.* (1998, Appendix B).

To enforce $x > 0$ so that the Weibull model may be applied, we also added 2.0 after applying the logarithm: $x = 2 + \log\{\text{concentration}\}$. At the control concentration of 0, we used consecutive-dose average spacing from (1.3) to define the predictor variable from the shifted logarithmic doses.)

The authors felt that a Weibull curve would provide a reasonable fit to these data. Following their suggestion, we apply model (2.20). For the initial estimates, a plot of the data (Exercise 2.20) suggests an increasing sigmoidal shape. This would be consistent with a Weibull curve where $\beta_1 > 0, \beta_2 > 0$, and $\beta_3 < 0$. Assuming this, we expect the mean response to approach β_0 as $x_i \to 0$, and to approach $\beta_0 + \beta_1$ as $x_i \to \infty$. Thus for initial estimates, we take $\beta_{00} = Y_{(1)} = \min_{i \in \{1,\dots,n\}}\{Y_i\}$ and $\beta_{10} = Y_{(n)} - \beta_{00} = \max_{i \in \{1,\dots,n\}}\{Y_i\} - \beta_{00}$. Next, we recognize that under (2.20) $\log\{[g(x_i; \boldsymbol{\beta}) - \beta_0]/\beta_1\} = -\beta_2 x_i^{\beta_3}$, so that $\log(-\log\{[g(x_i; \boldsymbol{\beta}) - \beta_0]/\beta_1\}) = \lambda_0 + \beta_3 \log(x_i)$ for $\lambda_0 = \log(\beta_2)$. Thus, if we construct the pseudo-variate

$$U_i = \log\left(-\log\left\{\frac{Y_i - \beta_{00}}{\beta_{10}}\right\}\right)$$

and regress this on $\log\{x_i\}$ (ignoring any term for which the logarithm is undefined), the resulting intercept and slope estimates can serve as $\log(\beta_{20})$ and β_{30}, respectively. Doing so for the data in Table 2.10 yields $\beta_{00} = -0.0437$, $\beta_{10} = 1.0763$, $\beta_{20} = 1.8850$, and $\beta_{30} = -1.2187$. Sample PROC NLIN code for performing this fit appears in Fig. 2.28, followed by the corresponding output (edited) in Fig. 2.29.

The output in Fig. 2.29 indicates reasonable convergence. A residual analysis of the fit is carried out as part of Exercise 2.20, where the possibility of a further model fit (with $\beta_0 = 0$) is explored. Here, we test simply whether the inhibition response differs significantly from a constant. We fit the reduced model with $g(x_i; \boldsymbol{\beta}) = \beta_0$ and

```
* SAS code to fit Weibull growth curve;
data pesticid;
  input logc  inhib  @@;
  datalines;
    0.3750532   0.00503433              0.3750532  -0.0437188
        ⋮                                    ⋮
    8.4997870   1.03250124              8.4997870   1.03262343
proc nlin method=marquardt;
  parameters    b0 =  -0.0437
                b1 =   1.0763
                b2 =   1.88497
                b3 =  -1.21866;
  model inhib = b0 + (b1*exp(-b2*(logc**b3))) ;
  der.b0 = 1;
  der.b1 = exp(-b2*(logc**b3));
  der.b2 = -(logc**b3)*b1*exp(-b2*(logc**b3));
  der.b3 = -(logc**b3)*b1*b2*exp(-b2*(logc**b3))*log(logc);
```

Figure 2.28 Sample SAS program to fit Weibull growth curve to pesticide ecotoxicity data

```
                          The SAS System
                          The NLIN Procedure

                       Method:  Marquardt
                       Iterative Phase
      Iter        b0          b1         b2         b3     Sum of Squ.
       0      -0.0437      1.0763     1.8850    -1.2187      2.5920
       1      -0.0244      1.1342     3.5772    -0.7812      1.4794
       2      -0.00368     2.4537     8.4842    -0.9300      0.6135
       ⋮         ⋮           ⋮          ⋮          ⋮           ⋮
      34       0.00813     2.3282    170.3     -2.5105       0.0702
NOTE:  Convergence criterion met.
```

Source	DF	Sum of Squares	Mean Square	F Value	Approx Pr > F
Regression	4	6.2625	1.5656	298.01	<.0001
Residual	17	0.0702	0.00413		
Uncorrected Total	21	6.3327			
Corrected Total	20	3.7605			

Parameter	Estimate	Approx Std Error	Approximate 95% Confidence Limits	
b0	0.00813	0.0212	-0.0366	0.0529
b1	2.3282	1.2174	-0.2402	4.8967
b2	170.3	279.4	-419.3	759.9
b3	-2.5105	1.0578	-4.7422	-0.2787

Figure 2.29 SAS output (edited) from fit of Weibull growth curve for pesticide ecotoxicity data

apply the discrepancy-measure approach based on (1.7). We find SSE(RM) = 3.7605 on 20 df. Since SSE(FM) = 0.0702, (1.7) yields $F_{calc} = \frac{1}{3}(3.6903/0.0041) = 3000.02$. When compared to an $F(3,17)$ reference distribution, the result is highly significant ($P < 0.0001$), and we conclude that the data exhibit a significant inhibition response. (Notice that the 95% confidence interval for β_3 given in Fig. 2.29 fails to contain zero, corroborating these results. By contrast, the 95% confidence interval for β_2 does contain zero, which might be used to argue that $g(x; \beta)$ is constant. This illustrates the irregular stability of the β_2 parameter estimate; see Ratkowsky, 1990, §5.4, for a further discussion.)

2.5 Rational polynomials

A group of highly nonlinear models that does not employ exponential forms is the class of *rational polynomial models*. Functions in this class are ratios of polynomials; the general expression is

$$g(x_i; \beta) = \frac{\beta_0 + \beta_1 x_i + \cdots + \beta_q x_i^q}{\beta_{q+1} + \beta_{q+2} x_i + \cdots + \beta_{q+r+1} x_i^r},$$

(2.21)

where the orders, q and r, of the two polynomials are positive integers such that $q \leq r$. (Models with $q > r$ are possible, but are less common in practice.) For simplicity, we usually do not operate with more than quadratic or possibly cubic terms in either the numerator or the denominator. Even when restricted to these simple cases, the rational polynomial model can provide a variety of flexible forms for nonlinear modeling.

2.5.1 Michaelis–Menten model

Perhaps the simplest of the rational polynomials is also one of the most famous: the *Michaelis–Menten model* (Michaelis and Menten, 1913). Based on equations for chemical enzyme or protein interactions, the Michaelis–Menten framework is related to models of compartmental phenomena such as those mentioned in §2.3. It is often employed to describe the velocity or rate of a chemical reaction, Y, as a function of the initial concentration, x, of a chemical substrate consumed by the reaction. The concepts are also useful for modeling toxicokinetic activity in organisms when recording blood or tissue concentrations (Becka *et al.*, 1993) after exposure to some environmental toxin, or for modeling predation over time in ecological field studies (Juliano, 1993).

As in (2.1), we let $Y_i = g(x_i; \beta) + \varepsilon_i, i = 1, \ldots, n$. Denote the maximum possible velocity of the reaction (achieved only asymptotically) as β_1, and the concentration at which $g(x_i; \beta)$ attains half this level as β_2. That is, $g(\beta_2; \beta) = \frac{1}{2}\beta_1$. Notice that β_2 is a form of median effective dose, as in (2.18). From this, the Michaelis–Menten model is derived as

$$g(x_i; \beta) = \frac{\beta_1 x_i}{\beta_2 + x_i},$$

(2.22)

which is (2.21) with $q = r = 1$, and where $\beta_0 = 0$ and $\beta_3 = 1$. (Note that the traditional notation in the chemometric literature for the unknown parameters is $\beta_1 = V_{\max}$ and $\beta_2 = K_{\mathrm{m}}$.) Defined only over $x_i \geq 0$, the curve starts at the origin and rises monotonically to the horizontal asymptote at β_1 (corresponding to saturation of the chemical and/or biological process).

Due to both its popularity and longevity, a number of different routines have been proposed to fit the Michaelis–Menten model; Ruppert *et al.* (1989) provide a good background. As a rule, most such methods are somewhat *ad hoc* and possess no underlying optimality other than their relative ease of implementation. Although useful in identifying excellent initial values for a more advanced iterative fit (see below), we do not recommend these older methods. Instead, we employ nonlinear LS. As above, when we also assume $\varepsilon_i \sim$ i.i.d. $N(0, \sigma^2)$, the LS estimators correspond to ML estimates for β_1 and β_2.

We continue to employ the Levenberg–Marquardt algorithm to implement the nonlinear LS routine. For the partial derivatives under (2.22) we find $\partial g/\partial \beta_1 = x/(\beta_2 + x)$, and $\partial g/\partial \beta_2 = -\beta_1 x/(\beta_2 + x)^2$. To find initial estimates for the iterative fit, we appeal to a linearization of (2.22) known as the *Lineweaver–Burk method* (Lineweaver and Burk, 1934): recognize that $1/g(x_i; \boldsymbol{\beta}) = \beta_1^{-1} + \beta_1^{-1}\beta_2 x_i^{-1}$. From this, construct the pseudo-variates $U_i = 1/Y_i$ and $z_i = 1/x_i$ and then regress U_i on z_i. Denote the resulting intercept and slope estimates as a_0 and a_1, respectively. In terms of (2.22), a_0 corresponds to an estimate of $1/\beta_1$ and a_1 corresponds to an estimate of β_2/β_1, so for the initial estimates use $\beta_{10} = 1/a_0$ and $\beta_{20} = a_1/a_0$.

Confidence intervals on the Michaelis–Menten parameters are available using the standard theory we noted above: an approximate, pointwise, $1 - \alpha$ confidence interval on β_j is $\beta_j \pm t_{\alpha/2}(n - p)se[b_j], j = 1, 2$. Here, $p = 2$. An approximate, joint, $1 - \alpha$ confidence region for both parameters (cf. §A.5.4) can be constructed as the two-dimensional ellipse defined by the inequality

$$\frac{(\beta_1 - b_1)^2 se^2[b_2] - 2(\beta_1 - b_1)(\beta_2 - b_2)\hat{\sigma}_{12} + (\beta_2 - b_2)^2 se^2[b_1]}{se^2[b_1]se^2[b_2]\hat{\sigma}_{12}^2} \leq 2F_\alpha(2, n - 2)$$

(2.23)

(Seber and Wild, 1989, §3.3.1), where $\hat{\sigma}_{12} = \mathrm{Cov}[b_1, b_2]$ is the estimated covariance between b_1 and b_2. One can also employ a Bonferroni adjustment, which yields $b_j \pm t_{\alpha/4}(n - 2)se[b_j]$, $j = 1, 2$. Extending this, simultaneous confidence bands for $g(x_i; \boldsymbol{\beta})$ under (2.22) are discussed by Tez (1991).

Example 2.10 (Puromycin) Bates and Watts (1988, §A1.3) present one of the best-known data sets in nonlinear regression analysis as an example of Michaelis–Menten modeling. The data are calculated initial rates (counts/min^2) of a radioactive product from a cellular enzymatic reaction after treatment with the peptidyl transfer inhibitor puromycin. Viewed here as a chemical substrate, puromycin was administered at initial concentrations ranging from 0.02 ppm to 1.10 ppm. The data appear in Table 2.11.

A plot of the data (Fig. 2.30) shows a monotone increasing pattern consistent with the Michaelis–Menten model. To fit the model we find initial estimates from the Lineweaver–Burk method to be $\beta_{10} = 195.8027$ and $\beta_{20} = 0.0484$. Sample PROC NLIN code for performing the fit appears in Fig. 2.31, followed by the corresponding output (edited) in Fig. 2.32. The output indicates adequate convergence. The quality of the fit is generally good; Exercise 2.23 explores this in detail.

A typical question of interest with such data would be, what is the maximum (asymptotic) reaction velocity? We estimate this as $b_1 = 212.68$ counts/min^2, with approximate 95% confidence limits from Fig. 2.32 given by $197.20 < \beta_1 < 228.16$ counts/min^2. Also, the concentration at which 50% of this maximum is reached is estimated by $b_2 = 0.0641$ ppm. Corresponding approximate 95% confidence limits are $0.0457 < \beta_2 < 0.0826$ ppm.

✪

Table 2.11   Chemical reaction velocities (radioactive counts/min^2) as a function of initial concentration of puromycin (ppm)

$x =$ Concentration	$Y =$ Reaction velocity
0.02	76
0.02	47
0.06	97
0.06	107
0.11	123
0.11	139
0.22	159
0.22	152
0.56	191
0.56	201
1.10	207
1.10	200

Source: Bates and Watts (1988, §A1.3).

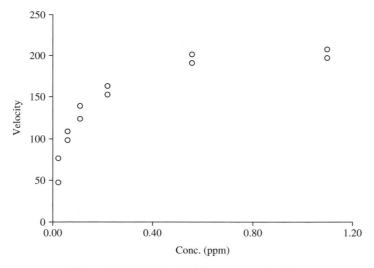

Figure 2.30   Scatterplot for puromycin data

Extension of the Michaelis–Menten model to more general forms of rational polynomial are common. For example, Bates and Watts (1988, §3.13.5) add a quadratic term to the denominator of (2.22) that allows the mean reaction velocity to rise to a true maximum and then drop:

$$g(x_i; \boldsymbol{\beta}) = \frac{\beta_1 x_i}{\beta_2 + x_i + \beta_3 x_i^2}. \tag{2.24}$$

```
* SAS code to fit Michaelis-Menten curve;
data puromycn;
  input conc vel @@;
  datalines;
  0.02    76      0.02    47      0.06    97      0.06    107
  0.11    123     0.11    139     0.22    159     0.22    152
  0.56    191     0.56    201     1.10    207     1.10    200
proc nlin method=marquardt;
  parameters  b1  = 195.80271
              b2  = 0.04841  ;
  model vel = b1*conc/(b2 + conc) ;
  der.b1 = conc/(b2 + conc);
  der.b2 = -b1*conc/((b2 + conc)**2);
```

Figure 2.31 Sample SAS program to fit Michaelis–Menten curve to puromycin data

```
                        The SAS System
                       The NLIN Procedure

                      Method:  Marquardt
                       Iterative Phase
            Iter        b1            b2      Sum of Squ.
            0         195.8        0.0484      1920.6
            1         210.9        0.0614      1207.9
            :           :            :           :
            4         212.7        0.0641      1195.4
            5         212.7        0.0641      1195.4
       NOTE: Convergence criterion met.

                            Sum of      Mean              Approx
       Source          DF   Squares    Square   F Value   Pr > F
       Regression       2   270214     135107   1130.18   <.0001
       Residual        10   1195.4     119.5
       Uncorrected Total 12  271409
       Corrected Total  11   30858.9

                             Approx       Approximate 95%
       Parameter  Estimate  Std Error   Confidence Limits
       b1          212.7      6.9471    197.20     228.16
       b2          0.0641     0.00828   0.0457     0.0826
```

Figure 2.32 SAS output (edited) from fit of Michaelis–Menten curve for puromycin data

Figure 2.33 compares the two Michaelis–Menten forms (2.22) and (2.24).

One can also modify the Michaelis–Menten curve to have a higher-power exponent, but still retain the two-parameter structure. This produces the *Hill equation* (Hill, 1910):

$$g(x_i; \boldsymbol{\beta}) = \frac{\beta_1 x_i^k}{\beta_2 + x_i^k}, \tag{2.25}$$

for some known $k > 0$. The Hill equation extends the Michaelis–Menten curve by allowing a sigmoid to describe the velocity or rate of the chemical reaction; see Heidel and Maloney (1999).

Alternatively, allowing β_0 to depart from zero in (2.22) yields a simple hyperbolic function

$$g(x_i; \boldsymbol{\beta}) = \frac{\beta_0 + \beta_1 x_i}{\beta_2 + x_i},$$

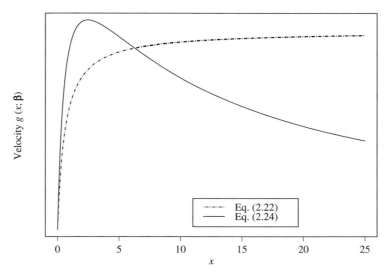

Figure 2.33 Michaelis–Menten curves (2.22) and (2.24)

which is often seen written in the equivalent form $g(x_i; \xi) = (\xi_0 + \xi_1 x_i)/(1 + \xi_3 x_i)$. Many other modifications of the Michaelis–Menten curve are possible; see Ratkowsky (1990, §4.3).

2.5.2 Morgan–Mercer–Flodin model

An additional complexity that can be incorporated into a rational polynomial model is to make one or more of the polynomial powers unknown, and hence add them to the regression parameters, β, for purposes of estimation. This drastically increases the intrinsic nonlinearity of $g(x_i; \beta)$, and we recommend such models only when the underlying subject-matter warrants their use.

One of the best-known unknown-order rational polynomials is the *Morgan–Mercer–Flodin (MMF) model*,

$$g(x_i; \beta) = \frac{\beta_0 \beta_1 + \beta_2 x_i^{\beta_3}}{\beta_1 + x_i^{\beta_3}}. \tag{2.26}$$

The MMF curve was developed as an alternative to the Michaelis–Menten and Hill models for describing nutritional response/growth in organisms faced with a limiting nutrient source (Morgan *et al.*, 1975). Its application has been shown to be far greater, however, since the model allows for a non-zero vertical intercept (β_0) and both hyperbolic and sigmoidal response patterns within its general structure. Note that some authors write (2.26) in its equivalent form $g(x_i; \xi) = (\xi_0 + \xi_1 x_i^{\xi_3})/(1 + \xi_2 x_i^{\xi_3})$.

As with the Michaelis–Menten model, we generally apply the MMF curve when $x \geq 0$. The function approaches a limiting horizontal asymptote of β_2 as $x \to \infty$, and reaches a 50% point of increase above its vertical intercept (a form of median effective dose) at $ED_{50} = \beta_1^{1/\beta_3}$. That is, $g(\beta_1^{1/\beta_3}; \boldsymbol{\beta}) = (\beta_0 + \beta_2)/2$.

To fit the model we again appeal to a nonlinear LS algorithm, such as Levenberg–Marquardt. Partial derivatives useful for implementing the algorithm are $\partial g/\partial \beta_0 = \beta_1/(\beta_1 + x^{\beta_3})$, $\partial g/\partial \beta_1 = (\beta_0 - \beta_2)x^{\beta_3}/(\beta_1 + x^{\beta_3})^2$, $\partial g/\partial \beta_2 = x^{\beta_3}/(\beta_1 + x^{\beta_3})$, and $\partial g/\partial \beta_3 = (\beta_2 - \beta_0)x^{\beta_3}\log(x)/(\beta_1 + x^{\beta_3})^2$. For the initial estimates, note that the minimum response is β_0, so start with $\beta_{00} = Y_{(1)} = \min_{i \in \{1,\dots,n\}}\{Y_i\}$. Similarly, since the upper asymptote is β_2, set $\beta_{20} = Y_{(n)} = \max_{i \in \{1,\dots,n\}}\{Y_i\}$. Lastly, Morgan *et al.* (1975) note that $g(x; \boldsymbol{\beta}) - \beta_2 = -\beta_1\{g(x; \boldsymbol{\beta}) - \beta_0\}/x^{\beta_3}$, so that $\log\{[\beta_2 - g(x; \boldsymbol{\beta})]/[g(x; \boldsymbol{\beta}) - \beta_0]\} = \log(\beta_1) - \beta_3\log(x)$. Thus, if we construct the pseudo-variate

$$W_i = \log\left\{\frac{\beta_{20} - Y_i}{Y_i - \beta_{00}}\right\}$$

and regress this on $-\log(x_i)$ (ignoring those terms for which any of the logarithms are undefined), the resulting intercept and slope estimates can serve as $\log(\beta_{10})$ and β_{30}, respectively.

When the errors satisfy the usual normal distribution/homogeneous-variance assumptions, point estimates and confidence intervals can be constructed from the nonlinear LS fit using the methods described above. To test if there is any effect due to the predictor variable, x_i, assess $\beta_1 = \beta_2 = \beta_3 = 0$ using the discrepancy-measure approach from (1.7).

One can also construct an approximate $1 - \alpha$ confidence interval for $ED_{50} = \beta_1^{1/\beta_3}$, using a combination of Fieller's theorem (§1.5) and the delta method (§A.6). Begin by taking the natural logarithm: $\log(ED_{50}) = \log(b_1)/b_3$. Clearly, this is a ratio of unknown parameters, estimated via $\log(b_1)/b_3$. Now, apply Fieller's result to build $1 - \alpha$ limits. This requires values for $se[\log(b_1)]$, $se[b_3]$, and $Cov[\log(b_1), b_3]$. Unfortunately, typical nonlinear regression outputs will not supply all of these terms; most give $se[b_3]$, but likely also provide only $se[b_1]$ and $\hat{\sigma}_{13} = Cov[b_1, b_3]$. This is where we apply the delta method. For $se[\log(b_1)]$, the delta method (Example A.1) gives $se[\log(b_1)] \approx se[b_1]/|b_1|$. Also, for $Cov[\log(b_1), b_3]$, a multivariate version of the delta method leads to $Cov[\log(b_1), b_3] \approx \hat{\sigma}_{13}/b_1$; cf. equation (A.25). With these quantities, we return to Fieller's theorem to generate approximate $1 - \alpha$ limits on $\log(ED_{50}) = \log(\beta_1)/\beta_3$. For $\log(\hat{ED}_{50}) = \log(b_1)/b_3$, the result becomes

$$\log(\hat{ED}_{50}) + \frac{\gamma}{1-\gamma}\left\{\log(\hat{ED}_{50}) - \frac{\hat{\sigma}_{13}}{b_1 se^2[b_3]}\right\}$$

$$\pm \frac{t_{\alpha/2}(\nu)}{(1-\gamma)|b_3|}\left\{\frac{se^2[b_1]}{b_1^2} + \log(\hat{ED}_{50})\left(se^2[b_3]\log(\hat{ED}_{50}) - \frac{2\hat{\sigma}_{13}}{b_1}\right)\right.$$

$$\left. - \frac{\gamma}{b_1^2}\left(se^2[b_1] - \frac{\hat{\sigma}_{13}^2}{se^2[b_3]}\right)\right\}^{1/2},$$

where $\gamma = t^2_{\alpha/2}(\nu)se^2[b_3]/b_3^2$ and $\nu = df_e$ from the nonlinear LS fit. With these limits on $\log(ED_{50})$, we exponentiate to arrive at corresponding $1 - \alpha$ limits on ED_{50}.

The MMF model can be difficult to fit, due to its highly nonlinear structure. Ratkowsky (1990, §5.4) gives some suggestions for reparameterizing (2.26) in order to increase its stability for nonlinear LS estimation. For example, replacing β_1 with e^{β_4} and operating with $\boldsymbol{\beta} = [\beta_0 \; \beta_2 \; \beta_3 \; \beta_4]^T$ can often improve behavior of the model fit.

2.6   Multiple nonlinear regression

Although our presentation has emphasized application of nonlinear models with a single predictor variable, it is also possible to incorporate multiple predictor variables, as in §1.2. Equation (2.1) becomes

$$Y_i = g(x_{i1}, x_{i2}, \ldots, x_{ip}; \boldsymbol{\beta}) + \varepsilon_i,$$

$i = 1, \ldots, n$, where $g(x_{i1}, x_{i2}, \ldots, x_{ip}; \boldsymbol{\beta})$ is some prespecified nonlinear function of $p > 1$ predictors. The possible forms and uses of such a *multiple nonlinear regression model* are quite varied, and a full treatment is beyond the scope here. We give a short example to illustrate the matter.

Example 2.11 (Dissolved oxygen rates)   How human activities impact landscape receiving waters such as lakes is often of interest to environmental scientists. An important monitoring endpoint in this regard is the level of dissolved oxygen (DO) at a lake's lower depths: unusually low DO levels are less supportive of aquatic life and indicate a lake whose ecosystem is in jeopardy.

A recent study of lake quality in the western United States compared two such lakes: Eagle Lake, a relatively remote lake with little human impact, and Tahoe Keys, an area of Lake Tahoe with high amounts of boating and other human activity. $Y = $ DO level was measured as a function of $x_1 = $ depth. Since DO levels tend to decrease exponentially with depth, the response takes on a nonlinear form: $E[Y_i] = \exp\{\beta_0 + \beta_1 x_{i1}\}$. In this particular study, it was expected that the rate of DO decline would be more extreme in Tahoe Keys than in Eagle Lake. To incorporate this feature into the nonlinear model, an additional predictor variable was defined that indexed whether or not the sample was from Tahoe Keys (i.e., $x_{i2} = 1$ if the ith DO sample was from Tahoe Keys and $x_{i2} = 0$ if the sample was from Eagle Lake). Also included was a third predictor variable to allow for a possible interaction between the depth and lake indicator variables: $x_{i3} = x_{i1}x_{i2}$. Thus the nonlinear model took the form $Y_i = \exp\{\beta_0 + \beta_1 x_{i1} + \beta_2 x_{i2} + \beta_3 x_{i3}\}$. Data from this study (supplied by Dr. J. T. Oris of Miami University) along with sample SAS code used to construct the indicator variables and fit the nonlinear regression model are given as part of Figure 2.34.

To fit this model via nonlinear LS we use the Levenberg–Marquardt algorithm. The partial derivatives of the nonlinear response function are $\partial g/\partial \beta_0 = e^{\beta_0 + \beta_1 x_{i1} + \beta_2 x_{i2} + \beta_3 x_{i3}}$, $\partial g/\partial \beta_1 = x_{i1} e^{\beta_0 + \beta_1 x_{i1} + \beta_2 x_{i2} + \beta_3 x_{i3}}$, $\partial g/\partial \beta_2 = x_{i2} e^{\beta_0 + \beta_1 x_{i1} + \beta_2 x_{i2} + \beta_3 x_{i3}}$, and $\partial g/\partial \beta_3 = x_{i3} e^{\beta_0 + \beta_1 x_{i1} + \beta_2 x_{i2} + \beta_3 x_{i3}}$.

```
*  SAS code to fit multiple nonlinear model;
   data lakes;
   input x1depth DO lakeid $ @@;
   x2TK = (lakeid='T');
   x3int = x2TK*x1depth;
   datalines;
    0 10.40   E        1  7.50   E        2  6.60   E
    3  6.10   E        4  5.70   E        5  5.40   E
    6  5.10   E       11  2.90   E       16  2.00   E
   21  1.20   E       26  1.00   E
    0  9.26   T        1  7.63   T        2  5.05   T
    3  2.52   T        4  1.95   T        5  1.47   T

   proc nlin  method=marquardt;
      parameters  b0 =   2.1243
                  b1 =  -0.0876
                  b2 =   0.1841
                  b3 =  -0.3122  ;
   model DO = exp(b0 + b1*x1depth + b2*x2TK + b3*x3int);
   der.b0 = exp(b0 + b1*x1depth + b2*x2TK + b3*x3int);
   der.b1 = x1depth*exp(b0 + b1*x1depth + b2*x2TK + b3*x3int);
   der.b2 = x2TK*exp(b0 + b1*x1depth + b2*x2TK + b3*x3int);
   der.b3 = x3int*exp(b0 + b1*x1depth + b2*x2TK + b3*x3int);
```

Figure 2.34 Sample SAS program to fit multiple nonlinear regression model to dissolved oxygen data

For initial estimates, we take the natural logarithm of Y_i and stratify the data by lake, then fit individual simple linear regressions on $U_i = \log(Y_i)$ for each lake. That is, we fit $U_i = \beta_{00} + \beta_{01}x_{1i}$ to only the Eagle Lake data to produce LS estimates b_{00} and b_{10}, and next we fit $U_i = \gamma_0 + \gamma_1 x_{1i}$ to only the Tahoe Keys data to produce LS estimates g_0

```
                       The SAS System
                       The NLIN Procedure

                     Method:  Marquardt
                      Iterative Phase
   Iter      b0         b1        b2         b3      Sum of Squ.
    0     2.1243    -0.0876    0.1841    -0.3122      6.6190
    1     2.1954    -0.1023    0.0809    -0.2645      5.2708
    :        :         :         :          :           :
    5     2.2019    -0.1053    0.0763    -0.2646      5.2450
    6     2.2019    -0.1053    0.0763    -0.2646      5.2450
   NOTE: Convergence criterion met.

                          Sum of    Mean                 Approx
   Source            DF   Squares   Square   F Value     Pr > F
   Regression         4    524.2    131.1    324.83      <.0001
   Residual          13    5.245    0.4035
   Uncorrected Total 17    529.5
   Corrected Total   16    136.1

                                Approx      Approximate 95%
   Parameter   Estimate      Std Error    Confidence Limits
   b0            2.2019        0.0460     2.1024      2.3013
   b1           -0.1053        0.0119    -0.1310     -0.0795
   b2            0.0763        0.0746    -0.0849      0.2376
   b3           -0.2646        0.0430    -0.3576     -0.1717
```

Figure 2.35 SAS output (edited) from multiple nonlinear regression fit

and g_1. The initial values for β_0 and β_1 are then simply b_{00} and b_{10}, while the initial values for β_2 and β_3 are $g_0 - b_{00}$ and $g_1 - b_{10}$. This leads to the values employed in Fig. 2.34.

Output (edited) from the code in Fig. 2.34 appears in Fig. 2.35, where good convergence is indicated. We find the final nonlinear LS estimates to be $b_0 = 2.202$, $b_1 = -0.105$, $b_2 = 0.076$, and $b_3 = -0.265$. A residual plot (not shown) shows a broadly reasonable fit, with perhaps a hint of heterogeneous variation at larger levels of DO.

Examining the results for the interaction parameter β_3, we see that the depth-related decrease in DO for Eagle Lake appears milder than that for Tahoe Keys, since the associated 95% confidence interval contains only negative values: $-0.36 \le \beta_3 \le -0.17$. Of additional interest is the parameter β_2, which can be viewed as a measure of the difference in surface DO between the two lakes. Here, the approximate 95% confidence interval is $-0.08 \le \beta_2 \le 0.24$; since this interval contains 0.00 we can infer that the surface DO rates between the lake lakes do not differ significantly.

Exercises

2.1. Show that under a continuity constraint, the segmented linear model in equation (2.2) simplifies to equation (2.3). [*Hint*: if continuity holds, then $\beta_0 = \beta_1 + \beta_2(x - \tau)$ at $x = \tau$.]

2.2. Kalf *et al.* (1996) study the effects of the environmental toxin benzene on human bone marrow and blood cells. As part of their study, the authors present data on the ability of a benzene metabolite, *p*-benzoquinone (BQ), to inhibit the activity of calpain, a component of human platelets. Increased inhibition suggests a debilitating effect due to the toxin. For this component of the study, we let $Y = $ percent calpain inhibition and $x = \log_{10}\{BQ \text{ concentration}\}(\text{in } \log_{10} \mu M)$. The data are:

$x = \log_{10}(BQ)$	−0.1852	0.0778	0.2296	0.3407
$Y = \%$ inhibition	32.3529	43.2941	54.1176	42.4706
$x = \log_{10}(BQ)$	0.6963	0.9556	1.5482	1.8259
$Y = \%$ inhibition	9.5294	95.1765	100.0	100.0

Critical to the model construction here is the fact that maximum inhibition must always be 100.0. Indeed, the last two data points exhibit this effect. Thus we have a right-plateau model where the height of the plateau is known.

(a) Express this model in a form similar to equation (2.2).

(b) Continuity at the join point τ is a natural assumption to make here. Update your model expression from part (a) to accommodate this. How many unknown parameters does the nonlinear model now possess?

(c) Find the partial derivatives of your model function from part (b). Express these in terms of indicator functions, as in Example 2.1.

(d) Plot the data. Does the plot seem to agree with the theoretical constructions you have made?

(e) Fit your model to the inhibition data. For initial estimates, we recommend the following: for the join point take the mean of the sixth and seventh log-BQ values: $\tau_0 = \frac{1}{2}(0.9556 + 1.5482) = 1.2519$. For the pre-plateau slope, take the slope between the first and last observations before this τ_0: $(95.1765 - 32.3529)/(0.9556 + 0.1852) = 55.0722$. Estimate the true join point and give an associated 95% confidence interval.

(f) Find the predicted values and from these calculate residuals from your fit in part (e). Construct a residual plot and comment on the fit.

2.3. Return to the polymer data from Table 2.2, as used in Example 2.2.

(a) By mimicking the approach in Fig. 2.6, corroborate the quantities given in Table 2.3. In particular, verify by direct calculation that the restricted LS point estimates $b_0(\tau)$ and $b_1(\tau)$ are simply the sample means from each relevant segment.

(b) Prove theoretically for the simple change-point model in this example that by restricting τ to lie in an interval between two consecutive predictor values, x_i and x_{i+1}, the restricted LS estimators $b_0(\tau)$ and $b_1(\tau)$ are the sample means of the corresponding segments defined over $x_1, \ldots, x_i$ and $x_{i+1}, \ldots, x_n$.

2.4. Mercuric compounds used in environmental pesticides can produce toxic effects in individuals exposed to these chemicals. For example, after an accidental phenyl-mercury exposure, data were recovered on infants' renal toxic response (as activity of the renal tubular enzyme γ-glutamyl transpeptidase, γ-GT, in U/L) as a function of log-mercury (Hg) concentration (in μg/L) in their urine. The data are:

log-Hg concentration	0.916	3.09	4.09	4.50	4.65	4.97
γ-GT activity	13.57	15.33	16.92	16.08	17.14	18.85
log-Hg concentration	5.18	5.35	5.45	5.55	5.70	5.99
γ-GT activity	18.63	19.30	25.77	28.44	35.46	45.26

(a) Since humans, even infants, may tolerate low levels of mercury with little toxicity, the toxic effects may exhibit themselves in a two-phase manner. Plot these data. Do you see a segmented effect?

(b) Fit a continuous, segmented linear model, (2.7), to these data. Use the plot from part (a) to determine your initial estimates. [*Hint*: we suggest setting the initial join point at x_7: $\tau_0 = 5.18$. Work from this to determine the other initial estimates.] Use the nonlinear LS fit to find a 95% confidence interval on the join point.

(c) Based on your fit in part (b), does the left segment deviate significantly from a horizontal line? Operate at $\alpha = 0.05$.

(d) Find the predicted values and from these calculate residuals from your fit in part (b). Construct a residual plot and comment on the fit.

(Piegorsch, 1987)

2.5. Return to the pond water level data in Example 2.3.

(a) Plot the pond water levels against aquifer level. Verify the values chosen as initial estimates for the nonlinear LS fit.

(b) Explore the suggestion in the example that the right linear segment may not deviate significantly from a horizontal plateau. Specifically, test $H_0: \beta_3 = 0$ by fitting a reduced model with $\beta_3 = 0$ and applying the discrepancy measure approach in (1.7). At $\alpha = 0.05$, what do you conclude?

2.6. Return to the pond water level data in Example 2.3.

(a) Find the predicted values and residuals from the fit in Fig. 2.9. Construct a residual plot and comment on the fit.

(b) In your residual plot from part (a), notice that there is a pattern of strictly decreasing residuals at low aquifer levels (below about $x = 38$). What might this indicate? To remedy the problem, modify the model in (2.4) to be continuous at both join points. Fit this continuous three-segment model to the original data and construct a new residual plot. Do you see any improvement?

2.7. An important measure of forest status and health is the ratio of single-sided leaf surface to ground surface area, known as the *leaf area index* (LAI). Leaf area index can be affected by or related to a number of factors. For example, Nilson *et al.* (1999) give LAI data for northern European coniferous forest canopies – so the index here might more correctly be called a needle area index – vs. age of the stand (in years). For Scots pine (*Pinus sylvestris* L.) the data are:

x = age	25	28	30	31	35	35	40
Y = LAI	3.8	12.8	6.3	7.2	3.6	3.7	4.9
x = age	45	50	55	56	60	70	120
Y = LAI	3.2	4.4	4.4	4.6	1.1	3.4	1.5

(a) Plot the data. Is a curvilinear relationship apparent?

(b) Fit the monoexponential model (2.8). (See Example 2.4 for suggestions with the initial estimates.) Do you see anything unusual with the estimated parameters?

(c) Test the null hypothesis that the horizontal asymptote is zero, $H_0: \beta_0 = 0$, using the discrepancy-measure approach from (1.7). At $\alpha = 0.05$, what do you conclude?

(d) If in part (c) you found that β_0 was insignificant, calculate the predicted values and residuals from a fit of the reduced model $Y_i = \beta_1 \exp\{-\beta_2 x_i\} + \varepsilon_i$. Use these to construct a residual plot. Comment on the fit.

(e) If in part (c) you found that β_0 was insignificant, proceed further and test the null hypothesis that age has no effect on LAI, $H_0: \beta_2 = 0$, using the discrepancy-measure approach from (1.7). At $\alpha = 0.05$, what do you conclude?

2.8. Verify the following indications from Example 2.5.

(a) Using differential calculus, show that under $g(x; \boldsymbol{\beta}) = \beta_1(e^{-\beta_2 x} - e^{-\beta_4 x})$, the maximum point occurs at $x^* = \{\log(\beta_2) - \log(\beta_4)\}/(\beta_2 - \beta_4)$ and the point of inflection occurs at $x^+ = 2\{\log(\beta_2) - \log(\beta_4)\}/(\beta_2 - \beta_4)$.

(b) Combine (2.11) and (2.12) to achieve the solution in (2.13).

2.9. Return to the nitrite utilization data in Example 2.5. From the residual plot in Fig. 2.17, notice that there is a pattern of possibly increasing variation as the predicted values increase. To address this, we can use nonlinear weighted LS to fit the biexponential model. The effort is facilitated here by the existence of replicated (here, triplicate) observations at the different levels of x_i. We calculate the sample variances for the three observations at each exposure level, and use as weights the reciprocals of these sample variances: $w_i = 1/S_i^2$. (Notice that for all observations within a triplet, the weights will be equal. We performed a similar operation in Exercise 1.3.) With these weights, invoke PROC NLIN as in Fig. 2.15, but include a _weight_=*name* statement, where *name* is the name of the SAS variable containing the w_is (cf. Fig. 2.26). After doing so, do your inferences change drastically from those in Fig. 2.16?

2.10. Return to the nonlinear function $g(x; \boldsymbol{\beta}) = \beta_1(e^{-\beta_2 x} - e^{-\beta_4 x})$. To gain a better understanding of this model, consider a Taylor series expansion for it. (A Taylor series expansion is an approximation of a nonlinear function by a polynomial of sufficiently high order; see §A.6.1). For instance, a third-order Taylor expansion of e^{-x} about $x = 0$ is $e^{-x} \approx 1 - x + \frac{1}{2!}x^2 - \frac{1}{3!}x^3$. Show that this result can be applied to $\beta_1(e^{-\beta_2 x} - e^{-\beta_4 x})$ to achieve the approximation $g(x; \boldsymbol{\beta}) \approx \beta_1\{(\beta_4 - \beta_2)x + (\beta_2^2 - \beta_4^2)\frac{x^2}{2!} + (\beta_4^3 - \beta_2^3)\frac{x^3}{3!}\}$. Show next that $g(x; \boldsymbol{\beta})$ increases linearly with a slope of $\beta_1(\beta_4 - \beta_2)$ for values of x close to zero (as noted in Example 2.5). Also, if $|\beta_2| < 1$ and $|\beta_4| < 1$, then the higher-order terms such as β_4^3 will be small and so $g(x; \boldsymbol{\beta})$ may be reasonably well approximated by a lower-order polynomial.

2.11. Similar to the data in Table 2.6, Bates and Watts (1988, §A1.12) also give results on nitrite utilization in *P. vulgaris* L. on a second day of experimentation. These are:

$x = \text{light}(\mu E/m^2 s)$	2.2	2.2	2.2	5.5	5.5	5.5
$Y = \text{nitrite (nmol/g-hr)}$	549	1550	1882	1888	3372	2362
$x = \text{light}(\mu E/m^2 s)$	9.6	9.6	9.6	17.5	17.5	17.5
$Y = \text{nitrite (nmol/g-hr)}$	4561	4939	4356	7548	7471	7642
$x = \text{light}(\mu E/m^2 s)$	27.0	27.0	27.0	46.0	46.0	46.0
$Y = \text{nitrite (nmol/g-hr)}$	9684	8988	8385	13505	15324	15430
$x = \text{light}(\mu E/m^2 s)$	94.0	94.0	94.0	170.0	170.0	170.0
$Y = \text{nitrite (nmol/g-hr)}$	17842	18185	17331	18202	18315	15605

(a) Plot the data. Does the pattern appear similar to that in Fig. 2.14?

(b) Mimic the calculations in Example 2.5 and fit the specialized biexponential model $g(x_i; \boldsymbol{\beta}) = \beta_1(e^{-\beta_2 x_i} - e^{-\beta_4 x_i})$ to these data. As in the example, use your results to estimate the level of light intensity at which the response reaches a maximum.

(c) Find the predicted values and residuals from the fit. Construct a residual plot and comment on the fit.

2.12. Seber and Wild (1989, §11.5) present chemometric data on the thermal isomerization (at 189.5 °C) of the monoterpene α-pinene into the turpentine oil allo-ocimene as a function of $x = $ time. (Allo-ocimene is used in rubber reclamation and also found as a phytochemical constituent of certain plants' leaves.) For $Y = $ concentration of allo-ocimene, the data are:

$x = $ time (hr)	20.5	51.0	82.0	130.0
$Y = $ conc. allo-ocimene	2.3	4.5	5.3	6.0
$x = $ time (hr)	178.0	250.5	377.0	607.0
$Y = $ conc. allo-ocimene	6.0	5.9	5.1	3.8

(a) Plot the data. Does the pattern appear nonlinear?

(b) Seber and Wild indicate that under a first-order kinetic model, the isomerization should proceed as per a triexponential model with zero intercept; that is, equation (2.10) with $\beta_0 = 0$. Fit this model to the data. (Initial estimates for a triexponential model can be difficult to determine. We suggest you work with values near the following: $\beta_{10} = 8.5$, $\beta_{20} = 10^{-3}$, $\beta_{30} = -6.5$, $\beta_{40} = 10^{-2}$, $\beta_{50} = -200$, and $\beta_{60} = 0.25$. Also, we found convergence occurred fastest using a derivative-free algorithm such as method=dud in PROC NLIN.)

(c) Find the residuals from the fit. Plot them against time and comment on the fit.

(d) Test if the triexponential components are necessary; that is, set $H_0: \beta_5 = \beta_6 = 0$ and fit this reduced model. From the fit, determine SSE(RM) so as to employ the discrepancy measure approach based on equation (1.7). Operate at $\alpha = 0.05$.

2.13. Verify analytically the following features of the Gompertz curve, $g(x; \boldsymbol{\beta}) = \beta_0 \exp\{-e^{-\beta_1 - \beta_2 x}\}$, from (2.14):

(a) For $\beta_2 > 0$, the upper asymptote is β_0; that is, $\lim_{x \to \infty} g(x; \boldsymbol{\beta}) = \beta_0$.

(b) For $\beta_2 > 0$, the lower asymptote is zero; that is, $\lim_{x \to -\infty} g(x; \boldsymbol{\beta}) = 0$.

(c) The vertical intercept is $\beta_0 \exp\{-e^{-\beta_1}\}$.

(d) The point of inflection occurs at $x^+ = -\beta_1/\beta_2$.

(e) The height at inflection is β_0/e.

(Winsor, 1932)

2.14. Return to the fruit diameter data in Example 2.6.

(a) Plot the fruit diameters against time. Verify the values chosen as initial estimates for the nonlinear LS fit.

(b) Plot the predicted Gompertz curve against time. Overlay this with the original data. Comment on the fit.

(c) Find the predicted values and residuals from the fit in Fig. 2.20. Construct a residual plot and comment on the fit.

2.15. Rhees and Atchley (2000) present data on the growth of male mice that were genetically selected for tendency toward increased late-term weight gain. This suggests a potential for asymmetric growth, and hence use of a Gompertz curve as in (2.14). The data, x_i = age (days) and Y_i = mean body weight (g) of five animals, are:

x = age	0	10	21	28
Y = body weight	1.5	8.0	14.6	24.4
x = age	35	42	49	56
Y = body weight	30.9	33.8	36.5	38.6

Leaving aside the issue of potential longitudinal or repeated measurements with these data, assume normal errors and homogeneous variances, and proceed with an analysis using the Gompertz model.

(a) Plot the data. Use it to help determine the initial estimates for the nonlinear LS fit.

(b) Fit the Gompertz curve (2.14) and test for a significant growth response. Operate at $\alpha = 0.01$.

(c) Estimate the point of maximal rate of growth, x^+. Include a 99% confidence interval, using Fieller's theorem as in (2.15).

(d) Find the predicted values and residuals from the fit in part (b). Construct a residual plot and comment on the fit.

2.16. Under the logistic growth curve, $g(x; \beta) = \beta_0/(1 + e^{-\beta_1 - \beta_2 x})$ from (2.16), verify analytically the following features:

(a) For $\beta_2 > 0$, the upper asymptote is β_0; that is, $\lim_{x \to \infty} g(x; \beta) = \beta_0$.

(b) For $\beta_2 > 0$, the lower asymptote is zero; that is, $\lim_{x \to -\infty} g(x; \beta) = 0$.

(c) The vertical intercept is $\beta_0/(1 + e^{-\beta_1})$.

(d) The point of inflection occurs at $x^+ = -\beta_1/\beta_2$.

(e) The height at inflection is $\beta_0/2$.

(f) The rate of growth at inflection is $g'(x^+; \beta) = \beta_0 \beta_2$.

2.17. In a study on the growth kinetics of the bacterium *Streptomyces lividans*, Daae and Ison (1998) recorded bacterial growth, measured as total biomass (g/L dry cell weight), in shake flask cultures over $x = $ time (hr). The data are:

$x = $ time (hr)	0	24	30	48
$Y = $ biomass (g/L)	0.500	0.503	0.600	1.113
$x = $ time (hr)	54	72	78	96
$Y = $ biomass (g//L)	1.391	2.069	2.049	1.924

(a) Plot the data. Is there a sigmoidal pattern?

(b) Assuming the data were taken from independent flasks over time, fit the logistic growth model (2.16). Test if there is a significant time effect. Operate at $\alpha = 0.01$.

(c) Find the predicted values and raw residuals from your fit in part (b). Construct a residual plot and comment on the fit.

2.18. Growth curve models may also be fitted via nonlinear weighted LS methods when some pertinent weight variable is identified *a priori*. For example, Kamakura and Takizawa (1994) report data on increasing mean body weights over time from rodents exposed to an environmental toxin. The individual data show an increase in variation as time progressed, questioning the assumption of homogeneous variance with these data. (The assumption of independence may also be in question, since the data are longitudinal; however, for purposes of illustration we will ignore this latter issue.) As $Y = $ mean body weight (in g, from 10 animals), $x = $ time (in days), and $s = $ the standard errors of the mean weights, the data are:

$x = $ time (days)	0	4	7	11	14	18
$Y = $ mean weight (g)	122.4	152.4	173.4	203.0	225.5	255.0
$s^2 = $ squared std. error	2.25	3.61	4.00	6.25	10.89	12.96
$x = $ time (days)	21	25	28	32	35	
$Y = $ mean weight (g)	275.9	297.4	309.0	330.4	337.8	
$s^2 = $ squared std. error	17.64	28.09	27.04	40.96	43.56	

(a) Plot the data. Include error bars as $\pm s_i$ to visualize the heterogeneous variation. Is there a sigmoidal pattern in the mean response?

(b) Fit the logistic model (2.16) to these data. To adjust for heterogeneous variance, include a weight of $w_i = 1/s_i^2$ for each observed mean. (To include a weight variable in PROC NLIN, use the _weight_= *name* statement; see Exercise 2.9.) Test if there is a significant effect over time. Operate at $\alpha = 0.01$.

(c) Find the predicted values and raw residuals from your fit in part (b). Construct a residual plot and comment on the fit.

2.19. To study the toxic effects in primates of exposure to heavy metals, Chen and Pounds (1998) gave data on renal cell damage in rhesus monkeys. The data, Y_i = fraction of a renal cell enzyme (lactate dehydrogenase, LDH) released in response to potential cell damage by exposure to cadmium (Cd), vs. x_i = Cd exposure (µm), are:

$x = $ Cd (µm)	1.0	1.0	1.0	2.0	2.0	2.0
$Y = $ LDH fraction	1.4	1.4	0.6	1.7	2.0	2.3
$x = $ Cd (µm)	3.0	3.0	3.0	4.0	4.0	4.0
$Y = $ LDH fraction	2.5	3.2	3.1	3.7	3.8	3.3
$x = $ Cd (µm)	5.0	5.0	5.0	6.0	6.0	6.0
$Y = $ LDH fraction	3.9	4.7	3.0	4.1	4.6	4.5
$x = $ Cd (µm)	7.0	7.0	7.0	8.0	8.0	8.0
$Y = $ LDH fraction	6.8	6.6	4.3	7.9	7.7	6.7
$x = $ Cd (µm)	9.0	9.0	9.0	10.0	10.0	10.0
$Y = $ LDH fraction	8.4	8.3	9.7	14.6	10.8	13.9
$x = $ Cd (µm)	15.0	15.0	15.0	20.0	20.0	20.0
$Y = $ LDH fraction	25.9	28.1	30.2	39.3	43.0	44.7

(a) Plot the data. Is there a curvilinear pattern?

(b) Assuming independent normal errors and homogeneous variances, fit the four-parameter logistic model (2.19) to these data.

(c) Calculate and plot the residuals from your fit. Does the pattern indicate any problems? If so, suggest an alternative to improve the fit. Does your alternative address the problem?

2.20. Return to the pesticide ecotoxicity data from Example 2.9.

(a) Plot the data. Do you agree with the indications in the example regarding the curvilinearity/shape of the response?

(b) Find the predicted values and raw residuals from the fit in Fig. 2.29. Construct a residual plot and comment on the fit.

(c) The method of calculating the inhibition response with these data allows for negative values, which in theory would suggest a protective effect of the toxin. In some settings, this is not unreasonable at low toxic exposures (a phenomenon known as *hormesis*), but with these data this appears to be more of a numerical anomaly. If so, we would expect β_0 to be zero. To assess this, test the null hypothesis $H_0: \beta_0 = 0$. What do you conclude at $\alpha = 0.05$?

2.21. Jolicoeur and Heusner (1986) display data on the growth of an experimental population of microorganisms (*Paramecium caudatum*) over time. We have $Y_i =$ organisms/ml and $x_i =$ days on study. The data are:

$x =$ time (days)	1	2	3	4	5	6
$Y =$ growth (orgs./ml)	0.080	0.386	0.571	0.943	1.441	1.471
$x =$ time (days)	7	8	9	10	11	12
$Y =$ growth (orgs./ml)	1.734	1.536	1.697	2.016	1.529	1.672
$x =$ time (days)		13	14	15	16	
$Y =$ growth (orgs./ml)		1.656	1.374	1.403	1.577	

(a) Plot the data. Do you see a nonlinear relationship?

(b) Fit the Weibull growth curve (2.20) to these data. Test for a significant growth response over time. Operate at $\alpha = 0.01$.

(c) Find the raw residuals from the fit in part (b). Plot these against time and comment on the fit.

2.22. Return to the biomass data of Exercise 2.17. The authors of that study found the logistic fit unsatisfactory. (Does this correspond with your own analysis?) As an alternative, repeat the steps of that exercise, now using (a) the Gompertz curve from (2.14), and (b) the four-parameter logistic growth curve from (2.19). Assess and comment on whether either model improves the fit.

2.23. Return to the puromycin data of Example 2.10.

(a) Find the predicted values and raw residuals from the fit in Fig. 2.32. Construct a residual plot and comment on the fit.

(b) Plot the predicted Michaelis–Menten curve against x. Overlay this with the original data. Comment on the fit.

(c) These data possess a feature that can be exploited for assessing model adequacy. Since the observations are replicated at the various levels of x_i, an estimate of the 'pure error' associated with σ^2 can be computed directly (Neter *et al.*, 1996, §3.7). To facilitate the operation, modify the notation to let Y_{ij} be the observed reaction velocities, where $i = 1, \ldots, I$ now indexes the *different* values of x_i, and $j = 1, \ldots, J_i$ indexes the replicated observations at each x_i. Let $\overline{Y}_i = \sum_{j=1}^{J_i} Y_{ij}/J_i$ be the observed mean value at x_i. Then the estimate of pure error is MSPE $= \sum_{i=1}^{I} \sum_{j=1}^{J_i} (Y_{ij} - \overline{Y}_i)^2 / (\{\sum_{i=1}^{I} J_i\} - I)$. Apply this operation to the data in Table 2.11 and compare the resulting value to the MSE estimate for σ^2 based on the nonlinear model fit in Fig. 2.32. If MSE greatly exceeds MSPE, lack of fit is evidenced. What do you find here?

2.24. Seber and Wild (1989, Table 3.1) present chemometric data on $Y_i =$ reaction velocity in an enzyme-catalyzed chemical reaction vs. $x_i =$ concentration of a substrate:

$x =$ concentration	0.2	0.2	0.222	0.222	0.286	0.286
$Y =$ reaction velocity	0.0087	0.0129	0.0169	0.0083	0.0183	0.0129
$x =$ concentration	0.4	0.4	0.667	0.667	2	2
$Y =$ reaction velocity	0.0258	0.0138	0.0258	0.0334	0.0527	0.0615

(a) Plot the data. What pattern do you see?

(b) Fit the Michaelis–Menten model (2.22) to these data. Estimate the maximum reaction velocity and give 99% confidence limits for this value.

(c) Find raw residuals from the fit in part (b). Construct a residual plot and comment on the fit.

2.25. Consider the three-parameter Michaelis–Menten curve from (2.24). For implementation via nonlinear LS, find the partial derivatives of the model function with respect to the three unknown parameters. Also suggest an approach for producing initial estimates. [*Hint*: can the Lineweaver–Burk method be modified for this purpose?]

2.26. Return to the nitrite utilization data from Example 2.5 and fit a three-parameter Michaelis–Menten curve, (2.24), to these data. (Your results in Exercise 2.25 may prove useful.) Compare the resulting fit with that in the example and those from Exercise 2.9.

(Bates and Watts, 1988)

2.27. Bioaccumulation factors (BAFs) describe the concentration of a contaminant in an organism at steady state relative to the concentration in the environment. In certain experiments, this characteristic is estimated by first placing a number of organisms in a test chamber at a fixed contaminant concentration, say C_w. A sample of $j = 1, \ldots, J_i$ organisms are selected at various times, x_i ($i = 1, \ldots, I$), and each organism's contaminant concentration is measured as, say, Y_{ij}. After a prespecified exposure interval of length τ, the remaining organisms are removed and placed in a clean chamber. A nonlinear model to describe data from the first phase of the experiment is $Y_{ij} = C_w(\kappa_u/\kappa_e)\{1 - \exp(-\kappa_e x_i)\} + \varepsilon_{ij}$, where κ_u is an uptake rate parameter and κ_e is an elimination rate parameter. After the organisms are placed into the clean chamber, we assume the contaminant is eliminated according to an exponential pattern of decay; that is, for $x_i > \tau$, organism contaminant concentration decreases exponentially at rate κ_e.

(a) A regression model that represents both phases of this experiment would involve a two-segment form where each segment is governed by a different nonlinear function. Express such a model assuming continuity at the join point $x_i = \tau$. [*Hint*: the expected concentration in the organism at the end of the exposure interval is $C_w(\kappa_u/\kappa_e)\{1 - \exp(-\kappa_e\tau)\}$.]

(b) Suppose we fit a set of data to the nonlinear model in part (a) and find ML estimators k_u and k_e for κ_u and κ_e, respectively. The BAF can be defined as κ_u/κ_e, and estimated as k_u/k_e. Use the delta method (§A.6) to construct an estimator for $se[k_u/k_e]$ and from this build a $1 - \alpha$ Wald confidence interval for BAF.

(Bailer *et al.*, 2000)

(c) Notice that since the estimated BAF is a ratio, some concern may arise as to whether its distribution is sufficiently symmetric for use in building a Wald interval (cf. Example 2.7). Thus, repeat part (b) by first constructing a delta-method interval for log(BAF) and then exponentiating to recover the original BAF scale.

(Bailer *et al.*, 2000)

(d) Repeat part (b) using Fieller's method from Exercise 1.16.

3

Generalized linear models

Fundamental to our presentations in Chapters 1 and 2 was the model formulation in equation (1.1), where the response variable Y_i is the sum of an additive error term and a possibly nonlinear function of a set of predictor variables. In particular, the additive errors were taken as normal with homogeneous variances. (We relaxed the homogeneous-variance assumption in select instances, but the normality assumption was never in question.) Indeed, the normal assumption made for the error terms automatically imposes a similar assumption on the data themselves: if $\varepsilon_i \sim$ i.i.d. $N(0, \sigma^2)$ then, under (1.1), $Y_i \sim$ indep. $N(g\{x_{i1}, x_{i2}, \ldots, x_{ip}; \boldsymbol{\beta}\}, \sigma^2)$. Thus for all the models and methods described in the previous two chapters, we assumed implicitly that the data, perhaps after transformation, were continuous measurements taken over a scale of values from $-\infty$ to ∞.

Clearly, while applicable to a wide variety of settings in the environmental sciences, this normal distribution assumption need not always hold. Examples of environmental data that cannot be normally distributed include continuous observations restricted to positive outcomes, data bounded over a fixed interval, and discrete data such as counts and proportions. In some of these cases, a further complication occurs when the variance and the mean response are related; For example, $\mathrm{Var}[Y_i]$ may increase as $\mathrm{E}[Y_i]$ increases. As we saw in Chapters 1 and 2, one can address this concern by employing some form of weighted least squares, or by transforming the observations into quantities that have roughly homogeneous variances and/or are at least approximately normal. The transformation approach can be useful in some settings – recall the logit and arcsine transforms mentioned in Example 1.5 – but it also can lead to problems of interpretation when the transformed results do not naturally translate back into the original scale of measurement. Modern environmetric practice seeks to make inferences wherever possible on the original scale, generating the need for extensions of the normal-based linear/nonlinear model in (1.1). In this chapter, we discuss a class of *generalized linear models* (GLiMs) that achieves this goal: it extends the linear model from Chapter 1 to include non-normal parent distributions for the data, and also incorporates concepts from Chapter 2 that allow for nonlinear relationships between the mean response and the linear predictor.

Analyzing Environmental Data   W. W. Piegorsch and A. J. Bailer
© 2005 John Wiley & Sons, Ltd   ISBN: 0-470-84836-7 (HB)

3.1 Generalizing the classical linear model

3.1.1 Non-normal data and the exponential class

As in Chapter 1, we assume independent observations $Y_i, i = 1, \ldots, n$, are collected alongside $p \geq 1$ predictor variables, $x_{i1}, x_{i2}, \ldots, x_{ip}$. GLiMs model the mean response $\mu_i = \mathrm{E}[Y_i]$ as a function of the predictor variables while accepting that the distribution of Y may differ from normal. In order to account for a variety of different possible parent distributions, they employ a class of probability functions that extends the normal probability function. (For a review of the basic notation and features of probability functions, see §A.1.) Specifically, we say that the probability density function (p.d.f.) or probability mass function (p.m.f.), $f(y)$, of a random variable Y belongs to the *exponential class* of probability functions if $f(y)$ takes the form

$$f(y) = \exp\left\{\frac{y\theta - B(\theta)}{A(\varphi)} + C(y, \varphi)\right\}, \tag{3.1}$$

where $A(\varphi)$, $B(\theta)$, and $C(y, \varphi)$ are functions of known form. The parameter θ is the (unknown) *natural parameter* of the distribution, and the parameter $\varphi > 0$ is an additional *dispersion parameter* (sometimes called a *scale parameter*). Equation (3.1) is actually a special case of a larger family of probability functions of the form $f(y) = \exp\{[T(y)\theta - B(\theta)]/A(\varphi) + C(y, \varphi)\}$. When $T(y) = y$, as in (3.1), we say the function is in *canonical form*. Also, for many models the function $A(\varphi)$ simplifies to the form φ/w, where $w > 0$ is a known constant.

An important, additional constraint on the class in (3.1) is that the support space, $\mathbb{S}$, of Y cannot depend upon θ. This is usually verified by incorporating some form of indicator function from §A.2.1 into $C(y, \varphi)$, such that the indicator does not depend on θ. We illustrate this in the examples below.

In (3.1), the mean $\mu = \mathrm{E}[Y]$ is related to the natural parameter θ via a partial derivative operation, $\mu = \partial B(\theta)/\partial\theta$, as is the variance, $\mathrm{Var}[Y] = A(\varphi)\partial^2 B(\theta)/\partial\theta^2$. When the dispersion parameter φ is known, we call $\partial^2 B(\theta)/\partial\theta^2$ the *variance function* of Y, since it incorporates all the unknown aspects of the variance term. We denote this as

$$V(\mu) = \frac{\partial^2 B(\theta)}{\partial\theta^2} = \frac{\partial}{\partial\theta}\left(\frac{\partial B(\theta)}{\partial\theta}\right) = \frac{\partial\mu}{\partial\theta}$$

to highlight the fact that the variance may be a function of the mean. The class in (3.1) also possesses a number of other interesting mathematical features that extend beyond its use as a p.d.f./p.m.f. for GLiMs; see, for example, the dedicated text by McCullagh and Nelder (1989, §2.2.2) for additional details.

Example 3.1 (Normal distribution) The normal p.d.f. is a member of the exponential class. We can see this as follows: recognize that the support space for Y is

$\mathbb{S} = (-\infty, \infty)$ and indicate this in $f(y)$ via an indicator function of the form $I_{\mathbb{S}}(y) = I_{(-\infty,\infty)}(y)$. Then, write the p.d.f. from (A.16) as

$$f(y) = \frac{1}{\sigma\sqrt{2\pi}} \exp\left\{ -\frac{(y-\mu)^2}{2\sigma^2} \right\} I_{(-\infty,\infty)}(y)$$

$$= \exp\left\{ -\frac{1}{2}\log(2\pi\sigma^2) + \log[I_{(-\infty,\infty)}(y)] \right\} \exp\left\{ -\frac{y^2 - 2y\mu + \mu^2}{2\sigma^2} \right\}$$

$$= \exp\left\{ \frac{y\mu - \frac{1}{2}\mu^2}{\sigma^2} - \frac{y^2}{2\sigma^2} - \frac{1}{2}\log(2\pi\sigma^2) + \log[I_{(-\infty,\infty)}(y)] \right\}.$$

In this form, we take the natural parameter to be $\theta = \mu$ and the dispersion parameter to be $\varphi = \sigma^2$ so that $A(\varphi) = \varphi$, $B(\theta) = \frac{1}{2}\theta^2$, and $C(y, \varphi) = -\frac{1}{2}[\varphi^{-1}y^2 + \log(2\pi\varphi)] + \log[I_{(-\infty, \infty)}(y)]$. Hence, the normal p.d.f. satisfies the class requirement given by (3.1).

Notice that the indicator function describing the support space for Y, $I_{(-\infty, \infty)}(y)$, was used as an explicit component in the function $C(y, \varphi)$. This was crucial for identifying the normal p.d.f. in exponential class form. Indeed, when we use the indicator function to write the p.m.f. or p.d.f. in its fully expressed form, we can check quickly whether the support of the p.m.f. or p.d.f. is dependent upon θ. If it is, then as noted above the random variable cannot be part of the exponential class. (For example, suppose Y follows a uniform distribution with parameters 0 and θ, from §A.2.7. Then the p.d.f. is $f(y) = \theta^{-1}I_{(0, \theta)}(y)$, and we cannot write it to satisfy (3.1). This random variable is not a member of the exponential class of distributions.) Notice also that for the normal p.d.f., $E[Y] = B'(\theta) = \theta = \mu$ and $Var[Y] = B''(\theta)A(\varphi) = \varphi = \sigma^2$, as expected. ✪

Recognition that the normal p.d.f. is a member of the exponential class of distributions is important. It shows that extensions of the classical linear model using (3.1) will serve as true generalizations, since they will include the normal model as a special case. The following examples present some other exponential class distributions; see also Exercise 3.1.

Example 3.2 (Exponential distribution) Perhaps the simplest distribution in the exponential class is the similarly named exponential distribution from equation (A.14), although the larger class is not named for this particular distribution. If $Y \sim \text{Exp}(\mu)$, then we have

$$f(y) = \frac{e^{-y/\mu}}{\mu} I_{(0,\infty)}(y) = \exp\left\{ -\frac{1}{\mu}y - \log(\mu) + \log[I_{(0,\infty)}(y)] \right\}.$$

In this form, we take the natural parameter to be $\theta = -1/\mu < 0$ and the dispersion parameter φ to be fixed at $\varphi = 1$. Then $f(y)$ does satisfy (3.1), with $A(\varphi) = 1$, $B(\theta) = \log(-1/\theta) = -\log(-\theta)$, and $C(y, 1) = \log[I_{(0, \infty)}(y)]$. Hence, the exponential p.d.f. is a member of the exponential class as given by (3.1).

A technical caveat: when y is positive the indicator function $I_{(0, \infty)}(y)$ equals 1 and the function $C(y, 1)$ is well defined. When y is negative or zero, however, the indicator function is 0, and thus the function $C(y, 1)$ here attempts to evaluate the natural logarithm of zero. Though this is technically impossible, we can appeal to a limiting argument for the evaluation: recognize that as its argument approaches 0, the natural logarithm approaches $-\infty$. Evaluated in the exponent of the p.d.f., this drives $f(y)$ to an infinitesimal value, the limiting value of which is itself 0. This is precisely what the probability density should be when y is not positive. ✪

Example 3.3 (Binomial distribution) Discrete distributions may also exist in the exponential class. Consider the binomial p.m.f. from (A.7),

$$f(y) = \binom{N}{y} \pi^y (1 - \pi)^{N-y} I_{\{0,1,\dots,N\}}(y)$$

$$= \exp\left\{y \log(\pi) + (N - y) \log(1 - \pi) + \log\left[\binom{N}{y} I_{\{0,\dots,N\}}(y)\right]\right\}$$

$$= \exp\left\{y \log\left(\frac{\pi}{1 - \pi}\right) + N \log(1 - \pi) + \log\left[\binom{N}{y} I_{\{0,\dots,N\}}(y)\right]\right\}.$$

In this form, we take the natural parameter to be $\theta = \log\{\pi/(1 - \pi)\} = \text{logit}(\pi)$ and the dispersion parameter φ to be fixed at $\varphi = 1$. (We also assume N is a known positive integer.) Then $f(y)$ does satisfy (3.1), with $A(\varphi) = 1$, $B(\theta) = -N\log(1 - \pi) = N\log(1 + e^\theta)$, and $C(y, 1) = \log\left\{\binom{N}{y} I_{\{0,\dots,N\}}(y)\right\}$. (Note that the technical caveat from Example 3.2 regarding the logarithm of an indicator function also holds here.) Hence, the binomial p.m.f. is a member of the exponential class. ✪

3.1.2 Linking the mean response to the predictor variables

A second extension GLiMs make to the classical linear model is to allow the relationship between the mean parameter, μ, and the linear predictor to deviate from strict equality. To distinguish the two, we introduce a separate notation for the linear predictor: η_i. For example, $\eta_i = \theta + \alpha_i$ is the linear predictor from a one-factor ANOVA model, while $\eta_i = \beta_0 + \beta_1(x_i - \bar{x}) + \beta_2(x_i - \bar{x})^2 + \cdots + \beta_p(x_i - \bar{x})^p$ is the linear predictor from a pth-order polynomial regression model. (In what follows, we omit the subscripts on μ and η unless the index is required for clarity.) In the classical linear model, the mean, μ, and the linear predictor, η, are identically related, $\mu = \eta$. GLiMs extend this into a functional relationship, say, $g(\mu) = \eta g(\cdot)$ is assumed monotone and is called the *link function* between μ and η. The link function defines the scale over which the systematic effects represented by η are modeled as additive.

In some cases the link is trivial; for example, $\mu = \eta$ is an *identity link*, $g(\mu) = \mu$. In other cases, the link can be used to model a necessary relationship between μ and η. For instance, in Example 3.2 the exponential mean must be positive, so we require the link to relate a strictly positive quantity, μ, to a linear predictor of any sign.

A common choice for such is the natural logarithm, $g(\mu) = \log(\mu)$, although of course other forms are possible.

When the choice of link is not motivated by the subject-matter, one can appeal to the natural parameterization in (3.1). The natural parameter, θ, is often viewed as the most compact representation of the unknown information in the data. Clearly, however, θ need not equal μ, and generally is some function of it. For instance, the natural parameter under an exponential p.d.f. is $\theta = -1/\mu$, and hence an alternative link function for this distribution is $g(\mu) = 1/\mu$ (coefficients such as -1 or $\frac{1}{2}$ typically are ignored). In general, if the relationship between θ and μ represents a monotone function of μ, it can be employed as a link function. In such a case, the link is called a *canonical link*. This special link function can be applied with any GLiM, and, because of its inherent elegance, it is a common choice when no other motivation is available. McCullagh and Nelder (1989, §2.2) and Piegorsch and Bailer (1997, Table 8.1) give some common GLiMs with their canonical links.

Since the link function is assumed monotone, it is possible to derive the *inverse link function*, $g^{-1}(\eta)$, which characterizes the mean as a function of the linear predictor. For simplicity, we write $h(\eta) = g^{-1}(\eta)$, so that $\mu = h(\eta)$. For example, with the exponential distribution, the inverse link to $g(\mu) = \log(\mu) = \eta$ is $\mu = h(\eta) = e^{\eta}$, while the inverse link to $g(\mu) = 1/\mu$ is $\mu = h(\eta) = 1/\eta$.

The exponential class in (3.1) and the link function $g(\cdot)$ relating μ to η are the two key generalizations that make up a GLiM. In effect, GLiMs comprise a class of nonlinear regression models defined for selected forms of possibly non-normal response. Within this context, a number of technical, theoretical developments are possible that allow for regression modeling of data from any member of the parent class in (3.1). We discuss these in the next section.

3.2 Theory of generalized linear models

We approach estimation and inferences for GLiMs in a manner analogous to that adopted in Chapters 1 and 2. Our primary estimation routine will be weighted least squares (see, §A.4.1). We employ weighting to account for the fact that most members of the exponential class possess variances which are non-constant functions of μ. Also, as in Chapter 2, the WLS solutions will require iterative calculation, via what is called an *iteratively (re)weighted least squares* (IRLS) algorithm (Jorgensen, 2002). A positive consequence of this strategy is that the IRLS results correspond to a maximum likelihood solution for the parameter estimates (Nelder and Wedderburn, 1972; McCullagh and Nelder, 1989, §2.5), and thus will possess all the optimal large-sample properties associated with the ML approach (§A.4.3).

The concepts and notation for fitting GLiMs are somewhat advanced, and the material we present in this section embodies this. Greater familiarity with differential calculus and matrix algebra will be necessary for better understanding of the technical theory behind GLiMs. Introductory readers may wish to skip ahead to §3.3 for applications of the methods before taking up the details that follow.

3.2.1   Estimation via maximum likelihood

Suppose data are given as $Y_i \sim$ indep. $f(y_i)$, $i = 1, \ldots, n$, with $f(y)$ defined by (3.1). The mean parameter is $\mu_i = E[Y_i]$, which is related to a set of predictor variables $x_{i1}, x_{i2}, \ldots, x_{ip}$ through the linear predictor $\eta_i = \beta_0 + \beta_1 x_{i1} + \cdots + \beta_p x_{ip}$; β_0 is used to model the zero response, intercept, background effect, etc. Thus the likelihood is a function of the $p + 1$ parameters in $\beta = [\beta_0 \ \beta_1 \ldots \beta_p]^T$, and we write $\ell(\beta)$ to denote its log-likelihood. Since the observations are independent, this is

$$\ell(\beta) = \sum_{i=1}^{n} \left\{ \frac{y_i \theta_i(\beta) - B[\theta_i(\beta)]}{A(\varphi_i)} + C(y_i, \varphi_i) \right\}, \tag{3.2}$$

where the notation $\theta_i(\beta)$ is used to emphasize that the natural parameters are modeled as functions of β. Note that we are assuming for the moment that the dispersion parameters φ_i are known.

Maximizing (3.2) with respect to β yields the ML estimates $\mathbf{b} = [b_0 \ b_1 \ldots b_p]^T$. This requires application of differential calculus. Find the $p + 1$ partial derivatives $\partial \ell(\beta)/\partial \beta_j$, set them equal to zero, and solve the resulting set of estimating equations. As in §A.3.2, these derivatives are referred to as *score functions*. Now, recognize that the derivative of $\ell(\beta)$ is the derivative of a finite sum, which can be written as a sum of derivatives. The individual summands are

$$\frac{\partial}{\partial \beta_j} \left(\frac{y_i \theta_i(\beta) - B[\theta_i(\beta)]}{A(\varphi_i)} + C(y_i, \varphi_i) \right) = \frac{1}{A(\varphi_i)} \left(y_i \frac{\partial \theta_i(\beta)}{\partial \beta_j} - \frac{\partial B[\theta_i(\beta)]}{\partial \beta_j} \right). \tag{3.3}$$

For the leading derivative on the right-hand side of (3.3), $\partial \theta_i(\beta)/\partial \beta_j$, apply the chain rule from differential calculus (Khuri, 1993, §4.1):

$$\frac{\partial \theta_i(\beta)}{\partial \beta_j} = \left(\frac{\partial \theta_i(\beta)}{\partial \mu_i} \right) \left(\frac{d\mu_i}{d\eta_i} \right) \left(\frac{\partial \eta_i}{\partial \beta_j} \right). \tag{3.4}$$

To address the first term in (3.4), we focus on $\partial \mu_i/\partial \theta_i(\beta)$. Recall that under (3.1), $\mu_i = B'[\theta_i(\beta)] = \partial B[\theta_i(\beta)]/\partial \theta_i(\beta)$, so $\partial \mu_i/\partial \theta_i(\beta) = B''[\theta_i(\beta)]$. But, $\mathrm{Var}[Y_i] = A(\varphi_i) V(\mu_i) = A(\varphi_i) B''[\theta_i(\beta)]$, so $V(\mu_i) = B''[\theta_i(\beta)]$. Hence, $\partial \mu_i/\partial \theta_i(\beta) = V(\mu_i)$, and from this we can write $\partial \theta_i(\beta)/\partial \mu_i = 1/V(\mu_i)$. For the second term in (3.4), recall that the inverse link relates μ_i and η_i via $\mu_i = h(\eta_i)$. Assuming $h(\cdot)$ is differentiable, $d\mu_i/d\eta_i$ is thus its first derivative, $h'(\eta_i)$. For the third term in (3.4), note that $\eta_i = \beta_0 + \beta_1 x_{i1} + \cdots + \beta_p x_{ip}$, so $\partial \eta_i/\partial \beta_j = x_{ij}$ (at $j = 0$ set $\partial \eta_i/\partial \beta_0 = 1$). Collecting the pertinent expressions together simplifies (3.4) into

$$\frac{\partial \theta_i(\beta)}{\partial \beta_j} = \frac{1}{V(\mu_i)} h'(\eta_i) x_{ij} = \frac{x_{ij} h'(\eta_i)}{V(\mu_i)}.$$

For the lagging derivative in (3.3), $\partial B[\theta_i(\boldsymbol{\beta})]/\partial\beta_j$, the chain rule yields

$$\frac{\partial B[\theta_i(\boldsymbol{\beta})]}{\partial\beta_j} = \left(\frac{\partial B[\theta_i(\boldsymbol{\beta})]}{\partial\theta_i(\boldsymbol{\beta})}\right)\left(\frac{\partial\theta_i(\boldsymbol{\beta})}{\partial\beta_j}\right).$$

Now $\partial B[\theta_i(\boldsymbol{\beta})]/\partial\theta_i(\boldsymbol{\beta})$ is $B'[\theta_i(\boldsymbol{\beta})]$, which we know under (3.1) is simply μ_i. Also, we found in the previous paragraph that $\partial\theta_i(\boldsymbol{\beta})/\partial\beta_j = x_{ij}h'(\eta_i)/V(\mu_i)$. Collected together, this all leads to a simple expression for (3.3): $(y_i - \mu_i)x_{ij}h'(\eta_i)/\{A(\varphi_i)V(\mu_i)\}$. We sum these terms over $i = 1, \ldots, n$ to find $\partial\ell(\boldsymbol{\beta})/\partial\beta_j$. Setting the derivatives equal to zero produces the $p + 1$ estimating equations

$$\sum_{i=1}^{n} \frac{x_{ij}h'(\eta_i)}{A(\varphi_i)V(\mu_i)}(y_i - \mu_i) = 0, \qquad j = 0, \ldots, p. \tag{3.5}$$

Solution of (3.5) yields the joint ML estimate of $\boldsymbol{\beta}$. Compare this to the multiple linear regression case from §1.2: there, $\mu_i = \eta_i$ and $A(\varphi_i) = \sigma^2$, so (3.5) simplifies to $\sigma^{-2}\sum_{i=1}^{n} x_{ij}\{y_{ij} - \mu_i\} = 0$ for each $j = 0, \ldots, p$. In effect, (3.5) generalizes the estimating equations that led to the LS/ML estimates from that section.

To solve (3.5) for all $j = 0, \ldots, p$ we typically employ computer iteration, which can be accomplished by most modern statistical computing packages. We will illustrate use of the SAS procedure PROC GENMOD in §3.3, below. (The similarly named procedure PROC GLM is useful for the classical linear model as described in Chapter 1, but it is not intended for use with GLiMs.) We should warn that PROC GENMOD fits GLiMs via a Newton–Raphson algorithm (similar to the Gauss–Newton method mentioned in Chapter 2; see Smythe, 2002), maximizing the log-likelihood directly (SAS Institute Inc., 2000, Ch. 29). Many other packages, such as the S-Plus® routine `glm` (S-Plus, 1997), use IRLS to solve (3.5).

Note that $E[b_j]$ need not equal β_j – i.e., the ML estimator of $\boldsymbol{\beta}$ is biased – and often it will only approach β_j as $n \to \infty$. (This is known as a form of *asymptotic consistency*.) Thus in small samples, analysts may wish to apply a bias correction. This will vary, depending on the nature of the linear predictor, the link function, any additional constraints on the model, and the underlying parent distribution (Cordeiro and McCullagh, 1991; Firth, 1993).

If the dispersion parameters are unknown, they may also be estimated. In the simplest case, the dispersion is homogeneous – i.e., $\varphi_i = \varphi$ – and there is only one additional unknown parameter. We estimate it via ML or via the method of moments (§A.4.2). In the later case, we find a quantity whose first moment is a simple function of φ and manipulate the relationship to produce the estimator, $\hat{\varphi}$. We present some quantities that can be useful in this regard in the next two sections.

3.2.2 Deviance function

A fundamental quantity useful for testing hypotheses, assessing model adequacy, and estimating dispersion is known as the deviance function. It is derived as a form of discrepancy measure based on the likelihood function $L(\beta; \mathbf{y})$, where $\mathbf{y} = [y_1 \ldots y_n]^T$ is the vector of observed data. Suppose we define the discrepancy of any model's fit by how much it deviates from the fullest possible model that can be fitted to the data. This latter quantity is achieved by fitting a separate parameter to each observation, giving a log-likelihood of the form $\log\{L(\mathbf{y}; \mathbf{y})\}$. The notation here is deceiving. Technically, we are estimating μ_i by Y_i wherever μ_i appears in the likelihood. But, since μ_i is related to θ_i via $\mu_i = B'(\theta_i)$ and since $\theta_i = \theta_i(\beta)$ is a function of β, we can in the most extreme case view β as being comprised of n 'parameters,' each corresponding to a value of Y_i. We say then that the model and hence the likelihood has been *saturated*, and we write $\hat{\theta}_i^{sat}$ to indicate that $\theta_i(\beta)$ is estimated under saturation. The corresponding value of $\ell(\beta)$ is denoted by $\ell_{sat} = \log\{L(\mathbf{y}; \mathbf{y})\}$.

Compared with the saturated model, any more parsimonious model with $p + 1 < n$ parameters in β will have a smaller log-likelihood; evaluated at the ML estimate of β, this is $\ell(\mathbf{b}) = \log\{L(\mathbf{b}; \mathbf{y})\}$. The quality of the model fit can be quantified by twice the difference of the two values: $2\{\ell_{sat} - \ell(\mathbf{b})\}$. (Multiplying by 2 has some useful consequences, which we illustrate shortly.) The closer this difference is to zero, the more the proposed model fit using only $p + 1$ parameters mimics that of a saturated model.

Under the exponential class in (3.1), we can make this more explicit. Suppose there is a single, known dispersion parameter φ, so that $A(\varphi)$ is a known quantity. Then

$$\ell_{sat} = \sum_{i=1}^n \left\{ \frac{y_i \hat{\theta}_i^{sat} - B(\hat{\theta}_i^{sat})}{A(\varphi)} + C(y_i, \varphi) \right\},$$

while

$$\ell(\mathbf{b}) = \sum_{i=1}^n \left\{ \frac{y_i \theta_i(\mathbf{b}) - B[\theta_i(\mathbf{b})]}{A(\varphi)} + C(y_i, \varphi) \right\}.$$

Twice their difference is

$$D^*(\mathbf{b}) = 2\{\ell_{sat} - \ell(\mathbf{b})\} = \frac{2 \sum_{i=1}^n \left\{ y_i[\hat{\theta}_i^{sat} - \theta_i(\mathbf{b})] - [B(\hat{\theta}_i^{sat}) - B\{\theta_i(\mathbf{b})\}] \right\}}{A(\varphi)}. \tag{3.6}$$

For simplicity, we denote the numerator of (3.6) as $D(\mathbf{b})$ and call it the *deviance function*. The entire quotient, $D^*(\mathbf{b})$, is called the *scaled deviance*, since $D(\mathbf{b})$ is being scaled by $A(\varphi)$. (A better moniker for (3.6) might be the *dispersion-adjusted deviance*, since the term 'scale' is used in many different ways for models satisfying (3.1).)

Specific forms of the deviance function for different exponential class p.d.f.s are given in numerous sources, including McCullagh and Nelder (1989, §2.3) and the PROC GENMOD documentation (SAS Institute Inc., 2000, Ch. 29). One special case is notable: for $Y_i \sim$ indep. $N(\mu_i, \sigma^2)$, the deviance is simply the residual sum of squares: $D(\mathbf{b}) = \sum_{i=1}^n (y_i - \hat{\mu}_i)^2$, where $\hat{\mu}_i = B'[\theta_i(\mathbf{b})]$ is the ith estimated mean response.

The deviance and scaled deviance are most useful when viewed within the context of testing discrepancies between models, as in equation (1.7). That is, suppose we have some full model (FM) represented by the $(p + 1)$-vector $\boldsymbol{\beta} = [\beta_0 \cdots \beta_q \ \beta_{q+1} \cdots \beta_p]^T$, and a reduced model (RM) in which only $q + 1 < p + 1$ of the elements of $\boldsymbol{\beta}$ are modeled (and the rest are set to zero). Put in terms of testing hypotheses, the null hypothesis often corresponds to the RM, where $p - q$ of the β_js are constrained to be zero. The alternative hypothesis corresponds to the FM, where no constraints are imposed on the β_js. Denote the reduced vector as $\boldsymbol{\beta}_{RM} = [\beta_0 \cdots \beta_q \ 0 \ldots 0]^T$. For consistency, also write the FM vector as $\boldsymbol{\beta}_{FM}$. If we fit both models to the data and find their scaled deviances $D^*(\mathbf{b}_{FM})$ and $D^*(\mathbf{b}_{RM})$, the difference between these is, for known φ, simply $\Delta D^* = D^*(\mathbf{b}_{RM}) - D^*(\mathbf{b}_{FM}) = 2\{\ell(\mathbf{b}_{FM}) - \ell(\mathbf{b}_{RM})\}$. Notice that this is twice the log of the ratio of the likelihoods. Recall from §A.5.3, however, that a ratio of likelihoods between a reduced (or null) and full model is the basis for constructing the likelihood ratio (LR) test of the reduced model. That is, to test if the $p - q$ parameters set to zero under the RM are significant, we can form the statistic

$$G_{calc}^2 = -2\log\left\{\frac{L(\theta(\mathbf{b}_{RM}); \mathbf{y})}{L(\theta(\mathbf{b}_{FM}); \mathbf{y})}\right\} =$$

$$-2[\log\{L(\theta(\mathbf{b}_{RM}); \mathbf{y})\} - \log\{L(\theta(\mathbf{b}_{FM}); \mathbf{y})\}] = -2\{\ell(\mathbf{b}_{RM}) - \ell(\mathbf{b}_{FM})\},$$

and this is just ΔD^*. Thus the difference in scaled deviances is an LR statistic for assessing the RM. Since we know under appropriate regularity conditions that $G_{calc}^2 \overset{\cdot}{\sim} \chi^2(p - q)$, we can say that $\Delta D^* \overset{\cdot}{\sim} \chi^2(p - q)$; the approximation improves as $n \to \infty$. This result is useful for testing the significance of reduced models in a nested hierarchy, as we formalize below in §3.2.4.

A χ^2 approximation also holds for the scaled deviances, $D^*(\mathbf{b})$. Given independent observations, Y_i, $i = 1, \ldots, n$, and an unknown parameter vector $\boldsymbol{\beta}$ of length $p + 1$ estimated via ML as $\mathbf{b}$, we have that $D^*(\mathbf{b}) \overset{\cdot}{\sim} \chi^2(n - p - 1)$. Unfortunately, the approximation here often depends on restrictive assumptions about the underlying distribution and/or the nature of the limiting operation, for example, the rate with which n approaches infinity; see McCullagh and Nelder (1989, §4.4.3). In small samples, the approximation can be poor near the tails of the distribution, and we do not generally recommend its use unless a careful investigation has suggested otherwise.

One specific operation with the χ^2 approximation for $D^*(\mathbf{b})$ does have some merit, however. When valid, the approximation can be used to estimate an unknown dispersion parameter, φ. Since $D^*(\mathbf{b}) \overset{\cdot}{\sim} \chi^2(n - p - 1)$, we know its expected value approximately equals the df: $E[D^*(\mathbf{b})] = E[D(\mathbf{b})/\varphi] \approx n - p - 1$, where $D(\mathbf{b})$ is the deviance function. By simple appeal to the method of moments (§A.4.2), we can equate the scaled deviance to its approximate mean and solve for φ. Doing so shows that the ratio of the deviance to the degrees of freedom, $\hat{\varphi}_D = D(\mathbf{b})/(n - p - 1)$, will be approximately unbiased for φ. This is a *deviance-based moment estimator* for φ.

3.2.3 Residuals

Assessing model adequacy is as important with a GLiM as it is with classical linear and nonlinear models. Residual analysis is a natural step in this process. With GLiMs, however, a number of possible definitions exist for what constitutes a residual. Start with the predicted values, $\hat{Y}_i, i = 1, \ldots, n$. As in §1.1, the *raw residual* is simply $r_i = Y_i - \hat{Y}_i$; this serves as an indicator of possible deviations or discrepancies in the model fit. An important concern with data whose parent distribution satisfies (3.1) however, is that the variance of Y_i is not necessarily constant. This can create spurious behavior and/or mask truly important patterns in residual plots and other diagnostic tools that use the r_is. To adjust for differential variation we scale r_i by the (estimated) standard deviation of Y_i:

$$\frac{Y_i - \hat{Y}_i}{\sqrt{\text{Var}(Y_i)}} = \frac{Y_i - \hat{Y}_i}{\sqrt{A(\varphi_i)V(\hat{\mu}_i)}},$$

where $\hat{\mu}_i = B'[\theta_i(\mathbf{b})]$ is the ith estimated mean response under the specific GLiM of interest. If, as is common, $A(\varphi_i) = \varphi/w_i$ for some known weights $w_i > 0$, this scaled residual becomes $\varphi^{-1/2}(Y_i - \hat{Y}_i)/\sqrt{V(\hat{\mu}_i)/w_i}$. For use as a graphical diagnostic, we can ignore the constant dispersion parameter φ, leading to $c_i = (Y_i - \hat{Y}_i)/\sqrt{V(\hat{\mu}_i)/w_i}$. These quantities are known as *Pearson residuals*. The term honors the statistician Karl Pearson, who proposed squaring and summing these residuals for use in analyzing tabular count data (Pearson, 1900). The result is a summary measure of residual variation, known as the *Pearson χ^2 statistic*:

$$X^2 = \sum_{i=1}^{n} c_i^2 = \sum_{i=1}^{n} \frac{w_i(Y_i - \hat{Y}_i)^2}{V(\hat{\mu}_i)}.$$

Many authors use X^2 to quantify goodness of fit, since scaling X^2 by φ leads to $X^2/\varphi \dot\sim \chi^2(n - p - 1)$ in large samples. (Large values of X^2/φ suggest possible inadequacy in the model fit. This can reflect a variety of factors, including inadequate specification of the linear predictor or incorrect selection of the underlying distribution for Y.) This *scaled Pearson χ^2 statistic* is simply a sum of the squared, scaled residuals introduced above: $X^2/\varphi = \sum_{i=1}^{n} c_i^2/\varphi = \sum_{i=1}^{n}(Y_i - \hat{Y}_i)^2/\text{Var}(Y_i)$.

Notice that since $X^2/\varphi \dot\sim \chi^2(n - p - 1)$, we know $E[X^2/\varphi] \approx n - p - 1$. Hence if φ is unknown, simple appeal to the method of moments leads to $\hat{\varphi}_P = X^2/(n - p - 1)$ for an approximately unbiased estimator of φ. This is a χ^2-*based moment estimator* for φ. $\hat{\varphi}_P$ is preferred over $\hat{\varphi}_D$ when $V(\mu)$ is not constant with respect to μ, since the latter estimate can exhibit instabilities in these instances.

An adjusted residual similar to c_i is based on the deviance function, $D(\mathbf{b})$. From (3.6), the ith contribution to $D(\mathbf{b})$ is $2[y_i\{\hat{\theta}_i^{\text{sat}} - \theta_i(\mathbf{b})\} - \{B(\hat{\theta}_i^{\text{sat}}) - B[\theta_i(\mathbf{b})]\}]$. Using this, a *deviance residual* can be defined as

$$d_i = \text{sgn}(r_i)\sqrt{2[y_i\{\hat{\theta}_i^{\text{sat}} - \theta_i(\mathbf{b})\} - \{B(\hat{\theta}_i^{\text{sat}}) - B[\theta_i(\mathbf{b})]\}]},$$

where $\mathrm{sgn}(r_i)$ equals -1 if $r_i < 0$, 0 if $r_i = 0$, and $+1$ if $r_i > 0$. The expression for d_i seems complex in its general form, but it simplifies once a specific parent distribution from (3.1) is identified. For example, under a Poisson specification for (3.1),

$$d_i = \mathrm{sgn}(r_i)\sqrt{2w_i\{Y_i\log(Y_i/\hat{Y}_i) - r_i\}},$$

while under a normal specification it is

$$d_i = \mathrm{sgn}(r_i)r_i\sqrt{w_i}.$$

Deviance residuals tend to be somewhat more stable than Pearson residuals (McCullagh and Nelder, 1989, §2.4), although they do posses a non-zero bias. Jørgensen (2002) suggests some modifications to d_i that can help correct for this.

In practice, we often standardize GLiM residuals so that they are approximately normal with constant variance. The calculations are best described using matrix notation: let $\mathbf{X}$ be a matrix whose columns are the $p+1$ vectors $\mathbf{J} = [1 \ldots 1]^{\mathrm{T}}$, $[x_{11} \ldots x_{n1}]^{\mathrm{T}}, \ldots, [x_{1p} \ldots x_{np}]^{\mathrm{T}}$. $\mathbf{X}$ is sometimes called a *design matrix*, where the columns correspond to the individual predictor variables in the model (and where the first column of ones is associated with the intercept); the rows correspond to observations. In this example, the matrix $\mathbf{X}$ has dimension $n \times (p+1)$. Also, let $\mathbf{M}$ be a diagonal matrix with diagonal elements $m_i = [w_i/V(\hat{\mu}_i)]^{1/2}/|g'(\hat{\mu}_i)|$, where $g'(\hat{\mu})$ is $\partial g(\mu)/\partial \mu$ evaluated at $\hat{\mu}$. (A diagonal matrix has zeros for all its off-diagonal elements; see §A.4.3.) Then, using matrix multiplication, construct the *hat matrix* $\mathbf{H} = \mathbf{M}^{\mathrm{T}}\mathbf{X}(\mathbf{X}^{\mathrm{T}}\mathbf{M}\mathbf{M}^{\mathrm{T}}\mathbf{X})^{-1}\mathbf{X}^{\mathrm{T}}\mathbf{M}$ and recover from this the diagonal elements, h_i, of $\mathbf{H}$. ($\mathbf{H}$ is called a hat matrix since it plays a role in computation of the predicted values $\hat{Y}_i$ – the 'Y-hat' values; see Lindsey, 1997, §B.2.) We divide d_i or c_i by $\sqrt{1 - h_i}$ to achieve a standardized residual with approximately constant variance (Jørgensen, 2002). For d_i, we can further divide by $\sqrt{\varphi}$ to achieve unit variance; Cordeiro (2004) gives equations to do the same for c_i. While computationally intensive, these efforts are easily handled via computer calculation.

As a technical aside, we should note that the inverse matrix $(\mathbf{X}^{\mathrm{T}}\mathbf{M}\mathbf{M}^{\mathrm{T}}\mathbf{X})^{-1}$ is assumed to exist and can be computed. If not, it can be replaced by what is known as a *generalized inverse*; see Searle (1982, §8.2).

Many other types of residuals can be constructed for GLiMs, including forms based on likelihood ratio and score statistics; see Pierce and Schafer (1986) or Lindsey (1997, §B.2), and, more generally, Davison and Tsai (1992).

3.2.4 Inference and model assessment

Inferences on the unknown parameters in a GLiM generally rely on large-sample features of the ML estimators (§A.5). For example, in many GLiMs the distribution of the Wald ratio $(b_j - \beta_j)/se[b_j]$ approaches standard normal: $W_j = (b_j - \beta_j)/se[b_j] \sim N(0, 1)$ in large samples $(j = 0, \ldots, p)$. Here, the standard error $se[b_j]$ is calculated as the square root of the estimated variance of b_j. The estimated variances are the diagonal elements of the inverse of the Fisher information matrix for **b**; see (A.18). In the SAS procedure PROC GENMOD, the information matrix is calculated using what is known as 'observed' information, where any quantities involving expected values are replaced by the actual observed data (Efron and Hinkley, 1978). This can improve the computational effort in some settings. To distinguish observed information from the more traditional Fisher information values in (A.18), we often say the latter represent 'expected' information. To force PROC GENMOD to use expected information rather than its default of observed information, use the `expected` option in the `model` statement. (By contrast, other software procedures that fit GLiMs may default to expected information. While the parameter estimates should not be affected, the standard errors may differ if a non-canonical link is used.) The choice between expected and observed information is not always an obvious one; their relative performance can vary, depending on the particulars of the GLiM being used. In order to present a consistent strategy, and unless previous research has suggested otherwise, we will default to use of expected information.

A (pointwise) $1 - \alpha$ Wald confidence interval for β_j is $b_j \pm z_{\alpha/2} se[b_j]$. Extensions to joint $1 - \alpha$ confidence ellipses – as in equation (2.23) – or to Bonferroni-adjusted $1 - \alpha$ simultaneous confidence intervals (§A.5.4) are also possible.

Hypothesis tests are similar. For example, a Wald test of $H_0: \beta_j = 0$ vs. $H_a: \beta_j \neq 0$ rejects H_0 in favor of H_a when $|W_{calc}| \geq z_{\alpha/2}$, or equivalently, when $W_{calc}^2 \geq \chi_\alpha^2(1)$, for $W_{calc} = b_j/se[b_j]$. The approximate P-value is $2P[Z \geq |W_{calc}|] = P[\chi^2(1) \geq W_{calc}^2]$, where $Z \sim N(0, 1)$. For a one-sided test against, say, $H_a: \beta_j > 0$, reject H_0 when $W_{calc} \geq z_\alpha$. The approximate P-value becomes $P[Z \geq W_{calc}]$.

Stability of Wald tests/intervals varies greatly among members of the exponential class. For example, Wald tests for most Poisson-based GLiMs are usually quite stable, while those for some binomial-based GLiMs can be so unstable as to be expressly contraindicated. (We comment in detail on this latter issue below, in §3.3.2.)

More complex hypotheses involving multiple predictors can be assessed via matrix-based extensions of the Wald test (Cox, 1988), or by likelihood ratio methods. The latter strategy is especially attractive here, since it corresponds to use of the scaled deviance function. Recall from §3.2.2 that the difference in scaled deviances between a full model and a reduced model nested within the full model corresponds to an LR statistic. Carrying forward this concept into a nested hierarchy of, say, $Q \leq p + 1$ models from the FM down to the simplest RM, we can imagine a series of LR statistics made up of these scaled deviance differences. Each scaled deviance difference/LR statistic will represent a test of one (or more) components of the FM after one (or more) previous components of the FM have been reduced away. (The degrees of freedom for the corresponding test will equal the number of parameters that are constrained to reduce the FM to the RM. Often, this is simply the difference in degrees of freedom between the two scaled deviances.) Arranged in tabular form, we have the result in Table 3.1. In the table, RM_k refers to the kth (sub)model being fitted, while 'Model df' refers to the number of non-zero parameters fitted for that particular (sub)model.

We call Table 3.1 an *analysis of deviance table*, analogous to the analysis of variance table from classical linear modeling. Using (3.6), one can sequentially assess the approximate significance of each nested component by referring the corresponding

Table 3.1 Schematic for an analysis of deviance table

Source	$D^*(\cdot)$	Model df	Δdf	$\Delta D^* = G^2_{\text{calc}}$	P-value
RM_1	$D^*(RM_1)$	df_1	–	–	–
RM_2	$D^*(RM_2)$	df_2	$df_2 - df_1$	$D^*(RM_1) - D^*(RM_2)$	$P[\chi^2(\Delta df) \geq \Delta D^*]$
$\vdots$	$\vdots$	$\vdots$	$\vdots$	$\vdots$	
RM_{Q-1}	$D^*(RM_{Q-1})$	df_{Q-1}	$df_{Q-1} - df_{Q-2}$	$D^*(RM_{Q-2}) - D^*(RM_{Q-1})$	$P[\chi^2(\Delta df) \geq \Delta D^*]$
FM	$D^*(FM)$	df_{FM}	$df_{FM} - df_{Q-1}$	$D^*(RM_{Q-1}) - D^*(FM)$	$P[\chi^2(\Delta df) \geq \Delta D^*]$

deviance difference/LR statistic to an appropriate χ^2 distribution. Reject the null hypothesis that the additional components in a larger nested model are insignificant at significance level α if $\Delta D^* = G^2_{\text{calc}} \geq \chi^2_\alpha(\Delta df)$. The corresponding P-value, derived from appeal to a large-sample χ^2 approximation, is the final column in Table 3.1.

As with classical linear modeling, it is important to note that in most cases this assessment must be made sequentially. Inferences on any nested (sub)component may depend on the terms that precede it in the model and on the sequential order in which they are fitted. In PROC GENMOD, this is labeled a `Type I` analysis, similar to the `Type I` sequential sums of squares from an ANOVA table in PROC GLM or PROC REG.

If φ is unknown in an analysis of deviance, the computations require adjustment. If estimation of φ proceeds via ML, we may still appeal to the large-sample asymptotics associated with ML, but the approximation may not be as stable. More often recommended is appeal to the method of moments (MOM) using a Pearson or deviance-based estimator. (As noted above, the Pearson form is usually preferred.) Then, given $\hat{\varphi}$, the difference in estimated scaled deviances between two nested models, RM_2 and RM_1 (with corresponding model degrees of freedom $df_2 > df_1$), is calculated as $\Delta D^* = \{D(RM_1) - D(RM_2)\}/\hat{\varphi}$. Although the MOM estimators also cause degradation in the χ^2 approximation, we can compensate: further divide ΔD^* by the corresponding $\Delta df = df_2 - df_1$, producing an F-statistic

$$F_{\text{calc}} = \frac{D(RM_1) - D(RM_2)}{(df_2 - df_1)\hat{\varphi}}.$$

Under certain regularity conditions on $\hat{\varphi}$, $F_{\text{calc}} \sim F(df_2 - df_1, n - df_2)$, and we can reject the null hypothesis that the additional components in RM_2 are insignificant at significance level α when $F_{\text{calc}} \geq F_\alpha(df_2 - df_1, n - df_2)$. The corresponding P-value is $P[F(df_2 - df_1, n - df_2) \geq F_{\text{calc}}]$.

An important question here is under which model (RM_1, RM_2, FM, etc.) $\hat{\varphi}$ should be determined. (As the model hierarchy changes from one RM to another, so may the point estimator.) Many different strategies are possible, each varying in quality based on the underlying parent distribution, linear predictor, link function, etc., and it is difficult to make an omnibus recommendation. The `Type 1` approach taken by PROC GENMOD is, however, a useful default: when prompted to apply an MOM estimator such as $\hat{\varphi}_P$, PROC GENMOD fits the maximal model (FM) to

determine the point estimator, and uses this same value throughout all its `Type I`
calculations of the estimated scaled deviances. This mimics the case with a linear
model when a set of (sub)models is compared using the general linear test of
hypotheses. In this latter situation, the MSE is used as a scale component in the
F-statistic of each model comparison: typically, the MSE from the maximal FM is
used for all comparisons.

Note that these scaled deviance/LR operations can also be used to construct approximate $1 - \alpha$ confidence intervals on the β-parameters, by inverting the corresponding
scaled deviance/LR test statistic (§A.5.1). It is not possible to write a closed-form
expression for the resulting intervals, but they can often be acquired as part of a GLiM
computer output. When estimating $\hat{\varphi}$ via ML, one can also estimate a standard error for
$\hat{\varphi}$, $se[\hat{\varphi}]$, and build an approximate $1 - \alpha$ Wald interval of the form $\hat{\varphi} \pm z_{\alpha/2} se[\hat{\varphi}]$. We
illustrate these methods via PROC GENMOD with the examples in §3.3.

3.2.5   Estimation via maximum quasi-likelihood

An interesting realization was made by Wedderburn (1974) regarding estimation of
the unknown β-parameters in a GLiM: the ML estimating equation for each β_j has
the same form. Specifically, suppose there is a single dispersion parameter, φ, and
that it is a known constant. From (3.5), the log-likelihood derivative (or 'score') for
β_j is proportional to $\sum_{i=1}^{n} (y_i - \mu_i) x_{ij} h'(\eta_i)/V(\mu_i)$. We denote this *score function* as U_j.
Wedderburn recalled from the derivation of (3.5) that $x_{ij} = \partial\eta_i/\partial\beta_j$ and
$h'(\eta_i) = \partial\mu_i/\partial\eta_i$. Thus their product is $x_{ij}h'(\eta_i) = (\partial\eta_i/\partial\beta_j)(\partial\mu_i/\partial\eta_i)$, which from the
chain rule of differential calculus simplifies to $\partial\mu_i/\partial\beta_j$. From this, one can write U_j as

$$U_j \propto \sum_{i=1}^{n} \frac{y_i - \mu_i}{V(\mu_i)} \left(\frac{\partial\mu_i}{\partial\beta_j} \right), \qquad j = 0, \ldots, p, \qquad (3.7)$$

(Zeger and Liang, 1992). To find the ML estimates of β_j we set $U_j = 0$ and solve the
$p + 1$ equations numerically. Wedderburn recognized that even if the underlying
distribution of Y_i is *not* of the exponential class form in (3.1), the estimating
equations in (3.7) can still yield valid estimates of β. To obtain these, the user must
specify the mean as a function of β, say $\mu_i = \mu_i(\beta)$, and the variance as a function of
μ_i, via $\text{Var}[Y_i] = \varphi V(\mu_i)$. Notice, however, that the model framework can extend
beyond a form of GLiM, since (3.7) is written to allow $\mu_i(\beta)$ to be any differentiable
function of β.

Wedderburn coined the term *quasi-likelihood* to describe a situation where (3.7) is
used to estimate regression coefficients with a specific form for the mean and
variance (or equivalent quantities), but where the underlying parent distribution is
left unspecified. The quantity U_j in (3.7) is the *quasi-score*, the anti-derivative of
which is called the *quasi-likelihood function* (McCullagh and Nelder, 1989, §9.2).
Hence, estimators, b_j, achieved by setting the U_js to zero are called *maximum
quasi-likelihood (MQL) estimators*. These possess large-sample features similar to
those of ML estimators (McCullagh, 1983); for example, the MQL vector $\mathbf{b}_{\text{MQL}}$ is

approximately multivariate normal with mean β and with variances and covariances given compactly in a *variance–covariance matrix* $\text{Var}[\mathbf{b}_{\text{MQL}}] = \varphi(\mathbf{D}^T\mathbf{V}^{-1}\mathbf{D})^{-1}$, where $\mathbf{V}$ is a diagonal matrix with diagonal elements $V(\mu_i)$ and $\mathbf{D}$ is an $n \times (p+1)$ matrix of the partial derivatives $\partial\mu_i/\partial\beta_j$.

If the dispersion parameter φ in a quasi-likelihood model is unknown, it must be estimated. It is typical to apply an MOM estimator similar to that described in §3.2.3. A Pearson statistic here is $X^2 = \sum_{i=1}^{n}(Y_i - \hat{Y}_i)^2/V[\mu_i(\mathbf{b}_{\text{MQL}})]$. Scaled, we again have $X^2/\varphi \stackrel{.}{\sim} \chi^2(n-p-1)$, from which an MOM estimator is available by equating X^2/φ to its df. This produces the Pearson-based dispersion estimator $\hat{\varphi}_{\text{P}} = X^2/(n-p-1)$. Concurrent estimation of φ does not adversely affect the normal approximation for $\mathbf{b}_{\text{MQL}}$. (Technically, the estimator $\hat{\varphi}_{\text{P}}$ must approach the true value of φ as $\sqrt{n} \to \infty$ (Moore, 1986). In most cases, the MOM estimator satisfies this requirement.)

One can extend the MQL approach by allowing the variance function to be more general, say $V[\mu_i(\beta), \varphi]$ (Nelder and Pregibon, 1987). To find the point estimators of β and φ, iterate between the obvious extension of (3.7) and a generalized Pearson estimating equation $\sum_{i=1}^{n}(Y_i - \hat{Y}_i)^2/V[\mu_i(\mathbf{b}_{\text{MQL}}), \varphi] = n - p - 1$ until convergence is achieved.

It is important to emphasize that the MQL approach requires the chosen forms for $\mu_i(\beta)$ and $V[\mu_i(\beta)]$ (or $V[\mu_i\{\beta\}, \varphi]$) to be correct for the actual effects they purport to model. This is especially true for $V[\mu_i(\beta)]$: if the variance function is misspecified, the MQL estimator $\mathbf{b}_{\text{MQL}}$ may not suffer greatly, but the associated standard errors and estimated covariances based on $\mathbf{b}_{\text{MQL}}$ will be incorrect. Inefficient and possibly incorrect inferences on β may result.

3.2.6 Generalized estimating equations

Liang and Zeger (1986) extended the quasi-likelihood estimating equations in (3.7) to GLiM settings where data are correlated. Their initial consideration was for longitudinal data, where correlations are induced by repeated measurements over time on the same subject (as mentioned in §2.4, above). They expected measurements on the same subject to be correlated, $\text{Corr}(Y_{ij}, Y_{ik}) \neq 0$, while assuming that measurements on different subjects are uncorrelated, $\text{Corr}(Y_{ij}, Y_{hk}) = 0$ when $i \neq h$. The Liang–Zeger construction has evolved to encompass more than just longitudinal forms of intra-observational correlation, however. Formally, let Y_{ij} be the jth correlated measurement on the ith subject, $i = 1, \ldots, n$; $j = 1, \ldots, m_i$. Thus the ith subject's observation vector is $\mathbf{Y}_i = [Y_{i1}\, Y_{i2} \ldots Y_{im_i}]^T$. There are $\sum_{i=1}^{n} m_i$ observations in total on n subjects. Also, assume that each Y_{ij} is observed along with a column vector of p concomitant predictor variables and a $(p+1)$st predictor associated with the intercept: $\mathbf{x}_{ij} = [1\, x_{ij1}\, x_{ij2} \ldots x_{ijp}]^T$. If the effect of $\mathbf{x}_{ij}$ is the same on all measurements, our interest is in estimation and inferences on an associated coefficient vector $\beta = [\beta_0\, \beta_1 \ldots \beta_p]^T$. (Note that it is virtually impossible to avoid vector and matrix notation with multivariate data of this sort. Readers unfamiliar with the notation may wish to study the short comments on matrices throughout Appendix A, or consult a text on matrix algebra such as Searle, 1982, or Harville, 1997.)

The regression model here exploits an important feature of the repeated/correlated data framework: the vector $\mathbf{x}$ can change with j, so that the value(s) of the predictor variable(s) can change across repeated/correlated measurements *within* a subject (Although this is not required). This provides the analyst with a wide, flexible variety of model options.

If the complete parametric details of the data – the exact form of (3.1), the exact intra-observation correlation structure, etc. – are known, then a log-likelihood can be constructed and ML estimators can be calculated. Under a quasi-likelihood format, however, only knowledge of the mean and variance–covariance structure of the ith subject's observation vector, $\mathbf{Y}_i = [Y_{i1} \, Y_{i2} \ldots Y_{im_i}]^T$, is required. The mean is an $m_i \times 1$ column vector and the variance is an $m_i \times m_i$ matrix,

$$\boldsymbol{\mu}_i(\boldsymbol{\beta}) = \begin{bmatrix} \mu_{i1}(\boldsymbol{\beta}) \\ \mu_{i2}(\boldsymbol{\beta}) \\ \vdots \\ \mu_{im_i}(\boldsymbol{\beta}) \end{bmatrix} \quad \text{and} \quad \mathbf{V}_i = \begin{bmatrix} \text{Var}[Y_{i1}] & \text{Cov}[Y_{i1}, Y_{i2}] & \cdots & \text{Cov}[Y_{i1}, Y_{im_i}] \\ \text{Cov}[Y_{i2}, Y_{i1}] & \text{Var}[Y_{i2}] & \cdots & \text{Cov}[Y_{i2}, Y_{im_i}] \\ \vdots & \vdots & \ddots & \vdots \\ \text{Cov}[Y_{im_i}, Y_{i1}] & \text{Cov}[Y_{im_i}, Y_{i2}] & \cdots & \text{Var}[Y_{im_i}] \end{bmatrix},$$

respectively. This is essentially a multivariate extension of the univariate quasi-likelihood formulation in §3.2.5. Indeed, Liang and Zeger (1986) recognized that a multivariate version of the vector of quasi-scores $\mathbf{U} = [U_0 \, U_1 \ldots U_p]^T$ based on equation (3.7) – and, for that matter, on the GLiM score equation (3.5) – is

$$\mathbf{U} = \sum_{i=1}^n (\nabla \boldsymbol{\mu}_i)^T \mathbf{V}_i^{-1} [\mathbf{Y}_i - \boldsymbol{\mu}_i(\boldsymbol{\beta})], \tag{3.8}$$

where $\nabla \boldsymbol{\mu}_i$ is the $m_i \times (p+1)$ gradient matrix of $\boldsymbol{\mu}_i(\boldsymbol{\beta})$, $\nabla \boldsymbol{\mu}_i = \partial \boldsymbol{\mu}_i(\boldsymbol{\beta})/\partial \boldsymbol{\beta}^T$, with (j, k)th element $\partial \mu_{ij}(\boldsymbol{\beta})/\partial \beta_k$. We assume that (i) the unknown parameters in $\boldsymbol{\beta}$ enter into the model via a linear predictor $\eta_{ij} = \mathbf{x}_{ij}^T \boldsymbol{\beta}$, and (ii) the linear predictor is related to the mean response via a monotone link function $g(\mu_{ij}) = \mathbf{x}_{ij}^T \boldsymbol{\beta}$. Extending the univariate relationship $\partial \mu_i/\partial \beta_j = x_{ij}/(\partial \eta_i/\partial \mu_i)$ and recognizing that $\partial \eta_{ij}/\partial \mu_{ij} = \partial g(\mu_{ij})/\partial \mu_{ij} = g'(\mu_{ij})$, we see $\partial \boldsymbol{\mu}_i(\boldsymbol{\beta})/\partial \boldsymbol{\beta}$ has elements of the form $x_{ijk}/g'(\mu_{ij}[\boldsymbol{\beta}])$, where $g'(\mu_{ij}[\boldsymbol{\beta}])$ is $\partial g(\mu_{ij}[\boldsymbol{\beta}])/\partial \mu_{ij}$ evaluated at $\boldsymbol{\beta}$. Thus the gradient matrix can be written as

$$\nabla \boldsymbol{\mu}_i = \begin{bmatrix} \dfrac{x_{i11}}{g'(\mu_{i1}[\boldsymbol{\beta}])} & \dfrac{x_{i12}}{g'(\mu_{i1}[\boldsymbol{\beta}])} & \cdots & \dfrac{x_{i1(p+1)}}{g'(\mu_{i1}[\boldsymbol{\beta}])} \\ \dfrac{x_{i21}}{g'(\mu_{i2}[\boldsymbol{\beta}])} & \dfrac{x_{i22}}{g'(\mu_{i2}[\boldsymbol{\beta}])} & \cdots & \dfrac{x_{i2(p+1)}}{g'(\mu_{i2}[\boldsymbol{\beta}])} \\ \vdots & \vdots & \ddots & \vdots \\ \dfrac{x_{im_i 1}}{g'(\mu_{im_i}[\boldsymbol{\beta}])} & \dfrac{x_{im_i 2}}{g'(\mu_{im_i}[\boldsymbol{\beta}])} & \cdots & \dfrac{x_{im_i(p+1)}}{g'(\mu_{im_i}[\boldsymbol{\beta}])} \end{bmatrix}.$$

Setting (3.8) equal to a column vector of $p+1$ zeros produces what Liang and Zeger called *generalized estimating equations* (GEEs):

$$\sum_{i=1}^n (\nabla \boldsymbol{\mu}_i)^T \mathbf{V}_i^{-1} [\mathbf{Y}_i - \boldsymbol{\mu}_i(\boldsymbol{\beta})] = \mathbf{0}, \tag{3.9}$$

the solution of which produces the GEE estimator $\mathbf{b}_{\text{GEE}}$.

The GEEs in (3.9) assume that the variance–covariance matrix, $\mathbf{V}_i$, of the ith subject's observation vector is known, which is unlikely in practice. Luckily, replacing it by a consistent estimator still results in asymptotically efficient estimators of $\boldsymbol{\beta}$ (Gouriéroux *et al.*, 1984). Under the usual GLiM assumption that $\mathrm{Var}[Y_{ij}] = \varphi V(\mu_{ij})$ for some variance function $V(\cdot)$, it is common to use $\mathbf{V}_i = \varphi \boldsymbol{\Delta}_i^{1/2} \mathbf{R}(\mathbf{a}) \boldsymbol{\Delta}_i^{1/2}$, where $\boldsymbol{\Delta}_i^{1/2}$ is a diagonal matrix with (diagonal) elements $\sqrt{V(\mu_{i1}[\mathbf{b}])}, \ldots, \sqrt{V(\mu_{im_i}[\mathbf{b}])}$ and $\mathbf{R}(\mathbf{a})$ is a *working correlation matrix* that models the correlation structure in the data. The vector $\mathbf{a} = [a_1 \ldots a_L]^{\mathrm{T}}$ is some additional set of L unknown parameters that allows $\mathbf{R}(\mathbf{a})$ to take on a variety of possible forms (see below). The working correlation matrix is estimated by first estimating $\mathbf{a}$, and from this constructing an estimate of $\mathbf{V}_i$ for use in (3.9). If unknown, the dispersion parameter φ can be estimated via a Pearson-based MOM estimator, similar to that seen in §3.2.5.

In large samples, the GEE estimator $\mathbf{b}_{\mathrm{GEE}}$ is approximately multivariate normal with mean $\boldsymbol{\beta}$ and variance–covariance matrix $\mathbf{C}$ given by inverting an analog of the expected Fisher information matrix. The information analog is $\mathbf{F} = \mathrm{E}[\mathbf{U}\mathbf{U}^{\mathrm{T}}]$, where $\mathbf{U}$ is given in (3.8). Using advanced matrix algebra, one can show (Exercise 3.3) that this is

$$\mathbf{F} = \mathrm{E}\left[\sum_{i=1}^{n} (\nabla \boldsymbol{\mu}_i)^{\mathrm{T}} \mathbf{V}_i^{-1} [\mathbf{Y}_i - \boldsymbol{\mu}_i(\boldsymbol{\beta})] \left(\sum_{k=1}^{n} (\nabla \boldsymbol{\mu}_k)^{\mathrm{T}} \mathbf{V}_k^{-1} [\mathbf{Y}_k - \boldsymbol{\mu}_k(\boldsymbol{\beta})] \right)^{\mathrm{T}} \right]$$

$$= \sum_{i=1}^{n} (\nabla \boldsymbol{\mu}_i)^{\mathrm{T}} \mathbf{V}_i^{-1} \nabla \boldsymbol{\mu}_i.$$

From this, take $\mathbf{C} = \mathbf{F}^{-1}$. If $\mathbf{C}$ depends on any unknown values in $\boldsymbol{\beta}$, $\mathbf{a}$, and/or φ, these are replaced by their own point estimates when calculating the variance–covariance elements.

The large-sample normality of $\mathbf{b}_{\mathrm{GEE}}$ facilitates construction of confidence intervals and hypothesis tests for $\boldsymbol{\beta}$. For example, a large-sample $1 - \alpha$ Wald interval on β_j is $b_j \pm z_{\alpha/2} se[b_j]$, where $se[b_j]$ is the square root of the jth diagonal element of $\mathbf{C}$.

An important feature of the GEE approach is that the quality of the estimator $\mathbf{b}_{\mathrm{GEE}}$ is not adversely affected if the form for the working correlation matrix $\mathbf{R}(\mathbf{a})$ is misspecified. If $\mathbf{R}(\mathbf{a})$ is only approximately correct, $\mathbf{b}_{\mathrm{GEE}}$ still approaches $\boldsymbol{\beta}$ in the limit. Unfortunately, the variance–covariance matrix does not share this large-sample robustness; under a misspecified correlation matrix, $\mathbf{C}$ is no longer a valid representation for $\mathrm{Var}[\mathbf{b}_{\mathrm{GEE}}]$. To adjust $\mathbf{C}$ in this case, a more robust, 'empirical' estimator is $\hat{\mathbf{C}} = \mathbf{F}^{-1} \hat{\mathbf{G}} \mathbf{F}^{-1}$, where

$$\hat{\mathbf{G}} = \sum_{i=1}^{n} (\nabla \boldsymbol{\mu}_i)^{\mathrm{T}} \mathbf{V}_i^{-1} [\mathbf{Y}_i - \boldsymbol{\mu}_i(\mathbf{b}_{\mathrm{GEE}})][\mathbf{Y}_i - \boldsymbol{\mu}_i(\mathbf{b}_{\mathrm{GEE}})]^{\mathrm{T}} \mathbf{V}_i^{-1} \nabla \boldsymbol{\mu}_i.$$

Due to Huber (1967), this construction for $\hat{\mathbf{C}}$ is often called an *information sandwich* (Pickles, 2002) since the empirical component is being 'sandwiched' by information quantities. $\hat{\mathbf{C}}$ has the property that it consistently estimates the true variance–covariance structure of $\mathbf{b}_{GEE}$ even when the correlation structure has been misspecified in $\mathbf{R}(\mathbf{a})$.

One particular feature of the GEE methodology that gives it great applicability is that the two sets of estimating equations for $\boldsymbol{\beta}$ and $\mathbf{a}$ are of very similar structure, hence they can be combined for efficiency's sake (Prentice and Zhao, 1991; Liang *et al.*, 1992). The resulting generalized score equations are solved simultaneously. If information in the data on $\boldsymbol{\beta}$ and $\mathbf{a}$ is separable (also called *orthogonal*), then large portions of the matrices making up the GEEs will have blocks of zeros. The result is called a set of *first-order estimating equations* (GEE1), and leads to greatly simplified calculations. Conversely, if there is pertinent information about $\boldsymbol{\beta}$ tied up in the information on $\mathbf{a}$ (or vice versa), then the associated matrix elements may be non-zero, leading to *second-order estimating equations* (GEE2). As noted above, applying GEE1 will yield consistent estimators for $\boldsymbol{\beta}$ even when $\mathbf{a}$ is misspecified. Although GEE2 does not share this robust feature, it can lead nonetheless to lower variances for the estimated regression coefficients. Liang *et al.* (1992) suggest that such improvement can be marginal, but it is still an issue that should be considered in the modeling process.

In most cases, solving (3.9) for the $p+1$ unknowns in $\boldsymbol{\beta}$ requires computer iteration. PROC GENMOD can accomplish this via its `repeated` statement. The user can control convergence conditions, output specification, and, perhaps most importantly, the link function $g(\cdot)$ and the structure of the working correlation matrix, $\mathbf{R}(\mathbf{a})$. Six different correlation structures are possible:

1. *independent*, with zero correlation assumed among the Y_{ij}s within a single $\mathbf{Y}_i$;

2. *fixed*, with $\mathrm{Corr}(Y_{ij}, Y_{ik})$ fixed at some user-specified values a_{jk};

3. *M-dependent*, with $\mathrm{Corr}(Y_{i,j}, Y_{i,j+t}) = a_t$ for $t = 1, \ldots, M$, and zero otherwise (except that, clearly, $\mathrm{Corr}[Y_{i,j}, Y_{i,j+t}] = 1$ at $t = 0$);

4. *exchangeable* (or constant), with $\mathrm{Corr}(Y_{ij}, Y_{ik}) = a$ for $j \neq k$ and $\mathrm{Corr}(Y_{ij}, Y_{ij}) = 1$;

5. *unstructured*, with $\mathrm{Corr}(Y_{ij}, Y_{ik}) = a_{jk}$ for $j \neq k$ and $\mathrm{Corr}(Y_{ij}, Y_{ij}) = 1$;

6. *first-order autoregressive*, AR(1), with $\mathrm{Corr}(Y_{i,j}, Y_{i,j+t}) = a^t$ for $t = 0, 1, \ldots, m_i - j$.

(Option 6 imitates a dampening effect often seen with temporally correlated data. The structure would make sense if the measurements were made at equally spaced time intervals. We discuss autoregressive models and temporally correlated data in Chapter 5.) For the first two structures no working correlation matrix is estimated, since the correlations are fixed by the user. For all other cases, PROC GENMOD builds an iterative scheme for evaluating the unknown values in $\mathbf{R}(\mathbf{a})$, based on standardized residuals from the current iteration's fit (SAS Institute Inc., 2000, Ch. 29).

3.3 Specific forms of generalized linear models

In this section, we illustrate uses of GLiMs for a variety of different members of the exponential class. Our examples use PROC GENMOD, although, as mentioned above, many other computer packages are available to fit GLiMs. In some cases, we also give some background on allied methods for performing the regression fit, depending on the nature of the parent distribution and the link function of interest. We start with the progenitor GLiM, the normal model.

3.3.1 Continuous/homogeneous-variance data GLiMs

When continuous data are observed with homogeneous variances and at least approximate symmetry about their mean, it is natural to consider use of the normal distribution as the parent p.d.f. for the probability model. As seen in Example 3.1, the normal p.d.f. is a member of the exponential class (3.1), and hence can be fitted as a GLiM. We do not necessarily recommend doing so, however. In general, dedicated routines for fitting normal models such as PROC REG or PROC GLM are much more efficient computationally and provide a much wider array of regression outputs and diagnostic options. Even with the case of nonlinear link functions (such as the natural logarithm), it is often more efficient computationally to apply a nonlinear routine such as PROC NLIN from Chapter 2. For illustration's sake, however, we give a brief example using PROC GENMOD to analyze the motor vehicle CO_2 data from Example 1.1.

Example 3.4 (Motor vehicle CO_2, cont'd) Consider again the data from Table 1.1 on motor vehicle use in the UK, seen originally in Example 1.1. There we modeled $Y = CO_2$ emission (as a relative index; $1970 = 100$) as normally distributed with constant variance and with a mean that was linearly related to $x =$ motor vehicle use. From Example 3.1, however, we know that a normal-distribution simple linear regression model is a special case of a GLiM, so we should expect both methods to produce identical results. (If, that is, they are asked to undertake the same model fit; see below.)

As in Example 1.1, we will fit the simple linear predictor, $\eta_i = \beta_0 + \beta_1 x_i$, to the data, but now using the two SAS procedures PROC REG and PROC GENMOD. PROC REG is dedicated to fitting simple and multiple linear regression models under a normal parent distribution (we could also use PROC GLM), while PROC GENMOD is dedicated to GLiMs. In either case, the code is fairly simple; Fig. 3.1 incorporates separate calls to each procedure. The code for PROC REG is straightforward and similar to other examples seen in Chapter 1. The code for PROC GENMOD looks almost identical; notable exceptions are the `dist=` and `link=` options in the `model` statement. These specify the exponential family distribution (`dist`) and the link function (`link`) of the GLiM. Here, `dist=normal` (or just `dist=n`) specifies a normal p.d.f. as in Example 3.1, while `link=id` specifies use of a simple identity link, $g(\mu) = \mu$, as is appropriate for a simple linear model.

```
*  SAS code to fit simple lin. regr. via REG and GENMOD;
data motorveh;
   input xveh yco2 @@;
datalines;
      105.742    104.619          110.995   109.785

          .            .                .         .
          .            .                .         .
          .            .                .         .

      225.742    194.667          229.027   193.438

*fit S.L.R. via PROC REG;
   proc reg;
   title "*****************PROC REG output*****************";
      model yco2 = xveh;
      run;

*fit S.L.R. via PROC GENMOD;
   proc genmod;
   title "**************PROC GENMOD output***************";
      model yco2 = xveh / dist=normal link=id;
```

Figure 3.1 Sample SAS programming to fit simple linear regression model to motor vehicle CO_2 data

Figure 3.2 gives output (edited) from the code in Fig. 3.1. The style and organization of the two (sub)outputs is clearly different, but closer inspection reveals many important similarities. First and foremost, the listing for Parameter Estimates in either case shows $b_0 = 28.3603$ and $b_1 = 0.7442$. This corroborates what we derived using direct calculations in Example 1.1.

Also under Parameter Estimates the **PROC REG** output gives the standard errors as $se[b_0] = 2.1349$ and $se[b_1] = 0.0127$. Again, this agrees with the results obtained in Example 1.1. Note, however, that the corresponding values under **PROC GENMOD** are slightly smaller: $se[b_0] = 2.0572$ and $se[b_1] = 0.0122$. What has happened?

A clue to the puzzling results given by **PROC GENMOD** is the following. The ratio of the squared standard errors (i.e., of the estimated variances) between each **PROC** does not depend upon the estimator: $(2.1349/2.0572)^2 = (0.01266/0.0122)^2 = 1.077$. Further, this ratio is equal to the ratio of sample size to df_e: $n/(n-2) = 1.077$ when $n = 28$. (In fact, this phenomenon is independent of the data. Any other data set modeled in this fashion would produce the same effect; see Exercise 3.4.)

The answer to this riddle lies in how **PROC GENMOD** estimates the population variance, σ^2. Recall from Example 3.1 that the dispersion parameter under the normal p.d.f. is $\varphi = \sigma^2$. When faced with estimating φ (**PROC GENMOD** calls $\sqrt{\varphi}$ the scale parameter here), the procedure defaults to maximum likelihood; see the NOTE under Analysis Of Parameter Estimates. For a normal-based linear regression model, it can be shown that the ML estimator of $\varphi = \sigma^2$ is $\frac{1}{n}$SSE, where SSE is the sum of squared errors $\sum_{i=1}^{n}(Y_i - \hat{Y}_i)^2$. This estimate of φ differs from the unbiased estimator, MSE $= \frac{1}{n-2}$SSE. Even though ML estimators possess a number of important optimal properties, when an unbiased estimator exists that differs from the ML estimator we typically prefer the unbiased form unless it is

```
                            The SAS System

            *****************PROC REG output*****************
    Model: MODEL1
    Dependent Variable: YCO2

                        Analysis of Variance
                           Sum of           Mean
    Source          DF    Squares          Square    F Value   Pr > F
    Model            1  26045.295       26045.295    3454.18   <.0001
    Error           26    196.046           7.540
    C Total         27  26241.341


            Root MSE        2.74595      R-square       0.9925
            Dep Mean      150.06999      Adj R-sq       0.9922
            C.V.            1.82978

                        Parameter Estimates
                      Parameter      Standard
    Variable  DF      Estimate         Error    t Value    Pr > |t|
    INTERCEP   1      28.36026       2.13490      13.28      <.0001
    XVEH       1       0.74419       0.01266      58.77      <.0001

            **************PROC GENMOD output**************
                        The GENMOD Procedure
                        Model Information
                    Distribution              Normal
                    Link Function            Identity
                    Observations Used            28

            Criteria For Assessing Goodness Of Fit
            Criterion            DF         Value      Value/DF
            Deviance             26      196.0457        7.5402
            Scaled Deviance      26       28.0000        1.0769
    Algorithm converged.

                    Analysis Of Parameter Estimates
        Parameter     DF    Estimate    Std Err   ChiSquare   Pr>Chi
        Intercept      1     28.3603     2.0572      190.04    <.0001
        xveh           1      0.7442     0.0122     3719.89    <.0001
        Scale          1      2.6461     0.3536          .         .
    NOTE:  The scale parameter was estimated by maximum likelihood.
```

Figure 3.2 SAS output (edited) from simple linear regression fit for motor vehicle CO_2 data

known to exhibit some other undesirable feature. Thus in **PROC REG**, the unbiased estimator is used to estimate the variance, while **PROC GENMOD** defaults to the ML estimator.

As seen in §1.1, the standard errors for both b_0 and b_1 are proportional to $\sqrt{MSE}$. **PROC GENMOD** is replacing the MSE with the biased estimator, which when squared will be a factor of $(n - 2)/n$ too small under a simple linear regression model. This explains the discrepancies seen in Fig. 3.2. More generally, for a linear regression

```
                        The SAS System
                     The GENMOD Procedure
                      Model Information
            Distribution                      NORMAL
            Link Function                     IDENTITY
            Observations Used                 28

                 Criteria For Assessing Goodness Of Fit
            Criterion              DF          Value       Value/DF
            Deviance               26        196.0457       7.5402
            Scaled Deviance        26         26.0000       1.0000
  Algorithm converged.

                     Analysis Of Parameter Estimates
                                   Standard        Chi-
      Parameter    DF    Estimate    Error        Square     Pr>ChiSq
      Intercept    1     28.3603    2.1349        176.47      <.0001
      xveh         1      0.7442    0.0127       3454.18      <.0001
      Scale        0      2.7459    0.0000
  NOTE:  The scale parameter was estimated by the square root
         of DEVIANCE/DOF.
```

Figure 3.3 SAS output (edited) from modified PROC GENMOD fit for motor vehicle CO_2 data

model with p predictor variables and an intercept, the standard errors from the biased estimator will be too small by a factor of $(n - p - 1)/n$.

To avoid use of the biased ML estimator here, we can recall from §3.2.2 that the deviance under a normal p.d.f. is simply the SSE (confirm this in the Fig. 3.2 output). Thus the deviance-based moment estimator of φ will be $\hat{\varphi}_D = \text{SSE}/(n - p - 1)$. Notice that the denominator here is df_e, so $\hat{\varphi}_D$ is just the MSE. To force PROC GENMOD to estimate the dispersion parameter with this form, one simply includes the dscale option in the model statement. In Fig. 3.1, this would be

```
model yco2 = xveh / dist = normal  link = id  dscale;
```

The corresponding output (edited) appears in Fig. 3.3, where we see that the standard errors are now being estimated as those under PROC REG. ✪

The concerns evidenced in Example 3.4 regarding normal-based regression models in PROC GENMOD extend to multiple linear regression as well. Thus, we caution readers to avoid haphazard use of PROC GENMOD when applying it to normal-based models (and in general!). Indeed, PROC REG and PROC GLM provide a wider variety of options and techniques for fitting such models, and they may be better choices for such analyses.

3.3.2 Binary data GLiMs (including logistic regression)

When discrete data are observed as binary observations (say, 'success' vs. 'failure') leading to proportion responses, it is natural to consider use of the binomial distribution as the parent p.m.f. for the probability model. Take Y as the number of successes out of N binary 'trials' for the outcome, so that $Y \sim \text{Bin}(N, \pi)$ where $\pi = \text{P[success]}$. (The proportion response is Y/N.) As seen in Example 3.3, this is a member of the exponential class in (3.1), and hence can be fitted as a GLiM. Recall from that example that the natural parameter is the log-odds of a success: $\theta = \text{logit}(\pi) = \log\{\pi/(1 - \pi)\}$. Although this can be written as a function of the binomial mean $\mu = \text{E}[Y] = N\pi$, it is more common in practice to direct attention to π rather than to μ. Thus we make a minor adjustment with the GLiM notation here and use π instead of μ as the argument of the link function; that is, we write and model $g(\pi)$ rather than $g(\mu)$.

The link function most often used with binomial data is the canonical form, the logit link: $g(\pi) = \text{logit}(\pi) = \log\{\pi/(1 - \pi)\}$. This transforms a value between 0 and 1 to a value between $-\infty$ and ∞. The corresponding inverse link is $h(\eta) = 1/(1 + e^{-\eta})$, which, since it is an inverse operation, takes a value between $-\infty$ and ∞ to a value between 0 and 1. This feature is particular to binary data: since π is a probability, any inverse link must return a value between 0 and 1. And, since $g(\pi)$ is monotone, so is its inverse $h(\eta)$. Notice, then, that if $h(\eta)$ is non-decreasing in η, it exhibits the properties of a cumulative distribution function (c.d.f., see §A.1): it is between 0 and 1, and it is a monotone, non-decreasing function. For example, with the logit link the inverse is the c.d.f. from what is known as the standard logistic distribution (Johnson *et al.*, 1995, Ch. 23); the corresponding GLiM is often called a *logistic regression*. Compare this function for $h(\eta)$ with the nonlinear logistic growth curve from §2.4.2: both are sigmoidal and continuous, but here the logistic regression model is restricted to $0 < h(\eta) < 1$. Since $h(\eta)$ is meant to model a probability, π, this is a necessary and reasonable restriction.

The correspondence between binary-data inverse link functions and c.d.f.s is not entirely coincidental. We can use this feature to motivate construction of the inverse link (and hence the link itself) by arguing for the existence of a latent, underlying (continuous) random variable that relates the response of each subject under study to its probability of expressing the binary outcome. That is, if the distribution of a subject's tolerance to an external stimulus takes on a certain form, the c.d.f. from that distribution will represent the inverse link function of a GLiM for the resulting binary data. This is known as a *tolerance distribution interpretation* for binary-data GLiMs. For example, if the tolerance distribution is thought to be standard normal, the corresponding inverse link is the standard normal c.d.f. $\Phi(\eta)$ (§A.2.10). Thus the link function is $g(\pi) = \Phi^{-1}(\pi)$, which is known as a *probit link*. The corresponding GLiM is called a *probit regression model*. Or, using an extreme-value c.d.f. (§A.2.9) for the tolerance distribution leads to $g(\pi) = \log\{-\log(1 - \pi)\}$, which is known as a *complementary log-log link*. We gave a further discourse on tolerance distributions for binary-data regressions in Piegorsch and Bailer (1997, §7.2).

Parameters estimated in the context of tolerance distributions can have specialized interpretations. For example, the intercept and slope under a linear probit model, $\Phi^{-1}(\pi) = \beta_0 + \beta_1 x$, are related directly to the mean and variance of a normal tolerance distribution (see Exercise 3.7). This interpretation becomes less clear in the case of multiple predictor variables.

In some cases, mechanistic or other environmental factors may motivate selection of a specific tolerance distribution. When present, this information simplifies selection of the (inverse) link function for the GLiM. Specification of tolerance distributions is not required to fit a GLiM to proportion data, however, and we recommend their use only when the subject-matter suggests appropriate forms.

When the subject-matter does not provide an obvious choice for the link function, many authors default to the logit, appealing to its status as the canonical link for the binomial model. In addition, the logit link has appeal due to the interpretability of its regression parameters. That is, $\log\{\pi/(1-\pi)\}$ represents the log-odds that a 'success' occurs. If we model π via this logit link and use, for example, the multiple linear predictor $\eta = \beta_0 + \beta_1 x_1 + \beta_2 x_2$, then β_2 represents the unit change in the log-odds – a *log-odds ratio* – for a unit change in x_2 while holding x_1 constant. Equivalently, $\exp(\beta_2)$ represents the multiplicative change in the odds ratio associated with a unit change in x_2 when x_1 is held constant.

One can also use the data to chose the link function empirically, by applying a series of candidate links and assessing the model adequacy of each. Select that link associated with the most favorable model fit. Useful in this regard is the scaled deviance function, $D^*(\mathbf{b})$ from §3.2.2. Recall that the scaled deviance measures the departure of a fitted model from the fullest possible model, so a smaller scaled deviance can indicate a better fit among candidate links. This assumes that each candidate link is fitted to the data using the same set of predictor variables, however. If this were not the case, we could adjust the measure by dividing by the degrees of freedom, $\mathrm{df} = n - p - 1$. That is, use $D^*(\mathbf{b})/\mathrm{df}$ instead of just $D^*(\mathbf{b})$. Indeed, since we know from the χ^2 approximation for $D^*(\mathbf{b})$ that $E[D^*(\mathbf{b})] \approx \mathrm{df}$, we expect that, for a reasonable model fit, $E[D^*(\mathbf{b})/\mathrm{df}] \approx 1$. When $D^*(\mathbf{b})/\mathrm{df}$ is much larger than 1, however, potential model inadequacy is indicated: a popular rule of thumb for this (under any GLiM) is

$$\frac{D^*(\mathbf{b})}{\mathrm{df}} > 1 + \frac{2.8}{\sqrt{\mathrm{df}}} \tag{3.10}$$

(McCullagh and Nelder, 1989, §4.4.3).

Example 3.5 (Groundwater contamination) In the late 1980s, the US Geological Survey (USGS) conducted a water quality study of land on Long Island, New York (Eckhardt *et al.*, 1989). As part of the study, results were presented on contamination of the groundwater by the industrial solvent trichloroethylene (TCE). We have $Y =$ number of wells where TCE was detected, out of N sampled wells in total. Also recorded for use as potential predictor variables were the qualitative variates $x_{i1} =$ land use (with 10 different categories, described below), and $x_{i2} =$ whether or not sewers were used in the area around the well, along with the quantitative variates $x_{i3} =$ median concentration (mg/L) of nitrate at the N well sites, and $x_{i4} =$ median concentration (mg/L) of chloride at the N well sites. Interest existed in identifying if and how these predictor variables affect TCE contamination. The data appear in Table 3.2.

Specification of an appropriate link function is unclear here, so we consider application of the three links mentioned above: the logit, probit, and complementary log-log (CLL). PROC GENMOD can fit all three of these links, so we apply it using the sample code in Fig. 3.4. Three critical features are evident in the code. First, special syntax must be used to inform PROC GENMOD which variable is Y_i and

Table 3.2 TCE groundwater contamination and potential predictor variables for Long Island, NY

$Y = \#$ wells contaminated	$N = \#$ wells tested	$x_1 =$ land-use category	$x_2 =$ sewer use	$x_3 =$ nitrate conc. (mg/L)	$x_4 =$ chloride conc. (mg/L)
0	17	Undeveloped	Yes	1.0	10
0	59	Undeveloped	No	0.8	17
0	7	Agriculture	Yes	0.9	9
2	48	Agriculture	No	7.0	20
0	5	Residential-L	Yes	1.6	6
2	21	Residential-L	No	3.5	14
12	43	Residential-M	Yes	3.9	18
5	86	Residential-M	No	2.6	15
33	76	Residential-H	Yes	5.2	33
1	17	Residential-H	No	4.2	18
3	26	Recreation	Yes	2.3	16
4	38	Recreation	No	1.7	11
8	32	Institution	Yes	3.3	36
0	22	Institution	No	0.9	11
7	29	Transportation	Yes	4.0	66
1	30	Transportation	No	1.3	12
20	42	Commerce	Yes	4.3	24
0	7	Commerce	No	1.1	30
17	33	Industrial	Yes	3.7	23
3	12	Industrial	No	2.2	19

Residential abbreviations: L = low-density (<2 dwellings/acre), M = medium-density (2–4 dwellings/acre), H = high-density (>4 dwellings/acre). Source: Eckhardt et al. (1989).

which is N_i in the proportion response. This is done in the model statement via the generic syntax

```
model Y/N= [predictor_variables_here] / dist=binomial;
```

Users should define each variable (Y_i and N_i) separately in an existing or new SAS data set prior to invoking PROC GENMOD. (Do *not* calculate the observed proportions $p_i = Y_i/N_i$ for use in the PROC. The first virgule (/) in the model statement does not stand for a division operation, nor does it imply that SAS programming options will follow. It is used syntactically to separate Y_i from N_i in the specification of the proportion response.)

As an aside, note that these binary models may be employed in problems where each subject possesses a unique profile of covariates. In this example, N wells possess a particular configuration of covariates $\{x_1, x_2, x_3, x_4\}$, called a *covariate class* (McCullagh and Nelder, 1989, §4.1.2). The data are essentially grouped binary observations with groups defined by the covariate classes. In cases where no covariate classes exist (so $N = 1$), the GLiM specification in PROC GENMOD simplifies to

```
model Y = [predictor_variables_here] / dist=binomial;
```

Next in Fig. 3.4 we define for each model a complete specification of the linear predictor, that is, we fit all four predictor variables to assess the quality of each potential link function. Notice also that we define x_1 (land use) and x_2 (sewer) as classification variates, using the `class` statement.

Third, notice in Fig. 3.4 that each different link is specified via the `link=` option in the `model` statement. We also include the `expected` option to ensure use of expected Fisher information when computing standard errors and estimated covariances. Since we wish to fit three different links (each under the same `dist=binomial` specification, and using the same linear predictor), we invoke PROC GENMOD three separate times. The corresponding output (edited) appears in Fig. 3.5.

We see from the output that the ratio of scaled deviance, $D^*(\mathbf{b})$, to df listed under `Value/DF` appears to be smallest for the logit link, although the differences are small, especially between the logit and probit. Nonetheless, if no other pertinent information were available to suggest the choice of link function, use of the logit would seem reasonable with these data, especially in light of the interpretability of the logistic regression parameters. Note also that under our informal rule of thumb (3.10) for

```
*  SAS code to fit logit, probit and CLL links;
   data grndwtr;
   input landuse $ sewer $ nitrate chloride ytce N @@;
   datalines;
   Unde yes  1.0 10 00 17        Unde no   0.8 17 00 59
    ⋮    ⋮    ⋮   ⋮  ⋮  ⋮          ⋮    ⋮    ⋮   ⋮  ⋮  ⋮
   Indu yes  3.7 23 17 33        Indu no   2.2 19 03 12

*  fit logit link via PROC GENMOD;
   proc genmod;
   title "*******************logit link*******************";
   class landuse sewer;
   model ytce/N = landuse sewer nitrate chloride
                  / dist=binomial link=logit expected;

*  fit probit link via PROC GENMOD;
   proc genmod;
   title "*******************probit link*******************";
   class landuse sewer;
   model ytce/N = landuse sewer nitrate chloride
                  / dist=binomial link=probit expected;

*  fit CLL link via PROC GENMOD;
   proc genmod;
   title "*******************CLL link*******************";
   class landuse sewer;
   model ytce/N = landuse sewer nitrate chloride
                  / dist=binomial link=cll expected;
```

Figure 3.4　Sample SAS program to fit logit, probit, and CLL links to groundwater contamination data

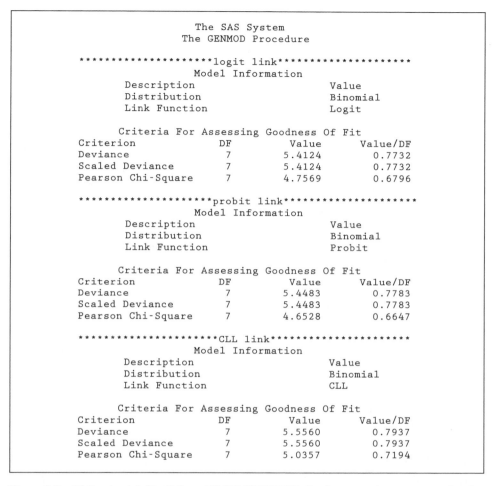

```
                        The SAS System
                      The GENMOD Procedure

        *********************logit link********************
                      Model Information
            Description                    Value
            Distribution                   Binomial
            Link Function                  Logit

            Criteria For Assessing Goodness Of Fit
        Criterion            DF        Value       Value/DF
        Deviance              7       5.4124         0.7732
        Scaled Deviance       7       5.4124         0.7732
        Pearson Chi-Square    7       4.7569         0.6796

        *********************probit link*******************
                      Model Information
            Description                    Value
            Distribution                   Binomial
            Link Function                  Probit

            Criteria For Assessing Goodness Of Fit
        Criterion            DF        Value       Value/DF
        Deviance              7       5.4483         0.7783
        Scaled Deviance       7       5.4483         0.7783
        Pearson Chi-Square    7       4.6528         0.6647

        ********************CLL link***********************
                      Model Information
            Description                    Value
            Distribution                   Binomial
            Link Function                  CLL

            Criteria For Assessing Goodness Of Fit
        Criterion            DF        Value       Value/DF
        Deviance              7       5.5560         0.7937
        Scaled Deviance       7       5.5560         0.7937
        Pearson Chi-Square    7       5.0357         0.7194
```

Figure 3.5 SAS output (edited) from PROC GENMOD fits for groundwater contamination data

model inadequacy, we would question the model fit here if $D^*(\mathbf{b})/\mathrm{df} > 2.058$. Since this is not the case with the logit – or, for that matter, with any of these three link functions – we operate under the supposition that the model fit is adequate.

Under a logit link, a more complete analysis would include an analysis of deviance, as in Table 3.1. To achieve this, we update the model statement in Fig. 3.4 to include a type1 option:

```
proc genmod; class  landuse   sewer;
 model ytce/n=landuse   sewer   nitrate   chloride
   /dist=binomial  link=logit  type1  expected;
```

Figure 3.6 presents the associated output (edited).

```
                        The SAS System
                        The GENMOD Procedure

                       Model Information
              Description                        Value
              Distribution                       Binomial
              Link Function                      Logit
              Dependent Variable                 ytce
              Dependent Variable                 N

                     Class Level Information
            Class      Levels   Values
            LANDUSE        10    Agrc Comm HigR Indu Inst LowR
                                 MedR Recr Tran Unde
            SEWER           2    no yes

                 Criteria For Assessing Goodness Of Fit
            Criterion             DF          Value       Value/DF
            Deviance               7         5.4124        0.7732
            Scaled Deviance        7         5.4124        0.7732
            Pearson Chi-Square     7         4.7569        0.6796
            Scaled Pearson X2      7         4.7569        0.6796
        Algorithm converged.

                  LR Statistics For Type 1 Analysis
            Source          Deviance      DF   Chi-Square   Pr > ChiSq
            Intercept       146.9559
            landuse          49.5025       9       97.45      <.0001
            sewer            15.2006       1       34.30      <.0001
            nitrate           5.4493       1        9.75      0.0018
            chloride          5.4124       1        0.04      0.8476
```

Figure 3.6   SAS output (edited) from logistic regression analysis of groundwater contamination data

To study the impact of the predictor variables on TCE contamination, we construct an analysis of deviance table using the results under LR Statistics For Type 1 Analysis. The PROC GENMOD output is almost identical to the schematic in Table 3.1, facilitating the effort. Table 3.3 displays the resulting calculations. We see that in the full model the first term studied (from the bottom up) is the chloride predictor. We find for it an insignificant result: the difference in scaled deviances between a full model containing the chloride term along with land use, sewer use and nitrate, and one without chloride is $\Delta D^* = 5.4493 - 5.4124 = 0.0369$. This is the LR statistic for testing $H_0: \beta_4 = 0$ vs. $H_a: \beta_4 \neq 0$ given that land use, sewer use, and nitrate are already included in the model. The approximate P-value is derived from a $\chi^2(1)$ reference distribution: $P \approx P[\chi^2(1) \geq 0.0369] = 0.8476$. At any reasonable α level this is insignificant, and we can move 'up' the table to assess the nitrate term (technically, testing $H_0: \beta_3 = 0$ vs. $H_a: \beta_3 \neq 0$ given that land use and sewer use are already included in the model). We find the LR statistic to be $\Delta D^* = 15.2006 - 5.4493 = 9.7513$, with $P \approx P[\chi^2(1) \geq 9.7513] = 0.0018$; clearly significant. If we were to remove chloride from the model and manipulate the order in which we fitted the remaining three predictor variables, we would find that after fitting the land use and nitrate terms, the categorical predictor for sewer use becomes insignificant ($P = 0.8933$; Exercise 3.6 studies this and other aspects of the logistic model

Table 3.3 Analysis of deviance table for groundwater contamination data

Source	$D^*(\cdot)$	Model df	Δdf	$\Delta D^* = G^2_{\text{calc}}$	P-value
Intercept	146.9559	19	—	—	—
Land use	49.5025	10	9	97.4534	<0.0001
Sewer use	15.2006	9	1	34.3019	<0.0001
Nitrate	5.4493	8	1	9.7513	0.0018
Chloride	5.4124	7	1	0.0369	0.8476

fit in more detail). Thus it appears that only land use and nitrite concentration significantly affect TCE contamination on Long Island. ✪

Besides **PROC GENMOD**, the logit link can be fitted directly using **PROC LOGISTIC**, or via **PROC PROBIT** with the `d = logistic` option in its `model` statement. Similarly, the probit link can also be fitted directly via **PROC PROBIT**, or via **PROC LOGISTIC** using the `link = normit` option in its `model` statement. The different procedures offer a variety of different options, and analysts may wish to study these options when deciding which procedure is most appropriate. This strategy obviously generalizes across statistical packages as well as across procedures within a package.

In some settings, only a single predictor variable, x_i, is under study, and the fundamental inference of interest can be distilled down to a single question: do increases in x_i lead to significant increases in the response probability, π_i, when the latter is viewed as function of x_i? This is often called the *quantal response problem*, and is a special case of the larger area of *stimulus-response/dose-response modeling*. (We touch upon dose-response issues when we discuss quantitative risk assessment in Chapter 4.) For the quantal response case, we write the basic, no-trend, null hypothesis as $H_0: \pi_1 = \cdots = \pi_n$. The goal is then to detect departures from H_0. A test procedure that addresses this is known as the *Cochran–Armitage (CA) trend test* (for proportions); its test statistic can be written as

$$Z_{\text{CA}} = \frac{\sum_{i=1}^{n}(x_i - \bar{x})Y_i}{\left\{\bar{p}(1-\bar{p})\sum_{i=1}^{n}N_i(x_i - \bar{x})^2\right\}^{1/2}}, \tag{3.11}$$

where $\bar{p} = \sum_{i=1}^{n} Y_i / \sum_{i=1}^{n} N_i$ is a pooled estimator of the common π under $H_0, \bar{x} = \sum_{i=1}^{n} N_i x_i / \sum_{i=1}^{n} N_i$, and it is assumed that the predictor variable has been ordered such that $x_1 < x_2 < \ldots < x_n$. In large samples the CA statistic is approximately standard normal. (The approximation is typically valid when every N_i is at least 10 and $\sum_{i=1}^{n} N_i \geq 50$.) One rejects H_0 in favor of an increasing trend when $Z_{\text{CA}} \geq z_\alpha$. The corresponding, approximate, one-sided P-value is $P \approx 1 - \Phi(Z_{\text{CA}})$. (To test against a decreasing trend, reject if $Z_{\text{CA}} \leq -z_\alpha$. The corresponding approximate P-value is $P \approx \Phi(Z_{\text{CA}})$. To test against any (two-sided) departure from H_0, reject if $|Z_{\text{CA}}| \geq z_{\alpha/2}$.)

The test based on (3.11) was developed by Cochran (1954) and Armitage (1955). It also was given in a more general form by Yates (1948). We introduce it here because the approach possesses a number of favorable properties. For instance, Z_{CA} is invariant to linear transformations of x_i, so that a change from x_i to $A + Bx_i$ does not change the value of (3.11); see Exercise 3.8. Furthermore, it has a form of omnibus optimality: it is locally most powerful against any form of twice-differentiable, monotone function for $\pi(x_i)$ (Tarone and Gart, 1980). With this, the test can be applied with some certainty whenever it is felt that π_i may be related to x_i via a smoothly increasing (or decreasing) inverse link.

The CA trend test is also optimally related to the logistic regression model. Suppose the logit link is valid, so that $\pi(x_i) = 1/(1 + \exp\{-\beta_0 - \beta_1 x_i\})$. To test for an increasing effect due to x_i we test $H_0: \beta_1 = 0$ vs. $H_a: \beta_1 > 0$. In this setting, the CA trend statistic is uniformly most powerful (§A.5.3) for testing H_0 against H_a. The same is true if we test $H_0: \beta_1 = 0$ vs. $H_a: \beta_1 < 0$ (Cox, 1958).

We should make one caveat: computation of (3.11) assumes that the x_is are reasonably symmetric about $\bar{x}$. When this is not the case, and if a simple transformation does not achieve symmetry, a skewness correction is advocated (Tarone, 1986): take as a measure of skewness in Z_{CA} the quantity

$$\hat{\tau} = \frac{(1 - 2\bar{p})\sqrt{N_+ - 1}}{\sqrt{\bar{p}(1 - \bar{p})(N_+ - 2)}} \frac{m_3}{m_2^{3/2}}, \tag{3.12}$$

where $N_+ = \sum_{i=1}^n N_i$ and $m_k = \sum_{i=1}^n N_i(x_i - \bar{x})^k / N_+$ ($k = 2, 3$). If (3.12) deviates too far from 0.0, reject H_0 in favor of an increasing trend when $Z_{CA} > z_\alpha + \hat{\tau}(z_\alpha^2 - 1)/6$.

Example 3.6 (Ozone exceedance)   A major health concern in many developed nations is the degradation of air quality by pollutants and other particulates. For example, while ozone (O_3) is an important constituent of the Earth's upper atmosphere, when present at ground level it is a major component of summertime smog and can become a health hazard (Cocchi and Trivisano, 2002).

Public health officials typically base ozone air quality standards on the number of exceedances of some measure of O_3 severity. For example, Huang and Smith (1999) discussed an air quality study of surface ozone air quality near Chicago, from which one can study whether increases in average surface O_3 concentrations are associated with increases in exceedances over an air quality standard. The standard at the time was that daily maximum surface O_3 should not exceed 120 parts per billion (ppb). Thus we have $Y =$ number of exceedances out of N observations taken against $x =$ average ozone concentration at each reporting location. The data appear in Table 3.4.

To address the issue of whether increasing average O_3 concentrations trend with higher exceedance rates, we apply the CA trend statistic from (3.11). The O_3 concentrations occur with weighted mean $\bar{x} = 71.9728$. There is minimal deviation from symmetry in the design ($\hat{\tau} = 0.0771$), but for illustration's sake we consider Tarone's skewness adjustment. The data show $\bar{p} = \frac{144}{2354} = 0.0612$. We find $\sum_{i=1}^n (x_i - \bar{x}) Y_i = 5352.1714$ and $\sum_{i=1}^n N_i(x_i - \bar{x})^2 = 1\,063\,342.4490$. Thus, from (3.11),

$$Z_{CA} = \frac{5352.1714}{\sqrt{0.0612 \times 0.9388 \times 1\,063\,342.4490}} = 21.6537.$$

Table 3.4 Exceedance rates over a threshold of 120 ppb in surface ozone near Chicago, as a function of average surface ozone concentration (ppb)

$x =$ avg. O_3 conc.	$Y =$ number of exceedances	$N =$ number of observations
39.31	0	184
49.42	2	230
56.89	0	315
64.20	2	285
65.10	2	191
74.67	8	292
80.46	11	309
85.05	3	95
87.62	7	110
99.43	27	148
106.30	15	62
114.00	28	77
125.40	18	33
143.60	16	18
196.40	5	5

Source: Huang and Smith (1999).

We reject H_0 at, say, $\alpha = 0.01$ when Z_{CA} exceeds $z_{0.01} + 0.0771(z_{0.01}^2 - 1)/6 = 2.326 + 0.0567 = 2.383$. The CA statistic clearly exceeds this cut-off point, so that a significantly increasing trend in exceedance is evidenced.

The CA statistic can be calculated directly in SAS using PROC FREQ with the `trend` option. This is illustrated in Fig. 3.7 (output suppressed). Note that the data must first be expanded so that the number of cases with exceedances and the number of cases with no exceedances are in separate data records. In the figure, this is performed in the data step defining the SAS data set `expand`. ✪

Due to its many favorable properties, the CA trend test is the method of choice for identifying an increasing (or decreasing) trend in a set of quantal response data. It is particularly important in one special case: logistic regression with a single quantitative predictor variable, $\pi(x_i) = 1/(1 + \exp\{-\beta_0 - \beta_1 x_i\})$. Under this model, the Wald test of $H_0: \beta_1 = 0$ against either a one- or two-sided alternative hypothesis suffers from a critical instability. The test statistic $W_{calc} = b_1/se[b_1]$ can decrease to zero as the true value of β_1 grows far from zero, that is, far away from H_0 (Hauck and Donner, 1977). As a result, the power of the Wald test can decrease with extreme departures from H_0, which is precisely what one wants *not* to happen. Væth (1985) gives general conditions for the Wald test to exhibit this poor performance. The CA trend statistic does not suffer from this instability (in fact, neither does the LR/ deviance statistic G_{calc}^2, although using G_{calc}^2 to test against one-sided departures is problematic). Hence, we recommend use of Z_{CA} over Wald statistics in simple linear logistic regression with large samples.

```
*  SAS code to perform C-A trend test;
data ozone;
 input avgO3conc Y N @@;
 datalines;
  39.31   0 184    49.42  2  230  ...  125.40  18  33
 143.60  16  18   196.40  5   5

data expand;
set ozone;
  exceed = "yes";  count = Y;    output;
  exceed = "no";   count = N-Y;  output;

proc freq;
  table avgO3conc*exceed / nocol nopct trend ;
  weight count;
```

Figure 3.7 Sample SAS program to fit Cochran–Armitage trend test to ozone exceedance data

All of the methods mentioned above will work best when the sample sizes grow large. (Typical recommendations are similar to those for use with the CA trend statistic: $N_i \geq 10$ for each proportion, and the total sample size should be upwards of 30. Also, a common suggestion includes limits for the actual responses, such as $Y_i \geq 5$ and $N_i - Y_i \geq 5$ for each i.) In smaller samples, the logistic model may still be favored, but the large-sample approximations can vary in their accuracy. One way to overcome this is to use exact estimation and testing methods for binomial data. Recall that the binomial p.m.f. is defined over a finite, discrete support space. Thus, given enough computing time, one could in principle enumerate all possible outcomes for any design under study, and assess how likely the outcome(s) actually observed are relative to all the others. Point estimates, test statistics, and P-values generated under such a schema are called *exact*, since no large-sample approximation is being used to derive the inferences (Mehta and Patel, 1995). For example, an exact version of the CA trend test is available by adding the statement

```
exact trend;
```

to the PROC FREQ sample code in Fig. 3.7. When the sample sizes are very large, we recommend also using the `maxtime=` option (in SAS ver. 8 or later).

The key issue for exact estimation and testing is access to adequate computing resources. In theory, performing an *exact logistic regression* is relatively straightforward, although a clever application of statistical conditioning helps make the operation practicable. By way of background, we can *condition* the probability of an event, $\mathcal{B}$, on any other pertinent random variable, say, W, in order to take into account the information in W. We write the probability that $\mathcal{B}$ occurs, *conditional* on the event $\{W = w\}$ as $P[\mathcal{B}|W = w]$, where the symbol '|' indicates the conditioning operator. Read it as 'given that' or 'conditional on'. (We encountered the conditioning notation in §A.5.3 when reviewing Type I error rates. It can also be extended to expected values and variances; for example, the conditional variance of Y given that $W = w$ is $\text{Var}[Y|W = w] = \text{E}[Y^2|W = w] - \text{E}^2[Y|W = w]$.) Details on how to perform conditional probability operations are given in texts on probability and statistics, such as Hogg and Tanis (2001, §2.3) or Casella and Berger (2002, §1.3).

When performing an exact logistic regression, it is advantageous to condition the underlying binomial probability statements on summarizing quantities – called 'sufficient statistics' – that remove the effects of any nuisance parameters. For example, when inferences concern the regression coefficient β_1 associated with an exposure variable x_{i1}, the logit intercept parameter β_0 often represents a nuisance quantity. Here, conditioning on terms associated with β_0 removes its impact on the exact probability calculations. A computing algorithm developed by Hirji *et al.* (1987) aids in the application of this *exact conditional logistic regression* approach. Once this is applied, conditional maximum likelihood estimation can proceed for any of the regression parameters (singly or in groups), associated inferences such as exact probability tests or $1 - \alpha$ confidence intervals/regions may be constructed, etc. In SAS ver. 8 or above, exact conditional logistic regression is available in PROC LOGISTIC, via its `exact` statement (Derr, 2000). For more on conditioning with sufficient statistics and use of conditional likelihoods in GLiMs, see Cox and Snell (1989) and McCullagh and Nelder (1989, Ch. 7).

3.3.3 Overdispersion: extra-binomial variability

A critical assumption with any of these methods for proportion data is that the underlying stochastic nature of the response is represented by a binomial p.m.f. This is typically true for most proportions or binary outcomes. Note, however, that the binomial model does impose one critical restriction: the variance of the proportions must take the form $\text{Var}[Y_i/N_i] = \pi_i(1 - \pi_i)/N_i$. If the environmental data under study are subject to some form of excess variation that cannot be accounted for as part of this variance assumption, the binomial model breaks down. We call this *extra-binomial variability*, or, more generally, *overdispersion*.

In the presence of extra-binomial variability, all of the methods described above can misrepresent the significance of a predictor variable. In effect, they underestimate the amount of variation in the data, and will reject true null hypotheses of no effect much more often than $100\alpha\%$ of the time or will produce confidence intervals whose true coverage is below $1 - \alpha$. When a statistical test rejects more frequently under the null hypothesis than nominally specified, the test is called 'anti-conservative' or 'radical.' The effect is considered undesirable.

To determine whether a set of proportions exhibits overdispersion, we require that the data consist of replicate proportions, say Y_{ij}/N_{ij} ($i = 1, \ldots, g$ groups; $j = 1, \ldots, m_i$ independent proportions per group). We test for excess variability in the proportions at the ith group level via a score statistic (§A.5.3) of the form

$$T_i^2 = \left(\sum_{j=1}^{m_i} \frac{(Y_{ij} - N_{ij}\hat{p}_i)^2}{\hat{p}_i(1 - \hat{p}_i)} - \sum_{j=1}^{m_i} N_{ij} \right)^2 \bigg/ \left(2 \sum_{j=1}^{m_i} N_{ij}(N_{ij} - 1) \right) \tag{3.13}$$

(Tarone, 1979), where $\hat{p}_i = \sum_{j=1}^{m_i} Y_{ij} / \sum_{j=1}^{m_i} N_{ij}$. In large samples, $T_i^2 \stackrel{\cdot}{\sim} \chi^2(1)$, so to test for significant extra-binomial variability, we calculate (3.13) and reject the null hypothesis of no overdispersion at each exposure level if $T_i^2 \geq \chi_\alpha^2(1)$. The associated P-value is, approximately, $P[\chi^2(1) \geq T_i^2]$. To assess excess variability across all i, we aggregate the individual T_i^2 statistics into $T^2 = \sum_{i=1}^g T_i^2$. Since a sum of independent χ^2 variables is itself χ^2, we have that $T^2 \dot\sim \chi^2(g)$. Thus we reject the global null hypothesis of no extra-binomial variability in favor of overdispersion at some level when $T^2 \geq \chi_\alpha^2(g)$. Extensions to settings where concomitant predictor variables are recorded with each observation are discussed by Dean (1992).

Example 3.7 (Litter effects) An important example of extra-binomial variability in environmental testing occurs when gestating animals are studied to determine if a toxic exposure affects their developing embryos or fetuses. This is an example of a *bioassay experiment*, where the biological activity of an environmental agent is studied for its effect(s) on an experimental organism (Coull and Ryan, 2002). For example, in environmental teratogenicity studies the proportion of offspring in an animal's litter that exhibit some congenital malformation is recorded after parental exposure to a hazardous environmental agent (Regan, 2002). A complication here is that embryos or fetuses are sampled from an individual female (called a 'dam'), and thus represent multiple observations from a single experimental unit. It is likely that the resulting per-fetus responses will be correlated. This intralitter correlation adds excess variability to the per-fetus observations, leading to what is known as a *litter effect* (Kupper, 2002).

Statistically, intralitter correlations generate variability in the observed proportions in excess of that assumed under the simple binomial model. Whenever presented with a set of per-litter proportions – or any set of replicate proportions – that may suffer from extra-binomial variability, we recommend applying the score statistic from (3.13) to assess if overdispersion is present.

Consider, for example, the data in Table 3.5 (plotted in Fig. 3.8), which are per-litter proportions of rodent fetuses dying *in utero* after maternal exposure to the industrial solvent and plasticizer di(2-ethyhexyl)phthalate, or DEHP (Ryan *et al.*, 1991). DEHP is often found as a contaminant in drinking water and hence its potential environmental toxicity is of interest.

To assess if the replicate proportions in Table 3.5 exhibit extra-binomial variability, we apply the score statistic from (3.13). The calculations are straightforward, and we illustrate them only on the control group ($i = 1$) in Table 3.5. We have $m_1 = 30$, and we find $\hat{p}_1 = \sum_{j=1}^{30} Y_{1j} / \sum_{j=1}^{30} N_{1j} = 65/395 = 0.1646$. For the numerator of T_1^2 we take $\sum_{j=1}^{30}\{[Y_{ij} - N_{ij}(0.1646)]^2/(0.1646 \times 0.8354)\} - \sum_{j=1}^{30} N_{1j} = 2055.8118 - 395.0 = 1660.8118$, square it, and divide by $2\sum_{j=1}^{30} N_{1j}(N_{1j} - 1) = 8690.0$. This yields $T_1^2 = 317.41$. Referred to a $\chi^2(1)$ distribution, this is highly significant ($P < 0.0001$). Hence we conclude that significant extra-binomial variability exists at the control level.

Similar calculations yield dispersion statistics (and P-values) for the other dose levels of $T_2^2 = 0.444$ ($P \approx 0.505$), $T_3^2 = 0.075$ ($P \approx 0.784$), $T_4^2 = 410.16$ ($P < 0.0001$), and $T_5^2 = 913.47$ ($P < 0.0001$). Extra-binomial variability also appears evident at the upper two dose levels. Aggregated, we find $T^2 = 1641.56$, which when compared to a $\chi^2(5)$ reference distribution also shows significant overdispersion ($P < 0.0001$).

Table 3.5 Per-litter data from a developmental toxicity study of DEHP as a function of exposure dose (mg/kg/day)

Dose: 0 (i = 1)		Dose: 44 (i = 2)		Dose: 91 (i = 3)		Dose: 191 (i = 4)		Dose: 292 (i = 5)	
dead fetuses	implants	dead fetuses	implants	dead fetuses	implants	dead fetuses	implants	dead fetuses	implants
3	12	1	11	4	11	19	19	11	11
1	16	1	15	0	12	1	1	4	13
3	11	2	14	2	8	2	14	9	14
10	16	1	14	2	15	6	14	11	11
2	9	1	14	2	14	4	4	13	17
5	14	3	13	3	14	16	18	7	7
0	16	1	13	1	15	7	12	7	11
1	13	3	8	3	14	9	13	10	12
0	16	0	13	2	14	9	15	13	13
1	14	1	16	0	11	3	15	10	16
11	18	2	9	1	11	10	10	14	14
0	13	1	12	1	11	1	11	11	11
0	15	1	13	3	14	12	12	8	13
1	10	1	13	2	12	2	14	12	12
0	13	0	11	3	14	2	2	13	13
0	13	1	13	2	15	10	12	10	10
0	12	1	15	0	14	2	10	12	12
4	12	2	13	2	13	3	11	13	13
2	14	1	12	1	13	2	11	4	13
1	17	1	15	0	12	6	6	10	10
0	12	0	13	2	16	5	14	13	13
1	13	1	12	2	11	6	14	12	12
0	13	0	9	0	8	2	11	6	12
0	11	4	15	0	12	0	13	12	12
0	14	1	4	3	6			9	9
0	12	1	10	1	9				
5	8								
1	10								
4	12								
9	16								

Source: Ryan *et al.* (1991).

Based on this analysis, we conclude that it would be inappropriate to analyze these proportions using a simple binomial likelihood. ✿

In the presence of overdispersion, analyses based on the binomial likelihood will operate in an unstable fashion. Alternative methods for analyzing overdispersed proportions are appropriate in this case. Many such methods exist, and space limits us from detailing them all. Interested readers should consult Pendergast *et al.* (1996), Hinde and Demetrio (1998), Chen (2002), and the references therein. Of the more

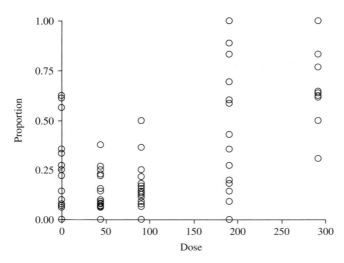

Figure 3.8 Scatterplot for litter effects data

stable methods, we favor two in particular: one for performing simple trend tests similar to the CA test, and another based on the quasi-likelihood estimation methods from §3.2.5.

For testing trend among a set of overdispersed proportions, $Y_{ij}/N_{ij}(i = 1, \ldots, g;$ $j = 1, \ldots, m_i)$, a generalization of the CA statistic (3.11) is

$$Z_{\text{GCA}} = \frac{\sum_{i=1}^{g} \sum_{j=1}^{m_i} (x_i - \bar{x}) Y_{ij}}{\left\{ \sum_{i=1}^{g} (x_i - \bar{x})^2 \sum_{j=1}^{m_i} (Y_{ij} - N_{ij}\bar{p})^2 \right\}^{1/2}}, \tag{3.14}$$

where now $\bar{p} = \sum_{i=1}^{g} \sum_{j=1}^{m_i} Y_{ij} / \sum_{i=1}^{g} \sum_{j=1}^{m_i} N_{ij}$ is a pooled estimator of π under H_0, and $\bar{x}$ is $(\sum_{i=1}^{g} x_i \sum_{j=1}^{m_i} N_{ij}) / \sum_{i=1}^{g} \sum_{j=1}^{m_i} N_{ij}$ under $x_1 < x_2 < \ldots < x_g$. In large samples this generalized Cochran–Armitage (GCA) statistic is approximately standard normal. Thus in all operational respects, one can employ Z_{GCA} exactly as one applies (3.11).

The GCA statistic can be derived as a generalized score statistic (Zhu and Fung, 1996), and also can be viewed as a quasi-likelihood test statistic (Lefkopoulou *et al.*, 1996). The latter authors note that the quasi-likelihood approach can spawn a variety of possible GCA-type statistics. In our experience, the statistic Z_{GCA} in (3.14) exhibits very stable false positive error rates and also good power to detect true increases in trend with overdispersed proportions, at least in large samples (Piegorsch, 1993; Carr and Gorelick, 1995). There can be stability problems with the test, however, when both the individual denominators, N_{ij}, are large and the numbers of observations per group, m_i, are small. In this case, one can adjust the denominator in (3.14) to make it slightly more robust. Lefkopoulou *et al.* (1996) gave a number of suggestions for this adjustment; among those they found to operate well, one standout resulted from only a slight change to (3.14):

$$Z_{\text{GNP}} = \frac{\sum_{i=1}^{g} \sum_{j=1}^{m_i} (x_i - \bar{x}) Y_{ij}}{\left\{ \sum_{i=1}^{g} (x_i - \bar{x})^2 \sum_{j=1}^{m_i} (Y_{ij} - N_{ij}\bar{p}_i)^2 \right\}^{1/2}}, \tag{3.15}$$

where $\bar{p}_i = \sum_{j=1}^{m_i} Y_{ij} / \sum_{j=1}^{m_i} N_{ij}, i = 1, \ldots, g$. As with (3.14), Z_{GNP} is approximately standard normal, so one employs Z_{GNP} in similar fashion to (3.14).

Example 3.8 (Litter effects, cont'd)   To illustrate use of this generalized trend test, return to the litter effect data of Example 3.7. We saw there that these data were overdispersed, so use of Z_{CA} from (3.11) is contraindicated. Instead, for assessing an increased trend in fetal mortality after exposure to DEHP, either (3.14) or (3.15) should be employed. Since the m_i values are reasonably large here, we apply (3.14). Doing so (Exercise 3.13) yields $Z_{\text{GCA}} = 6.71$, with approximate one-sided *P*-value given by $P[Z > 6.71]$ for $Z \sim N(0, 1)$. From Table B.1 we see this is well below 0.0001, so a significant increase in fetal mortality is indicated.   ⊛

Although highly useful for assessing trend against a single ordered predictor variable, the GCA trend test cannot be employed to analyze more complex GLiMs – such as a multiple regression model or an ANCOVA model – with overdispersed proportions. Among the many methods that could be used in this case, one general approach stands out for its wide applicability: the generalized estimating equations from §3.2.6. Applied to overdispersed proportions or binary data, the method requires only two basic specifications: a (regression) model for $\pi_i = E[Y_{ij}/N_{ij}]$ by assigning a link function, and a variance–covariance model for $\text{Var}[Y_{ij}]$ and $\text{Corr}[Y_{ij}, Y_{ik}]$ via designation of the working correlation matrix, **R(a)**. To illustrate, we consider again the litter effects data from Table 3.5.

Example 3.9 (Litter effects, cont'd)   Continuing with the litter effects data in Table 3.5, suppose we accept that a logit model for π_i is a reasonable way to model dependence on DEHP exposure. Denote the exposure dose by the single predictor variable x_i. Thus we take $\log\{\pi_i/(1 - \pi_i)\} = \beta_0 + \beta_1 x_i, i = 1, \ldots, 5$.

The significant extra-binomial variability seen with these data suggests that a litter effect is evident. Regarding each fetal observation as a simple dichotomy (i.e., whether or not it died *in utero*), the per-fetus responses represent *correlated binary data* and may be viewed as 'repeated' measurements within each exposed or control dam. Such a setting provides an excellent opportunity for implementation of GEEs. Operating with these individual binary outcomes, we model the overdispersion through specification of the working correlation matrix, **R(a)**. Possible options in PROC GENMOD are noted in §3.2.6; here, we choose the simple 'exchangeable' correlation matrix where all correlations are assumed equal among littermates. Figure 3.9 presents sample SAS code for this fit. Notice in the procedure call that we include the `descending` option. This ensures that PROC GENMOD will model π as the probability of a binary response ('Ydead = 1') rather than that of a non-response ('Ydead = 0'). If we failed to use the `descending` option here, the signs on the GEE parameter estimates would be reversed, possibly adding confusion to interpretation of the output.

In Fig. 3.9, the binary data must be entered in a format that allows SAS to identify: (i) each subject (here, the dams); (ii) any regression variables, which can be quantitative or qualitative (here dose, x, is quantitative); and (iii) the actual repeated measurements (here, the binary indicators of whether or not each fetus died *in utero*) up to the observed maximum of 19 fetuses per litter. This maximum is specified in the do command; dots indicate missing values in Fig. 3.9. The entries are employed as coding conveniences for entering the data via a looping device; in effect, we create a data set with a row of data for each dam–dose combination.

The GEEs are invoked via the repeated statement, which follows the model statement. Notice in the model statement that standard GENMOD syntax is used to identify the type of data (dist = binomial) and the link function for π_i (link = logit), although in fact the binomial likelihood will be used only to find initial estimates for the GEE iterative fit.

The nature of the working correlation matrix is defined in the repeated statement via the type= option. In Fig. 3.9, an exchangeable correlation matrix is generated via type = exch. The covb option requests output of the large-sample variance–covariance matrix for $\mathbf{b}_{GEE}$, in both its model-based ($\mathbf{C} = \mathbf{F}^{-1}$) and empirical ($\hat{\mathbf{C}} = \mathbf{F}^{-1}\hat{\mathbf{G}}\mathbf{F}^{-1}$) forms. Recall that if the working correlation matrix is misspecified, the latter form is preferred since it remains a consistent estimator of Var[$\mathbf{b}_{GEE}$]. Large-sample correlations are also printed.

The resulting output (edited) appears in Fig. 3.10. (The output begins by summarizing the binomial-based fit, then moves on to give the GEE model information and analysis.) From the GEE analysis, the Wald statistic for testing $H_0: \beta_1 = 0$ is $W_{calc} = b_{1GEE}/se[b_{1GEE}] = 9.37$. For testing against $H_a: \beta_1 \neq 0$, the approximate P-value is given under $Pr > |Z|$ as $2P[Z \geq |9.37|] < 0.0001$ for $Z \sim N(0, 1)$. Here, however, it is more pertinent to test against a strictly increasing teratogenic effect, $H_a: \beta_1 > 0$. To do so, we can continue to use W_{calc}, and now find the one-sided approximate P-value

```
*  SAS code to fit logit link via GEEs;
   data litter;
   input dam $ dose @@;
      do i = 1 to 19;
        input Ydead @@;
        output;
      end;
   datalines;
      d101   0   1 1 1 0 0 0 0 0 0 0 0 0 . . . . . . .
      d102   0   1 0 0 0 0 0 0 0 0 0 0 0 0 0 0 0 . . .
      d103   0   1 1 1 0 0 0 0 0 0 0 0 0 . . . . . . .
       ⋮     ⋮   1 1 1 1 0
      d523 292 1 1 1 1 1 1 0 0 0 0 0 0 . . . . . . .
      d524 292 1 1 1 1 1 1 1 1 1 1 1 1 . . . . . . .
      d525 292 1 1 1 1 1 1 1 1 1 1 . . . . . . . . .

   proc genmod descending;
      class dam;
      model Ydead = dose / dist=binomial link=logit expected;
      repeated subject=dam / type=exch covb;
```

Figure 3.9　Sample SAS program to fit logit link to litter effect data via GEEs

```
                        The SAS System
                        The GENMOD Procedure

                        Model Information
            Description                      Value
            Distribution                     Binomial
            Link Function                    Logit

                    Class Level Information
        Class        Levels   Values
        DAM             131    d101 d102 d103 d104 d105...

            Criteria For Assessing Goodness Of Fit
        Criterion               DF         Value         Value/DF
        Deviance               1612      1517.5021         0.9414
        Pearson Chi-Square     1612      1747.5833         1.0841

            Analysis Of Initial Parameter Estimates
    Parameter      DF    Estimate    Std Err    Chi-Square   Pr > Chi
    INTERCEPT       1     -2.3947      0.1117       459.58     <.0001
    DOSE            1      0.0128      0.0006       394.90     <.0001
    SCALE           0      1.0000      0.0000
NOTE:   The scale parameter was held fixed.

                    GEE Model Information
            Description                      Value
            Correlation Structure            Exchangeable
            Subject Effect                   DAM (131 levels)

Covariance Matrix (Model-Based)       Covariance Matrix (Empirical)
Parameter   INTERCEPT   DOSE          Parameter   INTERCEPT DOSE
INTERCEPT   0.05340   -0.000254       INTERCEPT   0.05136   -0.000263
DOSE        -0.000254  1.7929E-6      DOSE        -0.000263  1.9823E-6
Algorithm converged.

                Analysis Of GEE Parameter Estimates
                Empirical Standard Error Estimates
                            Standard      95% Confidence
    Parameter   Estimate     Error          Limits            Z      Pr>|Z|
    Intercept   -2.4120      0.2266    -1.9679 -2.8562      -10.64    <.0001
    dose         0.0132      0.0014     0.0159  0.0104        9.37    <.0001
```

Figure 3.10   Output (edited) from logit GEE fit to litter effect data

$P[Z \geq W_{calc}] = P[Z > 9.37]$. From Table B.1 we see this is well below 0.0001, so, as with the generalized trend analysis seen above, a strongly significant increase in fetal mortality is seen to associate with increasing DEHP exposure.   ✪

3.3.4   Count data GLiMs

When discrete data are observed as unbounded counts, it is natural to consider a Poisson distribution as the parent p.m.f. for the probability model. Take Y as the number of observed outcomes whose mean rate of occurrence is μ. As seen in Exercise 3.1(a), this is a member of the exponential class in (3.1), and hence can be fitted as a GLiM. The natural parameter is $\theta = \log(\mu)$. As with binomial-based GLiMs in §3.3.2, the dispersion parameter for a Poisson GLiM is constant, taken without loss of generality to be $\varphi = 1$.

The link function most often used with Poisson data is the canonical form, the natural logarithm $g(\mu) = \log(\mu)$. Clearly, this transforms a value between 0 and ∞ to a value between $-\infty$ and ∞. The corresponding inverse link is $h(\eta) = e^{\eta}$. We often call this a (Poisson) *log-linear regression* model, since under $\eta = g(\mu) = \log(\mu)$ the log of the mean is modeled as linear. Notice here that since $\mu > 0$, any inverse link, including e^{η}, must return a value greater than 0.

Example 3.10 (Oak tree associations)   Lindsey (1997, §8.1.4) discusses data on spatial associations between different species of oak trees in a hardwood forest. Here, the association is defined simply as whether or not any tree's nearest neighbor is of the same species of oak. (We discuss other, more complicated issues of spatial analysis in Chapter 6.) That is, if species type is unrelated to nearest-neighbor type, we say that no association exists as to how oak species distribute themselves spatially in this forest. Three different oak species were studied: white oak (*Potentilla fruiticosa*), red oak (*Quercus rubra*), and black oak (*Q. velutina*). We have $Y =$ number of times a given species was a nearest neighbor to itself or another species. The counts appear in Table 3.6.

The presentation in Table 3.6 represents the predictor variables as two separate factors that contribute to the observed counts: tree species and neighbor species. To assess the association between nearest-neighbor species, the table in effect asks whether the factors act as independent contributors, or if counts identified with one factor are contingent upon the other factor. Thus this sort of display is often called an $R \times C$ *contingency table* of count data, wherein the issue is one of association vs. independence between the R levels of the row factor and the C levels of the column factor.

Many statistical approaches exist for identifying association between two factors in an $R \times C$ contingency table. These include a form of the Pearson χ^2 statistic mentioned in §3.2.3, LR statistics, and a class of extensions known as phi-divergence statistics (Cressie and Pardo, 2002). A full survey exceeds the scope of this book; more details are available in sources such as Agresti (1990, 1996) and Imrey and Simpson (2002). For our purposes it is sufficient to recognize that associations in a contingency table may be modeled via a Poisson GLiM by making use of the table's $R \times C$ factorial structure. That is, view the row and column factors as qualitative predictors such that $i = 1, \ldots, R$ indexes the row factor's levels and $j = 1, \ldots, C$

Table 3.6   Counts of nearest-neighbor oak species in a hardwood forest

Oak species	Neighbor species		
	White oak	Red oak	Black oak
White oak	138	62	14
Red oak	59	104	20
Black oak	20	12	27

Source: Lindsey (1997, §8.1.4).

indexes the column factor's levels. Assuming each observed count is $Y_{ij} \sim$ indep. Poisson (μ_{ij}), the mean μ_{ij} can be related to a linear predictor via the two-factor ANOVA-type form $\eta_{ij} = \theta + \alpha_i + \beta_j + \gamma_{ij}$, where the α_is represent the row factor main effect, the β_js represent the column factor main effect, and the γ_{ij}s represent the row × column interaction. No association between row and column factors is indicated when $\gamma_{ij} = 0$ for all combinations of i and j. If we employ a logarithmic link, we have

$$\log(\mu_{ij}) = \theta + \alpha_i + \beta_j + \gamma_{ij}. \tag{3.16}$$

An important technical subtlety associated with use of (3.16) involves the linear parameters: in certain cases, the sampling scenario may force θ, and possibly also the α_is, to be included in the model. For the standard two-factor model as we describe it above, we assume each cell in Table 3.6 may be filled by any number of outcomes, leading to the Poisson counts, Y_{ij}. In this case, the only restrictions required on the model parameters in (3.16) involve estimability constraints (below). Suppose, however, that in Table 3.6 the total number of original trees (here 456) was fixed *a priori* by the sampling design or by some resource constraint. For instance, the forester may have been restricted to a maximum of 456 trees on which nearest-neighbor determinations could be made. (This seems unlikely here, but it might occur in other environmental sampling scenarios.) Restricting the total number of observations to 456 is, in effect, a constraint on the Y_{ij}s that must be incorporated into the log-likelihood. Restricting a set of Poisson counts to add to a fixed total leads to what is known as a *multinomial distribution* for the restricted counts (Fisher, 1922a). Fortunately, use of (3.16) together with a Poisson likelihood remains valid under multinomial sampling as long as any model fitted to the data always contains θ in the linear predictor (Agresti, 1990, §12.4.2). In most cases this is a transparent requirement, since it is common to include the constant term in an ANOVA-type linear predictor.

A slightly more specific requirement obtains if the sampling design fixes an entire margin of the $R \times C$ table. This can be either the row margin or the column margin; without loss of generality, suppose it is the former. (For instance, now suppose in Table 3.6 that the number of original oak trees was fixed by species, so that it was required *a priori* that exactly 214 white oaks be studied, then 183 red oaks, then 59 black oaks. Once again, this seems unusual here, but it can occur in other settings.) This restriction again changes the unrestricted Poisson likelihood, now into what is called a *product multinomial* likelihood (Agresti, 1990, §3.1). Here, use of (3.16) together with a Poisson likelihood remains valid as long as any model fitted to the data always contains both θ and the α_is. Although these various restrictions are motivated primarily by theoretical considerations, they nonetheless represent important aspects that must be built into in the GLiM.

Notice that the number of parameters under the full, two-factor predictor in (3.16) is greater than the number of observations available to fit the model, so some sort of estimability constraint is required on the model parameters (cf. §1.3.1). Examples include the corner-point constraints, $\alpha_R = \beta_C = \gamma_{iC} = \gamma_{Rj} = 0$, or the zero-sum constraints $\sum_{i=1}^{R} \alpha_i = \sum_{j=1}^{C} \beta_j = \sum_{i=1}^{R} \gamma_{ij} = \sum_{j=1}^{C} \gamma_{ij} = 0$. The choice between these, or of any other form of identifiability constraint, is usually arbitrary. Under (3.16), these constraints reduce a set of $1 + R + C + RC$ parameters down to a set of

$1 + (R-1) + (C-1) + (R-1)(C-1) = RC$ parameters, allowing for unambiguous estimation of all pertinent parameters in the linear predictor. Since there are RC independent observations in the $R \times C$ table, however, this will still *saturate* the table as described in §3.2.2: in effect, each independent datum fits one parameter. To test for association, we fit the saturated model (3.16) and also a model under the no-association null hypothesis $H_0: \gamma_{ij} = 0$ for all i and j. The difference in scaled deviances between the two models, ΔD^*, produces the LR test statistic of H_0. Under standard GLiM asymptotics, $\Delta D^* \overset{.}{\sim} \chi^2[(R-1)(C-1)]$, so we reject H_0 in favor of some association between row and column factors if $\Delta D^* \geq \chi^2_\alpha[(R-1)(C-1)]$. Notice that under a Poisson likelihood, the scale parameter is set to $\varphi = 1$, so the scaled deviance and unadjusted deviance functions are identical.

Unfortunately, we cannot advocate the LR/deviance statistic for testing association in this setting. The statistic exhibits poor small-sample stability, and can incur false-positive errors too often. Instead, we recommend an asymptotically equivalent approach for $R \times C$ contingency tables using Pearson's χ^2 statistic. (This was, in fact, the original use for which Pearson, 1900, developed the statistic, although of course he did not couch it in terms of GLiMs.) Start with the predicted values under the reduced, H_0-restricted model, $\hat{Y}_{ij}^{(0)}$, which can be shown to be

$$\hat{Y}_{ij}^{(0)} = \frac{\left(\sum_{i=1}^{R} Y_{ij} \right)\left(\sum_{j=1}^{C} Y_{ij} \right)}{\sum_{i=1}^{R} \sum_{j=1}^{C} Y_{ij}}.$$

Next, calculate $X_{calc}^2 = \sum_{i=1}^{R} \sum_{j=1}^{C} (Y_{ij} - \hat{Y}_{ij}^{(0)})^2 / \hat{Y}_{ij}^{(0)}$. In large samples, $X_{calc}^2 \overset{.}{\sim} \chi^2[(R-1)(C-1)]$. Guidelines for use of the large-sample approximation here vary; we favor the suggestion that every expected cell frequency, $\hat{Y}_{ij}^{(0)}$, be at least 1.0, such that the overall average expected cell frequency is at least 5.0 (i.e., $\sum_{i=1}^{R} \sum_{j=1}^{C} \hat{Y}_{ij}^{(0)} / RC \geq 5$). $R \times C$ tables that do not satisfy this rule of thumb are called *sparse*; we gave some guidance on how to analyze sparse contingency tables in Piegorsch and Bailer (1997, §9.3.4). If the table is not sparse, one can reject H_0 in favor of some association between row and column factors when $X_{calc}^2 \geq \chi^2_\alpha[(R-1)(C-1)]$.

In Table 3.6, $R = C = 3$. No association between row and column factors can be interpreted as a lack of any species clustering; that is, there is no reason to expect any one oak species to be nearer to its own or any other oak species in this forest. Rejection of H_0 implies that some ecological effect causes oak trees to cluster near each other or to avoid each other. (The former alternative seems more natural. Unfortunately, departures from H_0 detected by X_{calc}^2 are two-sided, so the test cannot in and of itself identify directional departures from H_0. We comment on ways to estimate directional effects below, using residuals from the model fit.)

Figure 3.11 gives sample SAS code, in which we invoke PROC GENMOD twice – once to fit the saturated model in (3.16), and a second time to fit the reduced model under H_0 in order to recover X_{calc}^2. (PROC GENMOD prints the Pearson statistic under Criteria For Assessing Goodness Of Fit.) In both cases, we use the class statement to identify the two main factors (species and neighbor) as qualitative classification variables. In the first procedure call, notice use of the bar symbol '|' in the model statement to fit a saturated model with all

```
* SAS code to fit Poisson log-linear regr.;
data oak;
input species $ neighbor $ Ycount @@;
datalines;
whitoak whitoak 138    whitoak redoak  62    whitoak blkoak  14
redoak  whitoak  59    redoak  redoak 104    redoak  blkoak  20
blkoak  whitoak  20    blkoak  redoak  12    blkoak  blkoak  27

proc genmod;   * saturated/full model;
title "****************** Saturated Model ******************";
  class species neighbor ;
  model Ycount = species|neighbor / dist=poisson
                              link=log type1 expected;

proc genmod;   * reduced model corresponding to independence;
title "**************** No-interaction Model ****************";
  class species neighbor ;
  model Ycount = species neighbor / dist=poisson
                              link=log expected;
```

Figure 3.11   Sample SAS program to fit Poisson log-linear model to oak tree association data as a 3×3 contingency table

possible terms (including the interaction species*neighbor). In the second call, the reduced model uses only the two main effect terms.

Output (edited) from the PROC GENMOD code in Fig. 3.11 appears in Fig. 3.12. There, the first component of the output reports the saturated model fit, where, as expected, the deviance is 0.00. (Since the model is saturated, it already *is* the most complete model one can fit. Thus the deviation from it, as measured by $D(\mathbf{b})$, must be zero.) If desired, we could use the information under LR Statistics For Type 1 Analysis to construct an analysis of deviance table. For example, the LR statistic for testing the null hypothesis $H_0: \gamma_{ij} = 0$ is $\Delta D^* = 88.1719$. Compared to $\chi^2(4)$, this is highly significant ($P < 0.0001$).

As noted above, however, we do not recommend use of the LR test to assess H_0. Instead, we find via the second call to PROC GENMOD that the Pearson statistic is $X_{\text{calc}}^2 = 104.4554$; see the output under No-interaction Model in Fig. 3.12 . Also compared to $\chi^2(4)$, this is highly significant ($P < 0.0005$ from Table B.3); a significant association appears to exist among nearest-neighbor relationships with these oak species.

As we noted above, neither the LR test nor the Pearson test for association in an $R \times C$ table can indicate in which direction(s) any significant association lies. Many approaches might be considered to overcome this. For instance, recall that to find the Pearson statistic, we square and sum the individual Pearson residuals, which here would be $c_{ij} = (Y_{ij} - \hat{Y}_{ij}^{(0)})/\sqrt{\hat{Y}_{ij}^{(0)}}$. PROC GENMOD can produce these values via the obstats option in the model statement when fitting the reduced model. The resulting obstats output will contain the Pearson residuals as the variable reschi. Standardized Pearson residuals are also given as the variable streschi if the residuals option is included. Doing so here (output not shown) and displaying the standardized residuals in the same format as the original $R \times C$ table leads to the values in Table 3.7.

```
                        The SAS System
                      The GENMOD Procedure

                     Model Information
              Description                    Value
              Distribution                   POISSON
              Link Function                  LOG

                  Class Level Information
            Class      Levels  Values
            SPECIES         3  blkoak redoak whitoak
            NEIGHBOR        3  blkoak redoak whitoak

       ****************** Saturated Model ******************
             Criteria For Assessing Goodness Of Fit
       Criterion                DF        Value      Value/DF
       Deviance                  0       0.0000          .
       Scaled Deviance           0       0.0000          .
       Pearson Chi-Square        0       0.0000          .
       Scaled Pearson X2         0       0.0000          .

               LR Statistics For Type 1 Analysis
       Source            Deviance    DF   ChiSquare    Pr>ChiSq
       Intercept         290.1899
       species           187.5099     2    102.68      <.0001
       neighbor           88.1719     2     99.34      <.0001
       species*neighbor    0.0000     4     88.17      <.0001

       **************** No-interaction Model ****************
             Criteria For Assessing Goodness Of Fit
       Criterion                DF        Value      Value/DF
       Deviance                  4      88.1719      22.0430
       Scaled Deviance           4      88.1719      22.0430
       Pearson Chi-Square        4     104.4554      26.1138
       Scaled Pearson X2         4     104.4554      26.1138
```

Figure 3.12 Output (edited) from Poisson log-linear model fit to oak tree association data

Table 3.7 Standardized Pearson residuals from reduced model, log-linear GLiM fit to oak tree association data

	Neighbor species		
Oak species	White oak	Red oak	Black oak
White oak	6.7945	−4.1424	−4.0321
Red oak	−5.3727	6.3778	−1.2574
Black oak	−2.2565	−3.1550	7.8319

The pattern in Table 3.7 is revealing. Strong positive residuals are seen along the diagonal, while moderate to strong negative residuals are seen off the diagonal. (Recall that standardization corrects the Pearson residuals to have unit variance, so values well past ± 2 are considered important.) The pattern indicates that many more trees were observed on the diagonals than would be expected under H_0, while the reverse is true off the diagonals. From this, it appears that trees of similar species tend to cluster near each other, at least for the three oak species studied in this hardwood forest. ✪

Of course, many environmental studies generate Poisson counts that are not as complex as the $R \times C$ table illustrated in Example 3.10. In some cases, the problem may be as simple as assessing the effect of a single quantitative predictor, x_i, via a simple log-linear regression: $\log(\mu_i) = \beta_0 + \beta_1 x_i$. Interest here can include inferences on the log-linear slope parameter β_1 (using ML estimates and $1 - \alpha$ confidence limits), or tests of the null hypothesis $H_0: \beta_1 = 0$ vs. a two-sided or one-sided alternative. Both Wald tests and LR tests can perform adequately in this case. All these operations are possible via PROC GENMOD.

When the predictor variable is ordered, say, $x_1 < x_2 < \cdots < x_g$, and when replicate observations are observed at each level of x_i, then it is also possible to employ a CA trend test, similar to that seen in §3.3.2. Suppose $Y_{ij} \sim$ indep. Poisson$[\mu(x_i)]$; $i = 1, \ldots, g; j = 1, \ldots, m_i$, and that we wish to assess the no-trend null hypothesis $H_0: \mu(x_1) = \cdots = \mu(x_g)$. At each x_i, the unrestricted ML estimate of $\mu(x_i)$ is the ith sample mean $\overline{Y}_{i+} = \sum_{j=1}^{m_i} Y_{ij}/m_i$. If H_0 is true, then we expect the $\overline{Y}_{i+}$s to approximate the pooled mean $\overline{Y}_{++} = \sum_{i=1}^{g} \sum_{j=1}^{m_i} Y_{ij}/\sum_{i=1}^{g} m_i$. The CA test quantifies departure from this pooled value, via the statistic

$$Z_{CA} = \frac{\sum_{i=1}^{g} m_i(x_i - \overline{x})\overline{Y}_{i+}}{\sqrt{\overline{Y}_{++} \sum_{i=1}^{g} m_i(x_i - \overline{x})^2}}, \tag{3.17}$$

where $\overline{x} = \sum_{i=1}^{g} m_i x_i / \sum_{i=1}^{g} m_i$. Reject H_0 in favor of an increasing (decreasing) trend when Z_{CA} is greater (less) than or equal to z_α ($-z_\alpha$). (Two-sided testing is also possible.)

As with its proportion-based cousin, the test based on (3.17) possesses many optimal properties for testing a monotone trend under a Poisson assumption on Y_{ij}. For instance, it is locally most powerful against any form of twice-differentiable, monotone function for $\mu(x)$. In particular, if the relationship between μ and x is log-linear, that is, $\log\{\mu(x)\} = \beta_0 + \beta_1 x$, then use of (3.17) approximates the uniformly most powerful test of $H_0: \beta_1 = 0$ (Tarone, 1982). The test is preferred for trend analysis of count data when the Poisson parent assumption is correct.

Example 3.11 (Drinking water toxicity) As part of a study on toxic substances in drinking water, Sakamoto *et al.* (1996) gave data on the genotoxic potential of water from the Katsura River in Kyoto, Japan. The authors used a microbial mutagenesis assay with the bacterium *Salmonella typhimurium* to identify genetic damage after exposure to the river's water. Recorded were $Y =$ numbers of mutant colonies on a standard Petri plate as a function of $x =$ river-water exposure (in ml equivalents); $m_i = 3$ replicate observations were taken at each x_i. Higher mutant yield indicated higher toxic potential of the water. The data appear in Table 3.8; they are plotted in Fig. 3.13.

The plot in Fig. 3.13 indicates a strong increase in mutant colonies over increasing dose. To assess this statistically we turn to (3.17). For the data in Table 3.8 we find $\overline{x} = 200.0$ and $\overline{Y}_{++} = 358.25$, such that $\sum_{i=1}^{4} m_i(x_i - \overline{x})\overline{Y}_{i+} = 748\,100.0$ and

Table 3.8   Counts of *Salmonella typhimurium* mutants (as colonies per plate) after exposure to water from the Katsura River, Kyoto

x = river water (ml equivalents)	Y = mutant colonies
0	38
0	44
0	40
100	170
100	167
100	152
200	305
200	341
200	304
500	853
500	929
500	956

Source: Sakamoto *et al.* (1996).

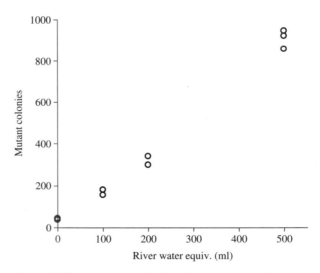

Figure 3.13   Scatterplot for drinking water toxicity data

$\sum_{i=1}^{4} m_i(x_i - \bar{x})^2 = 420\,000.0$. Thus (3.17) evaluates as $748\,100/\sqrt{358.25 \times 420\,000} = 60.9877$. Compared to a standard normal reference distribution, this is strongly significant ($P < 0.0001$ from Table B.1). Increasing exposure to the water of the Katsura River appears to significantly increase genetic damage in these organisms.

3.3.5 Overdispersion: extra-Poisson variability

An important feature of the Poisson distribution is that the observation's variance must be equal to its mean: if $Y_i \sim \text{Poisson}(\mu_i)$, then $\text{E}[Y_i] = \text{Var}[Y_i] = \mu_i$ (§A.2.4). As with the binomial distribution for proportions in §3.3.2, this is a fairly strict constraint. For environmental count data, it is not uncommon to see variation among homogeneously treated experimental units exceed that prescribed by the Poisson assumption. Such overdispersion – here we call it *extra-Poisson variability* – invalidates use of the statistical procedures we describe above. The effect is much the same as with extra-binomial variability: the methods underestimate the amount of variation in the data, and, for example, will reject a true null hypothesis more often than nominally specified.

Fortunately, we can test for the extra-Poisson effect. Similar to (3.13), suppose a set of independent, replicate counts is available in each group ($i = 1, \ldots, g$ groups; $j = 1, \ldots, m_i$ replicate counts per group). We test for excess variability in the counts in the ith group via what is known as Fisher's dispersion statistic,

$$X_i^2 = \frac{\sum_{j=1}^{m_i}(Y_{ij} - \overline{Y}_{i+})^2}{\overline{Y}_{i+}}, \tag{3.18}$$

(Fisher *et al.*, 1922), where $\overline{Y}_{i+} = \sum_{j=1}^{m_i} Y_{ij}/m_i$. In large samples, $X_i^2 \overset{\cdot}{\sim} \chi^2(m_i - 1)$, so to test for significant extra-Poisson variability, calculate (3.18) and reject the null hypothesis of no overdispersion at each exposure level if $X_i^2 \geq \chi_\alpha^2(m_i - 1)$. The associated *P*-value is, approximately, $\text{P}[\chi^2(m_i - 1) > X_i^2]$.

To assess excess variability globally, we aggregate the individual X_i^2 statistics into $X^2 = \sum_{i=1}^{g} X_i^2$. Since a sum of independent χ^2 variables is itself χ^2, we have that $X^2 \overset{\cdot}{\sim} \chi^2(m_+ - g)$, where $m_+ = \sum_{i=1}^{g} m_i$ is the total sample size. Thus we reject the global null hypothesis of no extra-Poisson variability in favor of overdispersion at some group level when $X^2 \geq \chi_\alpha^2(m_+ - g)$. Extensions to more complex regression settings are discussed by Collings and Margolin (1985) and Dean (1992).

Example 3.12 (Drinking water toxicity, cont'd) For the count data in Table 3.8, calculation of the Fisher dispersion statistic is straightforward. At each exposure level, we find that the sample means are $\overline{Y}_{1+} = 40.6667$, $\overline{Y}_{2+} = 163.0$, $\overline{Y}_{3+} = 316.6667$, and $\overline{Y}_{4+} = 912.6667$. With these, we arrive at the test statistics (and approximate *P*-values, each referenced to $\chi^2[2]$) $X_1^2 = 0.459$ ($P \approx 0.795$), $X_2^2 = 1.141$ ($P \approx 0.565$), $X_3^2 = 2.806$ ($P \approx 0.246$), and $X_4^2 = 6.250$ ($P \approx 0.044$). Only the last term is significant, and that barely so at $\alpha = 0.05$. Aggregated, we find $X^2 = 10.656$, which when referred to $\chi^2(8)$ is insignificant ($P \approx 0.222$). Thus we are reasonably well assured that no overdispersion exists with these data, and that the inferences achieved in Example 3.11 may stand as given. ✪

If a series of counts does exhibit overdispersion, the analysis must take this extra variability into account. When testing $H_0: \mu(x_1) = \cdots = \mu(x_g)$ against some monotone trend, a generalized score statistic similar to (3.17) can be constructed which remains valid in the presence of extra-Poisson variability. Recommended by Astuti and Yanagawa (2002), this is

$$Z_{GCA} = \frac{\sum_{i=1}^{g} m_i(x_i - \bar{x})\bar{Y}_{i+}}{\sqrt{\sum_{i=1}^{g}(x_i - \bar{x})^2 \sum_{j=1}^{m_i}(Y_{ij} - \bar{Y}_{++})^2}}, \tag{3.19}$$

where $\bar{x}$, $\bar{Y}_{i+}$, and $\bar{Y}_{++}$ are defined as in (3.17). In large samples, Z_{GCA} is distributed approximately as standard normal, thus we reject H_0 in favor of an increasing (decreasing) trend when Z_{GCA} is greater (less) than or equal to z_α ($-z_\alpha$). (Two-sided testing is also possible.)

Example 3.13 (Beetle species richness) From a study of how abiotic indicators affect biodiversity of animal species, Carroll (1998) gave data on the numbers of different tiger beetle species on the Indian subcontinent as a function of altitudinal relief. Altitudinal relief is the difference between the highest and lowest elevations above sea level in a study region of fixed size. Specifically, to predict Y = the number of beetle species in a $1\,km^2$ region, we let $x = \log\{relief\} + 1.75$. (The constant 1.75 is added to the log-relief values to make them all positive.) Grouped into non-overlapping categories, the species counts are given in Table 3.9.

In this study, it was thought that greater differential altitudes might give rise to greater species diversity, as measured by Y. Thus testing for increasing trend over levels of x_i was indicated. Calculation shows that the mean species numbers do appear to increase with increasing log-relief, but also that the sample variances change at a differential rate; see the last two columns of Table 3.9. This suggests possible extra-Poisson variability. Hence, we begin the analysis by checking for overdispersion in the data. Application of (3.18) to each of the four log-relief classes produces the dispersion statistics (and approximate P-values) $X_1^2 = 28.19$ ($P \approx 0.0004$), $X_2^2 = 193.11$ ($P < 0.0001$), $X_3^2 = 231.31$ ($P < 0.0001$), and $X_4^2 = 176.57$ ($P < 0.0001$). Thus significant overdispersion is evidenced at each log-relief class. (The aggregate dispersion statistic is also highly significant: $X^2 = 629.18$ on 57 df; $P < 0.0001$.) As a result, use of an adjusted trend statistic such as the generalized score in (3.19) is indicated.

To apply (3.19), we take as predictor values x_i the mid-points of the interval ranges in Table 3.9 (but, see also Exercise 3.19): $x_1 = 0.5$, $x_2 = 1.5$, $x_3 = 2.5$, and

Table 3.9 Counts of tiger beetle species in $1\,km^2$ regions on the Indian subcontinent, as a function of $x = \log\{altitudinal\ relief\} + 1.75$

Range of x	Numbers of beetle species	Mean	Variance
$0 < x < 1$	1, 14, 1, 3, 7, 6, 8, 11, 14	7.222	25.444
$1 \leq x < 2$	5, 8, 8, 8, 17, 3, 8, 8, 10, 24, 25, 3, 9, 15, 27, 29, 5, 20, 21, 3, 3, 4, 13, 20, 23, 24, 33, 39	14.821	106.004
$2 \leq x < 3$	1, 4, 1, 18, 30, 41, 1, 3, 2, 51, 48	18.182	420.564
$3 \leq x < 4$	5, 6, 25, 45, 29, 38, 6, 5, 58, 51, 3, 40, 41	27.077	398.410

Source: Carroll (1998).

$x_4 = 3.5$. This yields $\bar{x} = [(9 \times 0.5) + (28 \times 1.5) + (11 \times 2.5) + (13 \times 3.5)]/61 = 1.959$. The required values of $\bar{Y}_{i+}$ are given in Table 3.9. We also find $\bar{Y}_{++} = 16.918$. From these, we calculate $\sum_{i=1}^{4} m_i(x_i - \bar{x})\bar{Y}_{i+} = 365.295$ and $\sum_{i=1}^{4}(x_i - \bar{x})^2 \sum_{j=1}^{m_i}(Y_{ij} - \bar{Y}_{++})^2 = 13\,620.110$, so that (3.19) is $Z_{\text{GCA}} = 3.13$. Compared to a standard normal reference distribution, the one-sided P-value is $P \approx 0.0009$; a highly significant positive trend is evidenced. Thus it appears that increasing altitudinal relief is associated with significantly increasing species diversity for tiger beetles on the Indian subcontinent.

Overdispersion in count data may also be modeled directly, by incorporating a dispersion parameter into the GLiM. A number of possibilities exists for doing so. For example, one could appeal to GEEs as in Example 3.9. Or one might extend the Poisson p.m.f. via a dispersion parameter that helps account for the excess variability. That is, since PROC GENMOD can estimate an unknown scale parameter, φ, when fitting (3.1) – cf. Example 3.4 in §3.3.1 – one could call for such when fitting a GLiM to count data: specify a Poisson likelihood via the dist = poisson option, but then also include an option such as pscale to estimate φ via the method of moments. In effect, one fits a model where the variance remains proportional to μ but is no longer exactly equal to it. It is unclear how realistic such an assumption is in practice, but there may be cases where it represents a reasonable approximation to the overdispersion in the data.

One could also employ a different likelihood, such as the negative binomial (§A.2.5). Here, the mean parameter is still $E[Y_{ij}] = \mu_i$, but now the variance becomes quadratic in μ_i; $\text{Var}[Y_{ij}] = \mu_i + \delta\mu_i^2$, for some $\delta > 0$. Negative binomial GLiMs have been discussed by many authors, including Lawless (1987) and Young et al. (1999). Hilbe (1994) gives a user-defined macro for performing negative binomial regression in SAS ver. 6.08 or above; for SAS ver. 7 and above, PROC GENMOD can fit a negative binomial GLiM directly via use of the dist = nb option in its model statement.

An interesting modification for count data GLiMs occurs with environmental studies whose data are gathered over certain spatial regions or for different lengths of time. In this case, it is reasonable to expect that the mean response will be proportional to the area surveyed or the amount of time elapsed during sampling. For instance, in Example 3.13 the ecologist might have recorded the number of beetle species in regions of different areas. Let Y_i be the number of species observed in the ith region, and suppose that the expected number of species is proportional to the size of the area sampled, $\mu_i = E[Y_i] = \Omega_i\lambda_i$ where Ω_i is the area of the region and λ_i is the rate of species occurrence per unit area. As the rate of event occurrence, λ_i is the parameter of primary interest. Given a set of predictor variables, $x_1, \ldots, x_p$, one can model λ_i as a linear function of these variables, say, $\lambda_i = \beta_0 + \beta_1 x_{i1} + \cdots + \beta_p x_{ip}$. Under a log link, this becomes $\log(\mu_i) = \log(\Omega_i\lambda_i) = \log(\lambda_i) + \log(\Omega_i) = \beta_0 + \beta_1 x_{i1} + \cdots + \beta_p x_{ip} + \log(\Omega_i)$. Since the Ω_is are known in advance, the $\log(\Omega_i)$ terms are special to this model. They can be viewed as representing a predictor variable with a known coefficient, equal to one in this case. This type of term is called an *offset* in a regression model. Offsets can be incorporated in SAS by first defining the offset variable in the SAS DATA step and then referencing the offset in PROC GENMOD as an option in the model statement.

3.3.6   Continuous/constant-CV data GLiMs

A common GLiM for positive-valued continuous data where the variance is not constant involves a gamma p.d.f. The standard formulation for the gamma distribution is given in equation (A.15), with parameters $\lambda > 0$ and $\beta > 0$. As seen in Exercise 3.1(b), however, the gamma p.d.f. can be reparameterized for a GLiM by taking its mean parameter as $\mu = \lambda\beta$, and its exponential class dispersion parameter as $\varphi = 1/\lambda$. In this form, the population variance is quadratic in μ: $\mathrm{Var}[Y] = \lambda\beta^2 = \mu^2\varphi$.

When $\mathrm{Var}[Y] = \mu^2\varphi$, the corresponding coefficient of variation (CV) is $\sqrt{\mathrm{Var}[Y]}/\mathrm{E}[Y] = \sqrt{\varphi}$, which is independent of μ. That is, as μ changes, the CV remains fixed. For those situations where positive measurements are thought to have a constant CV, the gamma model is often a popular choice (McCullagh and Nelder, 1989, Ch. 8).

We should warn that some confusion exists on what to call φ under a gamma p.d.f. We call φ a dispersion parameter; in this sense, the parameter is viewed as extending the p.d.f. from some simple form to an overdispersed form. Although not often viewed quite this way for gamma models, our moniker remains valid since when $\varphi = 1$, the gamma p.d.f. reduces to the simpler exponential p.d.f. (§A.2.8). Exercise 3.21 explores this simplification in more detail.

Alternatively, many sources refer to φ as the exponential family *scale* parameter, while **PROC GENMOD** refers to $\lambda = 1/\varphi$ as the scale parameter (so analysts must proceed with caution when using information from **PROC GENMOD** outputs). This can be especially confusing, since in the general (non-GLiM) literature most authors refer to λ as the *shape parameter* of the gamma p.d.f. (Lawless, 1982, §1.3.3; Casella and Berger, 2002, §3.3) and to β (or $1/\beta$) as the scale parameter. Analysts must make sure they know precisely which form of parameter is being used when fitting gamma GLiMs to data.

Since the gamma mean must be positive, a useful link function to employ with gamma GLiMs is the natural logarithm: $g(\mu) = \log(\mu)$, with inverse link $h(\eta) = e^\eta$. This is a form of log-linear model for continuous measurements, since it relates a linear predictor to the mean response via a logarithmic link.

The canonical link for the gamma p.d.f. is the reciprocal: $g(\mu) = 1/\mu$. In order to satisfy the model constraint that $\mu > 0$, however, we must truncate η if it reaches or drops below zero. In practice, we often recommend the log link instead.

Under any appropriate link, estimation of the parameters in η proceeds via maximum likelihood, as in previous sections. To estimate the dispersion parameter, φ, a number of options present themselves. As discussed in §3.2, ML is a natural choice. The Pearson-based moment estimator is also useful, however, and use of either it or the ML estimator is generally reasonable for gamma GLiM data. (The deviance-based moment estimator is not recommended since it can exhibit instabilities when $V(\mu)$ varies with μ, as is true with the gamma model.)

Example 3.14 (Mutant frequencies)   Becker *et al.* (2001) report data on genetic damage to a population environmentally exposed to the chemical vinyl chloride through an accidental release of the material into the atmosphere. (Vinyl chloride is an odorless gas used in the manufacture of many electrical, industrial, and household products. The chemical can degrade air quality and contaminate ground-water supplies after use in manufacturing processes.) Recorded were mutant frequencies ($\times 10^6$) at the human *hprt* gene in peripheral blood lymphocytes after the accident, as a function of the subjects' exposure (exposed or unexposed), age (in years), and smoking status (yes or no). The latter two variables were included as adjustment variables: mutant frequencies (MFs) are known to increase with age, and smoking may also play a role. Of interest, however, is whether the MF response differs between exposed and unexposed groups. The data appear in Table 3.10. (Some data points were excluded from the analysis due to unusually low *in vitro* efficiencies when cloning the subjects' lymphocytes for MF determination. These do not appear in the table.)

Table 3.10 Mutant frequencies (MF) in human peripheral blood lymphocytes after exposure or no exposure to vinyl chloride

Exposed subjects				Control subjects		
Age	Smoking status	MF		Age	Smoking status	MF
57	No	5.6		57	Yes	1.9
57	No	3.9		56	No	1.4
60	Yes	1.1		57	No	2.4
29	Yes	2.1		33	Yes	0.3
50	No	2.8		55	No	1.5
53	No	1.8		23	No	1.6
34	Yes	1.9		43	No	1.9
50	No	2.5		30	Yes	1.8
53	Yes	3.7		28	Yes	2.3
21	No	1.2		39	No	6.4
25	Yes	0.3		34	No	2.9
34	No	2.9		45	No	5.2
48	No	3.4		58	No	10.4
58	No	5.7		57	No	4.4
43	No	5.1		46	No	4.2
45	No	5.2		31	Yes	0.7
54	No	1.5		20	Yes	2.3
56	No	6.6		44	Yes	6.1
34	Yes	2.5		53	No	2.7
31	No	5.8		55	No	4.4
33	Yes	2.6		60	Yes	1.4
51	Yes	3.4		51	Yes	3.4
46	No	2.0		30	Yes	5.4
46	Yes	1.7		50	No	1.5
37	No	2.8		42	Yes	1.5
44	Yes	7.2		54	No	6.4

Source: Becker *et al.* (2001).

A plot of the data distinguishing exposed from unexposed subjects (but ignoring the smoking indicator) appears in Fig. 3.14. As expected, an age-related increase in MF is evidenced. The difference between exposure groups is less clear, however. To assess the issue statistically, we turn to a GLiM. As in the figure, mutant frequencies often increase in variation along with increasing response (here, with increasing age), a candidate for a gamma GLiM. Our goal is to use the gamma GLiM to compare the exposed subjects' MF responses with the unexposed subjects' responses, while adjusting for age and also including smoking status. This has the flavor of an analysis of covariance (ANCOVA) from §1.3.2: we study the qualitative variables that classify exposure and smoking status, after adjusting for the quantitative variable represented by age. Recall that it is common in ANCOVA modeling to center the predictor variable: $x = $ age $- \overline{\text{age}}$, where $\overline{\text{age}}$ is the sample mean age across all subjects. In Table 3.10, $\overline{\text{age}} = 44.2308$ years.

Figure 3.15 presents sample PROC GENMOD code to fit such a model. In the code, notice the inclusion of: (i) the centered covariate, $x = $ age $- 44.2308$, entered as the first term in the `model` sequence to adjust the analysis for any age effects; (ii) an interaction term for smoking × exposure in the `model` statement (which, if significant, would be a critical component associated with the exposure effect); (iii) a logarithmic link specified in the `model` options; (iv) use of the `type1` option to call for an analysis of deviance; and (v) estimation of φ via the Pearson-based moment estimator, specified using the `pscale` option. The output (edited for brevity to highlight the `Type 1 Analysis`) appears in Fig. 3.16.

The output in Fig. 3.16 indicates that the smoking × exposure interaction appears to be insignificant: the calculated F-statistic is $F_{\text{calc}} = 0.01$. Compared to an F-reference distribution with (1,47) df, the approximate P-value is very near 1.0: $P \approx 0.94$. (When using a moment-based estimator for $\hat{\varphi}$, PROC GENMOD uses an F-approximation for the scaled difference in deviances; cf. §3.2.4. For completeness,

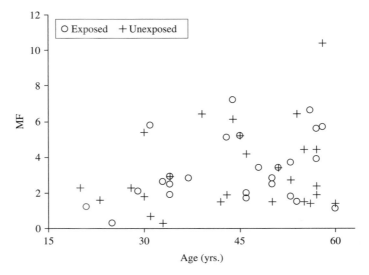

Figure 3.14   Scatterplot for mutant frequency data

```
*   SAS code to fit gamma GLiM;
    data mutantf;
    input expose $ age smoke mf @@;
    agebar = 44.2308;
    x = age - agebar;
    datalines;
        exposed 57 0   5.6              exposed 57 0   3.9
            :                              :
        control 42 1   1.5              control 54 0   6.4

    proc genmod;
    class expose smoke;
    model mf = x smoke expose smoke*expose
                / dist=gamma link=log type1 pscale expected;
```

Figure 3.15 Sample SAS program to fit gamma log-linear ANCOVA to mutant frequency data

```
                        The SAS System
                      The GENMOD Procedure

                       Model Information
              Description                      Value
              Distribution                     Gamma
              Link Function                    Log
              Dependent Variable               mf

             LR Statistics For Type 1 Analysis
                                                      Chi-
    Source        Deviance  NumDF DenDF  F Value  Pr>F  Square  Pr>ChiSq
    Intercept     22.7044
    x             20.8836     1    47    4.61  0.0370  4.61   0.0318
    smoke         20.0469     1    47    2.12  0.1523  2.12   0.1456
    expose        20.0469     1    47    0.00  0.9915  0.00   0.9915
    expose*smoke  20.0445     1    47    0.01  0.9389  0.01   0.9386
```

Figure 3.16 Sequential analysis-of-deviance results from SAS output (edited) under a gamma log-linear ANCOVA fit to the mutant frequency data

however, the unadjusted χ^2-based values are also included in the output.) Thus our first conclusion from this analysis is that, adjusted for any age effect, smoking status and exposure do not appear to interact in how they impact MF response in these subjects.

Removing the smoking $\times$ exposure interaction, we study next the exposure main effect. Since the interaction is insignificant, the analysis of deviance in Fig. 3.16 could be used for this effort by moving 'up' to the next difference in scaled deviances. Recall, however, that the deviances are scaled by a quantity estimated under a model containing the interaction. Removing the interaction term will change (slightly) the estimator of φ. Hence, we fit here a new model with interaction removed and the exposure main effect included last. The necessary SAS code (not shown) is essentially identical to that in Fig. 3.15; the only change is to remove the smoke*expose term in the model statement. The consequent output (edited) appears in Fig. 3.17. There,

```
                          The SAS System
                        The GENMOD Procedure

                         Model Information
              Description                      Value
              Distribution                     GAMMA
              Link Function                    LOG

                  Criteria For Assessing Goodness Of Fit
              Criterion            DF         Value        Value/DF
              Deviance             48        20.0469        0.4176
              Scaled Deviance      48        51.7873        1.0789
              Pearson Chi-Square   48        18.5808        0.3871
              Scaled Pearson X2    48        48.0000        1.0000

                    Analysis Of Parameter Estimates
      Parameter              DF    Estimate   Std Err  ChiSquare  Pr>Chi
      Intercept              1      0.9912    0.1673    35.10     <.0001
      X                      1      0.0130    0.0080     2.66     0.1032
      smoke      0           1      0.2800    0.1859     2.27     0.1320
      smoke      1           0      0.0000    0.0000      .         .
      expose     control     1     -0.0019    0.1727     0.00     0.9914
      expose     exposed     0      0.0000    0.0000      .         .
      Scale                  0      2.5833    0.0000
    NOTE: The Gamma scale parameter was estimated by DOF/Pearson's
          Chi-Squared

                    LR Statistics For Type 1 Analysis
                                                         Chi-
      Source       Deviance  NumDF DenDF  F Value   Pr>F Square  Pr>ChiSq
      Intercept    22.7044
      x            20.8836     1     48    4.70    0.0351  4.70   0.0301
      smoke        20.0469     1     48    2.16    0.1480  2.16   0.1415
      expose       20.0469     1     48    0.00    0.9914  0.00   0.9914
```

Figure 3.17 Output (edited) from gamma log-linear ANCOVA fit to the mutant frequency data, employing only main effects

under `Type 1 Analysis`, we see that exposure is insignificant: the calculated F-statistic is approximately 1.17×10^{-4}, which registers on the output as 0.00. Compared to an F-reference distribution with (1,48) df, the approximate P-value is again very near 1.0. Thus there does not appear to be a significant effect on these subjects' MF response due to vinyl chloride exposure. (Similar conclusions could be drawn about the smoking main effect as well; see Exercise 3.22.)

Figure 3.18 displays a residual plot using the predicted values from the fit in Fig. 3.17 and the standardized deviance residuals (obtained from PROC GENMOD by using the `obstats` and `residual` options in the `model` statement; output not shown). No substantial concerns are evidenced.

One additional aspect of the gamma GLiM worth mentioning is the possibility of making inferences on the dispersion parameter, φ. From Fig. 3.17 we see that the Pearson MOM estimate of φ is $\hat{\varphi}_P = 1/2.5833 = 0.3871$. (Recall that in PROC GENMOD the 'scale' parameter for a gamma GLiM is $\lambda = 1/\varphi$). Unfortunately, PROC GENMOD provides no capability for making inferences on φ via $\hat{\varphi}_P$. If, however, we allow the program to estimate λ and hence φ via its default procedure,

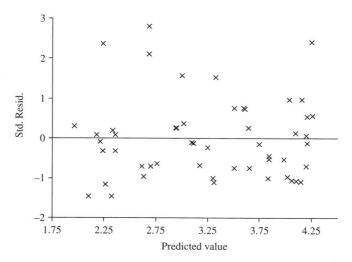

Figure 3.18 Residual plot for gamma log-linear ANCOVA fit from Fig. 3.17 for mutant frequency data

ML, then, using results from large-sample likelihood theory (§A.4.3), an approximate $1 - \alpha$ confidence interval for φ may be derived. That is, if **PROC GENMOD** can furnish two limits, say λ_L and λ_U such that $0 < \lambda_L < \lambda_U$ and $P[\lambda_L \leq \lambda \leq \lambda_U] \approx 1 - \alpha$, then by taking reciprocals the equivalent statement

$$P\left[\frac{1}{\lambda_U} \leq \varphi \leq \frac{1}{\lambda_L}\right] \approx 1 - \alpha \tag{3.20}$$

represents an approximate $1 - \alpha$ confidence interval for φ. The original interval can be of any valid form; PROC GENMOD can produce either Wald or LR-based limits (§A.5.1). (To ensure $\lambda > 0$, the Wald interval is actually constructed from $\log\{\hat{\lambda}\} \pm z_{\alpha/2} se[\log\{\hat{\lambda}\}]$, where $se[\log\{\hat{\lambda}\}]$ is found from a delta-method approximation; cf. Example A.1. Exponentiating the two limits guarantees a positive interval for λ.)

To do so here, return to the code in Fig. 3.15. Since smoking × exposure (Fig. 3.16), exposure (Fig. 3.17), and smoking (Exercise 3.22) are all insignificant, remove them from the `model` statement. (With all these terms removed, there is little reason to retain the `type1` option either.) To estimate φ using ML, remove `pscale` from the `model` options. The PROC GENMOD call becomes simply

```
proc genmod;
model mf = x / dist = gamma  link = log  expected  lrci;
```

where we have chosen the `lrci` option to call for LR intervals on the unknown parameters. The result (edited) appears in Fig. 3.19.

From Fig. 3.19, we see that the ML estimate of the gamma shape parameter (PROC GENMOD's 'scale') is $\hat{\lambda}_{ML} = 2.6448$, so that $\hat{\varphi}_{ML} = 1/2.6448 = 0.3781$, only

```
                        The SAS System
                      The GENMOD Procedure

                       Model Information
                Description                    Value
                Distribution                   Gamma
                link function                  Log
                Dependent Variable             mf

                  Analysis Of Parameter Estimates
                            Likelihood Ratio
                  Standard       95% Confidence        Chi-
  Parameter DF  Estimate   Error       Limits         Square   Pr>ChiSq
  Intercept  1    1.1653  0.0853   1.0001   1.3401     186.75    <.0001
  x          1    0.0175  0.0075   0.0016   0.0331       5.53    0.0187
  Scale      1    2.6448  0.4895   1.8031   3.7301
  NOTE:   The scale parameter was estimated by maximum likelihood.
```

Figure 3.19   Parameter estimate output (edited) from gamma log-linear ANCOVA fit to the mutant frequency data, employing only an age effect, x

slightly changd from $\hat{\varphi}_P$. Under `Likelihood Ratio 95% Confidence Limits` we find $1.8031 \le \lambda \le 3.7301$, so applying (3.20) yields $\frac{1}{3.7301} \le \varphi \le \frac{1}{1.8031}$, that is, $0.2681 \le \varphi \le 0.5546$. Notice, in particular, that this interval does not contain 1.0, which would correspond to the simpler exponential p.d.f. from Example 3.2.   ✪

Exercises

3.1. Show that the following probability functions are members of the exponential class given in (3.1). Identify the natural parameter θ and, if it exists, the dispersion parameter φ.

(a) The Poisson p.m.f. in (A.12):

$$f(y) = \frac{\lambda^y e^{-\lambda}}{y!} I_{\{0,1,\dots\}}(y), \qquad \text{with } \lambda > 0.$$

(b) The gamma p.d.f. in (A.15):

$$f(y) = \frac{1}{\Gamma(\lambda)\beta^\lambda} y^{\lambda-1} e^{-y/\beta} I_{(0,\infty)}(y), \qquad \text{with } \lambda > 0 \text{ and } \beta > 0.$$

(c) The geometric p.m.f.:

$$f(y) = \left(\frac{\mu}{1+\mu}\right)^y \frac{1}{1+\mu} I_{\{0,1,\dots\}}(y), \qquad \text{with } \mu > 0.$$

(d) The inverse Gaussian p.d.f.:

$$f(y) = \frac{1}{\sqrt{2\pi\xi y^3}} \exp\left\{-\frac{(y-\mu)^2}{2y\mu^2\xi}\right\} I_{(0,\infty)}(y), \qquad \text{with } \mu > 0 \text{ and } \xi > 0.$$

3.2. Verify that for the following probability functions in their exponential class parameterizations, $E[Y] = B'(\theta)$ and $Var[Y] = B''(\theta)A(\varphi)$.

(a) The Poisson p.m.f. in Exercise 3.1(a).

(b) The gamma p.d.f. in Exercise 3.1(b).

(c) The geometric p.m.f. in Exercise 3.1(c). [*Hint*: from §A.2.5, we know that the geometric p.m.f. is a special case of the negative binomial p.m.f.]

(d) The exponential p.d.f. in Example 3.2.

(e) The binomial p.m.f. in Example 3.3.

3.3. Verify the indication in §3.2.6 that when using GEEs to estimate parameters in a GLiM, the information matrix analog to $E[U\,U^T]$ is $\mathbf{F} = \sum_{i=1}^{n}(\nabla\mu_i)^T\mathbf{V}_i^{-1}\nabla\mu_i$.

3.4. Return to the following data sets and fit a linear model using computer routines that estimate the population variance using ML and using the residual sum of squares, such as PROC GENMOD and PROC GLM (or PROC REG), respectively. Verify that the ratio of squared standard errors for both b_0 and b_1 equals n/df_e, as in Example 3.4.

(a) The mercury toxicity data in Exercise 1.1 (simple linear model).

(b) The lead concentration data in Exercise 1.2(a) (simple linear model).

(c) The crop yield data in Exercise 1.3 (simple linear model; ignore the issue of variance heterogeneity).

(d) The soil pH data in Example 1.3 (multiple linear model).

(e) The streamflow rate data in Exercise 1.5(b) (multiple linear model).

3.5. In a study of the potential heritable toxic effects of exposure to dithiocarbamate fungicides (Härkönen *et al.*, 1999), Danish farm workers were examined for the rates of aneuploidy (chromosome loss) in their germ cells. The data are:

x_1 = prior aneuploid %	0.43	0.29	0.56	0.25	0.24	0.33
x_2 = cigarettes/day	30	20	20	2	15	12
x_3 = hours of exposure	38.5	10.0	23.5	87.0	56.0	12.0
Y = no. of aneuploid cells	48	25	47	26	27	17
N = no. of cells studied	10000	10000	10000	10000	10000	10000

x_1 = prior aneuploid %	0.20	0.29	0.36	0.43	0.32	0.48
x_2 = cigarettes/day	15	1	5	0	0	0
x_3 = hours of exposure	18.0	19.5	50.0	9.3	16.0	5.0

Y = no. of aneuploid cells	26	21	21	37	26	37
N = no. of cells studied	10000	10000	10000	10000	10000	10000

x_1 = prior aneuploid %	0.19	0.23	0.14	0.27	0.25	0.18
x_2 = cigarettes/day	0	0	0	0	0	0
x_3 = hours of exposure	53.0	71.0	18.3	66.5	29.0	17.0
Y = no. of aneuploid cells	15	15	19	22	16	26
N = no. of cells studied	10000	10000	10000	10000	10000	10000

x_1 = prior aneuploid %	0.19	0.21	0.24	0.25	0.18	0.27
x_2 = cigarettes/day	0	0	0	0	0	0
x_3 = hours of exposure	64.0	13.5	5.5	69.0	63.5	48.0
Y = no. of aneuploid cells	26	27	27	18	13	21
N = no. of cells studied	10000	10000	10000	10000	10000	10000

x_1 = prior aneuploid %	0.35	0.25	0.16	0.22
x_2 = cigarettes/day	0	0	0	0
x_3 = hours of exposure	41.5	31.0	106.0	36.0
Y = no. of aneuploid cells	39	31	24	25
N = no. of cells studied	10000	10000	10000	10000

(a) Determine which of the three basic link functions, logit, probit, or complementary log-log, fits the data best. Under that chosen link, build an analysis of deviance table to assess how the three predictor variables affect aneuploidy rates in these workers. Include in your analysis a study of the Smoking × Exposure interaction, using the product $x_4 = x_2 x_3$. Operate at $\alpha = 0.01$.

(b) Note that x_2 = no. of cigarettes per day was zero for 19 of the 28 farm workers. Assess the impact of x_1 and x_3 on aneuploidy rates only for non-smokers ($x_2 = 0$), using the same link chosen in part (a). Also consider an analysis of the impact of x_1, x_2 and x_3 on aneuploidy rates after recategorizing x_2 into three levels: 0, 1–12, and 12+. Do these variations on how to code smoking level affect the qualitative conclusions from this study?

3.6. Return to the groundwater contamination data in Example 3.5.

(a) Verify the indication given in the text that by fitting the predictor variables in the order land use, nitrate, sewer use, chloride, both the sewer use and chloride terms become insignificant.

(b) Update your model fit by dropping the sewer use and chloride predictors, but also by adding a term for land use × nitrate interaction. Is the interaction significant in the corresponding logistic regression?

(c) Construct a standardized residual plot from your fit in part (b) and comment on any anomalies. Use deviance-based residuals.

3.7. Let T be a random variable that quantifies the tolerance of a population of organisms to an environmental stimulus. Suppose it is believed that T follows a normal distribution with mean μ and variance σ^2, $T \sim N(\mu, \sigma^2)$. Further suppose that individuals in this population are exposed to the stimulus at varying doses, x. Show that the intercept and slope when fitting a linear probit model, $\Phi^{-1}(\pi) = \beta_0 + \beta_1 x$, are related to μ and σ^2. (*Hint*: work with the probability that a randomly selected individual has a tolerance less than or equal to the administered dose x, i.e., study $\pi(x) = P[T \leq x]$.) Consider how these parameters might be related to the resistance of a population to toxic insult.

3.8. Show that the Cochran–Armitage trend statistic for proportions in (3.11) is invariant to linear transformations of the predictor variable x_i.

3.9. In a study of cellular response to γ-ray exposure (Sorokine-Durm *et al.*, 1997), human blood cells in culture were scored for whether or not they exhibited chromosome damage, as a function of γ-ray dose. The data are:

$x = \gamma$-ray dose (Gy)	0.0	0.10	0.24	0.33	0.69	1.00
$Y = \#$ damaged cells	1	18	22	23	61	89
$N = \#$ cells studied	2305	2005	2028	2010	1501	869

$x = \gamma$-ray dose (Gy)	1.35	1.54	2.00	3.00	4.00
$Y = \#$ damaged cells	160	96	430	476	609
$N = \#$ cells studied	1005	505	1366	794	646

(a) Determine which of the three basic link functions, logit, probit, or complementary log-log, fits the data best. Use that chosen link to assess if the probability of cellular damage increases over increasing γ-ray dose. Operate at $\alpha = 0.01$.

(b) Use the Cochran–Armitage trend test from (3.11) to assess if the probability of cellular damage increases over increasing γ-ray dose. (Adjust for skewness in the dose variable, if necessary.) Operate at $\alpha = 0.01$. Does your result differ from that in part (a)?

3.10. Assessing a trend in proportions can be done with respect to any ordered predictor variable. For example, a common concern in environmental studies is assessing a trend over *time*. Assuming that the separate proportions are independent over the time variable (which may not always be true; see Chapter 5), application of (3.11) is perfectly valid. To illustrate, consider the data given by Burkhart *et al.* (1998) on proportions of structural malformations (thought to be due to embryonic deformities) over time in pond frogs (*Xenopus laevis*) taken from Minnesota wetlands in the summer of 1997. The data are:

Date	18 July	18 Aug.	25 Aug.	11 Sep.	19 Sep.	29 Sep.
$Y = \#$ malformed frogs	5	10	6	16	8	21
$N = \#$ frogs studied	104	144	90	245	61	115

(a) Plot the observed proportions, Y_i/N_i, against time, taken as $x_i = $ days from 15 July. Is there an increasing pattern?

(b) Assume that enough time passes between sampling events so that the proportions are statistically independent. Use the Cochran–Armitage trend test from (3.11) to assess if the probability of malformation increases over time. (Adjust for skewness in the time variable, if necessary.) Operate at $\alpha = 0.01$.

3.11. A setting where the complementary log-log (CLL) link occurs naturally is that of binary truncation of Poisson data. To show this, complete the following steps.

(a) Take data as $U_i \sim$ indep. Poisson(μ_i), $i = 1, \ldots, n$, to be modeled as a function of some linear predictor η_i. Under a log link for μ_i, we would write $\log(\mu_i) = \eta_i$. Show that this implies $\mu_i = \exp\{\eta_i\}$.

(b) Suppose that the U_is cannot be observed. Instead, all that is recorded is whether or not $U_i \geq 1$. Thus the Poisson counts have been truncated into binary observations Y_i, where

$$Y_i = \begin{cases} 1 & \text{if } U_i \geq 1, \\ 0 & \text{if } U_i = 0. \end{cases}$$

Let $\pi_i = P[Y_i = 1]$. Why must this also be $\pi_i = 1 - P[U_i = 0]$?

(c) Since $U_i \sim$ indep. Poisson(μ_i), show that $\pi_i = 1 - \exp\{-\mu_i\}$. Recall from part (a) that $\mu_i = \exp\{\eta_i\}$. Use this to write π_i as a function of η_i.

(d) Show that this relationship leads to the CLL link $\eta_i = \log\{-\log(1 - \pi_i)\}$.

(e) This effect occurs in a number of environmental settings. For example, in an environmental mutagenesis assay, test wells containing a bacterial suspension are exposed to a chemical toxin. If any of the bacteria in a well mutate, the well's suspension turns cloudy, indicating at least one mutational event has occurred. Thus Poisson-distributed mutations in the wells are truncated into binary indicators of genotoxicity. The following data are taken from such an assay; they are proportions of test wells indicating at least one mutagenic event after bacterial exposure to the industrial chemical cumene hydroperoxide (Piegorsch et al., 2000).

$x = $ dose(μg/ml)	0	0	0	1	1	1
$Y = $ # mutated wells	3	4	1	1	3	3
$N = $ # wells studied	144	144	144	144	144	144

$x = $ dose(μg/ml)	5	5	5	10	10	10
$Y = $ # mutated wells	6	15	11	21	17	13
$N = $ # wells studied	144	144	144	144	144	144

Fit a CLL link to these data and test if increasing doses of the chemical lead to increasing mutagenic response. Operate at $\alpha = 0.01$.

3.12. Return to the ozone exceedance data in Table 3.4.

(a) Plot the observed proportions, Y_i/N_i, against O_3 concentration, x_i. Does the pattern suggest any particular link function?

(b) Determine which of the three basic link functions, logit, probit, or complementary log-log, fits the data best. Use that chosen link to assess if the probability of ozone exceedance increases over increasing O_3 concentration. Operate at $\alpha = 0.01$. How do your results compare with those achieved in Example 3.6?

3.13. Confirm the values given in Example 3.8 for application of the generalized Cochran–Armitage statistic (3.14) to the litter effect data in Table 3.5. Also compute the robust statistic in (3.15) to these data. Comment on any difference you find between them.

3.14. Tinwell *et al.* (1996) report on the toxicity of the N', N'-dimethyl derivative (DMENU) of N-ethyl-N-nitrosourea (a chemical used in plant-growth research) in the bone marrow cells of male mice. Recorded were the proportions of bone marrow erythrocytes exhibiting micronuclei (MN) after exposure to increasing doses of the chemical. (Micronuclei are small portions of the cell's nucleus that have become detached, typically due to some intracellular damage.) Single oral exposures were applied, and after 24 hours the animals' bone marrow cells were sampled to check for MN formation. The exposure levels ranged from a zero-dose control, to the values 14.1, 70.4, 141, and 700 (all in mg/kg body weight). Since the exposure scale is heavily skewed to the right, we employ as dose levels the natural logarithms of the original exposures: $x_1 = 1.675$ [via (1.3)], $x_2 = 2.65$, $x_3 = 4.25$, $x_4 = 4.95$, $x_5 = 6.55$. The data are as follows:

$x_1 = 1.675$ ($i = 1$)		$x_2 = 2.65$ ($i = 2$)		$x_3 = 4.25$ ($i = 3$)		$x_4 = 4.95$ ($i = 4$)		$x_5 = 6.55$ ($i = 5$)	
MN cells	Total cells	MN cells	Total cells	MN cells	Total cells	MN cells	Total cells	MN cells	Total cells
1	2000	11	2000	52	2000	33	2000	38	2000
5	2000	12	2000	13	2000	95	2000	32	2000
6	2000	21	2000	47	2000	132	2000	116	2000
4	2000	22	2000	55	2000	96	2000	67	2000
2	2000	7	2000	26	2000	98	2000	71	2000
4	2000	9	2000	32	2000	111	2000	116	2000
3	2000	12	2000						
5	2000								
4	2000								

(a) Calculate the individual proportions, Y_{ij}/N_{ij}, and plot the results against x_i. Does there appear to be an increasing trend?

(b) Assess whether these replicate proportions exhibit extra-binomial variability by applying the T_i^2 statistic from (3.13). Also identify if an overall extra-binomial effect is evidenced by aggregating the individual test statistics. Operate at $\alpha = 0.10$.

(c) Test for an increasing trend in the MN proportion response by applying the generalized Cochran–Armitage statistic from (3.14). Operate at $\alpha = 0.10$.

(d) Test for an increasing trend in the MN proportion response by applying the robust generalized statistic from (3.15). At $\alpha = 0.10$, how does your result compare with that in part (c)?

(e) Test for an increasing trend in the MN proportion response by fitting a logistic regression model via GEEs. Use an exchangeable assumption on the working correlation matrix. (*N.B.*: Depending on your computer platform, the program may require a large amount of memory to perform the GEE calculations for these data.) At $\alpha = 0.10$, how does your result compare with those in parts (c) and (d)?

3.15. In a study of the relationship between an alligator's size and its primary type of prey, data were compiled leading to the following 3×3 contingency table:

Alligator size	Primary prey		
	Fish	Invertebrate	Other
size ≤ 1.75 m	8	13	3
$1.75 <$ size ≤ 2.5 m	14	6	1
size > 2.5 m	9	1	4

(a) Using the χ^2 method from §3.3.4, test the hypothesis that alligator size and primary type of prey are not associated. Set $\alpha = 0.05$.

(b) Application of the large-sample methods may be invalid here, since some of the expected counts under the invertebrate and other categories are so small. Is this the case with these data? If so, recalculate the test statistic after pooling across these two categories. Does your conclusion change from that in part (a)?

(c) As mentioned in Example 3.10, these analyses can also be conducted by appealing directly to the categorical nature of the data (Agresti, 1990). (In SAS, the primary procedure for analyzing categorical data is PROC FREQ.) If you are familiar with methods of categorical data analysis, repeat the test in part (a) using a categorical analysis and compare the results to the GLiM analysis.

(Agresti, 1996, §8.1.2)

3.16. Show that the Cochran–Armitage trend statistic for counts in (3.17) can also be written as

$$Z_{CA} = \frac{\sum_{i=1}^{g} m_i x_i (\bar{Y}_{i+} - \bar{Y}_{++})}{\sqrt{\bar{Y}_{++} \sum_{i=1}^{g} m_i (x_i - \bar{x})^2}}.$$

[*Hint*: show that the two numerators are equal. This alternate form is used by many authors; see Williams (1998).]

3.17. Pires *et al.* (2002) describe a study of how solid waste effluent from an industry source affects aquatic organisms. Working with the water flea *Ceriodaphnia dubia*, they report data as counts of *C. dubia* offspring produced 14 days after exposure to varying concentrations of the effluent. (Lowered offspring counts suggest an inhibitory effect on the organism's reproductive capacity.) The data are as follows:

x = effluent conc.	Y = # offspring	x = effluent conc.	Y = # offspring
0	46	1.3	38
0	41	1.3	39
0	58	1.3	28
0	40	1.3	41
0	58	1.3	42
0	42	1.3	39
0	56	1.3	46
0	49	2	6
0	61	2	12
0	42	2	13
0.6	42	2	13
0.6	43	2	15
0.6	42	2	23
0.6	50	3	18
0.6	51	3	21
0.9	43	3	17
0.9	38	3	22
0.9	42	3	10
0.9	40	3	21
0.9	44	3	20
0.9	39	3	22
0.9	39		
0.9	38		
0.9	39		

(a) Notice that there are replicate counts available at each of the different effluent concentrations studied. Calculate the per-concentration sample means, $\overline{Y}_{i+} = \sum_{j=1}^{m_i} Y_{ij}/m_i$. Plot the results against x_i. Does there appear to be a decreasing trend over x_i?

(b) Calculate the per-concentration sample variances. Do the ratios of sample mean to variance appear constant (as would be expected under a Poisson assumption on the Y_{ij}s)?

(c) Assess whether the per-concentration replicate counts exhibit extra-Poisson variability by applying the Fisher dispersion statistic from (3.18). Also identify if an overall extra-Poisson effect is evidenced by aggregating the individual test statistics. Operate at $\alpha = 0.10$.

(d) Based on your result in part (c), apply an appropriate trend statistic to determine if a significantly decreasing trend is evidenced in the offspring counts. Operate at $\alpha = 0.10$.

(e) Fit a log-linear model to these data under a Poisson likelihood. (Is this a valid operation, based on your results in part (c)?) Use a simple linear predictor in concentration: $\eta_i = \beta_0 + \beta_1 x_i$. Test for a decreasing trend; that is, test $H_0: \beta_1 = 0$ vs. $H_a: \beta_1 < 0$. At $\alpha = 0.10$, how does your result compare with that in part (d)?

(f) From your fit in part (e), find the ratio of scaled deviance to df: $D^*(\mathbf{b})/\mathrm{df}$. Is this value larger than the rule of thumb in (3.10) for assessing model adequacy? Does this agree qualitatively with your results from part (c)?

(g) Fit a log-linear model to these data under a negative binomial likelihood. Use a simple linear predictor in concentration: $\eta_i = \beta_0 + \beta_1 x_i$. Test for a decreasing trend; that is, test $H_0: \beta_1 = 0$ vs. $H_a: \beta_1 < 0$. At $\alpha = 0.10$, how does your result compare with those in parts (d) and (e)?

3.18. Return to the oak tree association data in Example 3.10. Lindsey (1997, §8.1.4) also reported nearest-neighbor data on other tree species in the forest under study. Including also Maple and Hickory trees, the larger data set is as follows:

Tree species	Neighbor species				
	White oak	Red oak	Black oak	Maple	Hickory
White oak	138	62	14	95	117
Red oak	59	104	20	64	95
Black oak	20	12	27	25	51
Maple	70	74	21	242	79
Hickory	108	105	48	71	355

Repeat the analysis from Example 3.10, now using this larger 5×5 contingency table.

3.19. Return to the beetle species richness data from Example 3.13.

(a) The original altitudinal relief values are in fact available for each observed species count. Thus instead of setting the predictor scores as the mid-points of the interval ranges in Table 3.9, we can operate with the actual average relief values from each interval. These are: $x_1 = 0.576$, $x_2 = 1.596$, $x_3 = 2.451$, and $x_4 = 3.609$. Repeat the calculations in Example 3.13 using these updated x_i values. Do your conclusions change from those in the example?

(b) Carroll (1998) also gave data on the numbers of bird species in each 1 km^2 region from this study. These were as follows:

Range	Mean relief	Numbers of bird species
$0 < x < 1$	0.576	129, 166, 137, 105, 155, 171, 174, 178, 181
$1 \leq x < 2$	1.596	165, 181, 184, 187, 193, 116, 158, 165, 169, 179, 203, 153, 173, 181, 187, 193, 175, 203, 211, 106, 114, 140, 177, 193, 196, 210, 231, 247
$2 \leq x < 3$	2.451	120, 145, 124, 225, 344, 430, 110, 118, 174, 193, 232
$3 \leq x < 4$	3.609	224, 171, 410, 518, 374, 387, 341, 326, 537, 533, 261, 460, 480

Repeat the calculations in Example 3.13 using this new outcome variable and x_i = mean relief. What do you conclude about the relationship between altitudinal relief and numbers of bird species?

3.20. Bailer and Oris (1997) display data on biomass (cells/mm) of green algae (*Selenastrum capricornutum*) after exposure to increasing levels of cadmium (Cd), in order to assess the growth-inhibitory effects of the heavy metal; cf. Exercise 2.19. The data are

x = Cd conc.	Y = biomass	x = Cd conc.	Y = biomass
0.0	1209	20	493
0.0	1180	20	416
0.0	1340	20	413
5.0	1212	40	127
5.0	1186	40	147
5.0	1204	40	147
10	826	80	49.3
10	628	80	40.0
10	816	80	44.0

(a) Notice that there are three replicate measurements per exposure group. Calculate the individual sample means, $\overline{Y}_{i+}$, and sample variances $S_i^2 = \frac{1}{2}\sum_{j=1}^{3}(Y_{ij} - \overline{Y}_{i+})^2$. From these find the coefficients of variation, $CV_i = S_i/\overline{Y}_{i+}$. Do the CV values appear roughly constant over changing Cd concentration?

(b) Plot the biomass measurements against Cd concentration. Does a pattern appear?

(c) Fit a gamma GLiM to these data, using a logarithmic link and a simple linear predictor in concentration, x_i. Fit the dispersion parameter via a Pearson-based moment estimator. Use the results to test if increasing concentration leads to a significant decrease in biomass. Set your level of significance to 5%.

(d) From the fit derived in part (c), calculate the standardized residuals (Pearson- or deviance-based), and construct a residual plot. Does the plot suggest any irregularities with the fit?

(e) Repeat part (c) but now fit the dispersion parameter via ML. Do your conclusions change?

3.21. In a study similar to that discussed in Example 3.14, Finette *et al.* (1994) reported data on possible gender and age differences among mutant frequencies (MF) at the human *hprt* gene in teenagers and children. The goal was to gain an understanding of the factors influencing spontaneous *hprt* MF response. Recorded were the individual subjects' MF values ($\times 10^6$), age, and sex. The data are as follows:

Sex	Age	MF	Sex	Age	MF	Sex	Age	MF
f	0.75	0.3	f	13.2	4.0	m	4.2	2.5
f	0.75	0.3	f	13.2	4.0	m	4.4	2.6
f	0.83	0.4	f	13.9	1.2	m	4.7	0.7
f	1.0	0.4	f	13.9	1.2	m	5.16	0.7
f	1.2	0.77	f	15.2	4.3	m	5.2	2.2
f	2.0	2.0	m	0.08	1.3	m	6.4	5.4
f	2.0	2.9	m	0.2	0.3	m	8.1	4.1
f	3.1	0.7	m	0.75	0.5	m	8.2	0.81
f	3.2	1.0	m	0.75	1.9	m	8.7	0.5
f	3.2	1.8	m	0.8	1.1	m	8.9	2.2
f	4.0	0.45	m	1.0	0.32	m	9.7	2.1
f	5.0	0.4	m	1.0	1.3	m	10.0	2.3
f	6.5	8.0	m	1.0	4.4	m	10.7	2.0
f	7.5	1.4	m	1.3	3.1	m	12.0	5.5
f	7.7	8.7	m	1.4	2.2	m	12.5	4.0
f	9.0	9.0	m	1.5	0.85	m	13.1	0.2
f	10.9	0.8	m	1.7	1.7	m	13.2	7.0
f	11.0	1.9	m	4.2	2.5			

(a) Fit an exponential GLiM to these data, that is, a gamma GLiM with the dispersion parameter set equal to $\varphi = 1$. (In PROC GENMOD, this is accomplished by including both `noscale` and `scale = 1` as options in the `model` statement.) Work with a logarithmic link function. Use in your linear predictor terms for age (centered, to represent a covariate as in

Example 3.14) and sex. Adjusting therefore for age, determine if there is a significant difference in MF response between sexes. Operate at $\alpha = 0.05$.

(b) Expand your fit in part (a) and allow φ to vary, as per the full gamma p.d.f. Use this to estimate φ via maximum likelihood, and from the results find a 95% confidence interval for φ. Does the interval contain 1.0? Why is this important?

3.22. Return to the mutant frequency data in Example 3.14. Continue with the analysis there and fit a gamma GLiM using a logarithmic link. Show that the smoking main effect is insignificant. (Do not include an exposure main effect, but continue to adjust for a possible age effect.) Operate at $\alpha = 0.05$.

4

Quantitative risk assessment with stimulus-response data

Quantitative risk assessment involves estimation of the severity and likelihood of adverse responses associated with exposure to hazardous stimuli. Within this context, we highlight in this chapter the use of stimulus-response data to assess environmental risks. Our emphasis is primarily on risks to biological (including human) or ecological systems, although one can also assess risk of economic loss due to natural disasters (Davenport and Kopp, 2002) or industrial or engineering risks such as nuclear power plant failures (Grimston, 2002).

As applied to human or ecological health, modern risk estimation is usually broken down into four fundamental stages (Stern, 2002). First is *hazard identification*, where an agent, substance, or other environmental stimulus induces detrimental outcomes in some industrial, occupational, public health, or ecological setting. To identify the damaging effects of a hazardous agent, investigators often consider data from biological assays on small mammals or other organisms, or epidemiological analyses of human populations at risk, in concert with any of the agent's structural similarities to other known, hazardous stimuli. Second is *stimulus/dose-response assessment*, where any quantifiable relationship between the agent and the detrimental outcome is modeled and estimated. Third is *exposure assessment*, where the extent and breadth of the population's or ecosystem's exposure to the agent is determined. This stage often requires specific stimulus/dose determination(s), and can employ models that synthesize the exposure information over time and/or space. Last is *risk characterization*, in which the risk analyst incorporates information from the previous three stages into a single assessment of the overall risk due to the agent. Included in this stage may be an additional effort to gauge the quality and confidence in the risk estimate as a whole. Note that we use the term 'dose' as a generic label for any quantification of a hazardous exposure. In some contexts, it might be the

Analyzing Environmental Data   W. W. Piegorsch and A. J. Bailer
© 2005 John Wiley & Sons, Ltd   ISBN: 0-470-84836-7 (HB)

concentration of a hazardous chemical, the delivered concentration of a toxic metabolite of some parent compound, the level of radiation to which an organism or ecosystem is exposed, etc.

At its core, risk is a multidimensional entity with a variety of quantitative and qualitative components. Managing these components in a risk assessment requires multiple inputs and expertise from a variety of disciplines (Griffiths, 2002), statistics being only one such contributor. Nonetheless, important quantitative considerations arise in each of the four stages, with perhaps the second, stimulus/dose-response assessment, requiring the greatest statistical input. A variety of issues impact the construction of a stimulus-response model, and these often lead to use of regression modeling. Thus in the spirit of the previous three chapters, we direct the majority of our attention here to quantitative methods for estimation and testing with stimulus-response data, concentrating on the risk-analytic aspects of the problem. We also introduce other allied components of the risk assessment, such as uncertainty analysis and comparative potency estimation.

4.1   Potency estimation for stimulus-response data

A common goal in environmental risk assessment is description of the hazardous agent's toxic or otherwise harmful effects, often for comparison with other agents of related structure or exposure. Summary measures are most useful in this regard, and these are viewed as describing the *potency* of the agent's stimulus on the organism or system under study. Using such a measure, a risk assessor can make comparative statements about a series of environmental agents as applied to a specific assay or system. When no significant response is evidenced after exposure to the agent, the potency is set equal to some boundary value that indicates no potent response. (Zero is common, although other boundary values are possible, depending on the measure used to describe potency; see Exercise 4.1.) For data exhibiting a significant response to the stimulus, a statistic departing from this boundary value is calculated.

Potency estimation has a long history, dating back to bioassays of pesticides, herbicides, and other poisons (Bliss, 1935; Finney, 1952). The simplest concern is how to measure or define the potency in an unambiguous manner: a number of quantitative solutions exists to answer this question, each based on a slightly different feature of the observed dose-response pattern. We consider a few of these in this section.

4.1.1   Median effective dose

A basic measure used to summarize a dose-response pattern is the *median effective stimulus* or *median effective dose*, denoted as ED_{50}. If the environmental agent is given as a concentration, we use the term *median effective concentration*, or EC_{50}. In either case, this is defined as the amount of agent required to produce a response at the median (50%) level (Trevan, 1927). For example, suppose we measure quantal response data as the proportion of times organisms respond to a toxic stimulus, as seen in §3.3.2. Then, since the proportions are bounded between 0 and 1, ED_{50} is the dose at which the response equals 0.50. Note here that ED_{50} is an inverse measure of potency: a higher value suggests a weaker agent, since a greater dose is required to produce the same median result.

Obviously, the ED_{50} depends upon the outcome variable under study, and the terminology has developed to indicate this. For instance, when simple lethality is the common outcome, the ED_{50} is more specifically the *median lethal dose*, or LD_{50}.

An extensive literature has developed on use of ED_{50}, EC_{50}, LD_{50}, etc.; see the early works by Irwin (1937) and Fieller (1940), and more contemporary discussions by Hamilton (1991), Soper (1998), and our own work (Bailer and Piegorsch, 2000). Our goal here is to distill from this material features of the ED_{50} useful for potency estimation, particularly as they pertain to stimulus-response risk analysis.

For example, suppose we observe quantal response data where the dose response takes the simple linear logistic form $\pi(x) = 1/(1 + \exp\{-\beta_0 - \beta_1 x\})$ from §3.3.2. In this case, the ED_{50} is found by setting $\pi(x)$ equal to $\frac{1}{2}$, and solving for x, that is, finding the value of x that solves $1/(1 + \exp\{-\beta_0 - \beta_1 x\}) = 0.5$. Assuming $\beta_1 \neq 0$, this is $ED_{50} = -\beta_0/\beta_1$ (Exercise 4.1). The ML estimate of this quantity is $\hat{ED}_{50} = -b_0/b_1$, where b_j is the ML estimate of β_j ($j = 0, 1$). Large-sample $1 - \alpha$ confidence limits on the ratio $-\beta_0/\beta_1$ are based on Fieller's theorem from Exercise 1.16. We find

$$\hat{ED}_{50} + \frac{\gamma}{1 - \gamma} \left\{ \hat{ED}_{50} + \frac{\hat{\sigma}_{01}}{se^2[b_1]} \right\}$$

$$\pm \frac{z_{\alpha/2}}{(1 - \gamma)|b_1|} \left(se^2[b_0] + 2\hat{\sigma}_{01}\hat{ED}_{50} + se^2[b_1]\hat{ED}_{50}^2 - \gamma \left\{ se^2[b_0] - \frac{\hat{\sigma}_{01}^2}{se^2[b_1]} \right\} \right)^{1/2},$$

$$(4.1)$$

where $\gamma = z_{\alpha/2}^2 se^2[b_1]/b_1^2$ measures departure from symmetry in the distribution of $\hat{ED}_{50}$ ($\gamma \to 0$ indicates greater symmetry), and $\hat{\sigma}_{01}$ is the estimated covariance between b_0 and b_1. The estimated covariance is available from most logistic regression computer outputs.

Example 4.1 (Copper toxicity)   We illustrate estimation of median effective dose with a toxicity study in which fish were exposed to potentially toxic concentrations of copper (Cu). In this study, $N = 20$ fish were exposed to the metal at each of seven concentrations (in µg/L). Recorded were proportions of fish dying after 4 days. The data, given by Wilhelm *et al.* (1998), are plotted in Fig. 4.1 and appear as part of the SAS code in Fig. 4.2. Of interest is determination of the median lethal concentration (LC_{50}) for this response via both point and interval constructions.

In Fig. 4.2, we use PROC PROBIT to perform the calculations. (Readers may wish to refer to §3.3.2 for background on SAS syntax when fitting regression models to proportion data.) The logit link is invoked via the `d=logistic` option in the `model` statement. The `inversecl` option calls for the LC_{50} estimate and 95% Fieller confidence limits. Notice also that we use the natural logarithm of concentration for

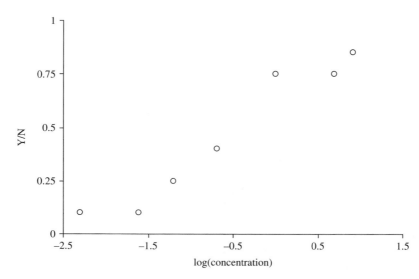

Figure 4.1   Scatterplot of observed mortality (as proportions Y/N) versus log-transformed Cu concentration for copper toxicity data

```
*  SAS code to find LC50;
data copper;
input conc ydead @@;
  N = 20;
  logconc = log(conc);
datalines;
  0.1  2     0.2  2     0.3  5     0.5 8
  1.0 15     2.0 15     2.5 17

proc probit;
   model ydead/N = logconc / d=logistic inversecl;
```

Figure 4.2   Sample SAS program to fit simple logistic regression model and find $\log(LC_{50})$ for copper toxicity data

the fit, due to the slight upper skew of the concentrations chosen in this study. This will require inversion via an exponential function to bring the final estimates back to the original scale. The (edited) output appears in Fig. 4.3.

From the **PROC PROBIT** output, we see that ML estimates of the logistic regression parameters are $b_0 = 0.5463$ and $b_1 = 1.3458$. A plot of the fitted logistic model overlaid on the data (Fig. 4.4) shows a reasonable fit.

To estimate LC_{50}, we begin with $\log(\hat{LC}_{50}) = -0.5463/1.3458 = -0.4059$. The large-sample 95% Fieller limits are given in Fig. 4.3, under 95% Fiducial Limits, as $-0.7199 \le \log(LC_{50}) \le -0.0734$. On the original Cu scale this is $\hat{LC}_{50} = e^{-0.4059} = 0.666\,\mu g/L$, with 95% Fieller limits $e^{-0.7199} = 0.487 \le LC_{50} \le e^{-0.0734} = 0.929\,\mu g/L$. Thus we are 95% confident that the Cu concentration at which half of the fish in this population will die is between 0.487 and 0.929 µg/L. ☯

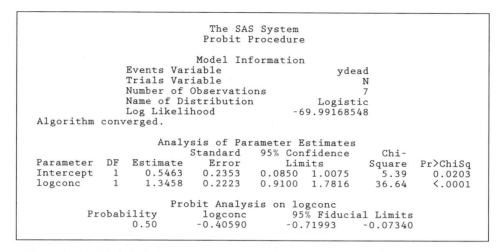

```
                        The SAS System
                       Probit Procedure

                     Model Information
           Events Variable                      ydead
           Trials Variable                          N
           Number of Observations                   7
           Name of Distribution              Logistic
           Log Likelihood                -69.99168548
  Algorithm converged.

                Analysis of Parameter Estimates
                         Standard    95% Confidence      Chi-
  Parameter   DF   Estimate    Error         Limits     Square   Pr>ChiSq
  Intercept    1     0.5463   0.2353   0.0850  1.0075      5.39     0.0203
  logconc      1     1.3458   0.2223   0.9100  1.7816     36.64    <.0001

                  Probit Analysis on logconc
          Probability         logconc      95% Fiducial Limits
             0.50            -0.40590     -0.71993    -0.07340
```

Figure 4.3 SAS output (edited) from simple logistic regression fit and $\log(\mathrm{LC}_{50})$ estimation for copper toxicity data

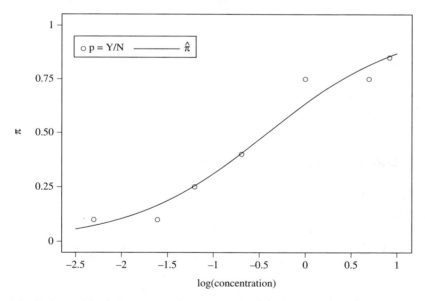

Figure 4.4 Estimated logistic response function and original proportions for copper toxicity data

Logistic functions are also useful for modeling continuous responses, as we saw with the growth curve models in §2.4. There, we noted that a logistic growth curve can be easily modified to represent an ED_{50}, as per equation (2.18).

For data in the form of counts, a simple log-linear model (§3.3.4) may be employed for stimulus-response analysis: $\log\{\mu(x)\} = \beta_0 + \beta_1 x$. If the response to the environmental agent is expected to produce a decrease in the mean rate of response, then we define the ED_{50} as that dose producing a 50% drop from the mean response at zero stimulus, $x = 0$. This contrasts with the quantal response setting above where we estimate a dose or concentration associated with a specified constant response in the population: now we are estimating the concentration or dose associated with an inhibition in response relative to the (unknown) control-level mean. Specifically, if $\mu(x) = e^{\beta_0 + \beta_1 x}$ then 50% of $\mu(0)$ is $\frac{1}{2} e^{\beta_0}$, so we solve for x in $2e^{\beta_0 + \beta_1 x} = e^{\beta_0}$. This yields $ED_{50} = -\log(2)/\beta_1$. Given stimulus-response data under a Poisson parent distribution, we find the ML estimate b_1 and then calculate $\hat{ED}_{50} = -\log(2)/b_1$. If we also find the approximate $1 - \alpha$ Wald limits $b_1 \pm z_{\alpha/2} se[b_1]$, a set of $1 - \alpha$ limits on $ED_{50} = -\log(2)/\beta_1$ is

$$\frac{-\log(2)}{b_1 - z_{\alpha/2} se[b_1]} \leq ED_{50} \leq \frac{-\log(2)}{b_1 + z_{\alpha/2} se[b_1]}.$$

An informal assumption being made here is that $b_1 \leq 0$, since we expect the stimulus to decrease the mean response. Notice, then, that if $z_{\alpha/2} se[b_1]$ is so large as to force $b_1 + z_{\alpha/2} se[b_1] > 0$, the interval calculation will lead to a nonsensical result. This is appropriate, however, since in this case an approximate $1 - \alpha$ confidence interval for β_1 contains zero. That is, there is no significant evidence for a dose effect and hence no reason to calculate a potency estimate.

Bailer and Oris (1997) consider the more general problem of estimating ED_{50}s for inhibition relative to control-level or baseline response under a generalized linear model (GLiM; cf. Chapter 3). They develop their linear predictors as polynomials in dose, and give examples with data in the form of binary survival rates, counts, and continuous measurements. They also describe confidence intervals for ED_{50} based on both Wald (§A.5.1) and bootstrap (§A.5.2) constructions.

4.1.2 Other levels of effective dose

Traditionally, the ED_{50} has been used as a summary measure of the dose effect due to its central location on the stimulus-response curve. Many other response levels are possible, however, such as ED_{01}, ED_{10}, ED_{90}, etc. Depending on the application, any of these impact levels may be relevant for a particular test system. In general, the *effective dose* ρ is the dose that yields a $100\rho\%$ effect over the stimulus-response curve, for $0 < \rho < 1$. We denote this by $ED_{100\rho}$. For example, with quantal response under the simple linear logistic model $\pi(x) = 1/(1 + \exp\{-\beta_0 - \beta_1 x\})$, the $ED_{100\rho}$ is the value of x that solves $\rho = 1/[1 + \exp\{-\beta_0 - \beta_1 x\}]$. This is

$$ED_{100\rho} = \frac{1}{\beta_1} \left\{ \log\left\{ \frac{\rho}{1 - \rho} \right\} - \beta_0 \right\} \tag{4.2}$$

(Exercise 4.5). At $\rho = \frac{1}{2}$, this collapses to the ED_{50} seen above. A similar construction is possible for any GLiM with quantal response data; for example, under the simple linear probit model $\pi(x) = \Phi(\beta_0 + \beta_1 x)$ we find $ED_{100\rho} = \{z_{(1-\rho)} - \beta_0\}/\beta_1$.

To estimate $ED_{100\rho}$, we substitute the ML estimates of the regression parameters, b_0 and b_1, into the pertinent expression for $ED_{100\rho}$. Tamhane (1986) reviews these and other estimation issues for measuring $ED_{100\rho}$. To construct large-sample $1 - \alpha$ confidence limits on $ED_{100\rho}$, we once again use Fieller's theorem. As long as the $ED_{100\rho}$ is of the form $(\delta - \beta_0)/\beta_1$ where δ is some known constant, the form of the confidence limits from (4.1) remains unchanged. For example, under the simple linear logistic model, the $ED_{100\rho}$ in (4.2) is of the form $(\delta - \beta_0)/\beta_1$, with $\delta = \log\{\rho/(1-\rho)\}$. Hence, a set of $1 - \alpha$ Fieller limits on (4.2) is

$$
\hat{ED}_{100\rho} + \frac{\gamma}{1-\gamma}\left\{\hat{ED}_{100\rho} + \frac{\hat{\sigma}_{01}}{se^2[b_1]}\right\}
$$

$$
\pm \frac{z_{\alpha/2}}{(1-\gamma)|b_1|}\left(se^2[b_0] + 2\hat{\sigma}_{01}\hat{ED}_{100\rho} + se^2[b_1]\hat{ED}^2_{100\rho} - \gamma\left\{se^2[b_0] - \frac{\hat{\sigma}^2_{01}}{se^2[b_1]}\right\}\right)^{1/2}.
$$

(4.3)

where $\gamma = z^2_{\alpha/2}se^2[b_1]/b_1^2$.

Example 4.2 (Copper Toxicity, cont'd) Returning to the copper toxicity data in Fig. 4.1, suppose now that we wish to calculate a lethal concentration level at, say, $\rho = 10\%$. To find this value we return to PROC PROBIT, as in Fig. 4.2. Recall that the output in Fig. 4.3 only gave information at $\rho = 50\%$. In fact, the `inversecl` option in PROC PROBIT provides estimates for a variety of ρ-values. (We edited the output in Fig. 4.3 for simplified viewing.) The additional output in Fig. 4.5 gives a larger picture. The $\log(LC_{100\rho})$ point estimates appear under `logconc`; the 95% Fieller limits are in the following two columns.

```
                        The SAS System
                        Probit Procedure

                     Model Information
              Events Variable                    ydead
              Trials Variable                       N
              Number of Observations                7
              Name of Distribution              Logistic

                    Probit Analysis on logconc
     Probibility         logconc        95% Fiducial Limits
        0.01            -3.82024      -5.45975      -2.96214
        0.05            -2.59373      -3.66883      -2.01269
        0.10            -2.03852      -2.86828      -1.57276
        0.20            -1.43597      -2.01864      -1.07613
        0.30            -1.03547      -1.47862      -0.72134
        0.50            -0.40590      -0.71993      -0.07340
        0.70             0.22367      -0.10488       0.71819
        0.80             0.62417       0.23966       1.26846
        0.90             1.22672       0.72934       2.12504
        0.95             1.78192       1.16652       2.92836
        0.99             3.00844       2.11360       4.72165
```

Figure 4.5 Additional SAS output (edited) for $\log(LC_{100\rho})$ estimation with copper toxicity data

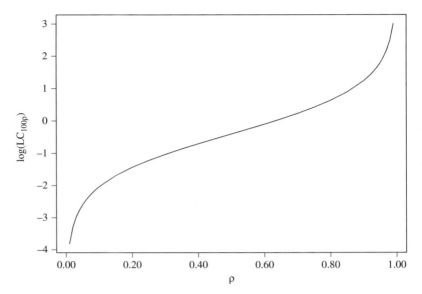

Figure 4.6　Estimated $\log(LC_{100\rho})$ function for copper toxicity data

From the output in Fig. 4.5, we find $\log(\hat{LC}_{10}) = -2.0385$, with 95% Fieller limits $-2.8683 \le \log(LC_{10}) \le -1.5728$. Converting to the original concentrations produces $\hat{LC}_{10} = e^{-2.0385} = 0.130\,\mu\text{g/L}$, with 95% Fieller limits $e^{-2.8683} = 0.057 \le LC_{10} \le e^{-1.5728} = 0.207\,\mu\text{g/L}$.

We can also use information in the complete output from **PROC PROBIT** (not shown) to plot the estimated $\log(LC_{100\rho})$ curve as a function of ρ for these data. The result appears in Fig. 4.6. ⊛

4.1.3　Other potency measures

Many other measures have been proposed to describe the potency of an environmental stimulus, taken within the context of risk estimation. Perhaps the simplest ones are known as observed effect levels. Specifically, the *no-observed-adverse-effect level* (NOAEL, sometimes written simply as NOEL), is the largest level of the stimulus or dose where no adverse effect is observed in the experimental units. If the stimulus is recorded as a concentration, we often write NOAEC. By extension, the *lowest-observed-adverse-effect level* (LOAEL) is the lowest level of the stimulus where a significant adverse effect is observed. If a significant increase over background occurs at the lowest level of the environmental stimulus, then the NOAEL will not exist while the LOAEL will correspond to that lowest tested level. The LOAEL is also called the *least effective dose* or *minimal effective dose* by some authors; the latter is especially common in the pharmacological and medical literature. For both NOAEL and LOAEL, estimates are determined statistically by performing a series of pairwise comparisons at each non-zero level of the stimulus with the response at the control or background level.

Numerous concerns have been raised regarding the NOAEL/LOAEL approach, since it suffers from a variety of instabilities. For instance, a NOAEL or LOAEL must by definition correspond to the actual levels of the stimulus under study. Thus, the spacing of stimulus levels is critical to these summary values, and ultimately to any risk assessments based upon them. Dose selection must be performed carefully, and include a dense enough grid of stimulus levels to derive a LOAEL and/or NOAEL with sufficient accuracy.

If the experiment is badly designed or poorly controlled, excessively high variability can result. This typically raises the values of NOAEL and LOAEL: high variability in the response will translate into assertions of low potency (large NOAELs), regardless of the true response effect. Also, a design that allocates too few experimental units to the stimulus levels may overestimate or even fail to identify a NOAEL or LOAEL. Indeed, it is unclear which parameter(s) of the dose–response relationship these quantities actually estimate. These difficulties have led to concerns that observed-effect levels such as NOAEL are poor summary statistics for stimulus-response data (Crump, 1984; Chapman *et al.*, 1996).

One feature that both the $ED_{100\rho}$ and the NOAEL share is that they are inverse measures of potency: a stimulus with a smaller $ED_{100\rho}$ or NOAEL is more potent than a stimulus with a larger $ED_{100\rho}$ or NOAEL. Some authors find this sort of 'inverse potency' to be counterintuitive, calling instead for transformations such as $1/ED_{100\rho}$ or $\log\{(1/ED_{100\rho}) + 1\}$ that recover direct proportionality, or for development of new, directly proportional measures. For instance, suppose the form of the stimulus-response function is known and that an increasing response indicates a detrimental effect. Since in this case a sharper increase indicates a more potent stimulus, a direct measure of potency is the rate of increase in the stimulus-response curve. Mathematically, this is the slope or tangent line of the curve at each level of the stimulus. Unfortunately, for most nonlinear models the slope varies with dose and so some specific dose must be chosen at which to measure the curve's slope. A common solution is to use the incremental rate of change in the dose response just past $x = 0$: this is typically taken as the first derivative at zero, $\mu'(0)$. This 'slope-at-zero' measure is motivated from low-dose estimation arguments: human exposures to a hazardous agent often approximate the lower dose levels of a laboratory animal study, so stimulus-response behavior observed at low doses in the animals may mimic human responses to the agent. As such, incremental change in dose response is of greatest interest near $x = 0$.

The use of a slope-at-zero measure is not without controversy, since for some stimulus-response functions slope measures may not exist or be sensible. For example, suppose $\mu(x)$ is an unknown-order polynomial such as $\beta_0 + \beta_1 x^{\beta_2}$. For $\beta_2 > 1$, the slope-at-zero is $\mu'(0) = 0$ (Exercise 4.7) and hence will always give a zero potency for any set of data. As we will see throughout this chapter, the choice of a stimulus-response function can influence the ultimate value of any functionally derived potency estimator, and so should be made carefully.

Measures such as NOAEL, LOAEL, $\mu'(0)$, etc., all refer to the background or control response, that is, to organisms or subjects in clean, uncontaminated conditions. When stimulatory or protective effects are observed at low concentrations of the environmental agent, however, alternate definitions of potency may be desired. For instance, suppose a protective effect appears at low doses, after which at higher doses a detrimental effect is exhibited as a reduction in response, such as inhibition of reproductive capacity. Then, the potency of the environmental inhibitor might be referred to the maximal observed reproductive response, in effect defining the detrimental impact relative to the optimal output of the organisms or subjects. Consider, for example, the log-quadratic model $\mu(x) = \exp\{\beta_0 + \beta_1 x + \beta_2 x^2\}$, which can allow for both low-dose increases and high-dose decreases in $\mu(x)$. Under this model, the relative inhibition (RI) concentration associated with a $100(1 - \rho)\%$ drop in response relative to the maximal level is

$$RI_{100\rho} = -\frac{\beta_1}{2\beta_2} + \sqrt{\frac{\log(1 - \rho)}{\beta_2}}, \qquad (4.4)$$

where we assume $\beta_2 < 0$ and $\beta_1 \geq 0$. Notice that the measure in (4.4) is a form of $ED_{100\rho}$.

Along with the control response level, the maximal response level is a common choice for the baseline from which an inhibition concentration is to be estimated. Bailer and Oris (2000) discuss these and several other possibilities for inhibitory risk modeling in more detail.

4.2 Risk estimation

4.2.1 Additional risk and extra risk

The potency measures described in §4.1 attempt to quantify the stimulus level(s) at which the subject's risk is increased or affected in some substantive manner. The actual risk to the subject or organism is defined implicitly, and estimation of this risk (or of the differential risk relative to background, or to maximal response, etc.) is left unspecified. In some settings, however, estimation of this risk is itself of interest.

Within this context, a necessary first step is development of a proper definition of risk. We define risk as the probability of some specific adverse effect, such as death, cancer, or growth inhibition. Assuming risk changes with increasing exposure at stimulus level x to the environmental agent, this is $R(x) = P[$exhibiting the adverse effect at level $x]$. Although seemingly straightforward, this definition contains an important, implicit feature: non-zero risk may exist, even change, for very small levels of x. This extends earlier concepts of risk where, at least for many non-cancer endpoints, one assumed that some dose threshold existed below which $R(x) = 0$. (The presence of a threshold is a debated concept in risk analysis, and risk management options can differ dramatically based on whether or not a threshold is believed to exist.) By modeling the risk more formally, however, a richer variety of possible stimulus-response functions and consequent statistical machinery becomes available (Krewski and van Ryzin, 1981). Indeed, one can even incorporate a threshold via some form of piecewise model for $R(x)$, as in §2.2.

Since $R(0)$ represents the background risk that all subjects in a population would be expected to encounter, a subject's differential risk is $R_A(x) = R(x) - R(0)$, which is known as the *additional risk* (also called *added risk*). Notice, however, that at stimulus level $x > 0$, $R(x)$ can be viewed as the proportion of individuals who would exhibit the response in the absence of exposure, $R(0)$, plus some stimulus-related fraction, π_x, of individuals who would not exhibit the response in the absence of exposure. This latter contribution is $\{1 - R(0)\}\pi_x$, producing $R(x) = R(0) + \pi_x\{1 - R(0)\}$. Solving for π_x yields $\pi_x = R_A(x)/\{1 - R(0)\}$, which in effect corrects the additional risk for non-response in the unexposed population. Since it also quantifies the risk of extra adverse effects in subjects exposed to a detrimental stimulus, we write this fraction as $R_E(x) = R_A(x)/\{1 - R(0)\}$ and call it the *extra risk* function. Together, $R_A(x)$ and $R_E(x)$ are known as forms of *excess risk*.

Example 4.3 (Multistage model)   Much of quantitative risk assessment is directed to the study of environmental carcinogenesis, that is, cancer induction by a hazardous environmental agent. A historically popular model for describing the risk function here is known as the *multistage model* of carcinogenesis. The model is based on the assumption that k transforming stages of damage are required for a normal cell to divide and progress into a cancerous state after exposure to the agent. In practice, it is common to set k equal to the number of non-zero doses tested minus 1. Another suggestion is to set $k = 2$ or $k = 3$, primarily for simplicity's sake. Developed in theory by Armitage and Doll (1954), the model has been studied by many authors; Whittemore and Keller (1978), Zeise *et al.* (1987), and Goddard and Krewski (1995) give useful reviews that also illustrate its evolution over time.

   In its more modern versions, the multistage model has developed in both complexity and biological validity to include multiple birth–death processes and elaborate stochastic pathways/stages for tumor onset and progression (Sherman and Portier, 1998; Moolgavkar, 1999). For our purposes, it will be sufficient to consider the basic multistage construction. Suppose data are taken as the number of identically aged subjects, Y_i, that exhibit a specific tumor, out of N_i total subjects all exposed for the same duration to the hazardous agent. This might occur, for example, in a carcinogenicity study with laboratory animals (Haseman, 1990; Ahn and Kodell, 1998). (Extensions exist that adjust for differential ages or exposure durations.) The index i is associated with differing doses or levels, x_i, of the carcinogenic stimulus, $i = 1, \ldots, n$. Assuming a binomial distribution for Y_i, we have $Y_i \sim$ indep. Bin$[N_i, R(x_i)]$, where $R(x)$ is the underlying risk that an individual subject will develop the cancer. The multistage supposition then leads to a formulation for $R(x)$ in what is essentially a GLiM for a binary regression. Under the 'same-age' assumption on all the subjects, $R(x)$ can be shown to take the form

$$R(x) = 1 - \exp\{-\beta_0 - \beta_1 x - \cdots - \beta_k x^k\}, \tag{4.5}$$

where we assume $x \geq 0$. For the expression in (4.5) to represent a probability, we require either (i) $\beta_j \geq 0$ for all $j = 0, \ldots, k$, or (ii) $R(x) = 0$ if $\beta_0 + \beta_1 x + \cdots + \beta_k x^k < 0$. The former constraint complies with certain biological requirements of the multistage assumptions and is more common. Under either constraint, (4.5) can be inverted into $\beta_0 + \beta_1 x + \cdots + \beta_k x^k = -\log\{1 - R(x)\}$. This is a GLiM with what is known as a *complementary log link*, here employing a kth-order polynomial predictor.

Under (4.5), $R(0) = 1 - \exp\{-\beta_0\}$, so the extra risk is simply $R_E(x) = [R(x) - R(0)]/[1 - R(0)] = [\exp\{-\beta_0\} - \exp\{-\beta_0 - \beta_1 x - \cdots - \beta_k x^k\}]/\exp\{-\beta_0\} = 1 - \exp\{-\beta_1 x - \cdots - \beta_k x^k\}$. ☉

Notice in (4.5) that the polynomial predictor is employed without any adjustment or centering of the x-variable. This contrasts with our previous recommendation in §1.5 that polynomial predictors should always be centered about $\bar{x}$. In fact, centering here can improve stability of the β_j estimates, and it remains a reasonable option. Since the dose metric is informative in most quantitative risk assessments, however, any dose manipulation(s) must be carefully reconstructed when performing inferences on risk. Indeed, centering about $\bar{x}$ can have awkward consequences since, for example, the extra risk and additional risk functions will become functions of it. This can lead to possible misinterpretations in the eventual risk analysis. Reversing the centering transformation is not onerous, of course, but in order to maintain simplicity in our presentation, we will not center the dose variables in this chapter. Interested readers are encouraged to experiment with dose manipulations such as centering or scaling in order to gain experience with their effects on excess risk functions and other measures of risk.

Example 4.4 (Extreme-value and Weibull models) The requirement in equation (4.5) that $\beta_0 + \beta_1 x + \cdots + \beta_k x^k$ must be non-negative can add complexity to the estimation process, since it forces us to impose some form of estimability constraint. One way to avoid this concern is to replace $\beta_0 + \beta_1 x + \cdots + \beta_k x^k$ with $\exp\{\beta_0 + \beta_1 x + \cdots + \beta_k x^k\}$, since e to any power is always positive. The result is

$$R(x) = 1 - \exp\{-\exp[\beta_0 + \beta_1 x + \cdots + \beta_k x^k]\}, \qquad (4.6)$$

where $k \geq 1$. This is known as an *extreme-value dose-response model* or as a *Gumbel dose-response model*, since it mimics the form of a c.d.f. from an extreme-value or Gumbel distribution (Beirlant and Matthys, 2002). Notice here that the inverse relationship is $\beta_0 + \beta_1 x + \cdots + \beta_k x^k = \log\{-\log[1 - R(x)]\}$, which is simply the complementary log-log link from a binomial-likelihood GLiM. Under (4.6) the extra risk is $R_E(x) = 1 - \exp\{e^{\beta_0}(1 - \exp[\beta_1 x + \cdots + \beta_k x^k])\}$ (Exercise 4.8).

Many authors employ the simplest case of (4.6) with $k = 1$: $R(x) = 1 - \exp\{-\exp[\beta_0 + \beta_1 x]\}$. When doing so, it is not uncommon to have x correspond to the natural logarithm of some original dose measure, d, so $x = \log\{d\}$. Then, following Chand and Hoel (1974), the risk function becomes $1 - \exp\{-\exp[\beta_0 + \beta_1 \log(d)]\} = 1 - \exp\{-e^{\beta_0} d^{\beta_1}\}$. (A small annoyance with the log transformation is its lack of existence at zero. Notice, however, that at $d = 0$, $R(0) = 0$ is well defined.) If for simplicity we write $\theta_0 = e^{\beta_0}$, this is

$$R(d) = 1 - \exp\{-\theta_0 d^{\beta_1}\}. \qquad (4.7)$$

Equation (4.7) is known as a two-parameter *Weibull dose-response model*. Some authors constrain $\beta_1 \geq 1$ in order to prevent (4.7) from representing certain unlikely biological responses, although this is not a fundamental requirement.

In many environmental epidemiology studies, a set of pertinent explanatory covariates, say $v_1, v_2, \ldots, v_m$, is often recorded along with the exposure dose variable, d. In this case the Weibull model can be expanded to also account for the covariates, via

$$R(d) = 1 - \exp\{-\theta_0 d^{\beta_0} - \exp[-\beta_1 v_1 - \cdots - \beta_m v_m]\}$$

(Crump, 2002).

Since in (4.7) $R(0) = 0$, we have $R_E(x) = R_A(x) = R(x)$. To broaden the model and allow for a non-zero background response, we can extend the two-parameter Weibull dose-response model into a three-parameter form:

$$R(d) = \varphi_0 + (1 - \varphi_0)(1 - \exp\{-\theta_0 d^{\beta_1}\}), \tag{4.8}$$

where $0 \leq \varphi_0 \leq 1$. This case is an application of *Abbott's formula* (Abbott, 1925) for incorporating a non-zero background effect into a dose-response model. Exercise 4.8 shows that under (4.8) the extra risk is $R_E(d) = 1 - \exp\{-\theta_0 d^{\beta_1}\}$ which, interestingly, recovers (4.7). ☉

In general, additional risk and extra risk functions are often quite complex. Because of this, in practice we may work with approximations to them. For example, the nonlinear multistage model from Example 4.3 can be simplified into an approximate linear form, as seen in the next example.

Example 4.5 (Linearized multistage model) To simplify the multistage R_E function, we can appeal to an approximation theorem from differential calculus. Known as a Taylor series expansion (§A.6.1), the approach approximates any smooth function with a polynomial of prespecified order. For instance, for any real number u near 0, it can be shown that the expression $1 - e^{-u}$ can be approximated to first order as simply $1 - e^{-u} \approx u$ (Exercise 4.10).

This is useful when combined with the following: note that as the positive value x approaches zero, the linear term in the sum $\beta_1 x + \cdots + \beta_k x^k$ dominates all the other x-related terms, since $x > x^2 > \cdots > x^k$. Thus near $x = 0, \beta_1 x + \cdots + \beta_k x^k \approx \beta_1 x$. Connect this with the Taylor approximation from above: as $x \to 0$, so does $u = \beta_1 x + \cdots + \beta_k x^k$. Thus, $R_E(x) = 1 - \exp\{-\beta_1 x - \cdots - \beta_k x^k\} \approx \beta_1 x + \cdots + \beta_k x^k \approx \beta_1 x$, which is linear in x. (When applied in cancer risk assessment, β_1 is sometimes referred to here as the 'unit cancer risk.') In this simplified form, the *linearized multistage model* can be used for various forms of risk estimation. ☉

Example 4.6 (Linearized extreme-value model) As given in Example 4.4, the extra risk from an extreme-value model is $R_E(x) = 1 - \exp\{e^{\beta_0}(1 - \exp[\beta_1 x + \cdots + \beta_k x^k])\}$. Using similar arguments to those in Example 4.5, this extra risk can also be linearized. That is, since $1 - e^{-u} \approx u$ near $u \approx 0$, we see that as x approaches zero, so does $\beta_1 x + \cdots + \beta_k x^k$. Thus $1 - \exp[\beta_1 x + \cdots + \beta_k x^k] \approx -(\beta_1 x + \cdots + \beta_k x^k)$ and so $R_E(x) \approx 1 - \exp\{-e^{\beta_0}(\beta_1 x + \cdots + \beta_k x^k)\}$. But, as x approaches zero, so does $-e^{\beta_0}(\beta_1 x + \cdots + \beta_k x^k)$, hence $R_E(x)$ can be further approximated as

$R_E(x) \approx e^{\beta_0}(\beta_1 x + \cdots + \beta_k x^k)$. Lastly, if $x \geq 0$, then $x > x^2 > \cdots > x^k$ as $x \to 0$. Thus near $x \approx 0$, we can write $R_E(x) \approx e^{\beta_0}\beta_1 x$. As in Example 4.5, this simplified form can make the extreme-value model more amenable to risk-analytic operations.　　☯

　　Given data on the detrimental phenomenon under study, the excess risk functions R_A and R_E are found by using estimates of the individual parameters on which they are based. Unless circumstances suggest otherwise, we favor maximum likelihood for acquisition of the parameter estimates. That is, suppose the excess risk function is itself a function of a set of unknown regression parameters, β. Given a parametric form for the p.d.f. or p.m.f. of the underlying data, we estimate β with its ML estimator $\mathbf{b}$, leading to the ML estimators $\hat{R}_A(x)$ or $\hat{R}_E(x)$ with β replaced by $\mathbf{b}$ wherever it appears.

　　Inferences on the β-parameters follow from large-sample likelihood theory (§A.5). For example, a test of whether β_k is required in the model – i.e., a test of $H_0\colon \beta_k = 0$ vs. $H_a\colon \beta_k \neq 0$ – employs the likelihood ratio statistic G^2_{calc} or the Wald statistic $W_{\mathrm{calc}} = b_k/se[b_k]$. Reject H_0 in favor of H_a if $G^2_{\mathrm{calc}} \geq \chi^2_\alpha(1)$ or if $W^2_{\mathrm{calc}} \geq \chi^2_\alpha(1)$, respectively.

Example 4.7 (Carcinogenicity of pentachlorophenol)　Return to the multistage model given in equation (4.5). Recall that the model was designed for use with proportion data, often in the context of environmental carcinogenesis studies. For example, the Integrated Risk Information System (IRIS; see http://www.epa.gov/iris/subst/0086.htm) of the US Environmental Protection Agency (EPA) gives data on carcinogenicity in female B6C3F$_1$ mice after chronic exposure to the fungicide/fumigant pentachlorophenol (PeCP). At human equivalent exposures of $x_1 = 0\,\mathrm{mg/kg/day}$, $x_2 = 1.3\,\mathrm{mg/kg/day}$, $x_3 = 2.7\,\mathrm{mg/kg/day}$, and $x_4 = 8.7\,\mathrm{mg/kg/}$ day, the mice exhibited the following proportions of hemangiosarcomas/hepatocellular tumors: $Y_1/N_1 = \frac{1}{34}$, $Y_2/N_2 = \frac{9}{49}$, $Y_3/N_3 = \frac{6}{46}$, and $Y_4/N_4 = \frac{42}{49}$. (The original carcinogenicity study was conducted in the USA by the National Toxicology Program, 1989.) The observed stimulus/dose-response pattern here is fairly flat for low doses, but then increases sharply by the highest dose group. Thus, some degree of nonlinearity is apparent in the response.

　　Consider estimating the extra risk function $R_E(x)$. Take $Y_i \sim$ indep. Bin$[N_i, R(x_i)]$ and let $R(x_i)$ satisfy the multistage model in (4.5). The number of non-zero dose levels here is 3, so we set $k = 2$. Figure 4.7 presents sample PROC NLMIXED code in SAS ver. 8 for this analysis.

　　The code in Fig. 4.7 specifies the functional form in (4.5) via SAS programming statements preceding the `model` statement: `eta` gives the $k = 2$ multistage predictor, while `prob` gives $R(x)$ as a function of `eta`. (The `parms` statement gives initial estimates to start the iterative fit.) The actual `model` statement here acts only to specify the binomial likelihood. In fact, PROC NLMIXED is designed to fit broader classes of statistical models than those we operate under here. Mentioned briefly in §1.4, these are known as *mixed models*, and they involve hierarchical specifications that can allow for random variation in the β-parameters of the linear predictor. We describe some specific versions of mixed models in §5.6 when we present analyses for correlated growth curve data. Here, however, the multistage formulation requires no mixed-model assumptions and our primary reason for employing PROC NLMIXED is its coding simplicity. To incorporate the constraint that $\beta_j \geq 0$ for all j, we employ

```
data PeCP;
input dose  ytumor  N  @@;
  d2 = dose*dose ;
datalines;
        0    1   34          1.3    6   49
        2.7  9   46          8.7   42   49

proc nlmixed;
    parms b0=.1  b1=.01  b2=.001;
  eta = b0 + b1*dose + b2*d2 ;
  prob = 1 - exp(-eta) ;
  model ytumor ~ binomial(N,prob) ;
    bounds b0 >= 0 , b1 >= 0 , b2 >= 0 ;
```

Figure 4.7 Sample SAS ver. 8 program to fit $k = 2$ multistage model to PeCP carcinogenicity data

the bounds statement. This requires PROC NLMIXED to institute a constrained maximization of the binomial likelihood. Other SAS procedures that can fit the multistage model include PROC NLIN, which also has bounds statement capability, and PROC GENMOD through use of a *user-defined link* (Piegorsch and Bailer, 1997, §9.5.3).

The output (edited) from PROC NLMIXED appears in Fig. 4.8. There, the ML estimates of the regression parameters are given as $b_0 = 0.035$, $b_1 = 0.025$, and $b_2 = 0.022$. The estimated extra risk is $\hat{R}_E(x) = 1 - \exp\{-0.025x - 0.022x^2\}$.

```
                         The SAS System
                       The NLMIXED Procedure

                            Specifications
        Dependent Variable                         ytumor
        Distribution for Dependent Variable        Binomial
        Optimization Technique                     Dual Quasi-Newton
        Integration Method                         None

                         Iteration History
     Iter   Calls   NegLogLike        Diff      MaxGrad        Slope
      1       6     9.33467932    42.52994     307.4632    -1065894
      2       8     8.30293054     1.031749    273.2638     -32.1928
      3      10     7.80485761     0.498073    122.1849      -5.95067
      4      11     7.04116726     0.763690     11.43309     -1.14515
      5      14     6.89549870     0.145669     60.98807    -301.278
      6      17     6.83368560     0.061813     17.09505      -0.15247
      7      19     6.81567528     0.018010      2.932736     -0.03776
      8      21     6.81516332     0.000512      0.5962       -0.00117
      9      23     6.81515025     0.000013      0.042285     -0.00003
     10      25     6.81515024     1.821E-8      0.004698     -4.15E-8
    NOTE: GCONV convergence criterion satisfied.

                         Parameter Estimates
                             Standard
        Parameter   Estimate     Error      DF   t Value   Pr > |t|
        b0          0.03519    0.03431       4     1.03     0.3631
        b1          0.02498    0.04056       4     0.62     0.5713
        b2          0.02174    0.007125      4     3.05     0.0380
```

Figure 4.8 SAS output (edited) from $k = 2$ multistage model fit for PeCP carcinogenicity data

Notice that $se[b_2] = 0.007$, producing a Wald statistic for testing $H_0: \beta_2 = 0$ of $|W_{calc}| = 3.05$. Referred to a large-sample standard normal distribution, the corresponding two-sided P-value is $2 \times 0.0011 = 0.0022$, from Table B.1. Thus there is significant evidence that the quadratic term is required in this multistage model. (The linear term may not be as significant. Recall, however, that when the linear and quadratic predictor variables in a polynomial regression have a high correlation – here it can be found as 0.978 – paradoxical significance patterns can emerge: a lower-order component when fitted last can appear insignificant but still represent an important contribution to the model. Hence, in keeping with our previous strategy from §1.5, we will retain all polynomial components whose orders are below the highest significant polynomial term.) Interestingly, in Fig. 4.8 PROC NLMIXED references the $t(4)$ distribution for this Wald P-value, producing $P = 0.038$. In effect, this acts as a conservative, small-sample adjustment to the usual standard normal reference for W_{calc}. Since the sample sizes are large enough here to lend credence to the large-sample approximation, this adjustment may be overly cautious. Regardless of the choice of reference distribution, however, the significance of the quadratic effect is clear for these data.   ✪

A slightly different approach to characterizing risk is often used when studying ecological impacts, based on ecosystem organization. The methods described above focus on the mean response (or the change in mean response) in a population of one species exposed to one hazardous agent. Risk assessors can also ask how an agent affects the individual organisms or how exposure to the agent affects a community of organisms. Thus the focus becomes one of studying the individual-level effect(s).

Example 4.8 (Individual-level added risk)   Environmental toxicity studies of single populations are often included in broader ecological risk assessments. For example, when assessing water quality, toxicologists often study response of the water flea *Ceriodaphnia dubia* to potential aquatic hazards. A common outcome is the count of *C. dubia* offspring produced a prespecified number of days after exposure to varying concentrations, x, of a potential ecotoxin. (Lowered offspring counts suggest an inhibitory effect on the organism's reproductive capacity; see Exercise 3.17.) It is natural to take the observed counts as Poisson distributed, and to model the mean number of offspring via a log-linear relationship $\log\{\mu(x)\} = \beta_0 + \beta_1 x$, or via a log-quadratic relationship $\log\{\mu(x)\} = \beta_0 + \beta_1 x + \beta_2 x^2$.

Under this model, equation (A.12) gives the probability of observing exactly M offspring from any individual organism as $P[Y = M] = e^{-\mu(x)}\mu(x)^M/M!$, while the probability of producing *at most* M offspring corresponds to the Poisson c.d.f.: $F_x(M) = \sum_{i=0}^{M} P[Y = i] = \sum_{i=0}^{M} e^{-\mu(x)}\mu(x)^i/i!$. For a value of $M > 0$ fixed at a sufficiently low level, this cumulative probability can be employed to represent the risk of detrimental response at exposure concentration x, that is, $R(x) = F_x(M)$. This defines adverse response at the organism-specific level as the probability that an individual organism produces no more than M offspring. The added risk is $R_A(x) = R(x) - R(0) = F_x(M) - F_0(M)$. Note that the choice of level, M, to represent affected reproductive output must be based on biological as well as risk-regulatory considerations.

For values of M less than about five, or when the control response rate $\mu(0)$ is larger than about 20, the control-level term, $F_0(M)$, in $R_A(x)$ is often trivial (Bailer and See, 1998). In such cases $R_A(x) \approx F_x(M)$, and we can operate under this simpler parametric structure. (For estimation without this simplifying assumption, refer to See and Bailer, 1998.) When $\mu(x)$ is modeled as a function of a set of regression parameters, say, $\mu(x) = \exp\{\beta_0 + \beta_1 x + \beta_2 x^2\}$, we can calculate the ML estimators of each β_j and then substitute these values into the appropriate equation to estimate R_A. ⊛

On a broader scale, the question of how best to set exposure levels for a collection or community of organisms remains open. Indeed, defining exposure limits so that, for example no more than a specified proportion of species in an ecosystem is adversely impacted by a hazardous agent is a somewhat different risk-analytic paradigm from that seen in human health risk assessment. Oris and Bailer (2003) summarize these issues and discuss other exposure assessment models and risk characterizations for ecological hazards.

4.2.2 Risk at low doses

When conducting predictive or environmental toxicity studies that generate data based on a stimulus/dose response, it is common for the dose levels to be taken at fairly high values. This is true primarily for laboratory experiments or controlled ecological studies conducted as screens for certain toxic effects; however, it can occur in human observational studies as well. Unfortunately, the dose-response pattern exhibited at high doses in such a study may not apply in the low-dose region. A number of candidate models may fit data equally well at high doses, but may also yield dramatically different potency estimators at lower dose levels (say, extreme impact levels near $x = \text{ED}_{0.001}$). This is one of the great challenges in quantitative assessment of possible human or ecological risks: so-called *low-dose extrapolation* from high dose levels to lower doses of regulatory interest (Brown and Koziol, 1983). Particularly with non-cancer responses, the goal is to identify, if possible, a 'safe' exposure level below which exposures to an environmental toxin or other hazardous agent do little if any harm. This is known in regulatory circles as a *reference dose (RfD)* or a *reference concentration (RfC)*, depending upon the application. To find it, one calculates a low-dose estimate of minimal effect, and then divides this by an *uncertainty factor* of between 10 and 1000 to account for interspecies extrapolations and other uncertainties in the risk estimation process (Calabrese and Baldwin, 1994; Gaylor et al., 1999); see §4.4.1, below.

Traditionally, low-dose estimation and calculation of RfDs/RfCs has relied on use of no-observed-adverse-effect levels (§4.1.3) to estimate a minimal effect level (Slob, 1999). This is based on the notion that a hazard's effect would likely be exhibited only after the organism passed a certain threshold of exposure, perhaps by overtaxing certain homeostatic controls or damage repair mechanisms. Used in this context, however, there is often a danger of misinterpreting the NOAEL as a true no-effect level, when in fact it is only the highest dose group whose responses do not differ statistically from the control. Indeed, given the concerns noted in §4.1.3 regarding its statistical limitations, use of the NOAEL in quantitative risk estimation remains an issue of debate.

A contemporary approach to low-dose estimation employs functional parametric models to represent the mean response $\mu(x)$ and the risk function $R(x)$. Known as *benchmark analysis*, the methodology uses the functional specification for $R(x)$ to provide low-dose estimates for risk and/or excess risk. By inverting the stimulus–response relationship, it estimates the benchmark dose level at which a predetermined level of risk is attained. We go into more detail on benchmark analysis in the next section. Here, we describe the inverse problem of making inferences on excess risk functions at low-dose or low-exposure levels of x.

Given a delineation for $R(x)$ under a specific parametric likelihood for the data, the excess risk can be taken as either the additional risk R_A or the extra risk R_E. For simplicity, assume the latter (although, the general operations we describe here can apply to R_A as well). Suppose that under the given likelihood/risk model, $R_E(x)$ can be written as a function of some linear combination of the regression parameters, β_j. Denote this linear combination as $\lambda(x)$ and its ML estimator as $\hat{\lambda}(x)$. For instance, in Example 4.3 use of a multistage model under a binomial likelihood produced the extra risk $R_E(x) = 1 - \exp\{-\lambda(x)\}$ for $\lambda(x) = \beta_1 x + \cdots + \beta_k x^k$. The ML estimators were then $\hat{R}_E(x) = 1 - \exp\{-\hat{\lambda}(x)\}$ and $\hat{\lambda}(x) = b_1 x + \cdots + b_k x^k$.

Assume further that $R_E(x)$ and $\lambda(x)$ are related such that each can be inverted into a function of the other. For example, $R_E(x) = 1 - \exp\{-\lambda(x)\}$ leads to $\lambda(x) = -\log\{1 - R_E(x)\}$. The standard error for $\hat{\lambda}(x)$ can be found using the standard errors and covariances of the ML estimators b_j. The details will vary from model to model, but the concept is the same throughout. For instance, under a $k = 2$ multistage model, $\hat{\lambda}(x) = b_1 x + b_2 x^2$. Then, $se[\hat{\lambda}(x)] = \{x^2 se^2[b_1] + 2x^3 \hat{\sigma}_{12} + x^4 se^2[b_2]\}^{1/2} = |x|\{se^2[b_1] + 2x\hat{\sigma}_{12} + x^2 se^2[b_2]\}^{1/2}$, where $\hat{\sigma}_{12}$ is the estimated covariance between b_1 and b_2.

Given $se[\hat{\lambda}(x)]$, an approximate $1 - \alpha$ Wald confidence interval for $\lambda(x)$ at any x is

$$\hat{\lambda}(x) \pm z_{\alpha/2} se[\hat{\lambda}(x)]. \tag{4.9}$$

One-sided limits are similar; for example, an approximate $1 - \alpha$ lower limit on $\lambda(x)$ is

$$\hat{\lambda}(x) - z_\alpha se[\hat{\lambda}(x)]. \tag{4.10}$$

An important caveat associated with these risk estimates – or, for that matter, with any predictive risk estimate – is that the underlying dose-response model used to construct the estimates must be correct, at least to a good approximation. If the model does not correctly represent the dose–response relationship under study, then the risk estimates may be suspect.

Equations (4.9) or (4.10) can be used for building $1 - \alpha$ confidence limits for $\hat{R}_E(x)$. Since we assume that $R_E(x)$ is related to $\lambda(x)$ in a one-to-one fashion, it is straightforward to manipulate this relationship into a confidence statement. We complete the details for the multistage model from Example 4.3 in the next example.

Example 4.9 (Multistage model, cont'd)   Recall for the multistage model that $R_E(x) = 1 - \exp\{-\lambda(x)\}$ for $\lambda(x) = \beta_1 x + \cdots + \beta_k x^k$. Given an ML estimator $\hat{\lambda}(x)$ with standard error $se[\hat{\lambda}(x)]$, suppose we calculate the two-sided limits in (4.9). Then, if we properly exponentiate the limits and subtract the results from 1.0, an approximate, two-sided, $1 - \alpha$ confidence interval for the multistage extra risk at any x is $1 - \exp\{-\hat{\lambda}(x) \pm z_{\alpha/2}se[\hat{\lambda}(x)]\}$. Note that if $-\hat{\lambda}(x) + z_{\alpha/2}se[\hat{\lambda}(x)] > 0$, we truncate the lower endpoint at 0.0.

In most risk estimation settings, it may be more appropriate to ask for a one-sided upper bound on $R_E(x)$, since regulatory interest often concerns only the severity of an adverse outcome. The corresponding analog to the two-sided interval at any x is then $1 - \exp\{-\hat{\lambda}(x) - z_{\alpha}se[\hat{\lambda}(x)]\}$. ✪

The approach we outline here for finding confidence limits on $R_E(x)$ – or, with proper modification, on $R_A(x)$ – holds for any stimulus/dose level, x, of interest. It is important to note, however, that these are *pointwise* $1 - \alpha$ limits, and hence are valid only for a single value of x. If confidence statements on $R_E(x)$ are desired over a set of, say, $L > 1$ prespecified values of x, $\{x_1^*, \ldots, x_L^*\}$, some correction for multiplicity must be incorporated into the calculations. A common choice is the Bonferroni adjustment from §A.5.4. This simply replaces the critical point in (4.9) with the adjusted value $z_{\alpha/(2L)}$ or the point in (4.10) with $z_{\alpha/L}$. Note that this adjustment may be quite conservative if L is very large.

Example 4.10 (PeCP carcinogenicity, cont'd)   Consider again the PeCP carcinogenicity data from Example 4.7, analyzed via a $k = 2$ multistage model in Fig. 4.8. For $\lambda(x) = \beta_1 x + \beta_2 x^2$ we have $\hat{\lambda}(x) = 0.025x + 0.022x^2$. As noted above, the corresponding standard error is $se[\hat{\lambda}(x)] = |x|\{se^2[b_1] + 2x\hat{\sigma}_{12} + x^2 se^2[b_2]\}^{1/2}$, where $\hat{\sigma}_{12}$ is the estimated covariance between b_1 and b_2. This latter quantity is available from PROC NLMIXED by including a `cov` option in the procedure call. Doing so (output not shown) produces the value $\hat{\sigma}_{12} = -2.3 \times 10^{-4}$ and also gives $se^2[b_1] = 0.0016$ and $se^2[b_2] = 5.1 \times 10^{-5}$. This yields $se[\hat{\lambda}(x)] = |x|\{0.0016 - 4.6 \times 10^{-4}x + 5.1 \times 10^{-5}x^2\}^{1/2}$. Thus an approximate $1 - \alpha$ upper bound on $R_E(x)$ at any single x is

$$1 - \exp\left\{-0.025x - 0.022x^2 - z_{\alpha}|x|\sqrt{0.0016 - 4.6 \times 10^{-4}x + 5.1 \times 10^{-5}x^2}\right\}.$$

For example, at $x = 1.3$ mg/kg/day, the ML estimate is $\hat{R}_E(1.3) = 1 - e^{-0.0325-0.0372} = 1 - e^{-0.0697} = 0.0673$, with corresponding 95% upper bound $1 - e^{-0.0325-0.0372-0.0705} = 1 - e^{-0.1402} = 0.1308$.

If interest exists in constructing simultaneous 95% upper confidence intervals at a set of $L = 4$ low doses, say, $\{0.1, 0.2, 0.5, 1.0\}$, we employ the Bonferroni-adjusted critical point $z_{\alpha/4} = z_{0.0125} = 2.2414$. This produces the following simultaneous 95% upper limits:

$$R_E(0.1) \le 1 - e^{-0.0116} = 0.0115; \qquad R_E(0.2) \le 1 - e^{-0.0233} = 0.0230;$$
$$R_E(0.5) \le 1 - e^{-0.0597} = 0.0579; \qquad R_E(1.0) \le 1 - e^{-0.1244} = 0.1169.$$

The confidence coefficient applies simultaneously to all four statements. ✪

Example 4.11 (Weibull model, cont'd)   Recall for the three-parameter Weibull model from (4.8) that $R_E(d) = 1 - \exp\{-\theta_0 d^{\beta_1}\}$. Write this as $R_E(d) = 1 - \exp\{-e^{\eta(d)}\}$

for $\eta(d) = \beta_0 + \beta_1 \log(d)$ and $\beta_0 = \log(\theta_0)$. Given ML estimators b_0 and b_1 such that $\hat{\eta}(d) = b_0 + b_1 \log(d)$, the standard error of $\hat{\eta}(d)$ is $se[\hat{\eta}(d)] = \sqrt{se^2[b_0] + 2\hat{\sigma}_{01} \log(d) + se^2[b_1] \log^2(d)}$ where $\hat{\sigma}_{01}$ is the estimated covariance between b_0 and b_1. Thus a pointwise $1 - \alpha$ Wald upper confidence limit on $\eta(d)$ is simply $\hat{\eta}(d) + z_\alpha se[\hat{\eta}(d)]$. From this, a pointwise $1 - \alpha$ upper confidence limit on $R_E(d)$ is $1 - \exp\{-\exp(\hat{\eta}(d) + z_\alpha se[\hat{\eta}(d)])\}$ (why?), or more fully

$$1 - \exp\left\{-\exp\left[b_0 + b_1 \log(d) + z_\alpha \sqrt{se^2[b_0] + 2\hat{\sigma}_{01} \log(d) + se^2[b_1] \log^2(d)}\right]\right\}$$

$$= 1 - \exp\left\{-\hat{\theta}_0 d^{b_1} \exp\left[z_\alpha \sqrt{se^2[b_0] + 2\hat{\sigma}_{01} \log(d) + se^2[b_1] \log^2(d)}\right]\right\},$$

where $\hat{\theta}_0 = \exp(b_0)$. ✺

One can also construct $1 - \alpha$ simultaneous confidence *bands* for an extra risk function, using Scheffé's (1953) S-method for confidence bands. The result is a $1 - \alpha$ confidence statement on $R_E(x)$ that holds for *all* values of x. Here again, the details will vary from model to model; Al-Saidy *et al.* (2003) describe the case for a binomial likelihood/multistage model, while Piegorsch *et al.* (2005) discuss a normal likelihood/quadratic model.

4.3 Benchmark analysis

4.3.1 Benchmark dose estimation

We can use $\hat{R}_E(x)$ for setting reference doses or other low-dose benchmarks. For example, given a *benchmark risk*, BMR, the estimated extra risk function can be inverted to find the level of x that produces this benchmark in the target population: set $\hat{R}_E(x) = \text{BMR}$ and solve for x. (Similar efforts are possible using $R_A(x)$.) This is a form of inverse regression, and it is essentially the same operation as used in §4.1 for finding effective doses such as the ED_{50}.

The result is called a *benchmark dose (BMD)* and it is a form of low-dose extrapolation. For use in quantitative risk assessment, BMD calculations proceed as follows:

1. Identify a parent distribution for the data and a model for the risk function $R(x)$. For example, with proportion data one might select a binomial likelihood and a multistage model, as in Example 4.3. Select an appropriate excess risk function, $R_E(x)$ or $R_A(x)$, for use in the benchmark analysis.

2. Set the benchmark risk, BMR. Common values include 0.1, 0.05, 0.01, or occasionally even 10^{-6}, depending on the nature of the adverse outcome, any regulatory considerations, and the quality of the data. In the latter case, for example, it may be unwise to extrapolate down to a very small level of BMR if the data

occupy much larger observed risk levels. Observed risks between 0.05 and 0.10 are common in many environmental studies, and as a consequence BMRs are often chosen in this range of response.

3. Acquire data and fit the model in step 1 to them. Use this to estimate the excess risk. If desired, construct confidence limits on the excess risk as described in §4.2.2.

4. Invert the relationship estimated in step 3 to estimate the BMD. For example, if extra risk, R_E, is under study, solve the relationship $\hat{R}_E(x) = $ BMR for x and use the result as $\widehat{\text{BMD}}$. (If more than one solution results from the mathematical operation, chose the smallest positive value.)

5. Estimate a $1 - \alpha$ lower limit on the dose associated with the specified BMR, and use this as the basis for any RfD calculations (cf. §4.4.1).

In some settings, the computations required to complete step 4 can become prohibitive, and approximations may be employed to simplify the process. The following examples provide some illustrations; see also Exercises 4.19–4.20.

Example 4.12 (k = 2 multistage model) Suppose $k = 2$ in the multistage model from Equation (4.5). For step 1, above, we will work with a binomial likelihood, and take the corresponding extra risk as $R_E(x) = 1 - \exp\{-\beta_1 x - \beta_2 x^2\}$. For step 2, suppose we fix some prespecified value for the BMR between 0 and 1. Given data from step 3, we calculate ML estimates of the regression parameters b_0, b_1, and b_2. The estimated extra risk is $\hat{R}_E(x) = 1 - \exp\{-b_1 x - b_2 x^2\}$.

To find the BMD as per step 4, above, note that the equation $R_E(x) = $ BMR is equivalent to $\beta_1 x + \beta_2 x^2 = -\log(1 - \text{BMR})$. For simplicity, let $C = -\log(1 - \text{BMR})$ and notice that $C > 0$, since $0 < \text{BMR} < 1$. Thus we can base the solution on a simple quadratic in x: $\beta_2 x^2 + \beta_1 x - C = 0$. As is well known, this has the two roots $\left(-\beta_1 \pm \sqrt{\beta_1^2 + 4\beta_2 C}\right)/2\beta_2$. If we exclude the trivial case $\beta_1 = \beta_2 = 0$, these roots will be distinct. Indeed, since $C > 0$ and, under the multistage model, we can require $\beta_j \geq 0$ for all j, the discriminant of the quadratic equation is $\beta_1^2 + 4\beta_2 C > 0$. Thus we see both roots are real.

To identify which root to use as the BMD, recall Descartes' famous rule of signs (see Borwein and Erdélyi, 1995, §3.2): the number of positive real roots for a polynomial is no larger than the number of sign changes in its coefficients. For $\beta_2 x^2 + \beta_1 x - C$, the number of sign changes is one ($\beta_2 \geq 0, \beta_1 \geq 0, -C < 0$), so the number of positive real roots is no larger than one. But since $4\beta_2 C \geq 0$, we see $\sqrt{\beta_1^2 + 4\beta_2 C} \geq \beta_1$, and thus the real root $\left(-\beta_1 + \sqrt{\beta_1^2 + 4\beta_2 C}\right)/2\beta_2$ must be positive. As such, there is exactly one real positive root and so we take it as the BMD.

Given the ML estimates from step 3, the corresponding ML estimate of BMD is

$$\widehat{\text{BMD}} = \frac{-b_1 + \sqrt{b_1^2 + 4b_2 C}}{2b_2}, \tag{4.11}$$

where $C = -\log(1 - \text{BMR})$.

Example 4.13 (PeCP Carcinogenicity, cont'd)   Returning to the PeCP carcinogenicity data from Example 4.7, recall that under a $k = 2$ multistage model the pertinent regression coefficients were estimated as $b_1 = 0.025$ and $b_2 = 0.022$. Thus for a benchmark risk of 10%, we take $\text{BMR} = 0.10$ and $C = -\log(1 - 0.1) = 0.1054$. The corresponding benchmark dose is estimated from (4.11) as

$$\hat{\text{BMD}} = \frac{-0.025 + \sqrt{0.025^2 + (4 \times 0.022 \times 0.1054)}}{2 \times 0.022}$$

or $\hat{\text{BMD}} = 1.69 \, \text{mg/kg/day}$. ✪

Example 4.14 (Linearized multistage model, cont'd)   For cases where $k > 2$ in the multistage model, the BMD operations become much more complex. For any $k \geq 1$, however, the linearization seen in Example 4.5 allows us to write $R_E(x) \approx \beta_1 x$. Thus we can set $\beta_1 x \approx \text{BMR}$, leading to $\text{BMD} \approx \text{BMR}/\beta_1$. Given an estimate of β_1 as b_1, we find $\hat{\text{BMD}} = \text{BMR}/b_1$. Note that the point estimate b_1 is calculated under the full k-order multistage model, but then used in the linearized approximation for BMD.

We should warn that the linearization here is notoriously inaccurate when the dose strays far from zero. Hence, it is best relegated to use only with very small BMRs, say, $\text{BMR} \leq 0.01$. ✪

Example 4.15 (Weibull Model, cont'd)   For the Weibull model in equation (4.8), the extra risk is $R_E(d) = 1 - \exp\{-\theta_0 d^{\beta_1}\}$. Once again, we write this as $R_E(d) = 1 - \exp\{-e^{\eta(d)}\}$ for $\eta(d) = \beta_0 + \beta_1 \log(d)$ and $\beta_0 = \log(\theta_0)$. To find the BMD at a fixed level of BMR, set $R_E(d) = \text{BMR}$ and solve for d. We find $\beta_0 + \beta_1 \log(d) = \log\{-\log(1 - \text{BMR})\}$, which leads to $\log(d) = [\log\{-\log(1 - \text{BMR})\} - \beta_0]/\beta_1$. Thus $\text{BMD} = \exp\{[\log\{-\log(1 - \text{BMR})\} - \beta_0]/\beta_1\}$. Given estimators b_0 and b_1, the estimated BMD becomes

$$\hat{\text{BMD}} = \exp\left\{\frac{\log\{-\log(1 - \text{BMR})\} - b_0}{b_1}\right\}.$$ ✪

4.3.2   Confidence limits on benchmark dose

To characterize uncertainty associated with the point estimator $\hat{\text{BMD}}$, it is standard practice to also determine a set of $1 - \alpha$ confidence limits for BMD (Crump and Howe, 1985). As noted in §4.2.2, risk-analytic considerations often favor only one-sided upper bounds on excess risk. This translates to one-sided lower $1 - \alpha$ bounds on BMD. Following Crump (1995), we refer to a $1 - \alpha$ lower confidence limit on the true BMD as a BMDL. To maintain consistency with the notation in the previous section, we denote this as $\hat{\text{BMDL}}$.

As was the case for $\hat{\text{BMD}}$, $\hat{\text{BMDL}}$ will vary from model to model. An example follows; see also Exercises 4.17 and 4.21–4.22.

Example 4.16 (Linearized multistage model, cont'd) For cases where $k \geq 1$ in the multistage model, we can use linearization to approximate a BMDL. In Example 4.14 we found the linearized BMD estimate as $\hat{\text{BMD}} = \text{BMR}/b_1$, where b_1 was calculated under the full k-order multistage model. For the BMDL, start with any $1 - \alpha$ upper confidence limit on β_1. Denote this as U_α, such that $P[\beta_1 \leq U_\alpha] \approx 1 - \alpha$; for example, we might take the upper $1 - \alpha$ Wald limit $U_\alpha = b_1 + z_\alpha se[b_1]$. Alternatively, Krewski and van Ryzin (1981) and Crump and Howe (1985) suggest that a likelihood ratio-based limit can be useful. Since standalone one-sided likelihood limits are often difficult to derive, they recommend constructing a $1 - 2\alpha$ two-sided interval for β_1 from the likelihood ratio statistic (§A.5.1), and then discarding the lower limit. Despite its *ad hoc* nature, Crump and Howe indicate that this approach can perform well in certain settings.

Using the upper-$(1 - \alpha)$ confidence limit on β_1, we can quickly find a lower limit on BMD. That is, since $P[\beta_1 \leq U_\alpha] \approx 1 - \alpha$, we have $P[\beta_1^{-1} \geq U_\alpha^{-1}] \approx 1 - \alpha$ and so, for any BMR > 0, $P[\text{BMR}/\beta_1 \geq \text{BMR}/U_\alpha] \approx 1 - \alpha$. But $\text{BMD} \approx \text{BMR}/\beta_1$ so this defines a BMDL; that is, take $\hat{\text{BMDL}} \approx \text{BMR}/U_\alpha$.

For instance, with the PeCP carcinogenicity data from Example 4.7 we saw under a $k = 2$ multistage model that $b_1 = 0.025$ with $se[b_1] = 0.041$. Thus an upper 95% Wald limit on β_1 is $U_{0.95} = b_1 + z_{0.05}se[b_1] = 0.025 + (1.645 \times 0.041) = 0.092$. As noted in Example 4.14, we do not recommend use of the linearization above BMR $= 0.01$. Applied here at that benchmark, we find $\hat{\text{BMDL}} \approx 0.01/0.092 = 0.107\,\text{mg/kg/day}.$ ✪

Here again, the BMDLs we describe hold for any single BMR of interest. As with the confidence limits for $R_E(x)$ discussed in §4.2.2, these represent *pointwise* $1 - \alpha$ limits on the BMD. If confidence statements are desired over a prespecified set of, say, $L > 1$ BMRs, a correction for multiplicity must be incorporated into the calculations. A common choice is again the Bonferroni adjustment from §A.5.4, which generally gives conservative bounds (Nitcheva *et al.*, 2005). For simultaneous BMDL bands, see Al-Saidy *et al.* (2003) or Piegorsch *et al.* (2005).

4.4 Uncertainty analysis

Uncertainty is a fundamental component in environmental risk assessment – indeed, in all areas of environmetric modeling – due to the broad variety of scientific inputs required for quantifying environmental risk. Probability and statistics are natural partners in this effort: probability distributions can characterize the stochastic components of and inputs to the phenomenon under study, while descriptive and inferential statistics provide information on the resulting outcome variable(s). Although we could apply the general concepts of uncertainty analysis to many of the modeling efforts described in other chapters of this text, we have chosen to highlight it here because of its important application in quantitative risk analysis.

For clarity, we follow convention and distinguish between uncertainty and variability. *Variability* describes the natural, underlying variation contained in the environmental process or phenomenon, such as biological variation in a population. This source of variation is inherent to the process and hence cannot be reduced or affected by any additional knowledge the investigator may obtain. By contrast, *uncertainty* describes those sources of variation in the process that can be affected by increasing the investigator's knowledge base. This can be, for example, the unknown value of a risk-function parameter or the number of stages in a multistage model. The goal in an uncertainty analysis is to understand and quantify contributions of the input variable(s) to statistical variation in the output variable when performing an environmental risk assessment.

Sources of uncertainty can be diverse. They include: (i) limited scientific knowledge of the environmental mechanisms or processes being studied, (ii) lack of sufficient data to provide the desired sensitivity to detect the environmental effect or (iii) to properly distinguish between competing models that represent the phenomenon, and (iv) inadequate background information on the values or ranges of a model's input parameters. In this section, we introduce some selected quantitative methods for approaching such uncertainties. Our presentation is necessarily brief; readers interested in more detail or a broader grounding will find the reviews by Griffiths (2002) or Smith (2002b) to be useful.

4.4.1   Uncertainty factors

At its simplest level, uncertainty analysis is used to bring the nature of uncertainty and its potential pitfalls to the risk assessor's and ultimately the risk manager's attention. For example, as described in §4.2.2 risk analysts often attempt to estimate reference doses or reference concentrations of acceptable levels of human exposure to an environmental toxin or other hazardous agent, using information from biological assays on test animals. For estimating an RfD or RfC, low-dose measures such as the BMD or BMDL from §4.3, or even the more limited NOAEL from §4.1.3, are common (US EPA, 1997). Calculation of a proper RfD or RfC is complicated, however, by a number of uncertainties. These include: (i) uncertainty in the form of model used to find the BMD or other statistical estimator; (ii) uncertainty in the differential effects of a hazardous agent when considered across different endpoints – e.g., neurotoxicity vs. developmental toxicity vs. carcinogenicity – if key data are absent on any endpoint; (iii) uncertainty in extrapolating results from limited-term, sub-chronic assays to longer-term, chronic exposures; (iv) uncertainty in extrapolating the low-dose results from the animal assay to humans; and (v) uncertainty in the differential sensitivity among members of the human population including, for example, extrapolation of adult-level results to children. Except perhaps for the last factor (intraspecies variability), all of these components deal primarily with uncertainties that could be reduced if more complete experimental information were available.

Some work has appeared that attempts to mitigate the uncertainty associated with extrapolating effects across species by using models of the uptake and elimination of hazardous environmental stimuli. These efforts include manipulation of physiologically based pharmacokinetic (PBPK) models to account for differences in species-specific tissue volumes, blood flows, metabolism, detoxification processes, etc. Bailer and Dankovic (1997) discuss uses of such PBPK models to assess uncertainty in a quantitative risk assessment; see also Becka (2002). While these methods may reduce uncertainty with across-species extrapolation, the uncertainties associated with unknown model parameters and with population variability in these traits remain. For more on problems with multiple sources of uncertainty, see Nair *et al.* (1995).

In the absence of such additional information, the simplest way to incorporate uncertainties when computing an RfD or RfC is to reduce the calculation via some form of downward scaling. That is, divide the BMDL by one or more *uncertainty factors* that represent substantive uncertainties in the computation. These are some-times called *safety factors*, since they provide a margin of safety in the eventual risk declaration (Dourson and Stara, 1983). In select cases, an additional *modifying factor* is applied to represent uncertainties not otherwise accounted for by the components described above. This might involve, for example, adjustment for low sample sizes in the experimental design or when selection of dose levels leads to a poor characteriza-tion of the exposure effect; see Crump *et al.* (1995, Table 1). Nominally, factors of up to 10 are used for each level of uncertainty felt to be pertinent in an individual RfD or RfC computation. Values other than 10 may be used if there is strong scientific judgment to support them (Gaylor and Kodell, 2000). For example, some recom-mendations exist for use of $\sqrt{10} = 3.163$ to account for dosimetric adjustments to the original data or for pharmacokinetic or other information on the agent's hazardous effects (US EPA, 1994, §4.3.9.1). (For simplicity, this is approximated as 3.0, although by convention when multiplying together two uncertainty factors of $3 \approx \sqrt{10}$ the product is taken equal to 10.) The total factor should not exceed 10 000, that is, no more than four to five full components of uncertainty (including a modifying factor, if necessary) should be present. Calculations requiring a total factor larger than this quite frankly question the validity of the study being used to support the risk assessment. Indeed, in practice the total uncertainty factor rarely exceeds 1000.

For example, suppose an RfD is desired on a pesticide used in agroforestry. A long-term, chronic-exposure assay is conducted in rats, and quantal data are recorded. Suppose also that from these data, a BMD at BMR $= 10^{-2}$ is calculated as $\hat{\text{BMD}} = 4.65$ ppm under a $k = 2$ multistage model. If it were felt that the database on this pesticide was fairly complete, and that the experimental design and data acquisition were of suitable quality, then the components of uncertainty pertinent to the RfD here would involve (i) interspecies extrapolation from rats to humans, (ii) intraspecies sensitivity among humans, and perhaps (iii) uncertainty in the possible simplicity of the $k = 2$ model. Thus the RfD might be calculated as

$$\text{RfD} = \frac{\hat{\text{BMD}}}{10 \times 10 \times 10} = \frac{\hat{\text{BMD}}}{10^3} = 0.0047 \text{ ppm.}$$

Readers may note that although the use of uncertainty factors incorporates true uncertainties in the RfD calculation, the issue of inherent variability is left unad-dressed. Indeed, any risk assessment will seem overly precise without some adjust-ment for both uncertainty and variability. At the simplest level, one can employ a statistical adjustment to the original point estimator that addresses variation. Thus, in the above illustration, suppose that the associated 95% $\hat{\text{BMDL}}$ is 3.19 ppm. Instead of employing the BMD (or, for that matter, an NOAEL) in the RfD calculation, we use this 95% lower limit to find $\text{RfD} = \hat{\text{BMDL}}/10^3 = 0.0032$ ppm.

For more complex adjustments one might attempt to generate an entire distribution of estimated BMDs or RfDs that reflect uncertainty in the larger risk estimation process (cf. §4.4.2, below).

For more on the use of uncertainty factors in human (and other) risk assessments, see Dourson and Derosa (1991), Gaylor *et al.* (1999), and US EPA (1997).

4.4.2 Monte Carlo methods

When the model used to characterize risk grows complex, simple corrections such as safety or uncertainty factors lose their ability to account for uncertainty in the analysis. In these cases, it is of interest to describe the entire statistical distribution of the environmental endpoint. Often, this is not difficult since the underlying statistical theory provides an analytical solution. For instance, suppose data are taken from a homogeneous-variance normal distribution under a simple linear regression model with mean response $\mu(x) = \beta_0 + \beta_1 x$, as in §1.1. Then computing a $1 - \alpha$ confidence interval on $\mu(x)$ at some environmentally pertinent value of x is a straightforward operation; see Example 4.17, below.

When the model involves a more complex combination of inputs, however, the distribution of the desired outcome variable may not support a simple or easily accessed solution. In these cases, it is common to approximate the unknown distribution by simulating the random outcome a large number of times via computer. This is an application of the *Monte Carlo method* mentioned in §A.5.2. Other names include *stochastic simulation, Monte Carlo simulation*, and *synthetic data generation*. When applied in a quantitative risk analysis, it may also be called *probabilistic risk assessment* (Carrothers *et al.*, 2002).

The Monte Carlo method uses the computer's ability to generate pseudo-random numbers as a surrogate for multiple repetitions of the actual process under study. (We use the term 'pseudo-random' to emphasize that no set of computer-generated numbers is truly random: some form of deterministic algorithm is used to make the generated variates appear random. Modern algorithms employ very clever, complex generators, however, so when properly applied they can mimic true randomness quite effectively.) That is, suppose an experimenter is interested in an environmental risk outcome, say, a BMD at BMR $= 10^{-2}$ for a suspected carcinogen. Given unlimited resources (and time), she could replicate her experiment M times, calculate each $\hat{\text{BMD}}$, and then plot all the M observed $\hat{\text{BMD}}$ values as a histogram. This would give an indication of the random variation in the $\hat{\text{BMD}}$, and allow her to report various percentiles of the empirical distribution of $\hat{\text{BMD}}$, give estimates of the distribution's central tendency, comment on the level of skew in the distribution, etc. When resources are limited, however, such an activity can only be simulated. Luckily, modern digital computers can undertake this effort in a rapid fashion.

Suppose more generally that the outcome of interest, U, can be written in generic form as $U = f(\mathbf{x}, \boldsymbol{\beta}, \varepsilon)$, where $\mathbf{x}$ is a vector of fixed explanatory variables and $\boldsymbol{\beta}$ is a vector of unknown parameters or other inputs that affect the outcome through the function f. The quantity ε is an additional stochastic error term which may itself depend upon unknown parameters; in many cases simple additive or multiplicative error models can suffice, $f(\mathbf{x}, \boldsymbol{\beta}, \varepsilon) = f(\mathbf{x}, \boldsymbol{\beta}) + \varepsilon$ or $f(\mathbf{x}, \boldsymbol{\beta}, \varepsilon) = f(\mathbf{x}, \boldsymbol{\beta}) \times \varepsilon$, respectively. If, say, $\boldsymbol{\beta}$ is the vector of (estimated) regression coefficients in a nonlinear model, and $\mathbf{x}$ is the value of the predictor variables at which a specific level of response, U, is to be estimated, then ε reflects the variability in the system. Of course, as in any model-dependent environmetric operation, such complex models may not truly represent reality, and often serve only to improve previous understanding or approximations of that reality.

For prespecified and fixed levels of $\mathbf{x}$, a Monte Carlo uncertainty analysis simulates $M > 0$ values of U, say $U_1, \ldots, U_M$, and from these makes risk-analytic statements about U, including any variation and uncertainty in its stochastic nature. The basic approach involves the following steps:

- Identify a quantifiable outcome, U, of environmental interest. Collect a pilot data set that provides information on this outcome.

- Formulate a model for the outcome in the form $U = f(\mathbf{x}, \boldsymbol{\beta}, \varepsilon)$.

- Specify the statistical distribution of ε and, in those settings where it too possesses stochastic components, of $\boldsymbol{\beta}$. Use information from the progenitor data set to select specific aspects of the distribution, such as the mean(s), variance(s), covariances, etc., of ε and $\boldsymbol{\beta}$.

- Generate pseudo-random data from these distribution(s) $M > 0$ times. Select M as large as is feasible given available computer resources. Values of $M \geq 10\,000$ are not unreasonable.

- Calculate a U_j from each of the $j = 1, \ldots, M$ pseudo-random data sets and collect these together into a empirical distribution for U.

- Construct summary measures and plots to describe pertinent risk-analytic aspects of the distribution of U. Common examples include histograms, empirical distribution plots, and summary statistics (Millard and Neerchal, 2001, Ch. 13).

In effect, the method attempts to explore some of the uncertainty in U that can be attributed to uncertainty in the input parameters $\boldsymbol{\beta}$. The following example gives a simple illustration.

Example 4.17 (Motor vehicle CO_2, cont'd) Consider again the data on motor vehicle use and CO_2 in the UK for the period 1971–1998, as seen in Example 1.1. We saw in that example that the ML estimates of the simple linear regression coefficients were $b_0 = 28.3603$ and $b_1 = 0.7442$. From Fig. 3.2, the corresponding MSE was 7.540.

As seen in Table 1.1, increases in the predictor variable, $x = $ vehicle use (relative to 1970), occur almost exactly in step with increasing time – over time, more vehicles are being used in the UK. For purposes of assessing risk to the environment of consequent increases in CO_2 emissions, one might be interested in predicting future CO_2 emissions at some high level of relative vehicle use. In fact, by the year 2000 data were available on vehicle use and, indeed, the recorded relative use increased to approximately $x = 233$. Thus we ask, what value will $Y = CO_2$ emissions (relative to 1970) take when vehicle use attains 233 relative units? This involves, in effect, prediction of Y at $x = 233$.

Of course, the exact solution to this problem is well known for a simple linear regression model. We estimate Y at $x = 233$ as $\hat{Y}(233) = b_0 + 233b_1 = 28.3603 + (233 \times 0.7442) = 201.759$ relative units. Also, an exact 95% prediction interval for $Y(233)$ is given by $\hat{Y}(233) \pm t_{0.025}(n - 2)se[\hat{Y}(233)]$, where

$$se[\hat{Y}(x)] = \sqrt{MSE\left\{1 + \frac{1}{n} + \frac{(x - \bar{x})^2}{\sum_{i=1}^{n}(x_i - \bar{x})^2}\right\}}$$

(Neter $et\,al.$, 1996, §2.5). Here, with $n = 28$, $\bar{x} = 163.5466$, and $\sum_{i=1}^{n}(x_i - \bar{x})^2 = 47028.472$, we find $201.759 \pm (2.056 \times 2.930) = 201.759 \pm 6.023$, or $195.74 \leq Y(233) \leq 207.78$.

A Monte Carlo uncertainty analysis may be appropriate here, however: the prediction occurs slightly outside the range of the data and it is well known that extrapolations outside the data range can lead to spurious inferences. A variety of Monte Carlo strategies are possible for implementing the analysis. We will choose the following approach: from realized values of b_0 and b_1, we reconstruct the Y_is to predict the response at $x = 233$. In effect, we are sampling from the joint distribution of the observations to predict the value of Y.

Our output value of interest is $U = \hat{Y}(233)$. This is modeled as $\hat{Y}(233) = b_0 + 233b_1 + \varepsilon$, where b_0, b_1, and ε are the stochastic input parameters for the model. We will replicate M entire sets of $n = 28$ data pairs (x_i, Y_i) and then for each replicated set fit a new b_0, b_1, and ε. To do so, we view the x_is as fixed constants at their observed values, that is, we set $x_1 = 105.742$, $x_2 = 110.995, \ldots, x_{28} = 229.027$, and then assign each Y_i a normal distribution with mean $b_0 + b_1x_i$ and variance σ^2. Here, we use the summary statistics obtained in Example 1.1 to specify these underlying input parameters: $b_0 = 28.3603, b_1 = 0.7442$, and $\sigma^2 = 7.540$ via the MSE. This is $Y_i \sim$ indep. $N(28.3603 + 0.7442x_i, 7.540)$ for $i = 1, \ldots, 28$. To generate the various normal variates for this Monte Carlo exercise, we first generate a standard normal deviate, Z, using a basic random generator (available in computer packages such as Microsoft® Excel, Minitab®, or S-Plus®). Then, we transform Z to achieve the desired mean and variance structure: $Y_i = b_0 + b_1x_i + Z\sqrt{MSE}$ (Gentle, 2003, §5.3).

Given the Monte Carlo pairs (x_i, Y_i), we fit a linear regression and find the estimated intercept and slope, b_0 and b_1, respectively. With these, we generate a new, independent, standard normal deviate Z and construct a new error term $\varepsilon = Z\sqrt{MSE}$. The result is $U = \hat{Y}(233) = b_0 + 233b_1 + \varepsilon$. Repeat this effort M times and base any desired inferences on the resulting Monte Carlo distribution of $\hat{Y}(233)$. We set $M = 10\,000$.

Figure 4.9 plots a histogram of the empirical frequency distribution from the resulting $10\,000$ pseudo-random predicted values. A clearly symmetric, unimodal shape is evidenced, corroborating the theoretical indication that the sampling distribution of $\hat{Y}(233)$ is itself normal. The empirical mean is $\bar{U} = 201.757$, mimicking the theory-based estimate of $\hat{Y}(233) = 201.759$. Also, the 2.5th and 97.5th percentiles of the empirical distribution are 196.03 and 207.43. Since these percentiles bracket 95% of the empirical outcomes, they approximate a 95% interval estimator for

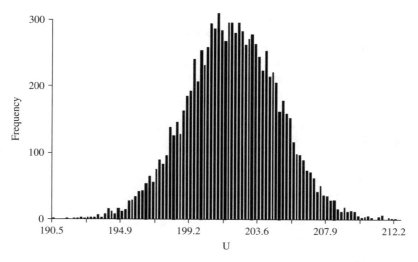

Figure 4.9 Histogram of 10 000 pseudo-random predicted values, $U = \hat{Y}(233)$, for motor vehicle CO_2 data

$Y(233)$. Although these limits are slightly tighter than the theory-based interval calculated above, the overall analysis still helps validate the inferences achieved using the basic theory.

Alert readers will notice that the statistical method employed in Example 4.17 is a form of parametric bootstrap (Dixon, 2002a), similar to that described in §A.5.2. Of course, the example serves as only an introduction to Monte Carlo methods for risk analysis. In more complex environmental risk assessments, model formulations can include large numbers of input variables, with diverse stochastic and non-stochastic characteristics. In such settings, the large number of input parameters requires the evaluation of a large number of Monte Carlo simulations to study the uncertainty fully. This can make the Monte Carlo approach difficult – or at least expensive – to implement. Instead, one typically selects a reduced subset of parameter inputs that still produces a representative series of model outputs. Iman and Helton (1988) studied a series of such reduced-computing methods, and concluded that an approach known as *Latin hypercube sampling (LHS)* exhibited the best adaptability to different levels of complexity. Coupled with its relative ease of implementation and its generally higher precision in estimating extreme features of the outcome (the 'tails' of the distribution), they found LHS to be an important tool for studying uncertainty with highly complex input–output models.

The LHS approach attempts to make more efficient use of the computer's operational power by stratifying the range of input variables into a series of $K > 1$ mutually exclusive intervals, each with equal probability. In each simulation run, one value is sampled randomly from each interval. The K values chosen for the first input variable are then paired randomly with the K values chosen for the second input variable. The resulting K pairs are next combined randomly with the K values of the third input variable to form K triplets, and so on. If $p \geq 1$ is the number of input variables under study, the result is a set of K p-tuples from which to assess the variation in the outcome variable(s). LHS was originally described by McKay *et al.* (1979) for use in industrial applications. It has since seen extensions to many areas of probabilistic risk analysis and stochastic simulation. For a full introduction to the approach, see Iman and Shortencarier (1984).

4.5 Sensitivity analysis

An important consideration when designing a Monte Carlo uncertainty analysis is whether or how an environmental stimulus or other input variable affects the outcome variable. To examine how different inputs impact the model outcome, we apply what is known as a *sensitivity analysis* to the risk assessment. Typically, this involves study of how the input variables affect the magnitude of the output variable; for example, calculating the first and second partial derivative(s) of the response function with respect to the inputs to quantify first- and second-order local sensitivity (Campolongo *et al.*, 2000), or computing correlations or other summary measures of association between the input and output variables (Helton and Davis, 2000). In some cases, actual output data are difficult or undesirable to acquire: when assessing risk of radioactive contamination from a nuclear plant accident or disposal mishap, say, the fewer outcomes that occur, the better! If so, and if adequate computing resources are available, we often use the Monte Carlo method to simulate and study the effects of different model inputs/assumptions on the output variable.

Similar to the allied case of uncertainty analysis from §4.4, methods of sensitivity analysis can and often should be applied to any statistical modeling effort. We give a brief introduction to some of the simpler concepts here in order to highlight their use when conducting quantitative risk assessments.

4.5.1 Identifying sensitivity to input variables

At its most basic level, a sensitivity analysis examines the input variables to find those that contribute heavily to the outcome variable and upon which to focus attention. Typically, variable plots and other graphical devices are used to identify important inputs, using simple measures of agreement. Natural quantities here include correlation(s) between the outcome and the various stimulus/input variables, standardized regression coefficients, and R^2 values (coefficients of determination) from multiple or stepwise regression analyses. For example, a useful graphic for summarizing the total impact of many input variables on a single outcome is known as the *tornado diagram*. This is simply a vertical bar chart of the pairwise product-moment correlations,

$$R_j = \frac{\sum_{i=1}^{n}(x_{ij} - \bar{x}_j)(Y_i - \bar{Y})}{\sqrt{\sum_{i=1}^{n}(x_{ij} - \bar{x}_j)^2 \sum_{i=1}^{n}(Y_i - \bar{Y})^2}}, \tag{4.12}$$

between the outcome, Y_i, and each input/regressor variable, $x_{ij}, j = 1, \ldots, p$, where $\bar{x}_j = \sum_{i=1}^{n} x_{ij}/n$ and p is the number of stimulus/input variables under consideration. If desired, rank correlations (Spearman, 1908) may be used instead of the product-moment correlations R_j. Or, to adjust the correlation between the outcome variable and each input variable for the impact of all other inputs, partial correlation coefficients (Neter *et al.*, 1996, §7.4; Helton and Davis, 2000) can be employed.

A tornado diagram charts the correlations from largest to smallest in absolute value (but, the sign is retained on the diagram), indicating visually which inputs have greatest correlative impact on the outcome. Ostensibly, the outcome variable should be more sensitive to these highly correlated inputs.

Example 4.18 (Brine flow sensitivity) Many geological and environmental factors impact radiation leakage from underground radioactive waste sites, and these factors must be considered carefully when developing such sites for long-term use. Towards this end, Helton and Davis (2000) gave summary data on how selected inputs affect predicted measures of fluid flow through a potential waste storage region. For example, for the outcome variable $Y =$ undisturbed-condition cumulative brine flow, $p = 8$ input variables were seen to have product-moment correlations (4.12) as given in Table 4.1.

A tornado diagram built from the correlations in Table 4.1 appears in Fig. 4.10. From it, we see that the first two variables, $x_1 =$ log anhydrite permeability and $x_2 =$ cellulose degradation, exhibit the greatest (linear) correlations with the predicted brine flow, followed by $x_3 =$ brine saturation of waste, $x_4 =$ steel corrosion rate, and $x_5 =$ anhydrite pore distribution. (Helton and Davis also gave the associated pointwise P-values for testing whether the true correlations with Y are zero for each of the x_js in Table 4.1, useful as adjunct measures of importance. These five variables were the only ones for which $P \leq 0.05$.) Thus, in terms of which variables to consider for more careful analysis of predicted fluid flow through the site, x_1 and x_2,

Table 4.1 Input variables and product-moment correlations (4.12) with cumulative brine flow at a radioactive waste storage location

Input variable	R_j
$x_1 =$ Log anhydrite permeability	0.5655
$x_2 =$ Cellulose degradation indicator	−0.3210
$x_3 =$ Brine saturation of waste	−0.1639
$x_4 =$ Steel corrosion rate	−0.1628
$x_5 =$ Anhydrite pore distribution	−0.1497
$x_6 =$ Microbial gas generation scaling	−0.1105
$x_7 =$ Waste residual brine saturation	−0.1080
$x_8 =$ Halite porosity	−0.0969

Source: Helton and Davis (2000).

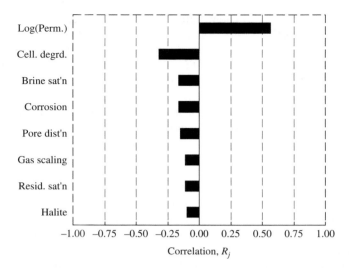

Figure 4.10   Tornado diagram of $p = 8$ input variable correlations with output brine flow, from Table 4.1

and perhaps also x_3 through x_5, provide the dominant contributions. We expect that predicted brine flow will be sensitive to their impact, and any risk assessments of the proposed site should be careful to include them.

Tornado diagrams can be modified to account for different types of importance measures. For instance, if the input variables have yet to be sampled or recorded, one cannot calculate correlation coefficients such as (4.12). But if the possible range of each variable is known, this can be presented instead in the tornado diagram: chart the variables, ordered in a decreasing fashion by range with the widest at top. (For more of a 'tornado' effect, center each at its median or, if known, its baseline or reference level.) If the variables range over different orders of magnitude, then chart their percentage change as [maximum/minimum] − 1. The resulting graphic is similar to a correlation-based tornado diagram, indicating potential output sensitivity to inputs with very large ranges.

Example 4.19 (Biofuel production)   Development of biofuels as a major alternative energy source is of growing worldwide interest. For instance, Chinese officials are studying *biogas*, a methane-rich fermentation product typically derived from animal dung, human sewage, or crop residues (Van Groenendaal and Kleijnen, 2002). Biogas fuel production aids in the recycling of waste products, does not typically lead to serious atmospheric pollution, and has the potential to be an important source of renewable energy. A major obstacle to increased biogas production is, however, its high production costs, particularly those of initial plant construction. Environmental economists must determine whether a biogas plant will recover its investment costs and achieve positive net present value (NPV). (NPV is the net return on investment, modified to reflect the effects of interest rates and other

Table 4.2 Input variables and their anticipated summary characteristics for a biogas plant's net present investment value

Input variable	Minimum	Maximum	% change
x_1 = Percentage waste as industrial byproducts	6.3	9.4	49.21
x_2 = Percentage waste as chicken dung	9.1	13.7	50.55
x_3 = Total raw waste (tons/yr)	25 000	37 000	48.00
x_4 = Total investment costs (yuan)	3 700 000	6 200 000	67.57
x_5 = Reduction in env. cleanup costs (yuan/yr)	460 000	670 000	45.65
x_6 = Desulferizer costs (yuan)	1 627	2 441	50.03
x_7 = Biogas price (yuan/m^3)	0.6	1	66.67
x_8 = Sludge processing costs (yuan/ton)	1.31	1.95	48.85
x_9 = Biogas plant production efficiency (m$^3_{gas}$/m$^3_{digest}$)	0.829	1.171	41.25

Source: Van Groenendaal and Kleijnen (2002).

economic factors. Higher NPVs indicate better investment opportunities.) That is, they view NPV as an outcome variable and assess if it is large enough to warrant the investment.

To identify which environmental and economic inputs have the greatest impact on the anticipated NPV for a major biogas plant, a sensitivity analysis was performed. Table 4.2 lists a series of potential input variables, along with their known or anticipated minima and maxima. Since the ranges differ greatly in their scales of magnitude, we will operate with the percentage change, 100[(maximum/minimum) − 1], also given as part of Table 4.2.

A tornado diagram of percentage change appears in Fig. 4.11. There, we see that the greatest impacts are associated with the two major price inputs, x_4 = total investment costs and x_7 = biogas price. (This is, perhaps, not surprising.) All other input variables exhibit a lesser change over their known or anticipated ranges, and thus may not have as great an impact on the plant's NPV. Future analyses should include x_4 and x_7, at a minimum, due to the NPV's expected sensitivity to their effects. ✸

Along with the tornado plot, a variety of graphical tools may be employed for plotting sensitivity indices and identifying important input variables, including devices to visualize complex, multidimensional effects. See Cooke and van Noortwijk (2000) for a general overview.

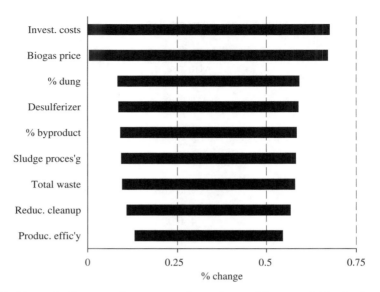

Figure 4.11 Tornado diagram of percentage change in $p = 9$ input variables for determining a biogas plant's net present investment value

4.5.2 Correlation ratios

Sensitivity measures must also incorporate any random variability to which the input variables are themselves subject. For instance, suppose we can ascribe a probability distribution to each X_j, $j = 1, \ldots, p$, with p.d.f. $f_{X_j}(x)$. (For simplicity, we assume X_j is continuous, although modifications to accommodate discrete input variables are also possible.) Using this, we study changes in variation of Y as X_j changes, relative to the total variation in Y. To adjust Y for variation in X_j, we operate with the conditional expectation: $E[Y|X_j = x]$. That is, when X_j is fixed at some realized value x, $E[Y|X_j = x]$ is the value we expect the random outcome Y to attain. (Notice that we are employing the conditional operator notation, as seen earlier in §3.3.2.) The variance of this conditional expectation (VCE) is

$$\text{VCE} = \text{Var}\{E[Y|X_j = x]\} = \int \{E[Y|X_j = x] - E[Y]\}^2 f_{X_j}(x)\mathrm{d}x.$$

Of interest is how the VCE compares to the unconditional variance of Y, $\text{Var}[Y]$. Helpful here is the fact that $\text{Var}[Y]$ can be decomposed into a sum of two terms, the first of which is the VCE and second of which can be viewed as the remaining variability in Y ascribable to other sources of variation when X_j is held fixed (Chan et al., 2000, §8.2; Casella and Berger, 2002, §4.4): $\text{Var}[Y] = \text{VCE} + E\{\text{Var}[Y|X_j = x]\}$. Thus, if VCE is large relative to $\text{Var}[Y]$, a large percentage of the variation in Y is related to variation in $E[Y|X_j]$ as X_j varies. This gives X_j a large impact on the random outcome.

A natural importance measure that incorporates these quantities is known as Pearson's (1903) *correlation ratio*,

$$\eta_j^2 = \frac{\mathrm{Var}\{E[Y|X_j]\}}{\mathrm{Var}[Y]}$$

for $j = 1, \ldots, p$. (Be careful not to confuse the symbol for the correlation ratio, η_j, with the similar symbol for a linear predictor, η_i, used throughout Chapter 3.) Since η_j^2 gives the proportion of variation in Y attributable to variation in $E[Y|X_j]$, it has interpretation similar to the squared correlation between Y and X_j,

$$\rho_j^2 = \frac{(\mathrm{Cov}[Y, X_j])^2}{\mathrm{Var}[Y]\mathrm{Var}[X_j]},$$

that is, the population analog to the square of (4.12). Indeed, if the relationship between Y and X_j is exactly linear, η_j^2 will equal ρ_j^2, while more generally $0 \le \rho_j^2 \le \eta_j^2 \le 1$. Kendall and Stuart (1967, §26.21) give other relationships between the two quantities, and also describe tests of hypotheses on η_j^2 based on analogous sample quantities. For use as a sensitivity/importance measure, large values of η_j^2 indicate potential sensitivity in Y to that particular input, X_j.

Example 4.20 (Nonlinear response) In ecological risk analysis the relatively simple nonlinear relationship $Y = X_2^4/X_1^2$ sometimes appears (Gardner *et al.*, 1981; Campolongo *et al.*, 2000). Suppose the input variables are themselves random, with independent uniform distributions (§A.2.7): $X_j \sim$ i.i.d. $U(1 - \varphi, 1 + \varphi)$, for $0 < \varphi < 1$. Given φ, we ask whether X_1 or X_2 has a greater impact on variation in Y, or if they appear to be equally important. That is, is η_1^2 substantively different from η_2^2?

Unfortunately, advanced methods of probability and integral calculus are required to find the exact values of the two correlation ratios under this model. (We explore this in detail as part of Exercise 4.28.) When $\varphi = \frac{1}{2}$, however, results given by Campolongo *et al.* (2000, §2.9) show that $\mathrm{Var}[Y] = 6.9012$, $\mathrm{Var}\{E[Y|X_1]\} = 1.8075$, and $\mathrm{Var}\{E[Y|X_2]\} = 3.5264$. Thus $\eta_1^2 = 0.262$ and $\eta_2^2 = 0.511$. Hence for $\varphi = \frac{1}{2}$, X_2 appears to have much greater impact on Y than does X_1. ✪

One can extend the sensitivity analysis to study how the η_j^2s change as pertinent parameters of the model vary. For instance, in Example 4.20, a plot of $\eta_2^2/\eta_1^2 = \mathrm{Var}\{E[Y|X_2]\}/\mathrm{Var}\{E[Y|X_1]\}$ against φ or against $\mathrm{Var}[X_j] = \frac{1}{3}\varphi^2$ gauges how the correlation ratios are affected by changes in φ; see Exercise 4.28.

Of course, many environmental risk scenarios exist where the model formulation is more complex than that seen in Example 4.20. In some of these cases, direct calculations to find η_j^2 may not be possible. To overcome this, risk analysts often appeal to the Monte Carlo method and simulate a large number of pseudo-random outcomes from the posited model. Then, using the simulated values, correlation ratios or other importance measures can be calculated in order to study the input variables. In this case, techniques such as LHS from §4.4 can prove useful (Iman and Helton, 1988). Methods for computing correlation ratios from LHS Monte Carlo studies and other forms of data are reviewed by Chan *et al.* (2000, §8.2).

4.5.3   Identifying sensitivity to model assumptions

When computing resources permit, one can also employ the Monte Carlo method and change the distribution used to model the stochastic features of either the input variable(s) or the outcome (or both). Sensitivity to such changes may indicate stability problems with the model under study. For example, the lognormal distribution (§A.2.10) is popular for modeling highly skewed variation in environmental variables. A possible alternative to the lognormal is the gamma distribution (§A.2.8), which also can accommodate a large skew but with generally lighter upper tails. Sensitivity analyses can be used to determine whether a skewed input or outcome variable is sensitive to this choice of distribution, and if so, how the sensitivity changes as pertinent features of the model change. This can aid in assessing the degree of risk associated with the environmental outcome.

Experienced readers will note that sensitivity analysis has close ties with the study of *robustness* in statistical inference (Hampel, 2002), and many interrelated themes exist between them. More complete, modern surveys of the method are available in the review by Fassó and Perri (2002) and the collection by Saltelli *et al.* (2000).

4.6   Additional topics

This chapter has provided a short introduction to topics in quantitative risk assessment for environmental analysis. The broad scope of the field of risk analysis is far too complex, however, for one single chapter to survey comprehensively. Our goal has been to highlight those issues of pertinence for quantitative analysis with single environmental exposures to a single population. Nonetheless, a few additional topics are worth mentioning. One important area of environmental risk assessment concerns *complex mixtures* of exposures, where two or more chemical agents make up the environmental exposure/hazard. The simultaneous effects of multiple chemical exposures can interact in a variety of ways, leading to increases, decreases, or even no change(s) in the response relative to that predicted from any separate, individual exposures. Environmetric analysis of exposure data from complex mixtures requires far more elaborate statistical methods than those we present above, and is an area of ongoing research (Kodell and Pounds, 1991; Hertzberg *et al.*, 1999).

Associated with the issue of complex mixtures is the problem of distinguishing the differential effects of a toxic environmental insult, say, inhibited reproductive output when coupled with reduced organism survival in an ecotoxicity study (Wang and Smith, 2000). Indeed, the general problem of how to model and analyze differential mortality/ survival due to hazardous environmental exposures is a broad area of study, extending well beyond the limited space available here. We gave a deeper discourse on statistical methods for survival analysis, and also provided some useful references for further reading, in Piegorsch and Bailer (1997, Ch. 11).

It is also worth mentioning that uncertainty analyses can be used for assessing risk in a broad variety of environmental settings. An interesting example is their use in pesticide exposure assessments to study the distribution of possible exposures. Since pesticides are often applied in a variety of contexts – e.g., different soil types, different proximities to aquifers, different application schedules – an uncertainty analysis can be used to incorporate variability associated with each or any combination of these applications. For example, one might incorporate variable meteorological conditions and soil conditions when estimating the distribution of pesticides reaching an aquifer. This is related to fate and transport modeling of the pesticide through the various water and soil media (Liu *et al.*, 1999). See the US EPA's website at http://www.epa.gov/ceampubl for more on exposure assessment modeling.

A final topic of note is how statistical considerations may apply to the risk characterization process. For instance, multiple species' sensitivities to a hazardous agent might be quantified in an ecological risk assessment by the distribution of ED_{50}s across the species. This sensitivity distribution could then be compared to an exposure distribution to estimate which level/concentration of the agent poses significant risk to a small fraction of species. *Expert judgment* plays an important role in this regard – indeed, in all of quantitative risk assessment – and must be employed in a thoughtful and effective manner (Sielken and Stevenson, 1997). A useful Internet site for learning more about ecological or human health risk assessment is the US EPA's National Center for Environmental Assessment at http://www.epa.gov/ncea/. This site includes guidelines for cancer and ecological risk assessment, along with software tools available for risk modeling and exposure characterization.

Exercises

4.1. Consider the simple linear logistic model $\pi(x) = 1/(1 + \exp\{-\beta_0 - \beta_1 x\})$.

(a) If $\beta_1 \neq 0$, verify that, as indicated in §4.1.1, the ED_{50} is $-\beta_0/\beta_1$.

(b) Comment on what value would be assigned if no dose-response pattern was observed – i.e., $\beta_1 = 0$ – in a set of data.

(c) Suppose that two chemicals are compared using their respective ED_{50}s. If chemical A is more potent than chemical B, what does this imply about their ED_{50}s? their β_1s?

4.2. Suppose $F(\eta)$ is an increasing, continuous function such that $0 < F(\eta) < 1$ and $F(\delta) = \frac{1}{2}$. Let $F(\eta)$ model a quantal response probability via $\pi(x) = F(\beta_0 + \beta_1 x)$. Show that the ED_{50} for this model is always of the form $ED_{50} = (\delta - \beta_0)/\beta_1$.

4.3. Fit a probit link to the following data sets. Use the fit to estimate LC_{50} or ED_{50}, as appropriate, and to find 95% Fieller confidence limits. How might you assess if the probit provides a better fit to the data than any other link used previously? [*Hint*: recall that the probit model is a special form of GLiM from §3.3.2.]

(a) The copper toxicity data in Example 4.1.

(b) The γ-ray exposure data in Exercise 3.9.

4.4. Recall that for the solid waste effluent data from Exercise 3.17, the counts did not appear overdispersed. As such, consider a Poisson parent assumption for their p.m.f. and consider a simple log-linear model in $x =$ effluent concentration, as in part (e) of that exercise. Over increasing levels of x, we might expect the mean offspring counts to decrease. From the fit in Exercise 3.17(e), estimate the EC_{50} (the concentration producing a 50% drop from the mean response at $x = 0$) and calculate a 99% confidence interval for this value.

4.5. Show that under the simple linear logistic model $\pi(x) = 1/(1 + \exp\{-\beta_0 - \beta_1 x\})$ with $\beta_1 \neq 0$, the $ED_{100\rho}$ is given by equation (4.2). Verify that at $\rho = 0.5$, this reduces to the ED_{50}.

4.6. Return to the following data sets and find the specified effective dose; also include 95% Fieller confidence limits. Be sure to apply an appropriate link function.

(a) LC_{01} for the copper toxicity data in Example 4.1.

(b) ED_{10} for the γ-ray exposure data in Exercise 3.9.

4.7. For the following stimulus-response functions, use differential calculus to find the slope at zero, $\mu'(0)$.

(a) $\mu(x) = \beta_0 + \beta_1 x$ for $\beta_1 \geq 0$.

(b) $\mu(x) = \beta_0 + \beta_1 x + \beta_2 x^2$ for $\beta_2 \geq 0$.

(c) $\mu(x) = \beta_0 + \beta_2 x^2$ for $\beta_2 \geq 0$.

(d) $\mu(x) = \exp\{-\beta_2 x^\theta\}(1 - \exp\{-\beta_0 - \beta_1 x\})$ for $\theta > 1$.

(e) $\mu(x) = \exp\{-\beta_2 x^\theta\}(\beta_0 + \beta_1 x)$ for $\theta > 1$.

(f) $\mu(x) = \exp\{\beta_0 + \beta_1 x + \beta_2 x^2\}$ for $\beta_2 \geq 0$.

(g) $\mu(x) = \beta_0 + \beta_1 x^{\beta_2}$ for $\beta_2 > 1$.

4.8. Verify the following from Example 4.4.

(a) Under the extreme-value model in (4.6), the extra risk is $R_E(x) = 1 - \exp\{e^{\beta_0}(1 - \exp[\beta_1 x + \cdots + \beta_k x^k])\}$.

(b) Under the three-parameter Weibull model in (4.8), the extra risk is $R_E(x) = 1 - \exp\{\theta_0 d^{\beta_1}\}$.

4.9. Suppose a risk function has the form $R(d) = 1 - \exp\{-\beta_0 - g(d)\}$ for some function $g(d)$ satisfying $g(0) = 0$. Let $\beta_0 = -\log(1 - \varphi_0)$ and show that $R(d)$ can be written in a form that mimics the three-parameter Weibull risk function in equation (4.8).

4.10. Show that for any real value u near $0, 1 - e^{-u} \approx u$ by applying a first-order Taylor series expansion (§A.6.1) to the function $f(u) = 1 - e^{-u}$ near the point $u = 0$. See also Exercise 2.10.

4.11. Crump et al. (1977) give data on the proportion of laboratory mice exhibiting hepatocellular tumors after exposure to the pesticide dieldrin over the following dose levels: $x_1 = 0.0$ ppm (control), $x_2 = 1.25$ ppm, $x_3 = 2.5$ ppm, and $x_4 = 5.0$ ppm. These are: $Y_1/N_1 = 17/156$, $Y_2/N_2 = 11/60$, $Y_3/N_3 = 25/58$, and

$Y_4/N_4 = 44/60$. Assume a multistage model is appropriate for these data, and estimate the extra risk function $R_E(x)$ via maximum likelihood. Assess if the $k = 2$ term in the model is required (operate at $\alpha = 0.05$).

4.12. Dunson (2001) gives data on the proportion of transgenic female Tg.AC mice exhibiting skin papillomas after 20-week dermal exposures to lauric acid diethanolamine (LADA) over the following dose levels: $x_1 = 0\,\text{mg}$ (control), $x_2 = 5\,\text{mg}$, $x_3 = 10\,\text{mg}$, and $x_4 = 20\,\text{mg}$. These are: $Y_1/N_1 = 1/18$, $Y_2/N_2 = 3/12$, $Y_3/N_3 = 10/13$, and $Y_4/N_4 = 12/13$. Assume a multistage model is appropriate for these data, and estimate the extra risk function $R_E(x)$ via maximum likelihood. Assess if the $k = 2$ term in the model is required (operate at $\alpha = 0.05$).

4.13. Suppose a risk assessment is conducted using continuous measurements such as body weight losses in laboratory animals, or time-to-response outcomes in neurotoxicity trials. Suppose the data satisfy $Y_i \sim$ indep. $N(\mu(x_i), \sigma^2)$, where x_i measures the environmental stimulus at the ith exposure level, $i = 1, \ldots, n$. Define the risk function as $R(x) = P[Y \leq \mu(0) - \delta\sigma]$, where $\delta > 0$ is a fixed tuning constant. (Thus a response which is more than δ standard deviations below the control mean is viewed as adverse. Common values for δ include 2.33 and 3.26.) Find the additional risk function $R_A(x)$ in terms of the standard normal c.d.f., $\Phi(\cdot)$, under the following models for $\mu(x)$:

(a) $\mu(x) = \beta_0 + \beta_1 x$.

(b) $\mu(x) = \beta_0 + \beta_1 x + \beta_2 x^2$.

(c) $\mu(x) = \beta_0 + \beta_1 x^\kappa$ for some fixed $\kappa > 0$.

(d) $\mu(x) = \beta_1 x + \beta_2 x^2$.

(Kodell and West, 1993)

4.14. Under the normal-distribution model employed in Exercise 4.13, let $R(x) = P[Y \leq \mu(0) - \delta\sigma]$ and view δ as a fixed constant. Show that $R_E(x)$ is proportional to $R_A(x)$. [*Hint*: find $R(0)$ in terms of the standard normal c.d.f., $\Phi(\cdot)$.]

4.15. Show how to construct a $1 - \alpha$ upper confidence limit on the extra risk R_E under a $k = 1$ multistage model.

4.16. Return to the following data sets and construct 99% upper confidence limits on $R_E(x)$ at the specified dose level(s). If $L > 1$ levels are given, adjust for multiplicity using a Bonferroni correction.

(a) The dieldrin carcinogenicity data from Exercise 4.11 under a $k = 2$ multistage model. Set $x = 0.25\,\text{ppm}$.

(b) The dieldrin carcinogenicity data from Exercise 4.11 under a $k = 2$ multistage model. Set $x \in \{0.25, 0.50, 0.75, 1.0, 1.25\}\,\text{ppm}$.

(c) The LADA carcinogenicity data from Exercise 4.12 under a $k = 1$ multistage model. (see Exercise 4.15). Set $x = 15\,\text{mg}$.

(d) The LADA Carcinogenicity Data from Exercise 4.12 under a $k = 1$ multistage model (see Exercise 4.15). Set $x \in \{5, 10\}\,\text{mg}$.

4.17. Mimic the construction in Example 4.12 and construct a BMD at some prespecified BMR for an extra risk function $R_E(x)$ under a $k = 1$ multistage model. (Must you make any assumptions on the model for the BMD to be valid?) Also show how to construct a $1 - \alpha$ BMDL for the given BMR.

4.18. Return to the following data sets and estimate a BMD from the extra risk function $R_E(x)$ at the given BMR.

   (a) The dieldrin carcinogenicity data from Exercise 4.11 under a $k = 2$ multistage model. Set BMR $= 10^{-1}$.

   (b) The LADA carcinogenicity data from Exercise 4.12 under a $k = 1$ multistage model (see Exercise 4.17). Set BMR $= 10^{-2}$.

4.19. Under a linearized extreme-value model, as in Example 4.6, set the extra risk, $R_E(x)$, equal to some prespecified benchmark risk, BMR, and solve for the corresponding BMD.

4.20. Return to the continuous-data setting described in Exercise 4.13.

   (a) For $\mu(x) = \beta_0 + \beta_1 x$, solve for the BMD by setting the additional risk, $R_A(x)$, equal to some prespecified benchmark risk, BMR.

   (b) Repeat part (a), but now use $\mu(x) = \beta_0 + \beta_1 x + \beta_2 x^2$. Assume the mean response is decreasing for large x, so that $\beta_2 < 0$.

4.21. Return to the three-parameter Weibull model in Example 4.15, where the BMD was estimated as $\hat{BMD} = \exp\{[\log\{-\log(1 - BMR)\} - b_0]/b_1\}$. Construct an approximate BMDL based on a delta-method approximation for $se[\hat{BMD}]$, as follows:

   (a) View $\hat{BMD}$ as a function of b_0 and b_1, say $h(b_0, b_1)$, and find the partial derivatives $\partial h/\partial b_0$ and $\partial h/\partial b_1$. For simplicity, denote the constant $\log\{-\log(1 - BMR)\}$ as C.

   (b) With the derivatives you found in part (a), apply the delta method from equation (A.23) to approximate $\text{Var}[h(b_0, b_1)]$. For the individual variance terms, take $\text{Var}[b_j] = se^2[b_j], j = 0, 1$. For the covariance, use the estimated value $\hat{\sigma}_{01}$.

   (c) Set $se[\hat{BMD}] = \sqrt{\text{Var}[h(b_0, b_1)]}$ and from this build the $1 - \alpha$ Wald limit as $\hat{BMDL} = \hat{BMD} - z_\alpha se[\hat{BMD}]$.

   (d) Repeat steps (a)–(c), but now operate with $h(b_0, b_1) = (C - b_0)/b_1$. In effect, this will produce a log(BMDL). To translate to the original dose scale, take $\exp\{\log(\hat{BMDL})\}$. Can you envision a setting where this approach might be favored over the more direct $\hat{BMDL}$ in part (c)?

4.22. Return to the $k = 2$ multistage model from Example 4.12, where the BMD was estimated via equation (4.11). For simplicity, let $C = -\log(1 - BMR)$. Construct an approximate BMDL based on a delta-method approximation for $se[\hat{BMD}]$, as follows:

(a) View $\hat{\mathrm{BMD}}$ as a function of b_1 and b_2, say $h(b_1, b_2)$, and show that the partial derivatives $\partial h/\partial b_1$ and $\partial h/\partial b_2$ are

$$\frac{\partial h}{\partial b_1} = -\frac{1}{2b_2}\left\{1 - \frac{b_1}{\sqrt{b_1^2 + 4b_2 C}}\right\} \quad \text{and} \quad \frac{\partial h}{\partial b_2} = \frac{1}{2b_2^2}\left\{b_1 - \frac{b_1^2 + 2b_2 C}{\sqrt{b_1^2 + 4b_2 C}}\right\}.$$

(b) With the derivatives from part (a), apply the delta method from Equation (A.23) to approximate $\mathrm{Var}[h(b_0, b_1)]$. For the individual variance terms, take $\mathrm{Var}[b_j] = se^2[b_j]$, $j = 1, 2$. For the covariance, use the estimated value $\hat{\sigma}_{12}$.

(c) Set $se[\hat{\mathrm{BMD}}] = \sqrt{\mathrm{Var}[h(b_1, b_2)]}$ and from this build the $1 - \alpha$ Wald limit as $\hat{\mathrm{BMDL}} = \hat{\mathrm{BMD}} - z_\alpha se[\hat{\mathrm{BMD}}]$.

4.23. Return to the following data sets and find a 99% BMDL at the specified BMR(s).

(a) The dieldrin carcinogenicity data from Exercise 4.11 under a $k = 2$ *linearized* multistage model. Set $\mathrm{BMR} = 10^{-2}$.

(b) The LADA carcinogenicity data from Exercise 4.12 under a $k = 1$ multistage model (see Exercise 4.17). Set $\mathrm{BMR} = 10^{-3}$.

(c) The LADA carcinogenicity data from Exercise 4.12 under a $k = 1$ multistage model. (see Exercise 4.17). Set $\mathrm{BMR} \in \{10^{-1}, 10^{-2}, 10^{-3}, 10^{-5}\}$. Adjust for the $L = 4$ risk levels using a Bonferroni correction.

4.24. The BMDL under a linearized multistage model may be constructed in a variety of ways. Consider the following alternative approach using the additional risk $R_A(x)$:

(a) Start with equation (4.5) and for simplicity set $k = 1$. Thus $R(x) = 1 - \exp\{-\beta_0 - \beta_1 x\}$. Show that a first-order Taylor series approximation for $R(x)$ about $x = 0$ is $1 - \exp\{-\beta_0 - \beta_1 x\} \approx (1 - \exp\{-\beta_0\}) + x\beta_1 \exp(-\beta_0)$. Simplify further by setting $\theta_0 = 1 - \exp\{-\beta_0\}$ and $\theta_1 = \beta_1 \exp\{-\beta_0\}$. Show that this yields $1 - \exp\{-\beta_0 - \beta_1 x\} \approx \theta_0 + \theta_1 x$.

(b) Under the approximation you found in part (a), find the additional risk $R_A(x)$. Set the BMD as that x satisfying $R_A(x) = \mathrm{BMR}$. Show that this is a function of only θ_1 (and BMR). How would you estimate the BMD here?

(c) Suppose U_α is a $1 - \alpha$ upper confidence limit on θ_1 such that $P[\theta_1 \le U_\alpha] \approx 1 - \alpha$. Manipulate the inequality under the probability statement to produce $P[(\mathrm{BMR}/U_\alpha) \le \mathrm{BMD}] \approx 1 - \alpha$. Thus we can take $\hat{\mathrm{BMDL}} = \mathrm{BMR}/U_\alpha$.

(d) How might you find U_α, as required in part (c)?

4.25. Smith (2002b) gives data on chlorophyll concentrations as a function of phosphorus load in a series of lakes. Both variables were transformed via the natural logarithm to approximate a linear relationship. A sample of $n = 21$ lakes produced the following data:

$x = $ log phosphorus	$Y = $ log chlorophyll
1.97	1.92
2.07	2.36
2.45	2.64
2.55	1.17
2.77	2.07
2.93	2.22
3.30	3.78
3.60	6.30
3.65	4.59
3.96	3.02
3.79	6.30
4.23	5.64
4.43	5.78
4.65	7.00
4.91	4.67
4.94	7.40
5.18	6.80
5.52	5.75
5.59	8.37
6.01	7.90
5.90	7.93

(a) Plot the data. Does a simple linear model seem reasonable for representing the relationship between x and Y?

(b) Apply the simple linear model $Y_i = \beta_0 + \beta_1 x_i + \varepsilon_i$, and find the ML estimators b_0 and b_1. Also, estimate the value of a new observation at an extreme value of x, say $x = 7$, $\hat{Y}(7)$, and give a 95% prediction interval for this predicted value.

(c) If you have access to pseudo-random number generators, perform a Monte Carlo uncertainty analysis similar to that in Example 4.17 on the outcome value $U = \hat{Y}(7)$. Specifically, produce a histogram of at least $M = 10\,000$ pseudo-random predicted values and comment on it. Also, comment on the quality of the exact prediction interval from part (b).

4.26. Similar to the data in Example 4.18, Helton and Davis (2000) give summary data on how selected input variables affect $Y = $ undisturbed-condition predicted repository pressure in the underground repository. The input variables, x_j, and correlations, R_j, are as follows:

Input variable	R_j
$x_1 = $ Log anhydrite permeability	0.1302
$x_2 = $ Cellulose degradation indicator	0.7124
$x_4 = $ Steel corrosion rate	0.2762
$x_8 = $ Halite porosity	0.4483
$x_9 = $ Initial brine pressure	0.0993

Construct a tornado diagram similar to Fig. 4.10 and identify which of these input variables impact Y in a substantive fashion.

4.27. Saltelli (2002) reports on a sensitivity analysis of human radionucleotide exposure from a potential radioactive waste disposal site. Among the potential inputs used to model $Y = $ annual human radionucleotide dose, the following variables (and their anticipated ranges) were considered:

Input variable	Minimum	Maximum
$x_1 = $ Containment time	100	1000
$x_2 = $ Chain nucleotide leach rate	10^{-6}	10^{-5}
$x_3 = $ Geosphere water velocity	0.001	0.1
$x_4 = $ Geosphere layer length	50	200
$x_5 = $ Geosphere layer retention factor	1	5
$x_6 = $ Geosphere layer retention coefficient	3	30
$x_7 = $ Stream flow rate	10^5	10^7

Construct a tornado diagram similar to Fig. 4.11 and identify which of these input variables impact Y in a substantive fashion. Use the natural logarithm of the percentage change as your charting variable.

4.28. Return to Example 4.20, where $Y = X_2^4/X_1^2$, with $X_j \sim$ i.i.d. $U(1 - \varphi, 1 + \varphi)$, for $0 < \varphi < 1$. There, we mentioned that the exact values of the correlation ratios could be determined using advanced methods of probability and statistics. If you are familiar with these methods, verify the calculations found in the example as follows:

(a) Show that the unconditional variance is

$$\text{Var}[Y] = \frac{\{(1 + \varphi)^9 - (1 - \varphi)^9\}\{(1 + \varphi)^3 - (1 - \varphi)^3\}}{216(1 - \varphi)^3(1 + \varphi)^3\varphi^3} - \frac{(\{1 + \varphi\}^5 - \{1 - \varphi\}^5)^2}{100(1 - \varphi)^2(1 + \varphi)^2\varphi^2}$$

using the fact that

$$E[X_j^m] = \int\limits_{1-\varphi}^{1+\varphi} \frac{x^m}{2\varphi} dx \qquad \text{for any } m \neq -1.$$

(b) Recognize that $E[Y|X_1] = E[X_2^4/X_1^2|X_1] = E[X_2^4]/X_1^2$ and find this as $\{(1+\varphi)^5 - (1-\varphi)^5\}/(10X_1^2\varphi)$. Thus, show that the first iterated variance is

$$\text{Var}\{E[Y|X_1]\} = \text{Var}\left[\frac{\{1+\varphi\}^5 - \{1-\varphi\}^5}{10X_1^2\varphi}\right] = \frac{\{(1+\varphi)^5 - (1-\varphi)^5\}^2}{75(1-\varphi)^3(1+\varphi)^3}.$$

Use the fact that $X_j \sim U(1-\varphi, 1+\varphi)$, so

$$E[X_j^{-2}] = \int\limits_{1-\varphi}^{1+\varphi} \frac{x^{-2}}{2\varphi} dx = \frac{1}{(1-\varphi)(1+\varphi)}.$$

(c) Recognize that $E[Y|X_2] = E[X_2^4/X_1^2|X_2] = X_2^4 E[X_1^{-2}]$ and find this as $X_2^4/\{(1-\varphi)(1+\varphi)\}$. Thus, show that the second iterated variance is

$$\text{Var}\{E[Y|X_2]\} = \text{Var}\left[\frac{X_2^4}{(1-\varphi)(1+\varphi)}\right]$$
$$= \frac{1}{2\varphi(1-\varphi)^2(1+\varphi)^2}\left\{\frac{(1+\varphi)^9 - (1-\varphi)^9}{9}\right.$$
$$\left. - \frac{\{(1+\varphi)^5 - (1-\varphi)^5\}^2}{50\varphi}\right\}.$$

(d) Set $\varphi = \frac{1}{2}$ and use the results in parts (a)–(c) to corroborate in Example 4.20 that $\eta_1^2 = 0.262$ and $\eta_2^2 = 0.511$, where $\eta_j^2 = \text{Var}\{E[Y|X_j]\}/\text{Var}[Y]$.

(e) Extend this analysis by finding $\eta_2^2/\eta_1^2 = \text{Var}\{E[Y|X_2]\}/\text{Var}\{E[Y|X_1]\}$ as a function of φ and plot this over $0 < \varphi < 1$. What do you conclude about the impact of X_2 relative to X_1 on Y?

(f) Show that $\text{Var}[X_j] = \frac{1}{3}\varphi^2$, and then plot η_2^2/η_1^2 against $\text{Var}[X_j] > 0$. How does this compare with the plot in part (e)?

5

Temporal data and autoregressive modeling

An important component of environmetric study concerns the analysis of data recorded over time or space (or both). This is referred to as *temporal* or *spatial* (or, if combined, *spatio-temporal*) statistical analysis. In this chapter we introduce and illustrate some basic statistical methods for temporal analysis, with emphasis placed on data from environmental studies. We discuss spatial analysis in Chapter 6.

5.1 Time series

When data are taken over time, they are referred to as a *time series*. The terminology reinforces the notion that the observations record a series of events indexed by time. Time series data are common in studies that monitor environmental conditions, especially in ecology, atmospheric science, oceanography, and natural resources. If the data are taken on a single subject or organism repeatedly over time – e.g., when measuring an organism's growth – the series is often referred to as repeated measures or longitudinal data. We discuss this specialized case in §5.6.1, below. More generally, an environmental time series is a record of response at specified observation stations or ecological locations over a large number of time points.

One of the most crucial aspects of a time series is *periodicity*, a harmonic repetition of some recurrent event or influence. (When applied to a seasonal influence, one often sees the term *seasonality*.) Periodicity occurs in many forms of temporal environmental data. To illustrate, consider the following example.

Example 5.1 (Water surface temperature) Data on water surface temperatures (in °C) at a monitoring station on North Inlet, South Carolina, from January 1981 to February 1992 were taken as part of the US National Science Foundation's Long-Term Ecological Research (LTER) project. The data (a sample of which appears in Table 5.1;

Analyzing Environmental Data   W. W. Piegorsch and A. J. Bailer
© 2005 John Wiley & Sons, Ltd   ISBN: 0-470-84836-7 (HB)

Table 5.1 Sample marsh water surface temperatures at North Inlet, SC

Date	Temp. (°C)
1/20/81	5.0
2/4/81	5.0
2/17/81	9.0
⋮	⋮
1/17/92	6.5
1/31/92	9.7
2/17/92	10.2

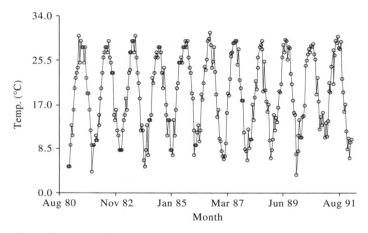

Figure 5.1 Marsh water surface temperatures at North Inlet, SC

see http://www.wiley.com/go/environmental for the full data set) are fairly dense, comprising $n = 274$ measurements taken as little as two weeks apart. A plot of the data appears in Fig. 5.1.

A clearly periodic pattern is evident in Fig. 5.1: the temperatures appear to cycle in a roughly annual pattern, although they do not exhibit any other form of underlying trend over time. (Such behavior is a form of *stationarity* in a time series, which we define formally in §5.3.2, below.) Questions of interest with such data include how to describe the periodic pattern and perhaps use this information to understand better the underlying phenomenon. To address these questions, we study these data further in the next section.

5.2 Harmonic regression

Strong periodic effects can have important impacts on a temporally recorded environmental outcome. In this chapter, we adopt the philosophy that most periodicity in environmental time series is at its source not stochastic – i.e., not due to random influences – and instead assume that it has some deterministic origin. This may be a response to the rotation of the Earth about its axis, the orbit of the Moon about the Earth, seasonal influences of one species upon another, etc. We view such periodicity as an effect to be modeled and removed before other effects can be properly studied, although we will also take advantage of the opportunity to estimate selected features of the periodic pattern. We begin with some simple models used to describe periodic data, known as *harmonic regression models*.

5.2.1 Simple harmonic regression

Formally, we define periodicity as the result of recurring or repeating influences on an experimental or observational unit. The building block we use to model it is the simple sine wave. Recall that the sine function of any angle θ varies over $-1 \leq \sin(\theta) \leq 1$, where θ is measured in radians. The function changes in a periodic fashion starting at 0 radians ($0°$) such that $\sin(0) = 0$, moving up to its maximum at $\pi/2$ radians ($90°$) with $\sin(\pi/2) = 1$, dropping back down through zero at π radians ($180°$), reaching its minimum at $3\pi/2$ radians ($270°$) such that $\sin(3\pi/2) = -1$, and then swinging back up to 0 at 2π radians ($360°$). The pattern repeats for $\theta > 2\pi$.

For a periodically oscillating observation Y_i, we use the sine function to build a regression model of the form

$$Y_i = \beta_0 + \gamma \sin\left(\frac{2\pi\{t_i - \theta\}}{p}\right) + \varepsilon_i, \qquad (5.1)$$

$i = 1, \ldots, n$, where t_i is an independent predictor variable that captures the time effect, π is the ratio of the circumference to the diameter of a unit circle ($\pi = 3.141\,592\,6\ldots$), ε_i is an additive error term, and the remaining quantities are parameters that affect the nature and shape of the sine wave. These are illustrated in Fig. 5.2 and summarized, along with some associated values, in Table 5.2.

If in (5.1) we assume no temporal correlation exists among the error terms, then it may be reasonable to set $\varepsilon_i \sim$ i.i.d. $N(0, \sigma^2)$ and view the model as a relatively straightforward nonlinear regression. This can be fitted using methods from Chapter 2. If we make the further assumption that the period p is known, then (5.1) can be reduced to an even simpler multiple linear regression model. That is,

$$Y_i = \beta_0 + \gamma \sin\left(\frac{2\pi\{t_i - \theta\}}{p}\right) + \varepsilon_i$$

$$= \beta_0 + \gamma \sin\left(\frac{2\pi t_i}{p} - \frac{2\pi\theta}{p}\right) + \varepsilon_i \qquad (5.2)$$

$$= \beta_0 + \gamma \left\{ \sin\left(\frac{2\pi t_i}{p}\right) \cos\left(\frac{2\pi\theta}{p}\right) - \cos\left(\frac{2\pi t_i}{p}\right) \sin\left(\frac{2\pi\theta}{p}\right) \right\} + \varepsilon_i,$$

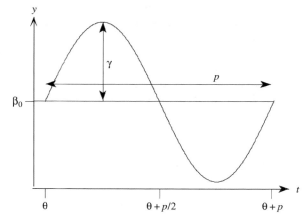

Figure 5.2   The anatomy of a sine wave: $\beta_0 + \gamma \sin(2\pi\{t - \theta\}/p)$

Table 5.2   Component parameters of the sinusoidal regression model in equation (5.1)

Parameter	Description
p	The *period* of the wave (the amount of time used in one cycle).
$1/p$	The *frequency* of the wave (no. of cycles/unit time); notation: $f = 1/p$.
β_0	The height of the wave's (horizontal) *centerline*.
γ	The *amplitude* of the wave (distance from the centerline to the wave maximum).
2γ	The *range* of the wave.
θ	The *phase angle* that lags the wave in a rigid fashion, either to the left or to the right.

the latter equality following from the well-known trigonometric 'double angle' formula $\sin(\psi - \varphi) = \sin(\psi)\cos(\varphi) - \cos(\psi)\sin(\varphi)$. For known p, each of the terms in this expansion can be written in a simpler form. Let $c_i = \cos(2\pi t_i/p)$ and $s_i = \sin(2\pi t_i/p)$ be two new (known) regression variables, and take $\beta_1 = -\gamma\sin(2\pi\theta/p)$ and $\beta_2 = \gamma\cos(2\pi\theta/p)$ as two new (unknown) regression coefficients. Then (5.2) simplifies to the multiple regression model

$$Y_i = \beta_0 + \beta_1 c_i + \beta_2 s_i + \varepsilon_i. \tag{5.3}$$

Equation (5.3) is known as a *simple harmonic regression model*, which can be fitted via the standard multiple regression methods discussed in Chapter 1.

Inferences under model (5.1) generally concern the unknown parameters γ and θ. Translating to model (5.3), the regression coefficients become functions of γ and θ, from which an inverse transform yields the relationships $\gamma^2 = \beta_1^2 + \beta_2^2$ – since $\sin^2(\psi) + \cos^2(\psi) = 1$ for any angle ψ – and $\theta = \frac{p}{2\pi}\cos^{-1}(\beta_2/\gamma) = \frac{p}{2\pi}\sin^{-1}(-\beta_1/\gamma)$.

As a result, the null hypothesis of no sinusoidal effect, $H_0: \gamma = 0$, can be tested via the equivalent null hypothesis $H_0: \beta_1 = \beta_2 = 0$. Simply fit the multiple regression model (5.3) and apply a standard F-statistic as in (1.7).

Parameter estimates also have specific analogs under (5.3): if b_1 and b_2 are the ML estimators for β_1 and β_2, then the ML estimator of γ is $\hat{\gamma} = \sqrt{b_1^2 + b_2^2}$ while that for θ is $\hat{\theta} = (2\pi)^{-1} p \cos^{-1}(b_2/\hat{\gamma})$ or, if desired, $\hat{\theta} = (2\pi)^{-1} p \sin^{-1}(-b_1/\hat{\gamma})$. Unfortunately, standard errors for these estimators and other associated inferences are somewhat more difficult to come by; one possibility would be to use the delta method from §A.6.2 to approximate the standard errors. Exercise 5.4 explores this approach.

Example 5.2 (Water surface temperature, cont'd) Return to the water surface temperature data of Example 5.1. It is natural to expect with these data that an annual period exists; that is, $p = 1$ if the time variable, t, is taken in years. Doing so here leads to the model $Y_i = \beta_0 + \gamma \sin(2\pi\{t_i - \theta\}) + \varepsilon_i$, which we re-express as the multiple linear regression model in (5.3). Here, c_i and s_i take the simple forms $c_i = \cos(2\pi t_i)$ and $s_i = \sin(2\pi t_i)$, while $\beta_1 = -\gamma \sin(2\pi\theta)$ and $\beta_2 = \gamma \cos(2\pi\theta)$. Sample SAS code for fitting this model to the data appears in Fig. 5.3. Notice that we have rescaled the time variable to be in years; most commercial software, including SAS, uses input time in days as the default temporal unit. We could also proceed by setting $p = 365.24$ and not rescaling the time variate.

The SAS output (edited for presentation) appears in Fig. 5.4. The output corroborates the indication in Fig. 5.1: statistically significant periodicity exists with these data. The overall F-statistic to test $H_0: \beta_1 = \beta_2 = 0$, or equivalently $H_0: \gamma = 0$, is $F_{\text{calc}} = 1426.12$ ($P < 0.0001$). We estimate the amplitude via

$$\hat{\gamma} = \sqrt{b_1^2 + b_2^2} = \sqrt{(-3.8298)^2 + (-9.1788)^2} = 9.946\,°C.$$

```
    data surface;
      input month day year y @@;
  *  create SAS date variable (in days from Jan. 1. 1960);
      sasdate = mdy(month,day,year);
  *  adjust MDY variable for units in years;
      dateyr = sasdate/365.24;
      pi = arcos(-1);              * pi = 3.1415926...;
      s1 = sin(2*pi*dateyr);   c1 = cos(2*pi*dateyr);
      datalines;
        1 20 81    5     2  4 81    5     2 17 81    9
        3  7 81   13     3 20 81   11     4  3 81   16
              :                :                :
        1 17 92  6.5     1 31 92  9.7     2 17 92  10.2

    proc reg;
      model y = s1 c1 ;
```

Figure 5.3 Sample SAS code for simple harmonic regression fit to the water surface temperature data

```
                         The SAS System

                    Dependent Variable: y
                    Analysis of Variance
                              Sum of            Mean
      Source            DF   Squares          Square     F Value   Pr > F
      Model              2   13617.3837      6808.6919   1426.122  <.0001
      Error            271    1293.8275         4.77427
      Corrected Total  273   14911.2112

                  Root MSE                 2.18501   R-Square   0.9132
                  Dependent Mean          18.82482   Adj R-Sq   0.9126

                        Parameter Estimates
                        Parameter      Standard
      Variable    DF    Estimate        Error       t Value   Pr > |t|
      Intercept    1    18.88290       0.13202      143.03     <.0001
      s1           1    -3.82978       0.18674      -20.51     <.0001
      c1           1    -9.17882       0.18665      -49.18     <.0001
```

Figure 5.4 SAS output (edited) from simple harmonic regression fit to the water surface temperature data

The corresponding estimated range of mean surface temperature is slightly less than 20 °C.

Further examination to assess the model fit via, for example, residual analysis indicates (Exercise 5.2) that the predicted values tend to overestimate mid-year temperatures; that is, during many of the summer months the predicted values are higher than the actual observed values. This is likely due to the extended South Carolina summers, where the warm season is generally longer-lasting than the winter season. As a result, the underlying seasonality is not well approximated by a simple harmonic. (We discuss more complex harmonic regression in the next section.) During the other seasons, however, the model fit appears to be adequate; for example, the calculated R^2 from Fig. 5.4 indicates that upwards of 91% of the variation in Y_i is explained by the simple harmonic regression terms. This is quite high given the lack of fit during the summer months.   ☉

We should warn readers to apply caution when employing the simple harmonic regression model. First, many authors use (5.3) only for the case of equally spaced time points so that an equal number of observations are recorded in each cycle. This leads to simplifications that can greatly facilitate the model fit. With the advent of modern regression software, however, the need for equally spaced time points with this simple model has diminished in importance. Of course, if certain regions of the time scale are regularly missing, such as winter months, the fitted curves may give unusual predictions or patterns during those missing periods. Second, and perhaps more importantly, the two predictor variables c_i and s_i are by nature related and may be highly collinear. Indeed, if the original t_is happen to take values completely in phase with the sine wave (e.g., $t_2 = t_1 + p$, $t_3 = t_1 + 2p$, etc.), then s_i is exactly collinear with c_i and the model cannot be fitted. Clearly, care is required when selecting or observing the original, temporal predictor variable. Finally, as we mention above, (5.1) and (5.3) assume that the measurements are independent and not serially correlated. If some form of temporal correlation is in evidence, the model fit is invalid. We discuss ways to incorporate temporal correlation in §5.3, below.

5.2.2 Multiple harmonic regression

In some settings the response may exhibit periodicity in excess of that captured by a single sine wave. If so, we can extend the simple harmonic regression model from (5.2) into a more flexible, multiple harmonic regression model: simply fit $H > 1$ different sine waves, each with specified, known periods $p_1, p_2, \ldots, p_H$. In this case we have

$$Y_i = \beta_0 + \gamma_1 \sin\left(\frac{2\pi\{t_i - \theta_1\}}{p_1}\right) + \gamma_2 \sin\left(\frac{2\pi\{t_i - \theta_2\}}{p_2}\right) + \cdots + \gamma_H \sin\left(\frac{2\pi\{t_i - \theta_H\}}{p_H}\right) + \varepsilon_i,$$

$$(5.4)$$

$i = 1, \ldots, n$. Since the p_hs are known, we can mimic the approach for the simple harmonic case and create a series of $2H$ regressor variables, that is, a pair of regressors for each period: $c_{ih} = \cos(2\pi t_i/p_h)$ and $s_{ih} = \sin(2\pi t_i/p_h)$. With these, an extension of (5.3) is

$$Y_i = \beta_0 + \beta_1 c_{i1} + \beta_2 s_{i1} + \beta_3 c_{i2} + \beta_4 s_{i2} + \cdots + \beta_{2H-1} c_{iH} + \beta_{2H} s_{iH} + \varepsilon_i. \qquad (5.5)$$

Notice that there are now $2H + 1$ unknown regression coefficients, so to allow for a proper model fit we require $n > 2H + 1$. The null hypotheses of no overall periodic effect becomes $H_0: \beta_1 = \beta_2 = \cdots = \beta_{2H} = 0$. Of added interest, however, is whether any of the individual periods are significant, that is, testing any of the 2 df hypotheses $H_0: \beta_{2h-1} = \beta_{2h} = 0$ for $h = 1, \ldots, H$.

Often, the periods $p_1, p_2, \ldots, p_H$ may be *harmonics* of a single, principle period, for example, $p, p/2, \ldots, p/H$. If so, the regression function oscillates with period p, but will not be a true sine wave. (It is a weighted average of sine waves.) Exercise 5.2 applies this approach to the water surface temperature data from Example 5.1, using additional harmonics of the annual period. Figure 5.5 illustrates how a sum of $H = 3$ different sine waves can look distinctly non-sinusoidal, yet still produce a periodic function.

5.2.3 Identifying harmonics: Fourier analysis

Harmonic analysis when the periods, p_h, are unknown is more complicated than when they are known. Standard methodology for estimating unknown p_hs from the data is known as *Fourier analysis* (Bloomfield, 2000), or sometimes as *spectral analysis*, *spectrum analysis*, or time series analysis 'in the frequency domain.' The term derives from the sine/cosine decomposition of (5.4) into (5.5): under certain conditions (defined in §5.3), a time series may be re-expressed as a sum of sine and cosine waveforms similar to the trigonometric series representation Fourier (1822) used in his famous treatise on the physics of heat. We give a simplified illustration of this in Fig. 5.5. Sinusoids are useful for describing periodically varying phenomena, and the sine/cosine *spectral representation* of a time series leads to powerful tools for understanding its periodic components (Brockwell and Davis, 1991, §4.2; Barnett, 2004, §10.6).

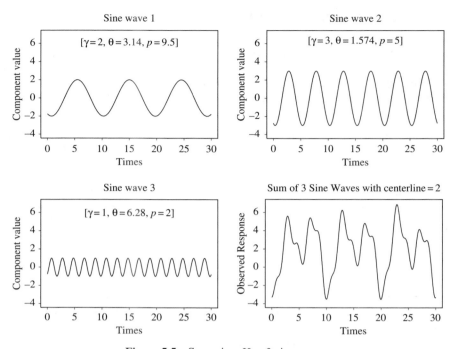

Figure 5.5   Summing $H = 3$ sine waves

In environmetric Fourier analysis, the goal is typically one of detecting cyclical phenomena in the time series. Indeed, detection and discovery of new periodicities are often of greater interest than confirmation of or inferences on previously identified periods. This has the flavor of an *exploratory data analysis (EDA)* (McLeod and Hipel, 1995); for example, one fits a multiple sine-wave model with many candidate periods to the data, and then compares the fitted periods' amplitudes to see if any of them stand out from the others.

To formulate the model, begin by requiring that the ordered time points $t_1, \ldots, t_n$ be equally spaced such that $t_{i+1} - t_i$ is a positive constant. This is used to improve interpretability of the various temporal parameters. Without loss of generality, we assume the time points are expressed in units of the sampling interval, $t_i = i$. We observe a response Y_i at each time point, and take the underlying error as independent, normally distributed, with zero mean and constant variance σ^2. An important restriction is that the mean of Y_i cannot itself trend over time. To model the periodic oscillations in Y_i, any underlying trend that interferes with or obscures the periodicity must first be removed. We go into more detail on this issue below.

Assuming the time series has at least three observation times, we denote by M the largest integer less than or equal to $n/2$. With this, we let the (maximum) number of possible sine terms in the multiple harmonic model be $H = M - 1$. For example, if $n = 15$ we would set $M = 7$ and consider a model with up to $H = 6$ possible sine terms.

A Fourier analysis takes these components and applies them in (5.4) with the known periods $p_1 = n$, $p_2 = n/2$, ..., $p_H = n/H$; the smallest of these candidate periods is slightly larger than 2. This is equivalent to modeling the H frequencies $f_1 = 1/n, f_2 = 2/n, \ldots, f_H = H/n$ as evenly spaced on the interval $0 < f < \frac{1}{2}$; cf. Table 5.2. Usually, Fourier analysis is described in terms of the frequencies rather than the periods, due to this even spacing. After detecting important frequencies, we invert these via $p_h = 1/f_h$ and identify the corresponding periods. Operationally, this is performed by translating to a multiple regression with the $2H$ regressors $c_{ih} = \cos(2\pi t_i / p_h)$ and $s_{ih} = \sin(2\pi t_i / p_h)$, $i = 1, \ldots, n$; $h = 1, \ldots, H$. Then, the corresponding multiple regression model in (5.5) is fitted.

A note of caution: many statistical software packages define the frequencies to be equally spaced on the interval $0 < f < \pi$, by multiplying what we call f_h by 2π. (See Example 5.3, below.) One must make certain of the frequency scale when identifying important frequencies and periods in this manner.

For our use, model (5.5) can be fitted via standard regression software such as PROC REG, but some simplification can be achieved in the multiple regression components. Recall from Chapter 1 that a linear regression model can be written in matrix notation, $\mathbf{Y} = \mathbf{X}\boldsymbol{\beta} + \boldsymbol{\varepsilon}$; here $\mathbf{Y}$ is a vector containing the observations, $[Y_1 \ldots Y_n]^T$, $\mathbf{X}$ is a matrix whose columns include the $2H + 1$ regressor variables via

$$\mathbf{X} = \begin{bmatrix} 1 & c_{11} & s_{11} & \cdots & c_{1H} & s_{1H} \\ 1 & c_{21} & s_{21} & \cdots & c_{2H} & s_{2H} \\ \vdots & \vdots & \vdots & \ddots & \vdots & \vdots \\ 1 & c_{n1} & s_{n1} & \cdots & c_{nH} & s_{nH} \end{bmatrix},$$

the vector $\boldsymbol{\varepsilon}$ contains the normally distributed errors $[\varepsilon_1 \ldots \varepsilon_n]^T$, and the superscript ^{T} indicates transposition of the vector or matrix. (Here, $2H + 1$ must be either $n - 1$ or $n - 2$.) We also have that the vector of ML estimators can be written in matrix form: $\mathbf{b} = (\mathbf{X}^T\mathbf{X})^{-1}\mathbf{X}^T\mathbf{Y}$, where $(\mathbf{X}^T\mathbf{X})^{-1}$ is the *inverse* of the matrix $\mathbf{X}^T\mathbf{X}$; cf. §A.4.3. As seen in §1.2, the corresponding individual ML estimators b_h cannot typically be written in convenient closed forms, hence the need for computer software to calculate these values. For the sines and cosines used in (5.5), however, a dramatic simplification occurs: the special structure between the trigonometric terms ensures that $\mathbf{X}^T\mathbf{X}$ is itself diagonal: $\mathbf{X}^T\mathbf{X} = \mathrm{diag}\{n, n/2, n/2, \ldots, n/2\}$. Since the inverse of a diagonal matrix is once again diagonal, the estimator $\mathbf{b} = (\mathbf{X}^T\mathbf{X})^{-1}\mathbf{X}^T\mathbf{Y}$ simplifies into the closed-form expressions

$$b_0 = \overline{Y}, \qquad b_{2h-1} = \frac{2}{n}\sum_{i=1}^{n} c_{ih} Y_i, \qquad b_{2h} = \frac{2}{n}\sum_{i=1}^{n} s_{ih} Y_i \qquad (5.6)$$

($h = 1, \ldots, H$); see Exercise 5.6. The coefficient b_{2h-1} may be viewed as a cosine transformation of the data, used to quantify the impact of the periodic cosine terms on Y_i. Similarly, b_{2h} is a sine transformation that quantifies the impact of the periodic sine terms.

The ML estimators (5.6) are used to construct the predicted values $\hat{Y}_i = b_0 + b_1 c_{i1} + b_2 s_{i1} + \cdots + b_{2H-1} c_{iH} + b_{2H} s_{iH}$, along with other pertinent statistical quantities such as the regression sum of squares $\text{SSR} = \sum_{i=1}^{n}(\hat{Y}_i - \overline{Y})^2$ with $2H$ df. Here again, the special structure of the sine and cosine regressors leads to some simplification:

$$\text{SSR} = \frac{n}{2}\sum_{h=1}^{H}(b_{2h}^2 + b_{2h-1}^2) = \frac{n}{2}\sum_{h=1}^{H}\hat{\gamma}_h^2.$$

(Recall that γ_h is the amplitude of the hth sine wave modeled in the harmonic regression.) In other words, variability in the response can be decomposed into contributions from each candidate period, p_h, quantified via $\frac{n}{2}\hat{\gamma}_h^2$. These contributions are often referred to as *intensities* of order h, and we denote them by $\mathfrak{I}(h) = \frac{n}{2}\hat{\gamma}_h^2$.

Under our i.i.d. normal assumptions on $\boldsymbol{\varepsilon}$, the ML estimators in (5.6) are independently distributed as $b_0 \sim N(\beta_0, \sigma^2/n)$ and $b_h \sim$ indep. $N(\beta_h, 2\sigma^2/n)$, $h = 1, \ldots, H$. With this, one can show that each $\mathfrak{I}(h)$ is a 2 df quantity, related to the $\chi^2(2)$ distribution (Exercise 5.7). Note, however, that when n is even, the last component $\mathfrak{I}(h) = \frac{n}{2}\hat{\gamma}_H^2$ has only 1 df (Box *et al.*, 1994, §2.2.2).

We use these various quantities to study the variation in $\mathbf{Y}$, by plotting the estimated contributions $\mathfrak{I}(h)$, or more often $\log\{\mathfrak{I}(h)\}$, against frequency $f_h = h/H$, $h = 1, \ldots, H$. The plot is known as a (raw) *sample periodogram* – a term originated by Schuster (1898) – or sometimes as a *raw spectral density estimate*. Traditionally, one inspects this plot in an EDA fashion, looking for peaks or spikes at specific frequencies in the periodogram. We expect that $\mathfrak{I}(h)$ will be large when the Y_is exhibit harmonic components near the corresponding frequency f_h, so peaks in $\mathfrak{I}(h)$ will indicate high impacts of those frequencies on the time series. Where peaks correspond to frequencies and hence periods that can be explained scientifically, there is the potential for discovery of unknown periodicities and, ostensibly, the phenomena responsible for them.

The sample periodogram often exhibits jagged structural features that can be difficult to distinguish from random noise. It is common therefore to *smooth* the plotted curve in order to clear away the noise (Buhlman, 2002). The effort shares similarities with construction of a histogram for a single-sample data set; when constructing histograms, we must choose class or bin lengths for grouping the data. If the class lengths are too narrow, the histogram will be too noisy to be useful; if they are too wide, important features in the data will be masked and possibly lost.

There are many different types of periodogram smoothers; reviews are given by Lee (2001), Solow (1993), or, for an earlier perspective, Wahba (1980). The simplest smoothers are moving averages of the raw periodogram, that is, weighted averages of the $\mathfrak{I}(h)$s or of their logarithms over some set of $K \geq 1$ indices around h. This set of indices is called a *spectral window*. Algebraically, the smoothed periodogram values are

$$\bar{\mathfrak{I}}_K(h) = \frac{\omega_0\mathfrak{I}(h) + \sum_{k=1}^{K}\{\omega_k\mathfrak{I}(h+k) + \omega_{-k}\mathfrak{I}(h-k)\}}{\omega_0 + \sum_{k=1}^{K}\{\omega_k + \omega_{-k}\}}, \tag{5.7}$$

where the weights ω_k are assumed non-negative. If the index on ω_k drops $h - k$ below 1 or raises it above H, one simply sets $\omega_k = 0$. (Other possibilities include reflecting the weights back at $h = 1$ and $h = n$, taking advantage of the cyclical nature of the model. Care needs to be taken when exercising this option, however, since instabil-ities can occur at the edges of the periodogram.) We often choose integer weights that are symmetric about h; for example, for $K = 2$, take the triangular weighting $\omega_{-2} = \omega_2 = 1$, $\omega_{-1} = \omega_1 = 2$ and $\omega_0 = 3$. By manipulating the window size, K, and the weights in (5.7), one can obtain varying degrees of smoothness. Plotting $\bar{\mathfrak{I}}_K(h)$ or $\log\{\bar{\mathfrak{I}}_K(h)\}$ against frequency f_h provides an EDA method for detecting important frequencies/periods in the time series.

In practice, assumptions underlying the periodogram plots must be met or the results may indicate spurious or incorrect periodicities. Chief among these is the requirement that observations be equally spaced in time. If there are a few gaps – say, fewer than 10% missing – it is generally acceptable to insert interpolated values. Next is the assumption of homogeneous variances: $\text{Var}[Y_i] = \sigma^2$ for all i. In many cases, observa-tions such as counts or concentrations will violate this assumption (apparent, for example, in the residual plot from a preliminary model fit), and if so the original series must be transformed to stabilize any variance heterogeneity; cf. our similar discussion for linear regression towards the end of §1.1. Perhaps the most common transformation with environmental data is the natural logarithm, although at times other functions can suffice as well. We illustrate this throughout the examples and exercises below.

Also, as mentioned above, Fourier analyses should be performed *only* on non-trending series, that is, series that do not exhibit a clear trend over time. In practice, we check for such trends by first plotting Y_i against t_i. A trending series will exhibit a discernible underlying 'signal' or pattern of monotone increase (or decrease), or perhaps some polynomial curvature. Any such trend in the data must be removed prior to periodogram construction. To do so, fit a simple linear model or low-order polynomial to the data via ordinary least squares and use the residuals from the fit for the Fourier analysis. The effort of pruning and refining the original time series in such a fashion is known as *prewhitening* or *detrending*.

SAS provides a simple program for performing Fourier analyses, via PROC SPECTRA. Given a prewhitened time series, PROC SPECTRA can produce raw and smoothed intensities, $\mathfrak{I}(h)$, or log-intensities, $\log\{\mathfrak{I}(h)\}$, which can then be graphed for the periodogram. We illustrate its use in the following examples.

Example 5.3 (Lynx trappings) Elton and Nicholson (1942) present data on num-bers of lynx (*Lynx canadensis*) caught by trappers on Canada's Mackenzie River. The data, yearly trappings from 1821 to 1934, are plotted in Fig. 5.6. (Table 5.3 gives a sample of the data; the full set is available online at http://www.wiley.com/go/environmental.) Clearly, some cyclical effect is evident; Elton and Nicholson felt that a 10-year period was possible, apparently related to a 'boom and bust' cycle between Canadian lynx and their main prey, the snowshoe hare, and complicated by the activities of the trappers.

Many authors have analyzed these or associated data; see, for example, Campbell and Walker (1977) or Solow (1993). Our interest here is not to supplant these previous analyses, rather to illustrate the use of Fourier methods and give some indication of the periodic characteristics in this time series.

Notice in Fig. 5.6 that no underlying trend appears in the series, so prewhitening is unnecessary. The raw data are in the form of counts, however, and it can be shown that their variation is heterogeneous (Exercise 5.8). Hence, some form of transformation

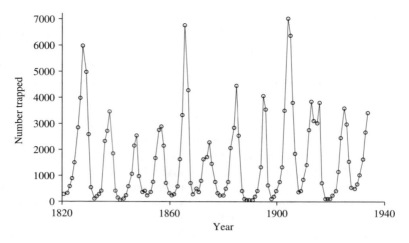

Figure 5.6 Annual lynx trappings on Canada's Mackenzie River

Table 5.3 Numbers of lynx trapped annually on Canada's Mackenzie River

Year	Number trapped
1821	269
1822	321
1823	585
⋮	⋮
1932	1590
1933	2657
1934	3396

Source: Campbell and Walker (1977).

is required to stabilize them. Most authors use the base-10 logarithm for these data, and we do likewise here: set $Y_i = \log_{10}$(number caught in year $1820 + i$), $i = 1, \ldots, 114$. For count data one could also consider a square root, $\sqrt{\text{number caught}}$, or a Freeman–Tukey, $\sqrt{\text{number caught}} + \sqrt{1 + \text{number caught}}$ (Freeman and Tukey, 1950), stabilizing transform. One advantage of the logarithm is the corresponding interpretation of the coefficients as order-of-magnitude changes in response over time.

Figure 5.7 gives sample SAS code for applying PROC SPECTRA to these data. The procedure call includes a number of important options. Among them, the center option centers the series before analysis begins in order to help stabilize the computations. *We highly recommend this option.* We also employ the out= option, which names a new SAS data set into which the periodogram values are to be stored. An additional, important variable also stored in the new data set is named

```
data lynx;
input year yloglynx @@;
datalines;
    1821   2.42975228      1822   2.50650503      1823   2.76715587
    1824   2.94001816      1825   3.16879202      1826   3.45040309
         :                      :                      :
    1932   3.20139712      1933   3.42439155      1934   3.53096768

proc spectra center out=out1 p s ;
   var yloglynx;
   weights 1 2 3 2 1;

proc print;            * prints all variables in data set;

data out1;   set out1;
   pi = arcos(-1);     * pi = 3.1415926...;
   fh = freq/(2*pi);
   lp = log10(p_01);   ls = log10(s_01);
```

Figure 5.7 Sample SAS code for periodogram construction with lynx trapping data

freq; this is the frequency, evenly spaced in radians from 0 to π, associated with each periodogram ordinate. For use in deriving the periods, we divide these first by 2π to work in units of cycles per observation. Then, the reciprocals of these divided quantities are the corresponding periods. (If we print out the new SAS data set, this forms a table from which one can easily identify the period corresponding to any important frequency; see Fig. 5.8.) For convenience, however, **PROC SPECTRA** also provides an output variable named period, which is precisely $2\pi/\text{freq}$. Lastly, the p and s options instruct **PROC SPECTRA** to include the raw and smoothed intensities, respectively, in the output data set.

Required in the call to **PROC SPECTRA** is the var statement, indicating which prewhitened data variable is of interest. For this example, no trend is evident in the

```
                        The SAS System

    Obs       FREQ        PERIOD        P_01          S_01
     1      0.00000          .          0.0000       0.03604
     2      0.05512       114.000       0.3813       0.05287
     3      0.11023        57.000       0.7035       0.08048
     4      0.16535        38.000       2.2845       0.10255
     5      0.22046        28.500       1.2788       0.09020
     6      0.27558        22.800       0.3990       0.06192
     7      0.33069        19.000       0.2942       0.04765
     8      0.38581        16.286       0.3754       0.05197
     9      0.44093        14.250       1.6787       0.07971
    10      0.49604        12.667       0.4064       0.08937
    11      0.55116        11.400       2.1214       0.28036
    12      0.60627        10.364       0.9127       0.44225
    13      0.66139         9.500      21.0264       0.60320
    14      0.71650         8.769       0.5771       0.39752
    15      0.77162         8.143       0.0398       0.20178
     :         :             :            :             :
    55      2.97625       2.11111       0.019186     0.00101
    56      3.03136       2.07273       0.010476     0.00069
    57      3.08648       2.03571       0.002954     0.00046
    58      3.14159       2.00000       0.000268     0.00030
```

Figure 5.8 SAS output (edited) for periodogram construction with lynx trappings data

log-transformed lynx counts, so the log-counts are entered as the data variable. If a smoothed periodogram is desired, the weights statement is used to specify the ω_ks for a moving average smoother. In Fig. 5.7, we employ the triangular weighting 1 2 3 2 1. Note that an odd number of weights must be specified, with the center weight corresponding to the center of the spectral window. The resulting SAS output (edited), which is simply the results from **PROC PRINT** in Fig. 5.7, appears in Fig. 5.8. We highlight in italics the largest value of S_01, which corresponds to a period of 9.5 years. As expected, this is close to the 10-year period mentioned above.

In Fig. 5.7 we follow the call to **PROC SPECTRA** with a **DATA** step transforming the raw periodogram (P_01) and smoothed periodogram (S_01) via a base-10 logarithm to make them easier to plot, and also include the desired frequency and period variables (output not shown). The resulting raw (using lp) and smoothed (using ls) $\log_{10}$-periodograms appear in Figs. 5.9 and 5.10, respectively. The plots use frequency as the horizontal variate over the scale $0 < f < 0.5$, in order to facilitate viewing.

As expected, the raw periodogram in Fig. 5.9 is difficult to read: besides the expected peak at $f = 0.105$ – i.e., $p = 1/0.105 \approx 9.5$ years – its jagged features make identification of other important periodicities difficult. By contrast, the smoothed periodogram in Fig. 5.10 not only highlights the 9.5-year period at $f = 0.105$, but also suggests an interesting series of two or three additional peaks. These occur near $f = 0.2$, 0.289, and perhaps 0.386, corresponding roughly to periods of $p \approx 5$, 3.5, and 2.5 years. (A small peak also loiters near $f = 0.465$, but it is so faint relative to the others that it has questionable validity.) There is also a peak at $f = 0.026$, corresponding to a *larger* period of approximately 38–40 years. Figure 5.6 tends to support this possibility, although one would then have to revisit the analysis and view $p = 10$ as a possible harmonic of this longer period.

Of course, some of these harmonics may be only statistical curiosities. To assess this, one could fit a model containing only certain specific periods and compare this reduced model to the full model via a formal test, as in (1.7). From what has emerged here, it is tantalizing to question whether and how any underlying ecological phenomena may be driving the more complex periodicities seen with these data. ✪

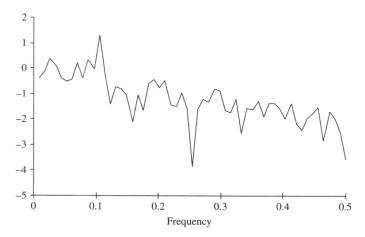

Figure 5.9   Raw periodogram for lynx trappings data

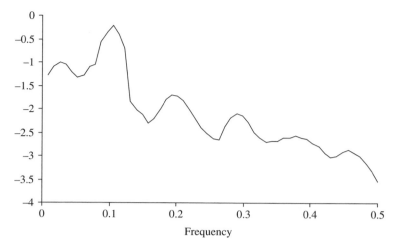

Figure 5.10 Smoothed periodogram for lynx trappings data

Example 5.4 (Mauna Loa CO₂ concentrations) A classic data set in the environmental sciences is the monthly recordings of atmospheric carbon dioxide concentrations (in ppm) taken at the Mauna Loa volcano in Hawaii (Bacastow *et al.*, 1985). The time series data were originally collected by the Scripps Institute of Oceanography; in mid-1974 collection was taken over by the US National Oceanic and Atmospheric Administration. Many sources exist for these data; we use those of Keeling and Whorf (2000), covering a period from March 1959 to December 1999. The data comprise $n = 490$ data points, a selection of which appear in Table 5.4. (The full set is available online at http://www.wiley.com/go/environmental. A few values are missing from the earlier years of the series; these have been interpolated in the online data set.)

A plot of the Mauna Loa CO_2 data is given in Fig. 5.11; an increasing trend in CO_2 concentration is evidenced, superimposed on a periodic cycle. The increasing trend has been taken by some observers to suggest the presence of potential *global*

Table 5.4 Atmospheric CO_2 concentrations (ppm) at Mauna Loa, Hawaii

Month	CO_2
Mar. 1959	315.56
Apr. 1959	317.29
May 1959	317.34
⋮	⋮
Oct. 1999	365.18
Nov. 1999	366.72
Dec. 1999	368.05

Source: Keeling and Whorf (2000).

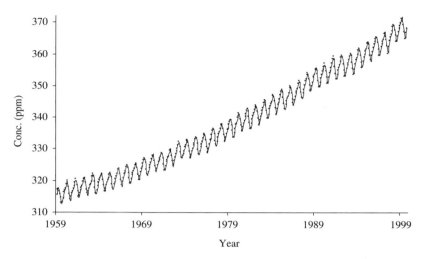

Figure 5.11   Atmospheric CO_2 concentrations at Mauna Loa, Hawaii

warming: increased atmospheric CO_2 may be associated with a *greenhouse effect* (North *et al.*, 2002) that leads to increased global atmospheric temperatures. (Of course, one should always be circumspect when using site-specific data to make statements about global phenomena.)

A fairly complete temporal analysis of these data was performed by Young and Pedregal (2002). Working from their results, and to gain a better understanding of the periodicity apparent in these data, we consider a Fourier analysis. Begin by noting in Fig. 5.11 that the increasing trend may have a curvilinear component. Thus we prewhiten the data by removing the trend via (centered) linear and quadratic terms. The analysis is then applied to the residuals from that fit. For the $\log_{10}$-smoothed periodogram, we use a moving average with symmetric weights 1 2 3 4 5 4 3 2 1. Figure 5.12 displays sample SAS code. In the code, notice use of the SAS internal variable _N_, which is the observation number. Since the data are equally spaced, we can use _N_ as a proxy for the cumulative month of observation in order to build the temporal predictor, t_i. Subtracting 245.5 from _N_ centers the predictor variable, since the mean of the first 490 integers is exactly 245.5.

The SAS output (not shown) provides values of the $\log_{10}$-smoothed periodogram (fh vs. ls from **PROC PRINT**) which are graphed in Fig. 5.13. Clear peaks are seen to occur at frequencies of $f = 0.083$, 0.167, 0.251, and 0.333, corresponding to periods of approximately $p \approx 12$, 6, 4, and 3 months. Thus a clear annual period is evidenced both in the data plot and in the periodogram, but additional harmonics at semi-annual, ternary, and quarterly periods are also possible. Exercise 5.10 studies periodicity in this time series in more detail; see also Lund and Reeves (2002) for a discussion of a possible change in the nature of the upward trend in the early 1990s.   ⊕

We conclude this section with some miscellaneous remarks. First, the largest frequency used in Fourier analysis, known as the *Nyquist frequency*, is $H/n \approx 1/2$.

```
      data maunaloa;
        input month year y @@;
 **   center month variable based on 490 monthly obs'ns;
        t = _N_-245.5;    tsq = t*t;
      datalines;
        03  1959  315.56      04  1959  317.29      05  1959  317.34
        06  1959  316.515     07  1959  315.69      08  1959  314.78
             :                     :                     :
        10  1999  365.18      11  1999  366.72      12  1999  368.05

 *    detrend series via quadr. regr.;
      proc glm;    model y = t tsq ;
        output out=prewhite r=resid ;

 *    apply Fourier analysis to residuals;
      proc spectra center out=out1 p s;
        var resid;      weights 1 2 3 4 5 4 3 2 1;

      data out1;    set out1;
      pi = arcos(-1);       * pi = 3.1415926...;
      fh = freq/(2*pi);
      lp = log10(p_01);     ls = log10(s_01);

      proc print; var fh period lp ls;

      proc gplot;
        plot (lp ls)*fh; run;
```

Figure 5.12 Sample SAS code for prewhitening and periodogram construction with Mauna Loa CO$_2$ data

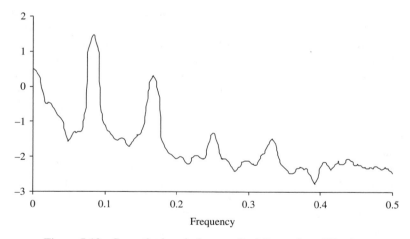

Figure 5.13 Smoothed periodogram for Mauna Loa CO$_2$ data

Some analysts interpret this to mean that one is unable to detect any cycle whose period is smaller than twice the sampling interval $t_i - t_{i-1}$. (That is, any f above the Nyquist frequency, H/n, cannot be detected.) A more accurate explanation is that such a cycle will not be directly recognized, but it may still appear as a cycle with a larger period. The larger period is called an *alias*. (Aliasing is a general concept for the masking of the effect of one factor by another factor.) For example, suppose we take observations daily, but an underlying phenomenon cycles at a period of every nine hours. In the daily data, a cycle of three days may appear that is actually caused by the nine-hour phenomenon (since both three days and 72 hours are divisible by the true nine-hour period and the one-day sampling interval).

Second, formal inferences for time series data in the frequency domain are still under development. Tests do exist that can assess whether the series is simply 'white noise' with no periodicities, but these are often sensitive to only one form of departure from the null, white noise effect. One such method involves Fisher's kappa statistic (Fisher, 1929), which is essentially the ratio of the largest periodogram ordinate to the sum of all the ordinates. The corresponding test will have good power if there is one, and only one, underlying, non-white noise frequency, but the presence of other frequencies can inflate the denominator of the ratio, limiting the test's power. Another alternative is Bartlett's (1955) cumulative periodogram, which can compare periodicity observed in the data to that expected under pure white noise. (The approach uses a form of goodness-of-fit test known as the *Kolmogorov–Smirnov (KS) test*. The KS test gauges the difference between the empirical c.d.f. of the data and the hypothesized c.d.f., here, under white noise; see Hogg and Tanis, 2001, §8.10.) PROC SPECTRA provides users with both of these test statistics, via the `whitetest` option. A more formal discussion on testing for white noise in Fourier analysis is given, for example, by Box *et al.* (1994, §8.2).

In practice, it is perhaps best to employ Fourier analysis in conjunction with the investigator's own subject-matter knowledge, keeping in mind the primary EDA character of the Fourier approach we outline here. For example, we should not be surprised if we find periods of 365 days or so in most environmental data. Often there will also be harmonics of this period – cycles with periods of six months, or four months, etc. These should not be interpreted as separate phenomena, but rather as a 12-month cycle which is not precisely sinusoidal in shape, with the smaller harmonics working together to approximate this 12-month period. Also common in data taken near the sea – e.g., in a salt marsh – are cycles and harmonics related to the lunar orbit (of approximately 29.5 days), the diel cycle (1 day), and/or the daily tides (24.84 hours or its first harmonic, 12.42 hours).

Lastly, we note an interesting historical feature associated with these trigono-metric/Fourier operations. The calculation of a long series of sines and cosines can become computationally burdensome, and an important area of study was for many years efficient calculation of Fourier-transformed quantities; see Press *et al.* (1992, §12.2). Readers may be familiar with the Cooley and Tukey (1965) *fast Fourier transform* (FFT), which is a numerical manipulation to improve the speed of such computations. For use with time series data, implementation of the FFT leads to especially efficient execution of the Fourier operations if the length of the series can be factored into a product of prime integers. For example, if $n = 2k$, Fourier analysis by the FFT involves an order of $n\log(k)$ operations, not n^2 as would otherwise be the case. Computational savings of such magnitude were once critical for implementing Fourier analyses on a digital computer, and analysts were often encouraged to set their sample sizes with this criterion in mind. With the advent of modern, high-speed computing, however, this issue has become considerably less important.

5.3 Autocorrelation

The preceding sections presented fairly simple temporal analyses, based primarily on studying cyclical or periodic features in a time series. Indeed, observations were assumed independent, and the cyclical or periodic features were viewed as deterministic and as such could be built into the model directly as a component of the mean, E[Y]. Departures from such simplifying assumptions are not uncommon with times series data, however, and many other kinds of serial effects can be encountered. For example, as environmental data on the same experimental or observation units are taken over time they may depend on their previous values: temporal observations on an animal's body temperature may vary due to some outside factor, but they will also vary according to internal factors specific to that particular animal. Or, a weather front can affect an ecosystem's ambient temperature level for many days after it passes through the region. When a stochastic phenomenon leads to this sort of effect, we say it induces *serial correlation*, *temporal correlation*, or *autocorrelation* in the time series. This reflects the association between observations in a time series not explained any deterministic component of the model, and measured via the correlation $\text{Corr}\{Y_i - \text{E}[Y_i], Y_{i+1} - \text{E}[Y_{i+1}]\}$. The terminology reinforces the notion that where observations are headed often depends on where they have been.

5.3.1 Testing for autocorrelation

Autocorrelation by its very nature induces serial dependencies in the error terms of a time series. This is a paradigm for which standard regression methods as used in §5.2.2 are not appropriate; if left unadjusted, autocorrelation can cause variance estimators, test statistics, and confidence intervals to be incorrectly calculated. Thus, for purposes of both statistical validity and basic scientific understanding, it is important to identify whether and to what extent a time series exhibits any autocorrelated effects.

A simple method to assess autocorrelation in a time series is known as the *Durbin–Watson test*. Given a postulated multiple regression model such as (5.5), one fits the model to the time series and finds the residuals from the fit, $r_i = Y_i - \hat{Y}_i$. The test statistic is a ratio of the squared successive differences to the squared residuals:

$$d_{\text{calc}} = \frac{\sum_{i=2}^{n}(r_i - r_{i-1})^2}{\sum_{i=1}^{n} r_i^2}.$$
(5.8)

Von Neumann (1941) described use of (5.8) for studying trends in a series of data. Durbin and Watson (1950, 1951) employed d_{calc} for the specific purpose of testing for autocorrelation in a time series. The test is generally quite powerful with respect to identifying at least first-order autocorrelation (see below), and can be used even in the presence of more complex forms of serial dependence (Weber and Monarchi, 1982).

Tables of critical points for (5.8) were given in Durbin and Watson's original papers and are also available in textbook sources such as Neter *et al.* (1996, Table B.7) or Manly (2001, Table B5). (An online source at the time of this writing is http://www.mhhe.com/business/opsci/aczel/textfigures/chap11/sld048.htm; an online calculator is available at http://home.ubalt.edu/ntsbarsh/Business-stat/otherapplets/Trend.htm.) A general rule of thumb is that values of d_{calc} near 2.0 suggest no significant autocorrelation. Values much larger than 2.0 suggest that a large residual tends to be followed by a small residual (or a large residual of opposite sign) and vice versa. This is indicative of a negative dependency among closely located data in the time series. By contrast, values much smaller than 2.0 suggest that a large residual is followed by a large residual (both positive or both negative) or that a small residual is followed by a small residual (both positive or negative). This is a positive dependency. Neter *et al.* (1996, §12.3) describe other features of the Durbin–Watson test.

Example 5.5 (Lynx trappings, cont'd) Return to the lynx trappings data of Example 5.3. To test if any autocorrelation is present in the $\log_{10}$-transformed data, we apply the Durbin–Watson statistic from (5.8) to residuals from the fit of a harmonic regression model. We saw in Example 5.3 that periods of approximately $p_1 = 38, p_2 = 9.5, p_3 = 5, p_4 = 3.5$, and $p_5 = 2.5$ years seemed to be evidenced in these data, so we fit the harmonic regression model $Y_i = \beta_0 + \beta_1 c_{i1} + \beta_2 s_{i1} + \beta_3 c_{i2} + \beta_4 s_{i2} + \cdots + \beta_9 c_{i5} + \beta_{10} s_{i5} + \varepsilon_i$, based on (5.5). Here, the predictor variables are sine and cosine terms of the form $c_{ih} = \cos(2\pi t_i / p_h)$ and $s_{ih} = \sin(2\pi t_i / p_h), i = 1, \ldots, 114; h = 1, \ldots, 5$.

Sample SAS code to perform the fit appears in Fig. 5.14. We employ a SAS procedure for fitting autocorrelated data, PROC AUTOREG. The syntax mimics

```
data lynxh;
input t yloglynx @@;
  pi = arcos(-1);      * pi = 3.1415296...;
  s1 = sin(2*pi*t/38);        * p = 38 sine regressor;
  c1 = cos(2*pi*t/38);        * p = 38 cosine regressor;
  s2 = sin(2*pi*t/9.5);       * p = 9.5 sine regressor;
  c2 = cos(2*pi*t/9.5);       * p = 9.5 cosine regressor;
  s3 = sin(2*pi*t/5);         * p = 5 sine regressor;
  c3 = cos(2*pi*t/5);         * p = 5 cosine regressor;
  s4 = sin(2*pi*t/3.5);       * p = 3.5 sine regressor;
  c4 = cos(2*pi*t/3.5);       * p = 3.5 cosine regressor;
  s5 = sin(2*pi*t/2.5);       * p = 2.5 sine regressor;
  c5 = cos(2*pi*t/2.5);       * p = 2.5 cosine regressor;
drop pi;
datalines;
  1821   2.42975228     1822   2.50650503     1823   2.76715587
          :                     :                     :
  1932   3.20139712     1933   3.42439155     1934   3.53096768

proc autoreg;
  model yloglynx  = s1 c1 s2 c2 s3 c3 s4 c4 s5 c5 / dwprob ;
```

Figure 5.14 Sample SAS code for harmonic regression and Durbin–Watson analysis with lynx trappings data

```
                        The SAS System
                     The AUTOREG Procedure

                  Dependent Variable      yloglynx
                  Ordinary Least Squares Estimates
        SSE            11.5650943        DFE                    103
        MSE               0.11228        Root MSE           0.33509
        Regress R-Square   0.6718        Total R-Square      0.6718
        Durbin-Watson      0.4875        Pr<DW              <.0001
        Pr>DW             1.0000

   NOTE:  Pr<DW is the p-value for testing positive autocorrelation, and
          Pr>DW is the p-value for testing negative autocorrelation.
```

Figure 5.15 SAS output (edited) for Durbin–Watson analysis with lynx trappings data

that for most SAS regression procedures, in that a `model` statement indicates the variables to be fitted in the regression. To call for the Durbin–Watson statistic, d_{calc}, we specify the `dwprob` option. This also provides the associated P-value for testing serial correlation. Since (5.8) is based on a simple LS fit to the data, however, one could use essentially similar code in **PROC REG** to find d_{calc}: simply apply the `dw` option in the **PROC REG** `model` statement.

Output (edited) from the code in Fig. 5.14 appears in Fig. 5.15. The Durbin–Watson statistic is given as $d_{calc} = 0.4875$, which is much smaller than 2.0. This indicates a strongly positive autocorrelation in the series. The (one-sided) P-value is given under `Pr<DW` as $P < 0.0001$, indicating a highly significant result. It appears that there is a strong serial correlation in these $\log_{10}$-transformed counts such that, from year to year, the number of lynx trapped on the Mackenzie River appears to depend on the numbers trapped in previous years, as well as on some periodic effect.

5.3.2 The autocorrelation function

Although it can identify the presence of autocorrelation, the Durbin–Watson test cannot quantify it directly. To do this, we turn to a more descriptive measure of serial dependence. Begin by supposing that the observations Y_i are taken at times t_i, $i = 1, \ldots, n$, and that the t_is are equally spaced such that $t_{i+1} - t_i$ is a positive constant. In the simplest case, we model $E[Y_i]$ as a constant, say β_0, along with $\mathrm{Var}[Y_i] = \sigma^2$, $i = 1, \ldots, n$. Notice that this can be written in the form of a very simple regression model

$$Y_i = \beta_0 + \varepsilon_i, \tag{5.9}$$

where the errors at time t_i satisfy $E[\varepsilon_i] = 0$ and $\mathrm{Var}[\varepsilon_i] = \sigma^2$. For similar sorts of regression models studied in Chapter 1, the errors were assumed uncorrelated. When a time series exhibits some form of serial correlation, however, a more elaborate error structure is required. We formalize this in the following definitions.

Definition 5.1 A time series with data Y_i is *mean stationary* if $E[Y_i]$ is constant with respect to t_i.

Definition 5.2 A time series with data Y_i is *second-order stationary* if the following conditions hold:

(i) Y_i is mean stationary.

(ii) $Var[Y_i] = \sigma^2$ is constant with respect to t_i.

(iii) Any serial correlation enters into the model independently of t_i; that is, where we start the series in time does not affect how the correlation progresses. Equivalently, we can require that $Cov[Y_i, Y_{i+j}]$ is independent of t_i for any t_{i+j}.

It is under (zero) mean stationarity that the spectral representation of a time series in the form of sinusoids mentioned in §5.2 applies.

Second-order stationarity is often also referred to as *weak stationarity* or *covariance stationarity* to indicate the constraints on the variance and covariances of Y_i. We will use the terms interchangeably. Other forms of stationarity are possible for defining a temporal process (Brockwell and Davis, 1991, §1.3; Barnett, 2004, §10.4.3), but for all the methods we discuss below only weak stationarity is required.

In practice, we check for weak stationarity by first plotting the series against time. A non-stationary series will show trends in the response or changes in variability. (See also the discussion on autocorrelation function plots on p. 237.) Defining formal test statistics for stationarity in a time series is a complex problem and at present represents an issue of ongoing statistical research; an extended exposition exceeds the scope of this book. SAS provides selected measures for assessing stationarity in its **PROC ARIMA**, via the `stationarity=` option in the `identify` statement.

If a series under study exhibits non-stationarity, various methods are available for filtering the effect. A common approach employs simple differencing to remove trends in the data. For instance, if an equally spaced series exhibits a simple linear trend, the first-order differences $C_i = Y_i - Y_{i-1}$ can be used to detrend the series. More complex trends can be removed by extending the differencing to, for example, second order, as in $D_i = C_i - C_{i-1} = Y_i - 2Y_{i-1} + Y_{i-2}$, etc. Or seasonal effects can be detrended by taking seasonal differences; for example, if the data are taken monthly, use differences over every four observations to remove a seasonal (four-month) fluctuation.

Alternatively, one can prewhiten a non-stationary time series, in similar fashion to that seen in §5.2.3 when constructing a periodogram. This is useful when multiple effects are present, for example both trend and cyclical seasonality, and if the effects are assumed wholly fixed and deterministic. Here, one might use LS to fit linear and where necessary, higher-order, centered, polynomial regression terms to detrend, and also fit a harmonic regression such as in (5.5) to deseason. The residuals from the LS fit are then examined for autocorrelation.

Serial correlation in a weakly stationary time series can be measured in a very intuitive way. The *first-order autocorrelation* is measured by calculating the ordinary sample correlation coefficient among the $n-1$ adjacent pairs $(Y_i, Y_{i+1}), i = 1, 2, \ldots, n-1$. Similarly, *second-order autocorrelation* uses the $n-2$ pairs (Y_i, Y_{i+2}), $i = 1, 2, \ldots, n-2$. In general, the kth-order sample auto-correlation is found via

$$\hat{\rho}_k = \frac{\sum_{i=1}^{n-k}(Y_i - \overline{Y})(Y_{i+k} - \overline{Y})}{\sum_{i=1}^{n}(Y_i - \overline{Y})^2} \qquad (5.10)$$

($k = 1, \ldots, n-1$), where $\overline{Y} = \sum_{i=1}^{n} Y_i/n$ is the sample mean of all n observations (Moran, 1948; Orcutt and Irwin, 1948). Common recommendations require n to be at least 30 (and preferably 40 or more) and to study k only over a range from $0, 1, 2, \ldots,$ up to about $n/4$. Some authors modify (5.10) by multiplying by $n/(n-k)$, but for sufficiently long series where $n \geq 40$ this typically will not make a major difference.

When a series of sufficient length has zero autocorrelation at all orders, (5.10) has expected value $\mathrm{E}[\hat{\rho}_k] \approx -1/(n-1)$ – which is near zero for large n – and standard error $se[\hat{\rho}_k] \approx n^{-1/2}$. (A more complex approximation for the standard error is available under the expanded assumption of exponentially decaying autocorrelation, that is, a series whose autocorrelation dampens down to zero at large lags; see Box *et al.*, 1994, §2.1.6.) Thus to assess the approximate significance of any kth-order autocorrelation, we compute $(\hat{\rho}_k + (n-1)^{-1})/se[\hat{\rho}_k]$ and view the autocorrelation as significant if in absolute value the ratio exceeds $z_{\alpha/2}$. This is equivalent to requiring that an approximate $1-\alpha$ Wald interval for ρ_k not contain $-(n-1)^{-1}$; i.e., that $\hat{\rho}_k$ exceeds $-(n-1)^{-1} + z_{\alpha/2}se[\hat{\rho}_k]$ or drops below $-(n-1)^{-1} - z_{\alpha/2}se[\hat{\rho}_k]$. Obviously, a great many such comparisons can be made if k is studied over a large range, and concerns over multiplicity of comparisons and false-positive error inflation become pertinent; cf. §A.5.4. When studying $\hat{\rho}_k$ in this fashion, we recommend using these limits more as an exploratory guide to the importance of a particular autocorrelation of order k than as a formal basis for inferences on the true, underlying correlations. If the latter are required, then use of an adjustment for multiplicity may be necessary.

A plot of $\hat{\rho}_k$ vs. the lag, k, can help gauge how the serial correlation is changing as observations lag apart in time. This is known as the (sample) *autocorrelation function (ACF) plot*, or sometimes as the *correlogram*. One important requirement when using an ACF plot in this fashion is, however, that it be applied only to stationary time series. If departure from stationarity is present the series should first be pre-whitened, and the plot based on the consequent residuals. (Indeed, the ACF plot can provide an informal indication that the time series is non-stationary: a non-stationary series will exhibit sample autocorrelations that decay very slowly over k. If this is evidenced, further prewhitening is necessary.)

SAS can produce ACF plots in PROC ARIMA: the ACF plot appears as part of the output from an `identify` statement or by invoking the `plot` option in PROC ARIMA's `estimate` statement. Approximate standard errors are also provided to help assess the strength of any estimated autocorrelation, as discussed above.

Example 5.6 (Lynx trappings, cont'd)   Continuing with the lynx trappings data from Example 5.3, we saw previously that the Durbin–Watson test identifies significant autocorrelation in the series (Example 5.5). To study this in more detail, we construct an ACF plot. Since we also saw that a non-stationary, periodic pattern appears in the data (Example 5.3), we first prewhiten by fitting a harmonic regression model (5.5) and recovering the residuals from that fit. For the regression, we follow our previous results and use periods of $p_1 = 38$, $p_2 = 9.5$, $p_3 = 5$, $p_4 = 3.5$, and $p_5 = 2.5$ years.

Sample SAS code to perform the fit and produce the ACF plot appears in Fig. 5.16. In the figure, PROC REG is used to fit the harmonic model via standard LS. Notice the use of a centered time variable for t to avoid numerical instabilities in the LS fit. The consequent residuals are exported to PROC ARIMA where we employ the `identify` statement to call for the ACF plot.

Output (edited) from PROC ARIMA that generates the ACF plot is given in Fig. 5.17. The plot shows a unit correlation at $k = 0$, which is an artifact: at $k = 0$ we are asking if Y_i is correlated with itself, and of course it is, perfectly so. More interestingly, the plot suggests strong serial correlation at lags $k = 1$ and $k = 2$, and then a damped cycling with perhaps a negative correlation at lag $k = 6$ for these log-transformed data. This is indicated by large values of $\hat{\rho}_k$ (see the column under `Corr`) that exceed twice their standard errors (the dotted markers down both sides of the plot). Rather than use the simple approximation $se[\hat{\rho}_k] \approx n^{-1/2}$ mentioned above, PROC ARIMA calculates values of $se[\hat{\rho}_k]$ via a slightly more complex approximation due to Bartlett (1946); see Box *et al.* (1994, §2.1.6). For these data, the ACF plot suggests that a substantial serial correlation is in evidence, at least at the smaller lag levels. Exercise 5.18 studies this in more detail.   ✴

```
data lynxctr;
input year yloglynx @@;
  t = year-1880;        * centered time variable;
  pi = arcos(-1);       * pi = 3.1415296...;
  s1 = sin(2*pi*t/38);      * p = 38 sine regressor;
  c1 = cos(2*pi*t/38);      * p = 38 cosine regressor;
  s2 = sin(2*pi*t/9.5);     * p = 9.5 sine regressor;
  c2 = cos(2*pi*t/9.5);     * p = 9.5 cosine regressor;
  s3 = sin(2*pi*t/5);       * p = 5 sine regressor;
  c3 = cos(2*pi*t/5);       * p = 5 cosine regressor;
  s4 = sin(2*pi*t/3.5);     * p = 3.5 sine regressor;
  c4 = cos(2*pi*t/3.5);     * p = 3.5 cosine regressor;
  s5 = sin(2*pi*t/2.5);     * p = 2.5 sine regressor;
  c5 = cos(2*pi*t/2.5);     * p = 2.5 cosine regressor;
drop pi;
datalines;
  1821   2.42975228      1822   2.50650503      1823   2.76715587
    :                      :                      :
  1932   3.20139712      1933   3.42439155      1934   3.53096768

proc reg ;
  model yloglynx = s1 c1 s2 c2 s3 c3 s4 c4 s5 c5 ;
  output out=prewhit  r=resid ;

proc arima data=prewhit; identify var=resid;
```

Figure 5.16   Sample SAS code for prewhitening and ACF plot with lynx trappings data

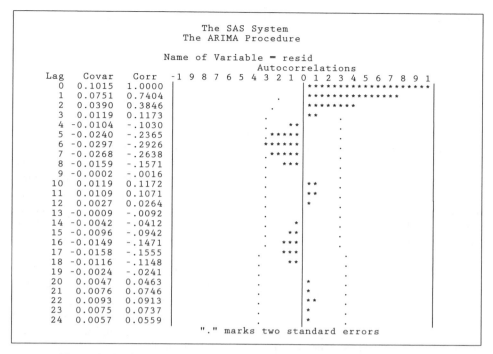

```
                        The SAS System
                      The ARIMA Procedure

                   Name of Variable = resid
                                       Autocorrelations
   Lag     Covar      Corr    -1 9 8 7 6 5 4 3 2 1 0 1 2 3 4 5 6 7 8 9 1
    0     0.1015    1.0000                      |********************
    1     0.0751    0.7404                      |***************
    2     0.0390    0.3846                 .    |********
    3     0.0119    0.1173                      |**    .
    4    -0.0104    -.1030                 .   **|      .
    5    -0.0240    -.2365                 .*****|      .
    6    -0.0297    -.2926                ******|       .
    7    -0.0268    -.2638                 .*****|      .
    8    -0.0159    -.1571                 .  ***|      .
    9    -0.0002    -.0016                 .     |      .
   10     0.0119    0.1172                 .     |**    .
   11     0.0109    0.1071                 .     |**    .
   12     0.0027    0.0264                 .     |*     .
   13    -0.0009    -.0092                 .     |      .
   14    -0.0042    -.0412                 .    *|      .
   15    -0.0096    -.0942                 .   **|      .
   16    -0.0149    -.1471                 .  ***|      .
   17    -0.0158    -.1555                 .  ***|      .
   18    -0.0116    -.1148                 .   **|      .
   19    -0.0024    -.0241                 .     |      .
   20     0.0047    0.0463                 .     |*     .
   21     0.0076    0.0746                 .     |*     .
   22     0.0093    0.0913                 .     |**    .
   23     0.0075    0.0737                 .     |*     .
   24     0.0057    0.0559                 .     |*     .
                            "." marks two standard errors
```

Figure 5.17   SAS output (edited) for ACF plot with lynx trappings data

5.4   Autocorrelated regression models

To incorporate autocorrelation into a regression model, harmonic or otherwise, we turn to a class of statistical models known as *autoregressive integrated moving average (ARIMA) models*, so named for reasons we explain below. The ARIMA class can combine many different forms of serial correlation. Other names often encountered in the literature are *Box–Jenkins models*, named for Box and Jenkin's groundbreaking work in this area (see Box *et al.*, 1994), and time series analysis 'in the time domain' (Barnett, 2004, §10.5).

As in §5.3 we assume the observations Y_i are taken at times t_i ($i = 1, \ldots, n$), and that the t_is are equally spaced such that $t_{i+1} - t_i$ is a positive constant. We impose the simple model from (5.9): $Y_i = \beta_0 + \varepsilon_i$, $i = 1, \ldots, n$, where the errors satisfy $\mathrm{E}[\varepsilon_i] = 0$ and $\mathrm{Var}[\varepsilon_i] = \sigma^2$. We also impose weak stationarity on the series.

5.4.1   AR models

The ARIMA class of time series models starts with what is known as the simple *autoregressive (AR) model*. Therein, each error term, ε_i, is taken as a linear combination of the most recent other errors, plus some additional, new sources of error. At its simplest level an AR model of order 1, denoted AR(1), stipulates that

$$\varepsilon_i = \varphi_1 \varepsilon_{i-1} + u_i \tag{5.11}$$

($i = 1, \ldots, n$). In (5.11), φ_1 is a constant known as the *autoregressive parameter* and u_i is a new, uncorrelated source of random error with zero mean and constant variance ζ^2. The new error term, u_i, is often called the *innovation*; when modeled in this way innovations are often referred to as *white noise* effects. We assume the autoregressive parameter satisfies $|\varphi_1| < 1$, which ensures that the time series will be stationary (Box *et al.*, 1994, §3.2.1). If φ_1 is positive and close to 1, the error ε_i at any given sampling time is very similar to the previous error; if it is negative and close to -1, the errors tend to alternate from positive to negative; if it equals zero, the model collapses to the uncorrelated case. Note that it is difficult to derive such a simple interpretation to the φ_1 parameter if the observations are not equally spaced in time. (However, see Diggle *et al.*, 1994, §5.2.1, for a possible exception using an exponential correlation model.)

Many authors assume the observations have been prewhitened such that their mean is zero; here take $Z_i = Y_i - \beta_0$ so that $E[Z_i] = 0$. Then one can write the AR(1) model directly in terms of these zero-mean variates: $Z_i = \varphi_1 Z_{i-1} + u_i$. One useful consequence of this re-expression is that the autoregressive effect on the data is more clearly illustrated. Indeed, one can recursively substitute for previous terms to find $Z_i = \varphi_1(\varphi_1 Z_{i-2} + u_{i-1}) + u_i = \varphi_1(\varphi_1\{\varphi_1 Z_{i-3} + u_{i-2}\} + u_{i-1}) + u_i$, etc. Mathematically, this can go on *ad infinitum*, leading to $Z_i = \sum_{j=0}^{\infty} \varphi_1^j u_{i-j}$. Consequently, $E[Z_i] = \sum_{j=0}^{\infty} E[\varphi_1^j u_{i-j}] = \sum_{j=0}^{\infty} \varphi_1^j E[u_{i-j}]$ and since $E[u_i] = 0$ for all i, we see $E[Z_i] = 0$, as desired. Similarly, $\text{Var}[Z_i] = \sum_{j=0}^{\infty} \text{Var}[\varphi_1^j u_{i-j}] = \sum_{j=0}^{\infty} \varphi_1^{2j} \text{Var}[u_{i-j}] = \zeta^2 \sum_{j=0}^{\infty} \varphi_1^{2j}$, the first equality following from the mutual independence of the u_is. Since we assume $|\varphi_1| < 1$, the sum represents a convergent geometric series and we find $\text{Var}[Z_i] = \zeta^2/\{1 - \varphi_1^2\}$ (Exercise 5.15). Now, since we took $Z_i = Y_i - \beta_0$, we also have $\text{Var}[Z_i] = \text{Var}[Y_i] = \text{Var}[\varepsilon_i] = \sigma^2$ (why?), so as a result we can move between the constants $\sigma^2 = \text{Var}[\varepsilon_i]$ and $\zeta^2 = \text{Var}[u_i]$ via the relationship $\sigma^2 = \zeta^2/\{1 - \varphi_1^2\}$. (A statistical technicality: at $i = 1$ (5.11) requires a specification for ε_0. Here, to retain the relationship $\sigma^2 = \zeta^2/\{1 - \varphi_1^2\}$ we apply the initiating assumption $\varepsilon_0 \sim N(0, \zeta^2/\{1 - \varphi_1^2\})$; see Exercise 5.15.)

Under the AR(1) error model the true, underlying serial correlation estimated by the $\hat{\rho}_k$s dampens exponentially with increasing lag indices, k. Specifically, $E[\hat{\rho}_k] \propto \varphi_1^k$ for $k = 1, 2, \ldots$ (Box *et al.*, 1994, §3.2.3). Recall that we encountered a similar sort of effect when modeling the correlation between two observations separated by a lag of t time units with the generalized estimating equations of §3.2.6: as the lag increases, raising a correlation-type parameter between -1 and 1 to that power will drive the overall correlation close to zero. Exponential dampening has an intuitive appeal, since in many cases environmental observations separated by long temporal periods exhibit weaker correlations than those separated by shorter periods. This is an important characteristic of the AR(1) process. To test the significance of φ_1 the

Durbin–Watson d_{calc} from (5.8) can be used. Indeed, d_{calc} is designed to identify autocorrelation of the AR(1) form (Neter *et al.*, 1996, §12.3).

We can also define higher-order AR models. For instance, an AR(2) model satisfies

$$\varepsilon_i = \varphi_1 \varepsilon_{i-1} + \varphi_2 \varepsilon_{i-2} + u_i \tag{5.12}$$

for constants φ_1 and φ_2. The correlation structures available under an AR(2) model are more complicated than those under AR(1) models. One property they share, however, is that under certain regularity conditions on φ_1 and φ_2 serial correlation still decays exponentially at large lags (Box *et al.*, 1994, §3.2). In general, an AR(k) model includes k autoregressive parameters, $k = 1, 2, \ldots$, such that, under certain regularity conditions, the true serial correlation decays exponentially at large lags. We explore use of AR(k) models to assess some common problems in environmental applications – such as trend analysis and effects of intervention in the presence of autocorrelation – in §5.5.

5.4.2 Extensions: MA, ARMA, and ARIMA

The second major class of autocorrelation models are called *moving average (MA) models*. These share the AR(k) assumption that (5.9) holds; however, they differ in their assumptions about the error structure. For example, an MA(1) model stipulates

$$\varepsilon_i = \theta_1 u_{i-1} + u_i \tag{5.13}$$

($i = 1, \ldots, n$), where θ_1 is a *moving average parameter*. Here again, if we prewhiten the data, or take $Z_i = Y_i - \beta_0$, so that $E[Z_i] = 0$, one often sees the MA(1) model written as $Z_i = \theta_1 u_{i-1} + u_i$. The innovations u_i are again white noise with $E[u_i] = 0$ and $Var[u_i] = \zeta^2$, but notice now how they enter into the errors: ε_i derives its stochastic nature from both current *and* past innovations, rather than from current innovations and past errors. As a result, the serial correlation in an MA(1) model is short-lived. We expect to see a spike in the ACF plot at lag $k = 1$, but nothing substantive at higher lags. MA models of higher order can also be constructed, allowing for more complicated but, again, short-lived serial correlation. We denote these as MA(q) models of order $q = 1, 2, \ldots$. These models all share the property that errors separated by a large enough lag become uncorrelated. (So, for example, the ACF plot will show zero autocorrelation for all lags greater than q.)

We can combine the AR(k) and MA(q) constructions into what is known as the class of *ARMA models*, creating even more complex error structures. ARMA models can be used to represent seasonal time series by lagging not only on recent errors/innovations, but on those, say, a year earlier.

The structure of the errors ε_i is crucial for specifying autoregressive and moving average models. Although the models' deterministic aspects based on (5.9) are rather simple, their stochastic components such as (5.11) or (5.13) can become quite intricate. As a result, a broad literature exists on how AR(k) and MA(q) model are derived theoretically from a larger class of statistical models known as *stochastic processes* (Ross, 1995; Bhattacharya, 2002). Readers interested in the mathematical and statistical theory underlying autoregressive, moving average, and other forms of ARMA processes can find further detail in sources such as Box *et al.* (1994, §1.2), Brockwell and Davis (1991), or, for a more historical viewpoint, Bartlett (1955), and the references therein.

As mentioned above, time series that exhibit an underlying trend often lend themselves well to study of their successive differences via first-order terms $C_i = Y_i - Y_{i-1}$, second-order terms $D_i = C_i - C_{i-1}$, seasonal differences, etc. Recall, for example, that if an equally spaced series exhibits a simple linear trend such as $E[Y_i] = \beta_0 + \beta_1 t_i$, the first-order differences $C_i = Y_i - Y_{i-1}$ can be used to detrend the series, since their expectations do not change with respect to i when $t_i - t_{i-1}$ is constant. (Second-order differences are constant if the trend is quadratic.) Indeed, differencing operations are an important tool in the analysis of time series data.

ARMA models can themselves be extended by *integrating* the effects on Y_i via summation over the differences: $C_i + C_{i-1} + \ldots$, etc. The resulting class of *ARIMA models* is flexible enough to approximate even some non-ARIMA correlation structures, at least for the sake of predicting a few lags into the future. Consequently, ARIMA models are used heavily in econometrics for forecasting time series; for similar reasons, there is growing interest in them for environmental applications (Young and Pedregal, 2002; Barnett, 2004, §10.5). Proper fit of a full-generality ARIMA model requires considerable expert judgment, however. Several plots are typically required, including the ACF plot and extensions called the *partial autocorrelation function (PACF) plot* and the *inverse autocorrelation function (IACF) plot*. The former is a plot of the 'partial' correlations between Y_1 and Y_{k+1} after correcting for the intervening observations $Y_2, \ldots, Y_k$ by regressing Y_1 on them, while the latter is based on inverting the roles of the $AR(k)$ and $MA(q)$ components in an ARMA model and building the ACF from this inverted model. Both can be useful for in-depth study of autocorrelation in a time series. For instance, the PACF plot is useful in diagnosing simple $AR(k)$ models, since the partial autocorrelations under an $AR(k)$ series should be at or near zero for all lags greater than k. Most computer packages make these graphical devices available, including, for example, PROC ARIMA.

An extended exposition on ARIMA modeling is beyond the scope of this book; interested readers should consult dedicated texts on time series such as Brockwell and Davis (1991) or Box *et al.* (1994).

5.5 Simple trend and intervention analysis

The issue of whether a series is trending upward or downward is often of interest with temporal environmental data. In what is commonly referred to as environmental *trend analysis* (Brillinger, 1994; Harvey, 2002), the investigator is interested in identifying the existence of a trend, the nature of that trend, and perhaps forecasting future temporal outcomes in the presence of the trend. Also of interest in a number of settings is whether some intervening event has caused a change in the overall level of the series; that is, an *intervention analysis* (Box and Tiao, 1975; Rasmussen *et al.*,

1993) of the response over time. This has important tie-ins with environmental *impact assessment*; see, for example, Stewart-Oaten (2002).

The general model considered for such problems takes a regression form with serially correlated errors. We extend (5.9) and write

$$Y_i = \beta_0 + g(t_i; x_{i1}, x_{i2}, \ldots, x_{iP}) + \varepsilon_i, \tag{5.14}$$

where $g(\cdot)$ represents any pertinent function of the time variable, t_i, and of other regressor variables, $x_{i1}, x_{i2}, \ldots, x_{iP}$, that are felt to influence Y_i in a deterministic fashion. These are typically some form of trend and/or seasonal and/or intervention terms and in the most general case may themselves be time-dependent. The errors, ε_i, follow an $AR(k)$ model of order $k \geq 1$ specified by the analyst; for example, (5.11) corresponds to $k = 1$ and (5.12) corresponds to $k = 2$. With strategic choice of the regressors, model (5.14) can be used to approximate a variety of phenomena underlying environmental time series. In what follows, we present a short selection.

5.5.1 Simple linear trend

If we let $g(t_i) = \beta_1 t_i$ in (5.14) we have the simple linear trend model

$$Y_i = \beta_0 + \beta_1 t_i + \varepsilon_i. \tag{5.15}$$

As usual, the slope coefficient β_1 is the change in $E[Y_i]$ per unit time. We can test the null hypothesis of no linear time trend $H_0: \beta_1 = 0$, find point estimates of the trend/slope parameter, or construct confidence intervals on β_0 and β_1. This is similar to the standard methods for analyzing a simple linear regression model seen in Chapter 1, except that in (5.15) the errors are assumed to follow an autoregressive structure such as (5.11) or (5.12).

5.5.2 Trend with seasonality

If we let $g(t_i; c_i, s_i) = \beta_1 t_i + \beta_2 s_i + \beta_3 c_i$ in (5.14), we can with proper choice of s_i and c_i impose a linear trend upon a harmonic regression, in effect extending (5.3). For example, if we let $s_i = \sin(2\pi t_i/12)$ and $c_i = \cos(2\pi t_i/12)$, we generate a trending sine wave with a 12-month period:

$$Y_i = \beta_0 + \beta_1 t_i + \beta_2 \sin\left(\frac{2\pi t_i}{12}\right) + \beta_3 \cos\left(\frac{2\pi t_i}{12}\right) + \varepsilon_i. \tag{5.16}$$

In (5.16) the centerline level of the sine wave now changes with a long-term trend of β_1 units per unit time. When fit using an autoregressive structure, we can test the hypothesis of no trend $H_0: \beta_1 = 0$, construct confidence intervals on β_1, or find point and interval estimates for the other regression parameters, all while adjusting for serial correlation.

If the single sine wave employed in (5.16) is inadequate to approximate complex seasonal cycles in the data, one can include additional sine and cosine terms. Often, these have periods which are harmonics of one basic period, such as 12 months, $12/2 = 6$ months, $12/3 = 4$ months, etc. A standard practice is to use periods which have been suggested by a Fourier analysis of the detrended series, as described in §5.2.3.

One can also extend (5.16) to include higher-order polynomial terms in t. These can be used to account for possible curvilinearity in the trend over time. As with any polynomial regression, however, some caveats are in order. First, it is often unlikely that the true underlying trend is exactly polynomial. Use of polynomial terms in (5.16) should be viewed as approximations to the true curvature for the time range under study. Also, we strongly recommend *centering* the time variate prior to including polynomial terms in the predictor; that is, use $(t_i - \bar{t})$, $(t_i - \bar{t})^2$, etc., in place of the corresponding terms t_i, t_i^2, etc. (If calculation of $\bar{t}$ is difficult for some reason, SAS can center variables directly by using **PROC STANDARD** with the `mean=0` option.) As seen in §1.5, centering reduces instabilities that occur when the polynomial terms are highly collinear. Lastly, we recommend use of very high-order polynomial terms only when one can argue for environmental or ecological mechanisms that support their use. Without these, and in most cases, an additional quadratic and possibly a cubic term will suffice to approximate any curvilinear departures from a simple linear trend.

We can fit $AR(k)$ regression models in SAS via **PROC AUTOREG**. The procedure's output begins with results from an LS fit assuming independent errors; however, if the $AR(k)$ error structure is found to be an important component of the data, inferences from the LS fit should be ignored. **PROC AUTOREG** can provide an ACF plot of the LS residuals, and will fit the $AR(k)$ model to these residuals. If desired, it can also perform a backward elimination search of $AR(k)$ parameters. This latter option is similar to backward search methods for variable selection in ordinary regression modeling (Neter *et al.*, 1996, §8.4). For terms that survive the backward elimination, **PROC AUTOREG** displays point estimates and standard errors. (A note of caution: **PROC AUTOREG**'s $AR(k)$ parameters are defined to be the *negatives* of the φ-parameters we define above.) It also gives a summary of the final fitted model, including approximate AR-corrected standard errors, t-tests, and approximate P-values for the regression parameters.

Example 5.7 (Ground-level ozone)   As noted in Example 3.6, ground-level ozone is a major component of summertime smog, and in high concentrations it can become a health hazard. Consequently, monitoring of ground-level ozone levels by environmental epidemiologists and other public health officials is a priority in many industrialized nations. Table 5.5 gives an example: a sample of $n = 96$ average monthly maximum concentrations (in ppb) of ground-level ozone between 1994 and 2001 recorded at the meteorological station in Scotland's Eskdalemuir Forest. (The full data are online at http://www.wiley.com/go/environmental.) Figure 5.18 plots the data, showing clear periodicity, but also a downward trend in the concentrations over time.

To identify specific harmonics in the series, we perform a Fourier analysis: using the residuals from a simple linear model fit, we construct a smoothed $\log_{10}$-periodogram. It can be shown that peaks in the plot occur at three frequencies: $f = 0.083$, 0.167, and 0.333, corresponding to periods of $p = 12$, 6, and 3 months (Exercise 5.21).

Table 5.5 A selection of monthly average maximum ground-level ozone concentrations (in ppb) in Eskdalemuir, Scotland

Month	Concentration
Jan. 1994	32.71
Feb. 1994	32.00
Mar. 1994	42.97
⋮	⋮
Oct. 2001	28.42
Nov. 2001	27.07
Dec. 2001	24.68

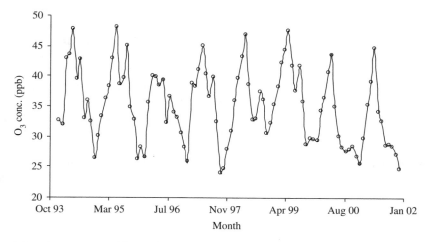

Figure 5.18 Average monthly maximum concentrations of ground-level ozone between 1994 and 2001 in Eskdalemuir, Scotland

The Durbin–Watson test statistic for these data is $d_{\text{calc}} = 1.342$ (one-sided $P = 0.0002$), suggesting significant positive autocorrelation. Thus we consider here an autoregressive model similar in form to (5.16), but using a centered linear predictor $t_i - \bar{t}$, for greater numerical stability, and the three periods mentioned above. The regression model has the form

$$Y_i = \beta_0 + \beta_1(t_i - \bar{t}) + \beta_2 \sin\left(\frac{2\pi\{t_i - \bar{t}\}}{12}\right) + \beta_3 \cos\left(\frac{2\pi\{t_i - \bar{t}\}}{12}\right)$$

$$+ \beta_4 \sin\left(\frac{2\pi\{t_i - \bar{t}\}}{6}\right) + \beta_5 \cos\left(\frac{2\pi\{t_i - \bar{t}\}}{6}\right) + \beta_6 \sin\left(\frac{2\pi\{t_i - \bar{t}\}}{3}\right)$$

$$+ \beta_7 \cos\left(\frac{2\pi\{t_i - \bar{t}\}}{3}\right) + \varepsilon_i,$$

where the ε_is are assigned an AR(k) error structure.

The ACF plot for these data is derived as part of Exercise 5.21. The plot exhibits strong positive autocorrelation at lag $k = 1$, and moderate positive autocorrelations at lags $k = 2$ and $k = 4$. This is followed by a cycling into negative autocorrelations. The negative values do not appear significant. Based on this information, we consider an AR(k) error structure of at least lag $k = 4$, and employ backward elimination to cull the AR component. Since we wish to ensure that no unusually long lags are overlooked, we set k to be fairly high; here we will start the procedure at lag $k = 8$.

Figure 5.19 gives sample SAS code for invoking PROC AUTOREG to fit this model. In the code, notice use of the SAS internal variable _N_, which is the observation number. Since the data are equally spaced, we can use _N_ as a proxy for month in order to build the temporal predictor, t_i. Subtracting 48.5 from _N_ centers the predictor variable, since the mean of the first 96 integers is exactly 48.5. To specify the AR(k) structure, we include the nlag=k option in the model statement. The code in Fig. 5.19 sets nlag=8. To cull unneeded lags from the model, we use the backstep option. As described above, this institutes an automated backward elimination process to remove insignificant AR terms. We also include an output statement, generating residuals and predicted values for purposes of model evaluation. (The predicted values and residuals are printed along with the observed values and the centered time variate using a call to PROC PRINT.)

The PROC AUTOREG fit (edited for presentation) appears in Fig. 5.20. Notice that the first two sets of output results correspond to an LS fit; as emphasized above, for an AR(k) model this output feature holds minimal value except perhaps to illustrate the impact of incorrectly assuming independence among the observations (see below).

The results in Fig. 5.20 support the autoregressive error assumption: backward elimination of the autoregressive terms results in a significant lag at $k = 1$ (see under Estimates of Autoregressive Parameters). Thus the final model includes

```
    data ozone;
      input y @@;
  *center month variable based on 96 monthly obs'ns;
      t = _N_  - 48.5;
      pi = arcos(-1);              * pi = 3.1415296...;
      s1 = sin(2*pi*t/12);         * p = 12 sine regressor;
      c1 = cos(2*pi*t/12);         * p = 12 cosine regressor;
      s2 = sin(2*pi*t/6);          * p = 6 sine regressor;
      c2 = cos(2*pi*t/6);          * p = 6 cosine regressor;
      s3 = sin(2*pi*t/3);          * p = 3 sine regressor;
      c3 = cos(2*pi*t/3);          * p = 3 cosine regressor;
    drop pi;     datalines;
      32.71    32.00    42.97    43.57    47.87    39.50    42.71
      ......
      32.65    28.65    28.77    28.42    27.07    24.68

    proc autoreg;
      model y = t s1 c1 s2 c2 s3 c3  /  nlag=8 backstep;
      output out=eval r=resid p=yhat;

    proc print; var t y yhat resid;
```

Figure 5.19 Sample SAS code for AR(k) regression with Eskdalemuir ozone data

```
                            The SAS System
                          The AUTOREG Procedure

                      Dependent Variable      y
                  Ordinary Least Squares Estimates
SSE                      931.719439    DFE                         88
MSE                        10.58772    Root MSE               3.25388
SBC                      527.128604    AIC                 506.613819
Regress R-Square            0.7419    Total R-Square          0.7419
Durbin-Watson               1.3418

                                       Standard                Approx
Variable          DF     Estimate        Error    t Value    Pr > |t|
Intercept          1      35.1408       0.3321     105.81      <.0001
t                  1      -0.0467       0.0121      -3.87      0.0002
s1                 1       4.7617       0.4720      10.09      <.0001
c1                 1      -4.8651       0.4697     -10.36      <.0001
s2                 1      -0.5768       0.4703      -1.23      0.2233
c2                 1      -1.7966       0.4697      -3.83      0.0002
s3                 1      -0.2874       0.4699      -0.61      0.5423
c3                 1      -0.9329       0.4697      -1.99      0.0501

                   Estimates of Autocorrelations
  Lag Covariance Correlation       Lag Covariance Correlation
   0      9.7054     1.0000
   1      3.1808     0.3277          5      1.3303     0.1371
   2      2.0003     0.2061          6      1.8109     0.1866
   3      1.4988     0.1544          7      1.6400     0.1690
   4      2.3096     0.2380          8      1.8020     0.1857

             Backward Elimination of Autoregressive Terms
  Lag   Estimate t Value     Pr>|t|     Lag   Estimate t Value      Pr>|t|
   3    -0.00640   -0.06     0.9557
   5     0.03327    0.29     0.7712       8    -0.08143    -0.77     0.4406
   7    -0.04268   -0.38     0.7044       6    -0.11221    -1.09     0.2767
   2    -0.05346   -0.49     0.6281       4    -0.19193    -1.90     0.0608

             Estimates of Autoregressive Parameters
                 Lag       Coefficient      Std Error     t Value
                  1         -0.327731        0.101290      -3.24

                                       Standard                Approx
Variable          DF     Estimate        Error    t Value    Pr > |t|
Intercept          1      35.1435       0.4670      75.26      <.0001
t                  1      -0.0475       0.0168      -2.83      0.0058
s1                 1       4.7493       0.6108       7.78      <.0001
c1                 1      -4.8607       0.6026      -8.07      <.0001
s2                 1      -0.5909       0.5074      -1.16      0.2474
c2                 1      -1.7939       0.5031      -3.57      0.0006
s3                 1      -0.3002       0.3744      -0.80      0.4249
c3                 1      -0.9321       0.3722      -2.50      0.0141
```

Figure 5.20 SAS output (edited) from autoregressive fit for Eskdalemuir ozone data

this lag at $k = 1$ in the error structure. Partial t-tests of the regressor variables at the end of the output indicate that all three periodic terms appear significant at $\alpha = 0.05$ (when either term of a sine–cosine pair for each modeled period is significant, we include both terms for that period), along with the linear trend term. Adjusting for multiplicity via a Bonferroni correction (§A.5.4) brings this down to perhaps only the $p_1 = 12$ and $p_2 = 6$ month periods and the linear trend.

It is instructive also to compare the autoregressive model fit to the independent-errors LS fit in Fig. 5.20. At first glance, the estimated regression coefficients exhibit minimal change between models; for example, the trend coefficient is estimated as

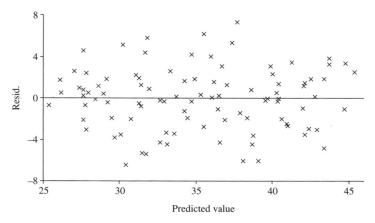

Figure 5.21 Residual plot from autoregressive fit for Eskdalemuir ozone data

$b_1 = -0.0467$ under independent errors and as $b_1 = -0.0475$ under autoregressive errors (in both cases, a decrease of about 0.05 ppb/month over the six-year period under study). A critical difference between the two occurs, however, with the standard errors. For example, $se[b_1] = 0.0121$ under independent errors increases by 39% to $se[b_1] = 0.0168$ under autoregressive errors. Increases in standard errors are not guaranteed: $se[b_7] = 0.4697$ under independent errors drops to $se[b_7] = 0.3722$ under autoregressive errors. Once again, sound inferences require correct specification of the error structure in a statistical model.

A residual plot based on the values provided from the call to **PROC PRINT** (output not shown) is displayed in Fig. 5.21. The plot shows a reasonably random spread of residuals.   ✪

5.5.3 Simple intervention at a known time

By employing more complex forms for $g(\,\cdot\,)$ in (5.14) we can model some rather sophisticated temporal effects. For example, we can employ indicator functions, $I_{\mathbb{A}}(x)$, from §A.2.1 to represent single, sudden changes in the series over time. (Recall that an indicator function has the form $I_{\mathbb{A}}(t) = 1$ if t is contained in the set $\mathbb{A}$, and $I_{\mathbb{A}}(t) = 0$ otherwise. Do not confuse the indicator function $I_{\mathbb{A}}(x)$ with the intensity function $\mathcal{I}(h)$ seen earlier.) For instance, to model a simple intervention that may cause some abrupt change in the overall level of the series at a known time τ, we can use

$$Y_i = \beta_0 + \xi I_{[\tau,\infty)}(t_i) + \varepsilon_i, \tag{5.17}$$

where t_i is the recorded time variable $(i = 1, \ldots, n)$. This corresponds to a simple two-segment plateau model

$$E[Y_i] = \begin{cases} \beta_0 & \text{if } t_i < \tau, \\ \beta_0 + \xi & \text{if } t_i \geq \tau. \end{cases}$$

A similar form was studied in Example 2.1, except there τ was unknown. In (5.17) the known change point τ denotes the time of some intervening phenomenon, such as a hurricane, an earthquake, an industrial site built upriver from a protected habitat, etc. (Appropriate modification is needed if the original time variable is centered about its mean.) For use in a regression model we define the simple predictor variable $I_{[\tau, \infty)}(t_i)$, set to one when $t_i \geq \tau$ and zero otherwise.

The model in (5.17) hypothesizes a non-trending time series both before and after the intervention at time τ, with an abrupt positive or negative jump in $E[Y_i]$ of size ξ. In this sense, ξ quantifies the intervention effect. We can also add appropriately constructed sine and cosine terms to account for seasonality.

By assuming an AR(k) model for the error terms in ε_i, hypothesis tests or confidence intervals on the jump size ξ will adjust for the serial correlation. Note that if we do *not* account for serial correlation in the data, inferences on ξ under model (5.17) are equivalent to those from a two-sample t-test, treating the data observed before and after time τ as separate samples. Thus (5.17) also can be viewed as an extension of this simpler statistical method to the temporally correlated data setting.

5.5.4 Change in trend at a known time

A straightforward extension to the simple model in (5.17) allows for an underlying trend that may change as the result of an intervention; take

$$Y_i = \beta_0 + \beta_1 t_i + \lambda(t_i - \tau)I_{[\tau, \infty)}(t_i) + \varepsilon_i, \tag{5.18}$$

$i = 1, \ldots, n$. The model in (5.18) is fitted by creating an additional regressor variable as the product of $I_{[\tau, \infty)}(t_i)$ and $(t_i - \tau)$, and then adding this to the simple trend model in (5.15). The coefficient λ is the change in trend at time τ, that is, after time τ the slope changes to $\beta_1 + \lambda$. Thus the mean response, $E[Y_i]$, is a continuous function in t with a change in slope at $t = \tau$. The model is similar to the continuous bilinear form in (2.7), except that in (5.18) the intervention point τ is assumed known.

If in (5.18) the errors are uncorrelated and τ is known, this model can be fitted as a multiple linear regression; see Neter *et al.* (1996, §11.5). If τ is unknown, however, (5.18) becomes a piecewise regression model. As discussed in §2.2, the model fit then requires application of nonlinear regression techniques.

To account for serial correlation, we employ an AR(k) structure for the ε_is. If k is unknown, one can specify a fairly large value for k and use backward elimination to cull the model back to an appropriate lag.

5.5.5 Jump and change in trend at a known time

We can combine (5.17) and (5.18) to allow for differential trend after the intervention point, that is, a trend of magnitude β_1 before the innovation, an abrupt jump of size ξ at time τ, and then a new trend with slope $\beta_1 + \lambda$ after time τ:

$$Y_i = \beta_0 + \beta_1 t_i + \xi I_{[\tau,\infty)}(t_i) + \lambda(t_i - \tau)I_{[\tau,\infty)}(t_i) + \varepsilon_i. \tag{5.19}$$

As with (5.18), when the intervention point τ is known (5.19) may be fitted using standard techniques: use either a multiple linear regression (e.g., via **PROC REG**) or an autoregressive model (e.g., via **PROC AUTOREG**), depending on the nature of the autocorrelated error structure. If τ is unknown, (5.19) becomes a nonlinear regression model. The latter case is similar to the bilinear form in (2.6).

Example 5.8 (River-water management)   North America's Colorado River is a major source of water for irrigation in the US states of Arizona and California, and in north-western Mexico. The last major US irrigation diversion structure on the lower Colorado River is the Imperial Dam, located 18 miles north of Yuma, Arizona. California and Arizona users receive water from the dam via two diversion canals; the Arizona side diversion is known as the Gila Gravity Main Canal. Table 5.6 gives a selection of data on the annual mean streamflow of water (in ft^3/s) into this canal between the years 1949 and 2000 (the full set is available online at http://www.wiley.com/go/environmental).

In 1964, the US Supreme Court issued a decree that settled a 25-year-old dispute between Arizona and California regarding supplies of Colorado River water. The Court established rights for various water users in the area that led to changes in how water from Imperial Dam was diverted and consumed. We will view the Court's decree as an intervention underlying the management and use of diverted river water.

Since the data are rates of flow, we anticipate that they possess a positive skew and that the variation in response may be heterogeneous. To correct for this, we apply a stabilizing transformation to the original streamflows, such as the square root $Y = \sqrt{\text{streamflow}}$. The full set of transformed data is plotted in Fig. 5.22; a change in response in the mid-1960s is clearly evident.

To assess and quantify the differential trend in these square-root-transformed data, we construct an ACF plot to examine the serial correlation. Since the strong signal in these data produces non-stationarity, prewhitening is required. Here, we

Table 5.6   A selection of Colorado River Imperial Dam annual mean streamflows (in ft^3/s) into the Gila Gravity Main Canal near Yuma, Arizona

Year	Streamflow
1949	186
1950	207
1951	254
⋮	⋮
1998	1055
1999	1057
2000	1124

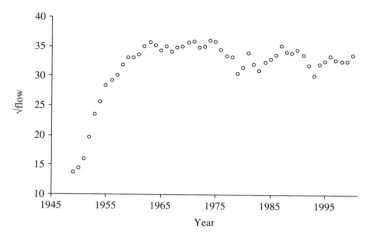

Figure 5.22 Square-root-transformed annual mean streamflow (in ft^3/s) into the Gila Gravity Main Canal near Yuma, Arizona

perform an initial LS fit of the model in (5.19) and use the residuals from the fit to construct the ACF plot. Sample SAS code for these operations is given in Fig. 5.23.

The ACF plot from PROC ARIMA appears in Fig. 5.24. There we see a strong positive autocorrelation indicated at lag $k = 1$, followed by a cycling down to potentially negative autocorrelation at lags $k = 6, 7$, and 8. To formalize this, we apply an AR(k) error structure in the differential trend model (5.19). To be conservative, we set $k = 10$ and employ backward selection. In PROC AUTOREG this corresponds to the following statements (used after the call to PROC ARIMA in Fig. 5.23):

```
proc autoreg;
model ysqflow=year jump upind/ covb nlag=10 backstep;
output r=resid p=yhat;
```

```
data river;
input year yflow @@;  yrtflow = sqrt(yflow);
  tau = 1964;  tadj = year - tau;
    if year >= tau then jump = 1;
    else jump = 0;
  upind = jump * tadj;
  datalines;
1949    186    1950    207    1951    254    1952    387
  :             :             :             :
1997  1070    1998  1055    1999  1057    2000  1124

proc reg;
  model yrtflow = year jump upind;
  output out=prewhit  r=resid;

proc arima data=prewhit; identify var=resid;
```

Figure 5.23 Sample SAS code for bilinear regression prewhitening and ACF plot of river-water management data

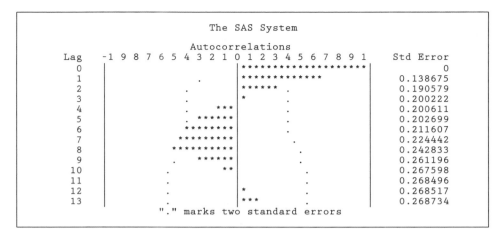

Figure 5.24 SAS output (edited) for ACF plot with prewhitened river-water management data

(the `covb` option in the `model` statement outputs the estimated variances and covariances of the final parameter estimates). The resulting output (edited) appears in Fig. 5.25. There, we see that through backward elimination a lag of $k = 5$ appears to accommodate the serial correlation. A residual plot appears in Fig. 5.26, where no gross discrepancies emerge. (The large negative residual in 1979 is of some secondary interest: the Colorado River basin experienced a major drought in 1977–1978, the impacts of which affected downstream river management in subsequent years.)

Estimates of the autocorrelation-adjusted regression parameters in Fig. 5.25 (under `Estimate`) indicate that the initial slope is important: $b_1 = 1.511$ ($P < 0.0001$). The jump at the intervention also appears significant: $\hat{\xi} = -3.254$ ($P = 0.0041$).

For inferences on the post-intervention response, we use the point estimate $b_1 + \hat{\lambda}$. This has standard error $\{se^2[b_1] + 2\text{Cov}[b_1, \hat{\lambda}] + se^2[\hat{\lambda}]\}^{1/2}$, where $\text{Cov}[b_1, \hat{\lambda}]$ is the estimated covariance between b_1 and $\hat{\lambda}$ (given under `Covariance of Parameter Estimates`). We calculate $b_1 + \hat{\lambda} = 1.511 - 1.607 = -0.096$, and $se[b_1 + \hat{\lambda}] = \{0.1138^2 + 2(-0.0134) + 0.1208^2\}^{1/2} = 0.027$. From these, we can build a Wald confidence interval for the post-intervention slope using a t-approximation for the critical point. Here, the PROC AUTOREG output suggests use of a $t(46)$ reference distribution (check `DFE` under `Yule-Walker Estimates`) so at $\alpha = 0.05$ we use $t_{0.025}(46) = 2.0129$, found from SAS code similar to that in Fig. A.4. The corresponding 95% limits for $\beta_1 + \lambda$, along with those for the other pertinent model parameters, appear in Table 5.7. There, we see that the post-intervention slope appears significant, since its 95% confidence interval fails to contains zero. Indeed, the data suggest a decreasing trend of between $0.04 \, \text{ft}^3/\text{s}$ and $0.15 \, \text{ft}^3/\text{s}$ per year since 1964. ✪

Obviously a great many variants on (5.17), (5.18), and (5.19) may be developed and fitted. For instance, the effects of an intervention can modulate, such as a gradually occurring change, or they can 'shock' the mean response at only

```
                              The SAS System
                           The AUTOREG Procedure

                     Estimates of Autocorrelations
      Lag  Covariance  Correlation       Lag  Covariance  Correlation
       0     2.2344      1.0000
       1     1.4894      0.6665            6    -0.8523     -0.3815
       2     0.6994      0.3130            7    -1.0562     -0.4727
       3     0.1423      0.0637            8    -1.0961     -0.4905
       4    -0.3306     -0.1480            9    -0.6629     -0.2967
       5    -0.6921     -0.3098           10    -0.2501     -0.1119

              Backward Elimination of Autoregressive Terms
    Lag  Estimate t Value   Pr>|t|    Lag  Estimate t Value   Pr>|t|
     3    0.0197    0.10    0.9194      7    0.1105    0.73    0.4666
     6   -0.0694   -0.37    0.7160      9   -0.1440   -0.97    0.3390
     4    0.0733    0.49    0.6299      2    0.2124    1.53    0.1342
    10    0.1060    0.69    0.4966      8    0.2277    1.96    0.0558

               Estimates of the Autoregressive Parameters
                  Lag    Coefficient      Std Error    t Value
                   1      -0.634640        0.106477     -5.96
                   5       0.215848        0.106477      2.03

                       Yule-Walker Estimates
        SSE           52.9083437      DFE                    46
        MSE            1.15018        Root MSE          1.07247
        Durbin-Watson  1.5004

                                  Standard                   Approx
        Variable    DF   Estimate    Error    t Value    Pr > |t|
        Intercept    1     -2929   222.6977   -13.15      <.0001
        year         1    1.5111     0.1138    13.28      <.0001
        jump         1   -3.2538     1.0763    -3.02      0.0041
        upind        1   -1.6074     0.1208   -13.30      <.0001

                    Covariance of Parameter Estimates
                    Intercept        year         jump         upind
        Intercept   49594.259    -25.34822    171.98224    26.123491
        year        -25.34822     0.0129558   -0.088017    -0.013352
        jump        171.98224    -0.088017     1.1585161    0.0766243
        upind       26.123491    -0.013352     0.0766243    0.0145979
```

Figure 5.25 SAS output (edited) from autoregressive fit for prewhitened river-water management data

one or a few particular observations (de Jong and Penzer, 1998). A novel approach used to study the nature of complex interventions in environmental settings is known as the *before–after control–impact (BACI) design*. One takes measurements before and after a known intervention, and also pairs observations taken at the impacted site with a control site to adjust for any site-specific variation. Popularized by Stewart-Oaten *et al.* (1986), the design and its associated analytic methodology are useful for ecological impact assessment. Smith (2002a) gives a practical review.

More generally, the issue of detecting changes in the deterministic features of a time series is contained in the larger study of *change-point analysis*. More details on this area of statistical theory and practice are available in, for example, Jandhyala *et al.* (1999, 2002), Esterby (2002), and the references therein.

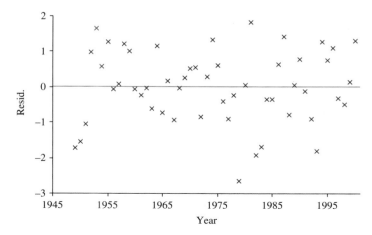

Figure 5.26 Residual plot from autoregressive fit for transformed river-water management data

Table 5.7 Approximate 95% Wald confidence limits for selected discontinuous intervention model parameters with transformed river-water management data

Parameter	Estimate	Approx. 95% limits	Parameter	Estimate	Approx. 95% limits
β_1	1.511	$1.282 < \beta_1 < 1.740$	ξ	−3.254	$-5.420 < \xi < -1.087$
λ	−1.607	$-1.851 < \lambda < -1.364$	$\beta_1 + \lambda$	−0.096	$-0.151 < \beta_1 + \lambda < -0.041$

5.6 Growth curves revisited

When modeling the growth of an organism or system over time, we saw in §2.4 that a variety of parametric regression functions were available. In that section we assumed the error terms were not subject to possible autocorrelation; however, for some growth curves that assumption is untenable. In this section, we revisit the issue of how to model and analyze growth curve data, now with the possibility of autocorrelation among the error terms.

5.6.1 Longitudinal growth data

Growth curve data taken over time are a form of *repeated measurements*, where the autocorrelation is induced among repeated observations on the same subject (Kenward, 2002). Repeated observations over time are also known as *longitudinal data*, although both these terms can refer to settings beyond the growth curve paradigm.

Indeed, the concept of 'growth' over a series of repeated measurements need not be restricted to observations taken over time; Potvin *et al.* (1990) give an example where repeated measurements were taken on plants to determine their photosynthetic response to changing CO_2 exposure. Increased CO_2 led to 'growth' in the plants' rates of photosynthesis.

Where interest exists in studying the time-course growth of a single subject, group, or system, the repeated measurements constitute observed values Y_i taken at time points t_i, $i = 1, \ldots, n$. Here again we assume the t_is are equally spaced such that $t_{i+1} - t_i$ is a positive constant. Since, by definition, a growth curve results from a process whose mean response changes over time, it represents a non-stationary series.

Consider model (5.14), where for simplicity the function $g(\cdot)$ is based solely on the time variate t_i:

$$Y_i = \beta_0 + g(t_i) + \varepsilon_i. \tag{5.20}$$

In the simplest cases, $g(t)$ is taken as a low-order polynomial in t; for example, a simple linear model corresponding to (5.15) or a quadratic model with $g(t_i) = \beta_1(t_i - \bar{t}) + \beta_2(t_i - \bar{t})^2$. Nonlinear functions such as those seen in §2.4 are also possible. (We comment briefly on nonlinear growth models with autocorrelated errors at the end of the chapter.) To accommodate the repeated measures effect, we can impose an $AR(k)$ error structure on the ε_is. Where necessary, the ACF plot and other diagnostic measures illustrated above may be required to specify the lag, k, for the autoregressive model.

5.6.2 Mixed models for growth curves

In many longitudinal growth curve studies, more than one subject or system contributes repeated responses over time. For such settings we denote the ith subject's observed responses as Y_{ij}, taken at equally spaced times t_j such that $t_{j+1} - t_j$ is a positive constant, $i = 1, \ldots, n; j = 1, \ldots, m$. Notice now that the notation must be expanded to accommodate the longitudinal structure. The number of time points, m, can be different from the number of subjects, n. We use this format in order to remain consistent with the similar GEE notation of §3.2.6. As was done there, we can collapse the Y_{ij}s into an observation vector for the ith subject: $\mathbf{Y}_i = [Y_{i1} \, Y_{i2} \ldots Y_{im}]^T$. In the simplest case, we assume each subject provides a complete set of m records, so that there are nm total observations.

When growth data are collected over multiple subjects it is not uncommon to observe variation between subjects in excess of that predicted under a simple model such as (5.20). That is, the potential for between-subject variability adds an extra component to the observed response. To model this, we extend (5.20) into

$$Y_{ij} = \beta_0 + g(t_j) + \delta_i + \varepsilon_{ij}, \tag{5.21}$$

$i = 1, \ldots, n; j = 1, \ldots, m$. Here we take $\varepsilon_{ij} \sim$ i.i.d. $N(0, \sigma^2)$, and view these as the usual experimental error terms. The additional, additive term δ_i is employed to model a subject-specific effect. We assume $\delta_i \sim$ i.i.d. $N(0, \psi^2)$ such that ε_{ij} and δ_i are independent for all i and j. Of interest is how this affects the correlation structure of the observation vectors $\mathbf{Y}_i$. That is, what are $\mathrm{Var}[Y_{ij}]$ and $\mathrm{Corr}[Y_{ij}, Y_{ik}]$? (We assume that observations between subjects are independent such that $\mathrm{Corr}[Y_{ij}, Y_{hk}] = 0$ for all $i \neq h$.)

We should note that the issue of modeling between-subject variability can be extended from the simple growth curves studied here to the larger class of linear models discussed in Chapter 1. These include simple or multiple linear regressions, ANOVA, or ANCOVA. As with (5.21), the extended model can be written as a sum of terms where (i) some unknown parameters do not vary randomly (called 'fixed effects') and (ii) others do (called 'random effects' in §1.4). In (5.21), each subject has its own intercept, $\beta_0 + \delta_i$, but the effect due to time is the same for all subjects. Since δ_i is random in $\beta_0 + \delta_i$, the intercept may be viewed as a random effect, while time is a fixed effect. The β_0 component reflects a population-averaged response at time zero (or, if the time variate is centered about its mean, at $t = \bar{t}$). The subject effect represents how a particular subject's intercept $\beta_0 + \delta_i$ differs from the population-averaged intercept β_0. We can also extend (5.21) to allow subjects to exhibit a different rate of growth in addition to a different intercept. For example, suppose the growth effect is linear such that $g(t_j) = \beta_1 t_j$ and write the linear model as $Y_{ij} = \beta_0 + \beta_1 t_j + \delta_{0i} + \delta_{1i}t_j + \varepsilon_{ij}$. Here, δ_{0i} and δ_{1i} are random intercept and slope effects with $\delta_{hi} \sim$ indep. $N(0, \sigma_h^2)$, $h = 0, 1$, while β_0 and β_1 are population-averaged intercept and slope parameters.

If a model contains both fixed effects and random effects, it is called a *mixed-effects model* or simply a *mixed model* (Drum, 2002). Repeated-measures growth curves are special cases of mixed models, and we can take advantage of this to analyze growth curve data using methods designed for the larger class. Rather than operate directly with the between-subject correlations, we will find it more expedient to work with the covariances, $\mathrm{Cov}[Y_{ij}, Y_{ik}] = \mathrm{Corr}[Y_{ij}, Y_{ik}]\sqrt{\mathrm{Var}[Y_{ij}]\mathrm{Var}[Y_{ik}]}$ $(j \neq k)$. Recall that $\mathrm{Cov}[Y_{ij}, Y_{ij}] = \mathrm{Var}[Y_{ij}]$. Under (5.21), it can be shown (Exercise 5.25) that $\mathrm{Var}[Y_{ij}] = \sigma^2 + \psi^2$ for all i, j, and that $\mathrm{Cov}[Y_{ij}, Y_{ik}] = \psi^2$ for all $j \neq k$. Thus our model induces constant variances for all the data *and* constant covariances between observations within a subject. This is perhaps the simplest non-independent covariance structure one could develop, since it leads to equal correlations among repeated measurements with a subject: $\mathrm{Corr}[Y_{ij}, Y_{ik}] = \psi^2/(\sigma^2 + \psi^2)$ for all $j \neq k$ (a form of intraclass correlation; cf. §1.4). The result is known as a *uniform correlation model* (Diggle *et al.*, 1994, §4.2.1). Note that with uniform correlations, the t_js can be taken over an unequally spaced grid and still allow for acceptable interpretations of the model parameters.

The uniform correlation model is also referred to as *compound symmetry (CS)*. The term derives from a matrix perspective: we can write out the variances and covariances for each $\mathbf{Y}_i$ into a variance–covariance matrix (§A.2.12), $\mathrm{Var}[\mathbf{Y}_i]$. Under uniform correlation, $\mathrm{Var}[\mathbf{Y}_i]$ has diagonal elements $\mathrm{Var}[Y_{ij}] = \sigma^2 + \psi^2$ and (symmetric) off-diagonal elements $\mathrm{Cov}[Y_{ij}, Y_{ik}] = \psi^2$. We can build this further into a variance–covariance matrix for the entire collection of observed vectors $[\mathbf{Y}_1 \mathbf{Y}_2 \ldots \mathbf{Y}_n]$, where now the individual $\mathrm{Var}[\mathbf{Y}_i]$ matrices are placed into the larger matrix in a block-diagonal fashion while the off-block diagonals are used to locate the inter-subject covariances. Since we assume independence between subjects, however, the off-block diagonals are all zero. Under compound symmetry, the resulting variance–covariance matrix is a 'compound' construction of separate, symmetric matrices (Littell *et al.*, 1996, §3.2.2). Note that in the GEE terminology from §3.2.6, compound symmetry corresponds to an 'exchangeable' correlation matrix.

Not surprisingly, the constant correlation induced by the CS assumption is likely to be unreasonable for growth curve data. Recall that under our AR(1) formulation from §5.4.1, as the lag between time indices grows the expected autocorrelation decays exponentially. That is, correlations between observations within a subject often decay as the time between them progresses. In terms of the covariances of the random error terms, we write this as $\mathrm{Cov}[\varepsilon_{ij}, \varepsilon_{ik}] = \sigma^2 \rho^{|k-j|}$ for $k \neq j$ and where $|\rho| < 1$. The corresponding variance–covariance matrix remains block diagonal – since we still assume that subjects are independent – but now each block term is patterned. The diagonal terms are constants based on $\mathrm{Var}[\varepsilon_{ij}] = \sigma^2$, while the covariances within the block-diagonal elements incorporate the heterogeneous, decaying correlation via $\mathrm{Cov}[\varepsilon_{ij}, \varepsilon_{ik}] = \sigma^2 \rho^{|k-j|}$. Since it mimics an AR(1) formulation, we call this a *first-order autoregressive model* for the within-subject covariance structure. It coincides with the similarly named correlation matrix in the GEE case.

For unequally spaced time points, the AR(1) formulation becomes $\mathrm{Cov}[\varepsilon_{ij}, \varepsilon_{ik}] = \sigma^2 \rho^{|t_k - t_j|}$ for $t_k \neq t_j$. This corresponds to an exponential correlation model described by Diggle *et al.* (1994, §4.2.2). Unequal spacing complicates analysis of such data, however, and different methods are required to accommodate them. For example, Littell *et al.* (1996, §3.5) suggest unequal-spacing modifications to the covariance structure in a general mixed model, while Jones and Boadi-Boateng (1991) describe unequal-spacing extensions in the special case of AR(1) models with longitudinal data.

To fit the mixed model we take advantage of the normality of δ_i and ε_{ij} and appeal to maximum likelihood (§A.4.3). Extra components such as ψ^2 or ρ add complexity to the log-likelihood, unfortunately, and a modification of the ML approach known as *residual maximum likelihood (REML)*, also called *restricted maximum likelihood*, is typically applied. REML was formally described by Patterson and Thompson (1971) for applications in the design of experiments. It maximizes the log-likelihood using residuals found after adjusting or 'filtering' the contributions of the fixed-effects terms. (Essentially, REML takes into account the loss in degrees of freedom due to estimation of the fixed-effect parameters.) Similar to the nonlinear LS/ML algorithms employed in Chapter 2, this requires iterative calculation and hence the issues of convergence to a stable maximum, choice of iterative algorithm and software, computational resources, etc., all come into play when calculating REML estimates.

Written as functions of the covariance terms σ^2 and ψ^2 or ρ, the resulting estimates for the β parameters are optimal in that they possess minimum mean squared error among all unbiased linear combinations of the original observations. These are called *best linear unbiased estimators (BLUEs)*. When the covariance parameters are unknown, as is common, we replace the unknown quantities with their REML estimates. This produces an empirical BLUE, or *EBLUE*, for each β parameter.

For inferences on the parameters in a mixed model we generally appeal to large-sample approximations based on the use of ML or REML estimators. Our primary concern is with estimation and testing of the trend component modeled via $g(t)$ in (5.21); for example, is there evidence for a significant trend over time? With the simple linear form $g(t) = \beta_1 t$, this amounts to testing $H_0: \beta_1 = 0$ or to constructing $1 - \alpha$ confidence limits on β_1. For a quadratic function, $g(t) = \beta_1(t - \bar{t}) + \beta_2(t - \bar{t})^2$, we might test $H_0: \beta_1 = \beta_2 = 0$, or perhaps also assess the pointwise hypotheses $H_{0j}: \beta_j = 0$ $(j = 1, 2)$. One can also estimate the covariance terms σ^2 and ψ^2 or ρ, test if ψ^2 is significantly greater than zero, or calculate confidence limits on ρ. (Under CS, a confidence interval for the intraclass correlation $\psi^2/\{\psi^2 + \sigma^2\}$ was given in §1.4.)

In some settings, inferences are desired on the random effects, δ_i, in (5.21). Common in this case are predictors of the actual subject-specific responses, since these can identify any subjects whose particular features differ dramatically from the population average. The effort is similar to the BLUE mentioned above, although since we are predicting a response, we call the estimator a *best linear unbiased predictor (BLUP)* of Y_{ij}, say $\hat{Y}_i(t_j)$. BLUPs assure the investigator that the prediction being made at a particular t_j is both unbiased and 'best' in the sense of attaining minimum mean squared error; see Robinson (1991) for a review on BLUPs or Littell *et al.* (1996, Ch. 6) for ways to construct BLUPs using SAS. We study use of BLUPs in more detail in §6.4 when we discuss spatial prediction.

Note that for many of these inferences, the error degrees of freedom may change, depending on what the correct denominator mean square (MS) is in the t- or F-statistics. Indeed, when the variance components are unknown the use of EBLUEs and of large-sample approximations can make determination of the correct df_e a matter of expert judgment. Wherever such input is unavailable, a useful default method is to modify the well-known Welch–Satterthwaite approximation for unequal-variance, two-sample t-tests (Welch, 1938; Satterthwaite, 1946). The method approximates the correct df_e using MS terms from the model fit, and hence is data-driven and computationally intensive. Nonetheless, it is typically quite accurate (Wang, 1971), and improves as the sample sizes get large; see Neter *et al.* (1996, §24.1) for details.

When inferences are performed on more than one parameter at a time, multiplicity adjustments may be necessary. For instance, when testing the two pointwise hypotheses $H_{0j}: \beta_j = 0$ $(j = 1, 2)$, we can adjust via a Bonferroni correction (§A.5.4) by reducing the pointwise α-level by a factor of 2. Or, for a simultaneous confidence ellipse on β_1 and β_2, we use the construction in (2.23) and apply a Welch–Satterthwaite approximation for df_e. One can also place approximate pointwise limits on the mean response at any t_i, or build simultaneous confidence *bands* on the entire repeated measures growth curve (Sun *et al.*, 1999). See Neter *et al.* (1996, Ch. 24), Littell *et al.* (1996), or Demidenko (2004) for more on these and many other statistical aspects of mixed-model analysis.

For repeated-measures growth curves, the CS and the AR(1) covariance structures are likely to be the most relevant for (5.21). (Others are possible, however; see the discussion at the end of Example 5.9.) In SAS, a number of procedures are available to fit such a mixed model. We suggest PROC MIXED, which is dedicated to mixed models. One can also appeal to PROC GLM to fit the CS covariance structure via its `repeated` statement; however, the required syntax and resulting output are often more complex than those from PROC MIXED (Littell *et al.*, 1996, §3.3).

Example 5.9 (Longspur growth rates) Hussell (1972) gives data on average body weights (in g) of hatchling Lapland longspurs (*Calcarius lapponicus*) across three Arctic nesting locations. We view the data as serial observations, since they are repeatedly taken on the same animals at each location over a series of days after hatching. In effect, each location is a 'subject' under our model. The data appear in Table 5.8; a corresponding plot appears in Fig. 5.27.

From the data plot, we see that a roughly linear pattern of growth is apparent over the limited time scale of the study, with perhaps a touch of curvilinearity. To assess this in more detail, we turn to a mixed model as in (5.21), and employ a quadratic predictor in (centered) time for the fixed-effect component. For the random-effect component, we employ an AR(1) covariance structure to accommodate potential serial correlation based on the repeated measurements over time. Sample PROC MIXED code for this appears in Fig. 5.28.

Certain features in the code in Fig. 5.28 deserve mention. Note first that we center the time variable about its mean $\bar{t} = 5$. Also, in the call to PROC MIXED the

Table 5.8 Growth in average weight (g) of Arctic longspurs

Location	$t = $ days since hatching				
	1	3	5	7	9
Igiak Bay, Alaska	3.22	7.20	11.72	15.99	20.95
Devon Island, Nunavut Territory	3.06	7.78	14.59	19.80	21.20
Churchill, Manitoba	2.98	6.96	12.24	17.52	21.57

Source: Hussell (1972).

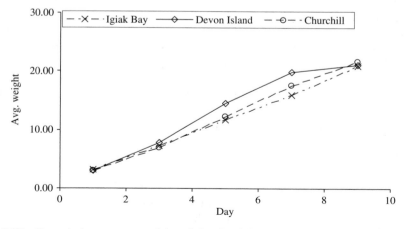

Figure 5.27 Growth in average weight of Lapland longspurs over time at three Arctic locations

```
      data longspur;
      input locatn $  y1 y2 y3 y4 y5;
      daybar = 5;
      yweight = y1; day = 1-daybar; output;
      yweight = y2; day = 3-daybar; output;
      yweight = y3; day = 5-daybar; output;
      yweight = y4; day = 7-daybar; output;
      yweight = y5; day = 9-daybar; output;
      drop y1-y5;
      datalines;
        Igiak      3.22   7.20   11.72   15.99   20.95
        Devon      3.06   7.78   14.59   19.80   21.20
        Churchil 2.98   6.96   12.24   17.52   21.57

    proc mixed covtest;
      class locatn;
      model yweight = day day*day/ solution ddfm=satterth htype=1;
          repeated / type=ar(1) sub=locatn;
```

Figure 5.28 Sample SAS code for AR(1) mixed model fit with longspur growth rate data

covtest option requests large-sample Wald tests – including point estimates and standard errors – for the covariance parameters. This is useful if interest exists in constructing confidence intervals or testing the variance components. (Note, however, that the Wald inferences may be unstable with the small sample size we have in this example. We include use of covtest here primarily for illustration's sake.) Also, PROC MIXED defaults to REML for performing the model fit. If another fitting criterion is desired, it must be specified via the method= option in the procedure call. For example, to employ full maximum likelihood use method=ml.

Much of the rest of the code is strikingly similar to the PROC GLM, PROC NLIN, or PROC GENMOD codes seen in Chapters 1–3. The class statement identifies locatn as a classification variable. In the model statement, the ddfm=satterth option calls for the Welch–Satterthwaite approximation to calculate df_e in any test statistics, while the htype=1 option calls for a table of sequential F-tests on the fixed-effect parameters based on the ordering chosen for the model terms. This latter option is similar to the type1 option seen in PROC GENMOD, and is appropriate when testing the pointwise importance of terms in a polynomial predictor such as we have here. (Contrastingly, the output given by the solution option provides 'partial' tests, similar to type3 inferences in PROC GENMOD or PROC GLM.)

To implement a repeated measures analysis for our growth curve, Fig. 5.28 invokes the repeated statement. The options specify the covariance structure: ar(1) specifies AR(1) covariances, while the sub= option identifies the 'subject' variable over which the repeated measurements are taken. sub= is a necessary option for our repeated-measures growth curve model. Here, the 'subjects' are the different locations, so we use sub=locatn. The corresponding output (edited) appears in Fig. 5.29.

The PROC MIXED output indicates that convergence was attained rapidly; see Iteration History. From the sequential (Type 1) analysis, we see that the quadratic term does not appear significant ($P = 0.351$). Given this, moving up the table shows the linear term is significant ($P < 0.001$). Thus as suggested by the plot in

```
                           The SAS System
                         The MIXED Procedure

                      Class Level Information
               Class      Levels   Values
               LOCATN         3    Churchil Devon Igiak

                       Iteration History
      Iteration     Evaluations     -2 Res Log Like        Criterion
              0               1          50.69384142
              1               2          47.57097852        0.00006369
              2               1          47.57016459        0.00000000
      Convergence criteria met.

                    Covariance Parameter Estimates
      Cov Parm    Subject    Estimate   Std Error   Z Value     Pr Z
      AR(1)       locatn       0.4504      0.2351      1.92    0.0554
      Residual                 1.2627      0.6138      2.06    0.0198

                         Fit Statistics
               -2 Res Log Likelihood               47.6
               AIC (smaller is better)             51.6

                    Solution for Fixed Effects
      Effect      Estimate   Std Error     DF    t Value    Pr>|t|
      Intercept   12.7463      0.5559     5.59     22.93    <.0001
      day          2.2998      0.1110     9.44     20.71    <.0001
      day*day     -0.03664     0.03773    11.7     -0.97    0.3511

                    Type 1 Tests of Fixed Effects
                          Num      Den
      Effect              DF       DF      F Value    Pr > F
      day                 1       9.44     428.95     <.0001
      day*day             1       11.7       0.94     0.3511
```

Figure 5.29   SAS output (edited) from AR(1) mixed-model fit with longspur growth rate data

Fig. 5.27, there is a significant trend in these growth data. The REML point estimate under `Solution for Fixed Effects` is $b_1 = 2.2998$, with $se[b_1] = 0.1110$. These values can be used to construct an approximate $1 - \alpha$ confidence interval on β_1: $b_1 \pm t_{\alpha/2}(\mathrm{df}_e)se[b_1]$. Here, the Welch–Satterthwaite approximation yields $\mathrm{df}_e = 9.44$ ≈ 9 under `DF`, so, for example, an approximate 95% confidence interval for β_1 is $2.2998 \pm (t_{0.025}(9) \times 0.1110) = 2.2998 \pm (2.262 \times 0.1110) = 2.2998 \pm 0.2511$, or $2.049 < \beta_1 < 2.551$. That is, we believe with 95% confidence that on average Lapland longspur hatchlings in these Arctic locations will grow between 2.05 and 2.55 g/day.

Estimates of the covariance parameters are given under `Covariance Para-meter Estimates`. The `Residual` component reports $\hat{\sigma}^2 = 1.2627$, while the estimate of `AR(1)` intralocation correlation is $\hat{\rho} = 0.4504$. (The estimate is associated with the 'subject' variable; here, `locatn`.) We also find $se[\hat{\rho}] = 0.2351$, thus an approximate 95% Wald confidence interval on ρ is $\hat{\rho} \pm z_{0.025}se[\hat{\rho}] = 0.4504 \pm (1.96 \times 0.2351) = 0.4504 \pm 0.4608$. Since this interval contains the point $\rho = 0$, the existence of a significant intralocation correlation with these data is called into question. The corresponding, two-tailed, Wald P-value for testing $H_0: \rho = 0$ is given under `Pr Z` as $2\{1 - \Phi(1.916)\} = 2 \times 0.0277 = 0.0554$; at $\alpha = 0.05$, this corroborates our indication that ρ is insignificant. (By contrast, the P-value in Fig. 5.29 for testing

$H_0: \sigma^2 = 0$ is one-tailed, $1 - \Phi(2.057) = 0.0198$, since by default σ^2 cannot be negative and so only positive departures from H_0 are considered.)

One can alternatively fit a CS covariance structure to the data. Although the AR(1) covariance structure seems far more pertinent for these data, we present the CS analysis for comparison purposes. The required SAS code is essentially unchanged from that in Fig. 5.28, except that in the `repeated` statement one uses the `type=cs` option. The resulting output (edited) appears in Fig. 5.30. From it, we see convergence was also attained rapidly. The sequential (`Type 1`) analysis of the quadratic predictor again shows an insignificant quadratic effect ($P = 0.3294$), while the linear term remains significant ($P < 0.001$). The REML point estimate under `Solution for Fixed Effects` is $b_1 = 2.3382$, with $se[b_1] = 0.0878$.

The REML estimates of the covariance parameters are $\hat{\sigma}^2 = 0.9249$ for the `Residual` component, and $\hat{\psi}^2 = 0.3847$ for the subject-specific `locatn` component. (Recall that locations are the 'subjects' in this mixed model.) The Wald statistic for testing $H_0: \psi^2 = 0$ is $W_{calc} = 0.3847/0.5756 = 0.668$, but the two-sided P-value listed under `Pr Z` is not useful. For a non-negative variance component only $H_a: \psi^2 > 0$ is of interest, so we should calculate a one-sided P. This is $P \approx 1 - \Phi(0.668) = 0.252$, and hence we cannot reject H_0. As with the AR(1) fit, the intralocation correlation is called into question.

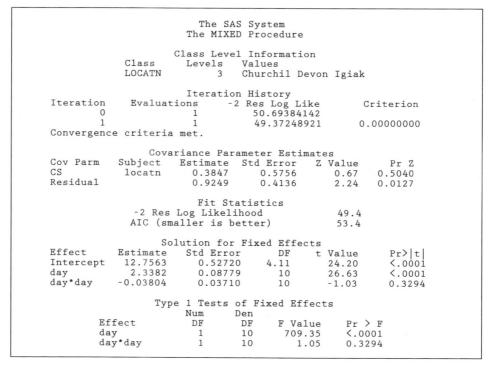

Figure 5.30   SAS output (edited) from CS mixed model fit to longspur growth rate data

Note that we could have fitted a repeated-measures CS model in PROC MIXED, using its `random` statement in place of its `repeated` statement. The appropriate code for use in Fig. 5.28 would be

```
random intercept / sub = locatn;
```

See Exercise 5.26.

Taken as a whole, the CS fit is almost identical to the earlier AR(1) fit. Indeed, one might question which fits better. A measure for assessing this known as *Akaike's information criterion (AIC)* (Akaike, 1973). The AIC gauges the quality of a model fit by subtracting from the maximized log-likelihood a penalty term for the number of unknown parameters included in the model. In this form, the larger the AIC, the better the model fit. (Many sources compute this as an added penalty to minus-twice the log-likelihood – as in Figs. 5.29 and 5.30 – so that smaller quantities are better. PROC MIXED can provide either form via its `IC` option.) Here, listed in both outputs under `Fit Statistics` is the smaller-is-better form: AIC{AR(1)} = 51.6 and AIC{CS} = 53.4. As might be expected, the measure indicates a slightly better fit for the AR(1) structure. For more on use of information quantities for comparing covariance structures, see Wolfinger (1993).

Lastly, we should note that PROC MIXED makes available an extensive variety of possible covariance structures; the AR(1) and CS versions we describe barely scratch the surface! For use with growth data, or for any form of repeated-measures temporal data, other possibilities include:

- `type=ARMA(1,1)`: an autoregressive moving average covariance structure that combines features of the AR(1) and MA(1) error models;

- `type=toep`: the so-called Toeplitz structure where the covariance between two errors depends only on the lag between them, and each of these covariances is allowed to differ in any way – this can mimic an AR(m) structure where m is the number of repeated measurements per subject;

- `type=toep(q)`: a 'banded' Toeplitz structure with all covariances greater than $q < m$ units apart set to zero;

- `type=sp(pow)(list)`: a spatial power law that can be used to mimic an AR(1) covariance structure with variably spaced time points (Littell *et al.*, 1996, §3.5); `list` is a quantitative SAS variable that lists the ordered time points.

There is also an 'unstructured' option, `type=un`, which places no restrictions on the covariances. This corresponds to the similarly named 'unstructured' correlation in a GEE analysis (§3.2.6), and can be very computationally intensive. ⊛

One limitation both the AR(k) models in §5.6.1 and the mixed models in this section share is that they impose a linear structure on the parameters in $g(t)$. As seen in §2.4, however, environmental growth curves can take on a variety of nonlinear forms. Analyses for repeated-measures growth curves using a nonlinear function of t continue to be redeveloped and refined, and a sizable amount of work has appeared in the literature; see, for example, the texts by Davidian and Giltinan (1995) or Vonesh and

Chinchilli (1996) and the various references therein. Used in environmetric applications, the most common nonlinear forms are the logistic curve (or modifications thereof) in (2.16) and the various exponential curves in §2.3 such as $g(t) = \beta_0 + \beta_1 e^{-\beta_2 t}$.

As might be expected, the added complexity in fitting a nonlinear growth function can be extensive. We do not go into detail on this here, but mention only that SAS users may appeal to the ver. 8 procedure PROC NLMIXED. For earlier versions, the separate SAS macro NLINMIX can prove useful; see Littell *et al.* (1996, §12.2). Also, for data of an appropriate form the `repeated` statement in PROC GENMOD can fit a number of nonlinear link functions via its appeal to GEE methodology; cf. §3.2.6.

Exercises

5.1. Thiébaux (1994, §2.5) gives data on daily maximum temperatures (in °F) at Little Rock, Arkansas, from January 1979 to December 1986. Collected into monthly averages, the data are available online at http://www.wiley.com/go/environmental (the data are in two columns: month | temperature); a sample follows:

Month	Temperature
Jan. 1979	36.4
Feb. 1979	46.7
Mar. 1979	65.6
⋮	⋮
Oct. 1986	73.6
Nov. 1986	56.7
Dec. 1986	50.0

(a) Plot the average maximum temperatures against time. Do you see a periodic effect?

(b) Assume a known period of $p = 12$ months and fit a simple harmonic regression (5.3) to the data. Test if the periodic effect is significant at the $\alpha = 0.05$ level. Use your results to give a point estimate for the amplitude, γ.

(c) Compute the residuals from your fit and plot them against time. Do you see any problems with the plot? Also compute the predicted values from your fit and overlay the predicted curve on the data plot. Comment on the quality of the fit.

5.2. Return to the water surface temperature data in Example 5.2, available online at http://www.wiley.com/go/environmental.

(a) Verify the indication in the example that the simple harmonic regression fit apparently overpredicts the data in many summer seasons. Do this by finding the predicted values and overlaying the predicted curve on the data plot. Notice that the troughs appear near the summer months, and that the predicted curve is typically well above the observed values at these troughs. (Alternatively, graph the residuals vs. time and observe a similar effect.)

(b) Add a second, half-harmonic to the regression; that is, given $p_1 = 1$, incorporate a second period of $p_2 = p_1/2$ via addition of two appropriately constructed regressor variables. Test to see if the additional harmonic is significant at the $\alpha = 0.10$ level using a single 2 df test statistic via (1.7).

(c) Extend the results from part (b) by adding a further harmonic. That is, fit three periods: $p_1 = 1$, $p_2 = p_1/2$, and $p_3 = p_1/3$. Test to see if the third harmonic is significant at the $\alpha = 0.10$ level using a single 2 df test statistic via (1.7).

5.3. Return to the water surface temperature data in Example 5.2 and fit the sine wave in (5.1) directly, via a nonlinear regression. (Use a software procedure such as PROC NLIN.) Use Fig. 5.1 in conjunction with Fig. 5.2 to identify initial estimates for the parameters γ, β_0, and θ.

(a) Does your estimate of γ correspond to that found in the example? What about θ?

(b) Give (approximate) 95% confidence intervals for γ and θ. Does the corresponding inference on γ agree with the result in Example 5.2?

5.4. Apply the delta method from §A.6.2 to approximate the variances of $\hat{\gamma} = \sqrt{b_1^2 + b_2^2}$ and $\hat{\theta} = (2\pi)^{-1} p \cos^{-1}(b_2/\hat{\gamma})$ under the simple harmonic regression model (5.3). Given these variances, replace any unknown parameters with their MLEs and then take square roots to find standard errors for the point estimates.

5.5. Illustrate the claim that when

$$\mathbf{X} = \begin{bmatrix} 1 & c_{11} & s_{11} & \cdots & c_{1H} & s_{1H} \\ 1 & c_{21} & s_{21} & \cdots & c_{2H} & s_{2H} \\ \vdots & \vdots & \vdots & \ddots & \vdots & \vdots \\ 1 & c_{n1} & s_{n1} & \cdots & c_{nH} & s_{nH} \end{bmatrix}$$

with $c_{ih} = \cos(2\pi ih/n)$ and $s_{ih} = \sin(2\pi ih/n)$, then $\mathbf{X}^\mathsf{T}\mathbf{X} = \mathrm{diag}\{n, n/2, n/2, \ldots, n/2\}$. Specifically, let $H = 2$ and $n = 8$. Show that $\mathbf{X}^\mathsf{T}\mathbf{X} = \mathrm{diag}\{8, 4, 4, 4, 4\}$.

5.6. Extend the results from Exercise 5.5 and show that for $\mathbf{Y} = [Y_1 \ldots Y_n]^\mathsf{T}$, estimating $\boldsymbol{\beta}$ via $(\mathbf{X}^\mathsf{T}\mathbf{X})^{-1}\mathbf{X}^\mathsf{T}\mathbf{Y}$ leads to the closed-form solutions in (5.6). [*Hint:* $\mathbf{X}^\mathsf{T}\mathbf{X} = \mathrm{diag}\{n, n/2, n/2, \ldots, n/2\}$, so $(\mathbf{X}^\mathsf{T}\mathbf{X})^{-1} = \mathrm{diag}\{1/n, 2/n, 2/n, \ldots, 2/n\}$.]

5.7. Let $\mathcal{J}(h) = \frac{n}{2}\hat{\gamma}_h^2 = \frac{n}{2}(b_{2h}^2 + b_{2h-1}^2)$, $h = 1, \ldots, H$, where b_h is given via (5.6).

(a) Verify the claim that $\mathcal{J}(h)$ is related to a $\chi^2(2)$ distribution. Specifically, suppose $b_h \sim$ indep. $N(\beta_h, 2\sigma^2/n)$, $h = 1, \ldots, H$, and that H_0: $\beta_{2h} = \beta_{2h-1} = 0$ holds. Show then that $\mathcal{J}(h)/\sigma^2$ is distributed as $\chi^2(2)$. To simplify your effort, assume n is odd and ignore b_0.

(b) Given the result in part (a), suppose we have an estimator of σ^2, say $\hat{\sigma}^2$, which is mutually independent of the b_h values such that $v\hat{\sigma}^2/\sigma^2 \sim \chi^2(v)$.

(Typically, $v = n - 2H - 1$.) What is the distribution of $\mathcal{I}(h)/2\hat{\sigma}^2$ under $H_0: \beta_{2h} = \beta_{2h-1} = 0$?

5.8. Return to the lynx trappings data of Example 5.3, available online at http://www.wiley.com/go/environmental.

(a) To examine the suggestion in Example 5.3 that the original count data may exhibit variance heterogeneity, fit a non-trending harmonic regression with a period of 9.5 years (as suggested by the example and by the previous literature) to the original counts. Store the residuals from the fit and plot them against the predicted values. Does variability in the residuals increase as the predicted values increase?

(b) Plot the base-10 logarithms of the number trapped against time. That is, mimic the display in Fig. 5.6 using the log-transformed data. Does your plot show a similar pattern?

5.9. A classic data set in the history of harmonic analysis is that from the 11-year solar sunspot cycle. In 1849 J. R. Wolf of the Berne Observatory proposed and later published (Wolf, 1856) an index, with which he intended to quantify the relative pattern of dark prominences ('spots') seen on the surface of the Sun. Wolf used previous data and his own recordings to amass a database of relative sunspot indices. The database has been updated continuously since then; a full series from 1749 to 1960 – most authors agree that data earlier than 1749 are questionable – was published by the Swiss Federal Observatory in Zurich in 1961 (Waldmeier, 1961). Different versions of the series exist, depending apparently on the data source to which one appeals; we will use a data set described by Andrews and Herzberg (1985). To keep things manageable, we limit our study to the monthly indices from 1900 to 1977. The corresponding data are available online at http://www.wiley.com/go/environmental (the data are in two columns: month | spot index); a sample follows:

Month	Spot index
Jan. 1900	9.4
Feb. 1900	13.6
Mar. 1900	8.6
⋮	⋮
Dec. 1976	16.4
Jan. 1977	23.1
Feb. 1977	8.7

Sunspots and more generally solar activity can affect Earth's climate and environment, and hence this time series is worthy of study. Plot the series vs. month. Is the (now famous) 11-year cycle evident? Is there any other trend component? If so, prewhiten the data to detrend them. After prewhitening, or if no trend is present, perform a Fourier analysis on the relative sunspot indices and report your findings. Use a moving average smoothing of the periodogram, with triangular weights 1 2 3 . . . 18 19 18 . . . 3 2 1, and plot the base-10 log of the smoothed periodogram.

5.10. To study periodicity in the Mauna Loa CO_2 data from Example 5.4 (available online at http://www.wiley.com/go/environmental), perform the following:

(a) Plot the raw periodogram and compare it to the $\log_{10}$-smoothed version in Fig. 5.13. (For example, plot `lp` vs. `fh` from Fig. 5.12.) Are the periods as easy to identify?

(b) Compare the $\log_{10}$-smoothed periodogram from Fig. 5.13 with those produced using the following symmetric weighting schemes. Comment on the effect the change in weights has on the ability of the plot to identify likely periods.

(i) 1 2 3 2 1

(ii) 1 2 3 4 5 6 5 4 3 2 1

(iii) 1 2 3 4 5 6 7 8 9 8 7 6 5 4 3 2 1

5.11. Isaacson and Zimmerman (2000) give temporal data on yearly snow water equivalent (SWE, a measure of the amount of water in snow) at a particular watershed in Yellowstone Park. The data are available online at http://www.wiley.com/go/environmental (the data are in two columns: year | SWE); a sample follows:

Year	SWE
1935	8.20
1936	11.70
1937	9.30
⋮	⋮
1994	12.92
1995	14.28
1996	18.13

(a) Show that an upward trend is evident in these data by plotting SWE vs. year.

(b) Fit a simple linear regression to the data over time in order to prewhiten the series. Collect the residuals from this fit and plot them against time. Does the plot reveal any important anomalies?

(c) Using the residuals from the linear fit in part (b), perform a Fourier analysis on the series. Take the base-10 logarithms of the raw periodogram ordinates and plot them against frequency. How easy is it to identify periodicity from this plot?

(d) Apply a moving average smoother to the raw periodogram ordinates, using the triangular weighting seen in Example 5.3. Take the base-10 logarithms of the smoothed periodogram ordinates and plot them against frequency. Show that periodicities emerge at approximately $p = 21, 7, 4.8, 3.5$, and 2.4 years.

5.12. Return to the following data sets. For each set, fit the specified regression model and test for autocorrelation in the residuals via the Durbin–Watson statistic (5.8). Operate at $\alpha = 0.05$.

(a) The Zurich sunspot data from Exercise 5.9. Fit a simple harmonic regression (5.3) with period $p = 11.4$ years.

(b) The Mauna Loa CO_2 data from Example 5.4. Fit (centered) linear and quadratic terms in time along with harmonic terms corresponding to periods of $p = 12$, 6, 4, and 3 months.

(c) The snow water equivalent data from Exercise 5.11. Fit a linear term in time along with harmonic terms corresponding to periods of $p = 21, 7, 4.8, 3.5$, and 2.4 years.

5.13. Helgesen and Pribble (2001, Fig. H-11) present data on ice loss from retreating (melting) Eurasian glaciers over the period 1980 to 1999. Recorded was the annual mean cumulative loss (in mm) for a localized group of 30 glaciers:

Year	Ice loss	Year	Ice loss	Year	Ice loss	Year	Ice loss
1980	201.7	1985	581.9	1990	1956.1	1995	3403.5
1981	298.2	1986	701.7	1991	1903.5	1996	3907.9
1982	745.6	1987	1105.3	1992	2197.4	1997	4390.4
1983	598.5	1988	1495.6	1993	2690.5	1998	5201.8
1984	701.8	1989	1403.5	1994	2796.8	1999	5600.9

Take as the response, Y, the natural logarithm of the cumulative ice loss.

(a) Plot the log-transformed data against time. Do you see a trend? If so, of what kind?

(b) Fit a polynomial regression with (centered) linear and quadratic terms in time to the log-transformed data. If the logarithms are independent and normally distributed with constant variance, is the quadratic term necessary? Is the linear term necessary? Operate at $\alpha = 0.05$.

(c) Test for autocorrelation in the residuals via the Durbin–Watson statistic (5.8). Operate at $\alpha = 0.05$. What do you conclude about the autocorrelation in the log-transformed data?

5.14. Return to and/or consider the following data sets. Find the sample autocorrelations and construct an ACF plot to identify if and to what extent autocorrelation is evidenced in each series. If necessary, prewhiten the data by removing trends and/or periodicities.

(a) The Little Rock temperature data from Exercise 5.1.

(b) The snow water equivalent data from Exercise 5.11. For greater numerical stability, operate with the centered time variable $t - \bar{t} = t - 1965.5$.

5.15. For the AR(1) model, complete the details from §5.4.1 that establish the relationship $\mathrm{Var}[Z_i] = \zeta^2/\{1 - \varphi_1^2\}$. Also, show that by modeling $\varepsilon_0 \sim N(0, \zeta^2/\{1 - \varphi_1^2\})$ in (5.11) we recover $\mathrm{Var}[\varepsilon_1] = \sigma^2 = \zeta^2/\{1 - \varphi_1^2\}$.

5.16. (a) How would you extend the AR(2) error structure in (5.12) to an AR(3) model? to any general AR(k) model?

(b) How would you extend the MA(1) error structure in (5.13) to an MA(2) model?

5.17. Return to the snow water equivalent data from Exercise 5.11 and fit an autoregressive model to the data. Include a term for increasing trend over time (for greater numerical stability, operate with the centered time variable $t - \bar{t} = t - 1965.5$) and harmonic regression terms corresponding to periods of $p = 21$, 7, 4.8, 3.5, and 2.4 years. Use the results from Exercise 5.14(b) to determine whether an AR(k) error structure is necessary, and if so, adopt an appropriate lag value, k, for the AR(k) model. Use backward elimination to cull the AR error component. Plot the residuals from your model against time. Comment on the quality of the fit.

5.18. Return to the lynx trappings data in Example 5.3, available online at http://www.wiley.com/go/environmental, and fit an autoregressive model to the base-10 logarithm of the number trapped. Use for your predictor variables harmonic regression terms to account for the periodicity seen in the example; employ periods of $p_1 = 38$, $p_2 = 9.5$, $p_3 = 5$, $p_4 = 3.5$, and $p_5 = 2.5$ years. Also, for the AR(k) error component, set $k = 12$. (In Example 5.6 we found potential lags of $k = 1$, 2, and possibly 6, so $k = 12$ adds some conservatism to this possibility.) Use backward elimination to cull the AR error component. Also plot (i) the residuals and (ii) an overlay of the observed and predicted values. Comment on the quality of the fit.

5.19. Schipper and Meelis (1997) report data on annual numbers of breeding pairs of Arctic tern (*Sterna paradisaea*) for Griend Island in the Dutch Wadden Sea. The data, taken over the period 1964–1995, are as follows:

Year	Count	Year	Count	Year	Count	Year	Count
1964	300	1972	500	1980	350	1988	400
1965	300	1973	500	1981	700	1989	700
1966	300	1974	450	1982	350	1990	630
1967	350	1975	800	1983	600	1991	410
1968	300	1976	800	1984	500	1992	800
1969	600	1977	500	1985	300	1993	1100
1970	600	1978	500	1986	600	1994	1200
1971	500	1979	375	1987	500	1995	1000

To correct for possible variance heterogeneity, apply a Freeman–Tukey transform (Freeman and Tukey, 1950) to the original counts: $Y = \sqrt{\text{count}} + \sqrt{1 + \text{count}}$.

(a) Plot the transformed data against time. Is a trend evident?

(b) Prewhiten the transformed data by fitting a term for linear trend over time and use the residuals from this fit in a periodogram analysis. Determine if any periodic effect is present in the time series.

(c) Find the sample autocorrelations and construct an ACF plot to identify if and to what extent autocorrelation is evidenced. Prewhiten the transformed data first by fitting a term for linear trend over time and any harmonic regression terms (paired sines and cosines) from your analysis in part (b).

(d) Fit an autoregressive model to the transformed data. Include a term for increasing trend over time and test to see if this term is significant at $\alpha = 0.05$. Also include appropriate harmonic regression terms (paired sines and cosines) from your analysis in part (b). Use the results from part (c) to determine whether an AR(k) error structure is necessary, and if so, adopt an appropriate lag value, k, for the AR(k) model. Use backward elimination to cull the AR error component.

(e) Plot the residuals from your model against time. Comment on the quality of the fit.

5.20. Return to the Zurich sunspot data in Exercise 5.9 and fit an AR(2) model to the data, including also any pertinent harmonic terms. Compare this fit to one without the AR(2) error structure. Does the assumption of AR(2) errors improve the fit?

(Yule, 1927)

5.21. Verify the indication in Example 5.7 that for the Eskdalemuir ozone data a Fourier analysis suggests periods of $p = 12$, 6, and 3 months after prewhitening with a simple linear trend. (Download the data from http://www.wiley.com/go/environmental.) Also (i) verify the calculation of the Durbin–Watson test statistic given in the example and (ii) construct an ACF plot after prewhitening with a simple linear trend and the three periods mentioned above.

5.22. Consider the functional forms in the following models and show how the various parameters in each model relate to each other:

(a) The continuous bilinear models (5.18) and (2.7).

(b) The discontinuous bilinear models (5.19) and (2.6).

5.23. In a study of environmental remediation after a leak at a jet-fuel tank farm near Charleston, South Carolina, Vroblesky et al. (1997) present data on groundwater concentrations of hydrocarbons. In particular, they questioned whether benzene concentrations were reduced after an extraction well was put into service at the farm. The data, in µg/L over time, are:

Date	Winter 1991	Spring 1991	Summer 1991	Fall 1991
Conc.	225.0	300.0	375.0	392.5
Date	Winter 1992	Spring 1992	Summer 1992	Fall 1992
Conc.	410.0	240.0	97.0	170.0
Date	Winter 1993	Spring 1993	Summer 1993	Fall 1993
Conc.	87.0	170.0	160.0	120.0
Date	Winter 1994	Spring 1994	Summer 1994	Fall 1994
Conc.	150.0	46.0	41.0	12.0
Date	Winter 1995	Spring 1995	Summer 1995	
Conc.	9.7	19.0	11.0	

(The data points at spring and fall 1991 have been interpolated to allow for equally spaced time points.) The extraction well was added in early 1992, after the observation for winter 1992 was taken. View this as a known intervention with $\tau =$ spring 1992.

(a) Plot the data. Do you see any obvious effect due to the intervention?

(b) Consider the autoregressive intervention model in (5.19) and fit it to these data. Identify whether the pre- and post-intervention slopes are significant. Also assess whether the discontinuous jump parameter is necessary. In all cases, operate at $\alpha = 0.05$.

(c) Find the predicted values and residuals from your model fit in part (b). Plot each against time. Comment on the quality of the fit.

5.24. Recall that in Exercise 2.18 we studied growth in body weights over time in rodents exposed to a toxin. In that exercise, we ignored the longitudinal nature of the data, in effect assuming no autocorrelation over time. Return to those data and complete the following:

(a) To adjust for the heterogeneous variation seen with those data, work with the natural logarithm of the observed means. Plot the log-transformed values against time and verify that a shallow curvilinearity is present.

(b) Apply an AR(k) model with (centered) linear and quadratic predictors in time. Notice that $\bar{t} = 17.7273$ with these data. To determine the lag, k, construct first an ACF plot by prewhitening the data under your quadratic model (use the residuals from the fit for the ACF plot). Is significant autocorrelation indicated? If not, then our analysis in Exercise 2.18 was in fact reasonable.

(c) Using your fit from part (b), determine if the quadratic term is required in the model. Operate at $\alpha = 0.05$.

5.25. Assume a general compound symmetry model $Y_{ij} = \mu_{ij} + \delta_i + \varepsilon_{ij}$ ($i = 1, \ldots,$ n; $j = 1, \ldots, m$) such that μ_{ij} is a fixed effect, $\delta_i \sim$ i.i.d. $N(0, \psi^2)$ is a random effect, and $\varepsilon_{ij} \sim$ i.i.d. $N(0, \sigma^2)$ is a random error term. Also assume that ε_{ij} and δ_i are independent for all i and j. This is a generalization of the compound symmetry growth curve described in §5.6.2.

   (a) Show that $\text{Var}[Y_{ij}] = \sigma^2 + \psi^2$ for all i, j. [*Hint*: use features of the variance operator described in §A.1.]

   (b) Show that $\text{Cov}[Y_{ij}, Y_{ik}] = \psi^2$ for all $j \neq k$. [*Hint*: use the fact that for any random variables T, U, V, and W, $\text{Cov}[T + U, V + W] = \text{Cov}[T, V] + \text{Cov}[U, V] + \text{Cov}[T, W] + \text{Cov}[U, W]$.]

   (c) Verify the indication in §5.6.2 that under this model $\text{Corr}[Y_{ij}, Y_{ik}] = \psi^2/(\sigma^2 + \psi^2)$.

5.26. Return to the longspur growth rate data from Example 5.9.

   (a) If you have access to PROC MIXED, compare the fit under a compound symmetry covariance structure attained using the `repeated` statement in Fig. 5.30 to that using the `random` statement, as suggested in the example. Are there any substantial differences?

   (b) Since these data were sample averages, there is information available on the variability at each time point. In particular, the sample variances, s_i^2, associated with the means in Table 5.8 are:

Location	$t =$ days since hatching				
	1	3	5	7	9
Igiak Bay, Alaska	0.078	0.436	1.877	4.709	0.941
Devon Island, Nunavut Territory	0.058	0.608	1.300	1.416	3.063
Churchill, Manitoba	0.073	1.166	7.563	5.760	2.190

Consider again a repeated-measures mixed model based on an AR(1) covariance structure, as in the example. To adjust for the variance heterogeneity seen here, include a weight of $w_i = 1/s_i^2$ for each observed mean. (To include a weight variable in PROC MIXED, use the `weight` statement.) Is the AR correlation significant? Is the quadratic term necessary?

5.27. Von Ende (1993) presents repeated growth measurements on sizes of seedling plants raised under a low-nutrient regime. Recorded were $t =$ weeks on study and $Y =$ number of leaves per plant. The data are:

Week	Plant no. 1	Plant no. 2	Plant no. 3	Plant no. 4	Plant no. 5
1	4	3	6	5	5
2	5	4	7	7	6
3	6	6	9	8	7
4	8	6	10	10	8
5	10	9	12	12	10

(a) To correct for possible variance heterogeneity, apply a Freeman–Tukey transform (Freeman and Tukey, 1950) to these count data: $Y = \sqrt{\text{count}} + \sqrt{1 + \text{count}}$. Assess the growth pattern in the data by fitting a mixed model as in (5.21) with a single linear predictor in time. Employ an AR(1) covariance structure. Is the time trend significant at $\alpha = 0.05$?

(b) Instead of the AR(1) covariance structure in part (a), analyze the transformed data using a compound symmetry covariance structure. Do the results change drastically? Which of the two structures shows the better fit?

(c) Since the data are counts, we could also analyze them using GEEs from §3.2.6. Do so, using the original counts. Employ a log-linear link function and base your preliminary estimates on a Poisson likelihood. (But base all final inferences on the GEE results.) For the working correlation matrix, return to an AR(1) assumption. Is the time trend significant at $\alpha = 0.05$?

6

Spatially correlated data

As mentioned at the start of Chapter 5, when environmental data are recorded over time or space, the analyses must incorporate potential temporal or spatial correlation. Chapter 5 was concerned with the temporal aspects of this issue. Here, we consider the spatial aspects, and give an introduction to some basic statistical methods for spatial data analysis.

6.1 Spatial correlation

In a sense, the analysis of temporally correlated data as described in Chapter 5 is relatively simple, since the correlation is induced in a unidimensional fashion. That is, serial dependency in a univariate time series derives most often from a single source, the temporal autocorrelation. This can be thought of as existing on a line, going backwards and forwards but not also 'up and down.' (Indeed, most temporal correlation only exists backward in time.) In fact, any environmental study that generates correlated data equally spaced over a single dimension, not only time, may be amenable to the autocorrelated temporal analyses we introduced in Chapter 5.

When observations are made in space instead of time, however, the data can exhibit more complex correlation structures. The correlation can be two-dimensional if the data are taken only over a spatial surface, three-dimensional if the data are taken over true three-dimensional space – longitude, latitude, and altitude, say – or even four-dimensional if the data are taken over three-dimensional space *and* in time; we call this a *spatio-temporal model* (Barnett, 2004, §11.7). Indeed, in theory one can extend the correlation to any number of dimensions, $d \geq 1$.

In multiple dimensions the notation is necessarily more complex. Suppose that data $Z(\mathbf{s}_1), Z(\mathbf{s}_2), \ldots, Z(\mathbf{s}_n)$ are recorded at known 'sites' $\mathbf{s}_i, i = 1, \ldots, n$, where $\mathbf{s}_i$ is

Analyzing Environmental Data   W. W. Piegorsch and A. J. Bailer
© 2005 John Wiley & Sons, Ltd   ISBN: 0-470-84836-7 (HB)

a d-dimensional vector of spatial site coordinates. We use Z instead of Y to represent an observed response here, since in the typical case of $d = 2$, x and y will represent spatial coordinates as the pair $\mathbf{s}_i = (x_i, y_i)$. For instance, $\mathbf{s}_i$ might be the longitude, x_i, and the latitude, y_i, at which $Z(\mathbf{s}_i)$ is observed. Similar to the case for temporally correlated data, we can connect this to a stochastic process, say $Z(\mathbf{s})$, operating on a subset of d-dimensional Euclidean space $\mathbb{R}^d$; in the spatial setting this is also known as a *random field* (Adler, 2002). The underlying stochastic process contributes stochastic variation in the response. This basic structure can underlie a wide variety of environmental scenarios. Since this chapter is intended as an introduction to spatial analyses, we will confine the discussion to only a limited assortment of such. Our concern is primarily with the simple case of spatial data in the plane, $d = 2$.

6.2 Spatial point patterns and complete spatial randomness

Perhaps the simplest case of spatial data in the plane occurs when $Z(\mathbf{s}_i)$ is a discrete measurement of some environmental phenomenon, such as a count of nests from an endangered avian species in a protected study region. This can be as simple as a binary response indicating whether or not the phenomenon occurs/exists at a particular location, $\mathbf{s}_i$; for example, whether or not a nest is sighted at site $\mathbf{s}_i$ in the study region. Similar in nature are data that indicate simply *where* certain events have occurred. That is, the data are the $\mathbf{s}_i$ values themselves. In such a setting, we may question whether the phenomenon is occurring in a purely random fashion or if it appears to be spatially clustered. In the former case, the random process underlying the phenomenon is said to satisfy *complete spatial randomness* (CSR). To wit:

Definition 6.1 A spatial stochastic process $Z(\mathbf{s})$ satisfies *complete spatial randomness* if:

(i) the number of events in a d-dimensional region Ω_d follows a Poisson distribution (§A.2.4) with mean rate parameter $\mu = \lambda \|\Omega_d\|$, where λ is the *intensity* – i.e., the expected number of points per unit area – of the spatial process under study and $\|\Omega_d\|$ denotes the size (area, volume, etc.) of the region; and

(ii) any set of n event locations $\mathbf{s}_1, \ldots, \mathbf{s}_n$ observed in the region represents a random sample from a d-dimensional uniform distribution on Ω_d.

For example, if we are studying bird nests in a region Ω_d, λ is the density of nests per unit area while $\|\Omega_d\|$ is the total area of the region.

The Poisson distribution enters into the operation to create a CSR process in a way analogous to how uniform probabilities are used in the postulates that generate a Poisson random variable; we gave an exposition of this in Piegorsch and Bailer (1997, §1.2.5). As a result, under CSR the probability of encountering a point event in a spatial region behaves proportionally to the area of the region (Barnett, 2004, §11.1).

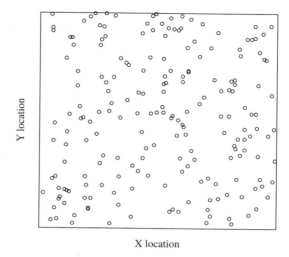

Figure 6.1 Realization of a CSR process on the unit square; $n = 200$

A set of spatially distributed points is called a *spatial point pattern*. When examining such a pattern, it is not always easy to identify CSR. For example, and in what follows, set $d = 2$. Figure 6.1 displays 200 points in the unit square randomly generated under CSR. Viewing it, one gets a broad sense of spatial randomness, but there are also some clear clumps and gaps, as well as some regular patterns. Clearly, to assess CSR in practice we must turn to more formal, objective environmetric approaches. We detail some of these in the next few sections.

6.2.1 Chi-square tests

The simplest way to assess CSR in a two-dimensional spatial region is to partition the region into contiguous, equal-area units – called *quadrats* – and count the number of points occurring in each quadrat. Suppose we denote the region by Ω_2 and partition it into an $I \times J$ grid of contiguous quadrats. Let the number of points in the (i,j)th quadrat be Y_{ij} such that the total number of points is $n = \sum_{i=1}^{I} \sum_{j=1}^{J} Y_{ij}$. Under CSR, the Y_{ij}s should show no deviation from homogeneous Poisson variation across the cells. This can be assessed via simple modification of the Pearson goodness-of-fit statistic for count data described in §3.3.4. Here, to test against deviation from homogeneity/CSR the statistic takes the form

$$X_{\text{calc}}^2 = \frac{\sum_{i=1}^{I} \sum_{j=1}^{J} (Y_{ij} - \overline{Y})^2}{\overline{Y}}, \tag{6.1}$$

where $\overline{Y} = n/IJ$ is the average number of points per quadrat. We reject CSR in favor of spatial clustering in the point pattern if X_{calc}^2 exceeds the $\chi^2(IJ - 1)$ upper-α critical point, $\chi_\alpha^2(IJ - 1)$, from Table B.3; the associated P-value is $P \approx P[\chi^2(IJ - 1) \geq X_{\text{calc}}^2]$.

Of course, the usual regularity conditions apply for use of the approximating χ^2 reference distribution: (i) the overall effect under CSR should not be too sparse, measured as, say, $\overline{Y} > 1$, and (ii) the number of quadrats should not be too small, say, $IJ > 5$ (Kathirgamatamby, 1953).

Note that we use the term *clustering* here to represent greater variability in quadrat counts than would be expected under CSR: the greater variation causes points to cluster together spatially. One can also encounter smaller spatial variability than would be expected under CSR if some underlying factor causes the points to scatter in a regular fashion; for example, bird nests might spatially locate in a regular pattern if they were being actively protected. We use the term *regularity* to describe such an effect. To test against regularity in the data at significance level α, reject CSR if $X^2_{calc} \leq \chi^2_{1-\alpha}(IJ - 1)$. To test both the clustering and regularity alternatives simultaneously at significance level α, reject CSR when $X^2_{calc} \geq \chi^2_{\alpha/2}(IJ - 1)$ or when $X^2_{calc} \leq \chi^2_{1-\alpha/2}(IJ - 1)$.

Notice, by the way, that the X^2_{calc} statistic in (6.1) is simply Fisher's statistic (3.18) for testing overdispersion with count data. (Spatial regularity would correspond to *underdispersion* in the terminology from Chapter 3.) This is not surprising, given the interrelationships between the Poisson distribution and the effects underlying CSR.

Example 6.1 (Redwood seedlings) Diggle (2003, §1.1) presents a classic point pattern data set, originally studied by Ripley (1977) and taken from Strauss (1975), on the locations of $n = 62$ redwood seedlings (*Sequoia sempervirens*). The data are pairs of coordinates in meters from the origin of a square region of sides approximately 23 m long; a selection appears in Table 6.1 (see http://www.wiley.com/go/environmental for the full data set). Figure 6.2 plots the locations of all 62 seedlings, showing a clear clustered effect. (Often, young redwoods sprout in groupings around older, progenitor stumps, taking advantage of the previously established root system and producing so-called 'fairy ring' clusters of young trees.) The figure also overlays a 3×3 grid of square quadrats, following Diggle (2003, §2.5.2).

Table 6.1 A selection of paired locations from a sample of $n = 62$ redwood seedlings over a 23 m × 23 m square

Site index	X-location	Y-location
1	1.886	8.372
2	1.886	11.109
3	1.886	19.297
4	2.254	10.143
5	2.254	11.500
6	2.254	20.654
⋮	⋮	⋮
60	20.746	17.158
61	20.746	19.872
62	22.057	22.218

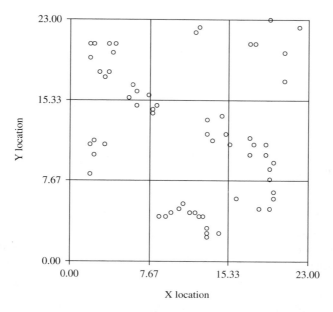

Figure 6.2 Locations of $n = 62$ redwood seedlings over a 23 m × 23 m square, with 3 × 3 grid overlay

To assess if indeed the clustering in Fig. 6.2 deviates from CSR, we count the number of seedlings lying in each of the nine quadrats. The resulting 3 × 3 table is

13	2	6
6	8	9
0	13	5

We have $\overline{Y} = 62/9 = 6.889$. For these data, one can show (Exercise 6.1) that the 3 × 3 format satisfies both regularity conditions to validate the χ^2 approximation. Hence, we find $X^2_{\text{calc}} = \{(13 - 6.889)^2 + \cdots + (5 - 6.889)^2\}/6.889 = 156.889/6.889 = 22.774$. At $\alpha = 0.01$, we reject CSR in favor of a clustered effect if $X^2_{\text{calc}} \geq \chi^2_{0.01}(8) = 20.09$. Since this is indeed the case, we conclude that significant non-CSR clustering is evident in these data.

One can perform these calculations in SAS PROC GENMOD or PROC FREQ. For instance, in PROC GENMOD the Pearson statistic for homogeneity can be computed by modifying the code in Fig. 3.11: use the same basic GLiM instructions, but do not include any variable terms in the model statement. (Under homogeneity, there are no such terms in the corresponding linear predictor of the GLiM.) Sample code appears in Fig. 6.3. The output (not shown) quickly verifies our computation of $X^2_{\text{calc}} = 22.774$.

To investigate how and where CSR fails with these data, we use the individual Pearson residuals from the X^2 statistic to examine each quadrat's contribution to X^2_{calc}. (Pearson residuals were introduced in §3.2.3.) Here, the Pearson residuals are

```
data redwood;
input ygrid xgrid Ycount @@;
datalines;
1  1 13      1  2  2        1  3  6
2  1  6      2  2  8        2  3  9
3  1  0      3  2 13        3  3  5

proc genmod;   * reduced model corresponding to homogeneity;
  model Ycount = / dist=poisson link=log ;
```

Figure 6.3 Sample SAS program to find Pearson X^2 for testing homogeneity in a 3×3 contingency table using a Poisson log-linear model

$c_{ij} = (Y_{ij} - \overline{Y})/\overline{Y}^{1/2}$ such that $X^2_{\text{calc}} = \sum_{i=1}^{I} \sum_{j=1}^{J} c_{ij}^2$. Residuals very far from zero in absolute value indicate strong departures from CSR in the corresponding quadrat.

We can also recover the Pearson residuals from the PROC GENMOD code in Fig. 6.3 by using the `obstats` option in the model statement. (Exercise 6.1 applies this to the 3×3 configuration for these data.) Adjusting to standardized Pearson residuals (also discussed in §3.2.3) can give a more precise pattern of departure, if one exists. These can be found in PROC GENMOD by also including the `residuals` option in the model statement; cf. Example 3.10.

More complex methods for identifying departure from CSR are based on associated indices of clustering/dispersion. We describe one such method, Ripley's K function, in §6.2.3. ◉

Notice in Example 6.1 that the 3×3 grid of quadrats leading to $X^2_{\text{calc}} = 22.774$ was somewhat arbitrary. While an expanded 2×2 grid would not satisfy the χ^2 conditions – since $IJ = 4 < 5$ – a tighter 4×4 grid might be perfectly valid (or it might not; see Exercise 6.1). This arbitrary aspect can have a critical effect on the χ^2 test: even with the same set of data, no guarantees exist that one analyst's grid will generate the same statistical inferences as another's if they choose different quadrat configurations. This concern is not entirely due to use of the simple Pearson statistic in (6.1). More complex test statistics are available for testing CSR with contiguous quadrat counts (see Cressie, 1993, §8.2.3), but even with these the choice of quadrat configuration can affect the final inferences. Also, depending on how sparsely the quadrats are chosen, classification of the spatial observations into discrete quadrat counts can represent a loss of information. (In effect, we start with data representing n distinct locations in two dimensions but then analyze counts over $IJ < n$ quadrats.) For reasons such as these, it is more common in modern analyses to base inferences on the precise spatial distances between data points, using what are called *distance methods*. These are introduced in §6.2.2.

We should note that the effort to employ quadrats for studying spatially varying environmental data does have practical benefits. Use of quadrats is prevalent, for example, in field ecology, soil science, and forestry, but there it is more common to *sample* the spatial area of interest using quadrats that are non-contiguous and, at that point, not necessarily rectangular. *Quadrat sampling*, as it is called, is an important tool in environmental survey sampling and we briefly discuss its use in §8.3.2. Indeed, many of the methods discussed here for enumerative, contiguous-quadrat data can also be applied to the quadrat sampling scenario (Diggle, 2003, §3.2).

6.2.2 Distance methods

When accurate spatial information is available, use of arbitrary quadrats sacrifices the accuracy for the sake of simplicity. With only minor increases in complexity, however, more advanced methods are available for precisely mapped data. These are known as *distance methods*, since they attempt to take into account the actual map distances observed between points in the spatial pattern.

Perhaps the best known of the variety of distance methods for spatial analysis is the class of *nearest-neighbor methods* (Cressie, 1993, §8.2.5; Dixon, 2002b). These quantify the pattern(s) of spatial association within a cloud of points in d-dimensional space using the distances between all pairs of neighboring points. There are many ways to define these distances; we limit ourselves to the single best-known way: Euclidean distance ('as the crow flies') between two points. Specifically, for $d = 2$ suppose two points s_i and s_j have individual spatial locations (x_i, y_i) and (x_j, y_j), respectively. Then, we denote the Euclidean distance between them by

$$\delta_{ij} = \sqrt{(x_i - x_j)^2 + (y_i - y_j)^2}.$$

A common, alternative notation for δ_{ij} is $\|s_i - s_j\|$. A point s_i has nearest neighbor s_j when $\|s_i - s\|$ is minimized at $s = s_j$ among all possible observed points $s \neq s_i$ in the spatial region under study, Ω_2. For simplicity, we denote this minimized, nearest-neighbor distance at each point s_i by $W_i = \min_{j \neq i}\{\delta_{ij}\}$.

A nearest-neighbor analysis summarizes information in the distances W_i by comparing their mean, $\overline{W}$, to that expected under CSR (more precisely, by comparing $\overline{W}$ to its expected value under a homogeneous Poisson process with intensity λ). Under spatial clustering, the average nearest-neighbor distance $\overline{W}$ should be smaller than expected under CSR, while under spatial regularity $\overline{W}$ should be larger than expected under CSR. If we adjust for sampling variation by dividing this difference by the standard error of $\overline{W}$, $se[\overline{W}]$, the result is a Wald statistic which can be referenced to a standard normal distribution. We reject CSR when the statistic is too large in absolute value.

While easy to describe, this methodology is somewhat more difficult to put into practice. The mean and standard error of $\overline{W}$ depend upon the intensity λ, and hence an estimator of λ must be employed in the analysis. A simple choice is the corresponding sample quantity $\hat{\lambda} = n/A_2$, where $A_2 = \|\Omega_2\|$ is the spatial area of Ω_2. An additional consideration when working with nearest-neighbor distances, however, is the problem of *edge effects*. That is, if the spatial region Ω_2 is cordoned off from a larger spatial continuum that may also contain possible point outcomes of interest, the 'neighbor' nearest to a point s_i in Ω_2 might be another point *outside* Ω_2. By missing these true nearest neighbors we can induce bias in $\overline{W}$

and inaccuracies in $se[\overline{W}]$. This must be accounted for in any inferences made on the spatial pattern.

Donnelly (1978) presents an adjustment for edge effects that works well with small sample sizes *if* Ω_2 is at least approximately rectangular. (Donnelly suggests sizes as low as $n \geq 8$ can yield reasonable results.) Modify the Wald statistic to account for possible edge-effect bias via

$$Z_{calc} = \frac{\overline{W} - 0.5\hat{\lambda}^{-1/2} - n^{-1}P_2(0.0514 + 0.041n^{-1/2})}{se_{adj}[\overline{W}]}, \qquad (6.2)$$

where the adjusted standard error is calculated as

$$se_{adj}[\overline{W}] = \sqrt{\frac{0.0703}{n\hat{\lambda}} + \frac{(0.037)P_2}{n^2\hat{\lambda}^{1/2}}}$$

and P_2 is the total perimeter of Ω_2. In large samples, reject CSR when $|Z_{calc}|$ exceeds the standard normal upper-$(\alpha/2)$ critical point $z_{\alpha/2}$.

Example 6.2 (Redwood seedlings, cont'd) For the redwood seedlings data in Fig. 6.2 a number of the $n = 62$ trees sit near the border of the test region. To correct for possible edge effects in our previous analysis, we can apply nearest-neighbor calculations. Here $A_2 = 23^2 = 529\,m^2$ and $P_2 = 92\,m$, so that $\hat{\lambda} = n/A_2 = 0.1172$ trees per m^2.

The nearest neighbors may be computed from the full data set; for example, point $s_1 = (1.886, 8.372)$ in Table 6.1 can be shown to have nearest neighbor $s_5 = (2.254, 10.143)$, producing $W_1 = \{(1.886 - 2.254)^2 + (8.372 - 10.143)^2\}^{1/2} = 1.809$, while point $s_2 = (1.886, 11.109)$ can be shown to have nearest neighbor $s_6 = (2.254, 11.5)$, producing $W_2 = 0.537$. Collected together, we find $\overline{W} = 0.8995$. We adjust $\overline{W}$ in the numerator of (6.2) by subtracting the edge-adjusted, expected, nearest-neighbor distance under CSR, $0.5\hat{\lambda}^{-1/2} + n^{-1}P_2(0.0514 + 0.041n^{-1/2}) = 1.5445$. We also find the adjusted standard error to be $se_{adj}[\overline{W}] = 0.1107$. This produces the test statistic $|Z_{calc}| = 5.83$, with corresponding P-value well below 0.0001 from Table B.1. Once again, significant departure from CSR is evidenced in this spatial pattern of young trees. Indeed, the sign of Z_{calc} is negative, $Z_{calc} = -5.83$, so that nearest-neighbor distances are lower on average than expected under CSR. This implies an overall clustering effect, corroborating the earlier indication in Example 6.1.

Example 6.3 (CSR process) Applying (6.2) to the point pattern data in Fig. 6.1 would produce $\overline{W} = 7.3674/200 = 0.0368$. The region in the figure was taken as the unit square, so $A_2 = 1$ and $P_2 = 4$. We find $\hat{\lambda} = n/A_2 = 200$ and adjust $\overline{W}$ in the numerator of (6.2) by subtracting $0.5\hat{\lambda}^{-1/2} + n^{-1}P_2(0.0514 + 0.041n^{-1/2}) = 0.0364$. Also, the adjusted standard error is $se_{adj}[\overline{W}] = 1.421 \times 10^{-3}$. This produces the test statistic $|Z_{calc}| = 0.279$, with a corresponding P-value of 0.78. Since these data were artificially generated under a CSR process, the insignificant result is not unexpected.

Another approach to adjust for edge effects is to pull the study region's operational boundary away from the original border by a fixed distance – a *buffer* – and then ignore points in the buffer when calculating statistical quantities. By choosing a large enough buffer, one can reduce or even eliminate edge effects, but, of course, at the cost of possible loss of data.

Although use of (6.2) is typically valid with even moderately sized samples, one still suffers a loss of information when collapsing the entire sample of nearest-neighbor quantities, W_i, into a single summary statistic. We can overcome this deficiency by analyzing the empirical distribution of the W_is, studied via a modified Kolmogorov–Smirnov test (mentioned near the end of §5.2.3) or other form of advanced test statistic to see if the pattern of response differs from CSR (Mugglestone and Renshaw, 2001). Given appropriate computer resources, one might also appeal to Monte Carlo approaches for testing CSR (Besag and Diggle, 1977; Cressie, 1993, §8.4.1), similar to those described in §4.4.2: generate a large number of pseudo-random configurations of n different spatial locations under CSR in the study region Ω_2. Then calculate, say, Z_{calc} as per (6.2) for each simulated configuration. The proportion of simulated configurations resulting in a calculated $|Z_{calc}|$ larger than that observed from the original data is a Monte Carlo-based estimated of the P-value. Reject CSR if this P-value drops below a preassigned level of α.

For more on methods for testing CSR, see Cressie (1993, §§8.2–8.3), Dixon (2002b), Diggle (2003, §2.3), and the references therein.

6.2.3 Ripley's *K* function

Once a spatial point pattern has been seen to deviate from CSR, it is natural to question how and/or where the spatial randomness fails. As noted in §6.2.1, many methods are available for approaching this question. We introduce here a graphical method for studying spatial point patterns known as *Ripley's K function*. The function shares certain similarities with estimation and prediction for continuous spatial processes, which we will discuss in §6.4, below.

To calculate the K function, we make the following assumptions: (i) the region under study, Ω_2, is two-dimensional (although it is possible to expand the construction to $d > 2$); (ii) the distance between any two points, $\mathbf{s}_i = (x_i, y_i)$ and $\mathbf{s}_j = (x_j, y_j)$, in Ω_2 is given by the Euclidean distance $\delta_{ij} = \sqrt{(x_i - x_j)^2 + (y_i - y_j)^2}$; (iii) Ω_2 is *completely mapped*, that is, the spatial locations of all n events in Ω_2 are identified; (iv) the data are *isotropic*, that is, the separation from point $\mathbf{s}_i$ to point $\mathbf{s}_j$ imparts the same information as that from $\mathbf{s}_j$ to $\mathbf{s}_i$ (cf. Definition 6.4, below); and (v) the intensity, λ, is constant throughout Ω_2. Recall that λ is the expected number of points per unit area.

The K function was originated by Bartlett (1964) and later developed by Ripley (1976, 1977). It is defined by conceptualizing a random variable V_h as the number of events within a constant distance $h > 0$ of any randomly selected point in Ω_2. Then

$$K(h) = \frac{\mathrm{E}[V_h]}{\lambda}$$

is the expected number of events within h of any randomly selected point, scaled by the expected number of points per unit area. Defined in this manner, the K function serves as a measure of spatial dependence between any two points separated by h units. By incorporating the ability to vary h, $K(h)$ quantifies how the spatial pattern varies over different distances within Ω_2.

Since $K(\cdot)$ is defined in terms of unknown quantities in both its numerator and denominator, we estimate it from the spatial data. For λ we use the simple estimator $\hat{\lambda} = n/A_2$, where A_2 is the area of Ω_2. For $E[V_h]$, we use an empirical measure based on how often two points are a distance $h > 0$ or less apart. The result can be written as

$$\hat{K}(h) = \frac{1}{n\hat{\lambda}} \sum_{i=1}^{n} \sum_{\substack{j=1 \\ j \neq i}}^{n} I_{[0,h]}(\delta_{ij}) \tag{6.3}$$

for any $h > 0$. (Recall that the indicator function from §A.2.1 has the form $I_{\mathbb{A}}(t) = 1$ if t is contained in the set $\mathbb{A}$, and $I_{\mathbb{A}}(t) = 0$ otherwise. In (6.3), the indicator is 1 if the points s_i and s_j are within a distance h of each other, 0 otherwise.) To view how the spatial variation changes, plot $\hat{K}(h)$ against h. For rectangular study areas, a standard recommendation is to range h no higher than about one-third to one-half the shortest dimension of the rectangle.

If the spatial pattern has been generated under CSR, the true $K(h)$ function will be $K_{\text{CSR}}(h) = \pi h^2$. Thus, a plot of $\hat{K}(h)$ vs. h that mimics πh^2 is consistent with CSR, while deviations from πh^2 indicate departures from CSR. Other spatial point processes can lead to different shapes for the K function; a plot of $\hat{K}(h)$ similar in shape to one of these forms can indicate that the corresponding, alternative process might underlie the observed spatial response. Dixon (2002c) describes some of these other models for $K(h)$.

In practice, it is useful to work with

$$\hat{L}(h) = \sqrt{\frac{\hat{K}(h)}{\pi}}, \tag{6.4}$$

since $\text{Var}[\hat{L}(h)]$ is approximately constant under CSR (Ripley, 1979). As such, under CSR a plot of $\hat{L}(h) - h$ will approximate a horizontal line of zero height (Exercise 6.7). One can also plot upper and lower quantile curves, obtained under the CSR null hypothesis via Monte Carlo simulation (§4.4.2) or bootstrapping methods (§A.5.2). If the estimated K (or L) function deviates substantially from the quantile curves over much of the range of h, departure from CSR is indicated; see, for example, Moser (1987, Fig. 3), Cressie (1993, Fig. 8.14), or Dixon (2002c, Fig. 1).

When edge effects are present (6.3) is contraindicated, since it can exhibit substantial bias as h grows large. A simple edge-effect adjustment is possible by applying a buffer to the region under study and then calculating (6.3) and (6.4) over the buffered region. More elaborate edge-effect adjustments also exist, however; perhaps the most common is due to Ripley (1976):

$$\hat{K}_R(h) = \frac{1}{n\hat{\lambda}} \sum_{i=1}^{n} \sum_{\substack{j=1 \\ j \neq i}}^{n} \frac{I_{[0,h]}(\delta_{ij})}{\omega(s_i, s_j)}, \tag{6.5}$$

where $\omega(\mathbf{s}_i, \mathbf{s}_j)$ is a weight function that adjusts the contribution each pair of points makes to the overall empirical measure. The weight is based on centering a circle at $\mathbf{s}_i$ and extending it to $\mathbf{s}_j$ such that the circle has radius δ_{ij}. If this circle lies entirely within Ω_2, then set $\omega(\mathbf{s}_i, \mathbf{s}_j) = 1$. If not, then set $\omega(\mathbf{s}_i, \mathbf{s}_j)$ equal to the proportion of the circumference of the circle that lies within Ω_2; cf. Diggle (2003, Fig. 4.1). This refinement adds complexity to the $K(h)$ estimator, but it can be well worth the effort. To facilitate matters, a SAS macro exists (Moser, 1987), available at the time of this writing from http://www.stat.lsu.edu/faculty/moser/spatial/spatial.html.

Note that some authors use $(n-1)\hat{\lambda}$ instead of $n\hat{\lambda}$ in (6.5), to serve as an adjustment for small sample sizes. Clearly, the difference between the two forms diminishes as n grows large.

Example 6.4 (Redwood seedlings, cont'd)   For the redwood seedlings data in Fig. 6.2, recall that a Pearson test of CSR indicated significant departure. To study this graphically, $\hat{K}(h)$ is straightforward, if tedious, to compute using (6.3). The study region is a $23\,\mathrm{m} \times 23\,\mathrm{m}$ square, with $A_2 = 529\,\mathrm{m}^2$. From these, we calculate $\hat{\lambda} = 62/529 = 0.117$. The square study area has sides of length $23\,\mathrm{m}$, so we limit our calculations to no higher than $h = \frac{23}{3} \approx 7.67$. The result is plotted in Fig. 6.4, along with the CSR reference curve πh^2; Fig. 6.5 gives the corresponding plot of $\hat{L}(h) - h$. In both figures departure from CSR is evidenced: the curves rise above the CSR reference over small values of h, then drop below the CSR reference as h grows. The effect is particularly clear in the $\hat{L}(h) - h$ plot.

Rises above the CSR standard are indicative of spatial clustering, while falls below it indicate spatial regularity. Here, the figures suggest that clustering is prevalent for trees located close together, but that as h grows a more regular pattern may be evident. This is consistent with the 'fairy ring' phenomenon mentioned earlier: young redwoods often cluster around old stumps. Whether such clustering appears at increasing separations is less certain throughout a grove.

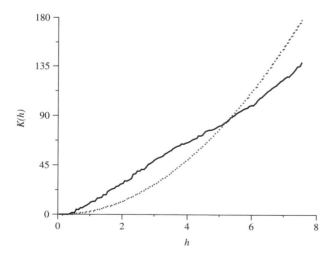

Figure 6.4   Estimated K function (thick curve) from (6.3) and CSR reference (dotted curve) for redwood seedlings data

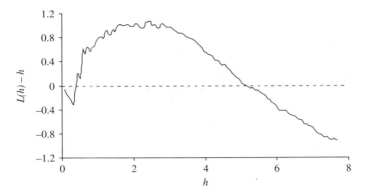

Figure 6.5 $\hat{L}(h) - h$ plot for redwood seedlings data. Horizontal dashes indicate CSR reference line

Recall from Example 6.2 that a number of trees sit near the border of the test region, raising concerns about possible edge-effect biases. The adjusted nearest-neighbor analysis in that example still indicated significant departure from CSR. Another way to correct for edge effects is to apply Ripley's edge-adjusted estimator from (6.5). For these data, we invoked Moser's (1987) SAS macro `%spatial` under the specifications `maxx = 23`, `maxy = 23`, `konly = 1`, `ktcrit = 7.67`, `kmax = 200`. These indicate to `%spatial` that the spatial area ranges over $0 < x < 23$ and $0 < y < 23$, that only a Ripley K analysis is requested, that the range of h is $0 < h < 7.67$, and that the K function should be estimated at 200 distinct points over the specified range for h, respectively. The results (not shown) lead to the edge-adjusted estimator $\hat{K}_R(h)$ graphed in Fig. 6.6.

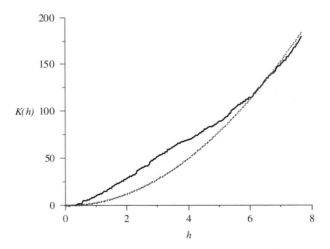

Figure 6.6 Edge-adjusted K function (thick curve) from (6.5) and CSR reference (dotted curve) for redwood seedlings data

The figure based on the edge adjustment pulls $K(h)$ slightly higher away from the CSR standard as h grows large, suggesting that some bias away from localized clustering exists in the data near the border. The evidence nonetheless corroborates our earlier inferences of departure from CSR with these data.   ✍

6.3 Spatial measurement

Environmental investigators often measure more than just the location of items or events when they study spatial phenomena. It is common in many studies to record specific outcomes on the observations made at each spatial location (for point process data, this is called *marking*). For example, one might measure diameters of trees after observing their spatial pattern in a forest (Exercise 6.15) or biomass of a faunal species in a specialized ecosystem (Example 6.5). One can also undertake an environmental study at well-defined spatial locations, such as determining concentrations of a toxic contaminant in a waste site (Example 6.9), or of ground-level ozone at city centers across a county (similar to Example 3.6).

Thus we return to our earlier notation and let $Z(\mathbf{s}_1), Z(\mathbf{s}_2), \ldots, Z(\mathbf{s}_n)$ be measured, continuous observations over a sample of spatial sites $\mathbf{s}_i$, $i = 1, \ldots, n$, in a predefined study region Ω_2. Unless otherwise specified, we continue to restrict attention to the case of $d = 2$. We suppose that the data are generated from some underlying model, say

$$Z(\mathbf{s}_i) = \mu(\mathbf{s}_i) + \varepsilon(\mathbf{s}_i), \tag{6.6}$$

where both the deterministic term $\mu(\mathbf{s}_i)$ and the stochastic term $\varepsilon(\mathbf{s}_i)$ may depend on the spatial location at which $Z(\mathbf{s}_i)$ is recorded. We assume the stochastic errors have zero mean, $E[\varepsilon(\mathbf{s}_i)] = 0$, and that variation between spatial points is determined by some *covariance function* $C(\mathbf{s}_i, \mathbf{s}_j) = \text{Cov}[\varepsilon(\mathbf{s}_i), \varepsilon(\mathbf{s}_j)]$. Recall from §A.1 that a covariance measures the joint variability between two random variables: $\text{Cov}[X, Y] = E[(X - \mu_X)(Y - \mu_Y)]$, where, in particular, $\text{Cov}[Y, Y] = \text{Var}[Y]$. We will vary the specificity of the model in (6.6) as we move through the rest of this chapter.

When measurements are taken at arbitrary locations over a continuous region, we often refer to the recorded outcomes as *geostatistical data* (Webster and Oliver, 2001), referring to the geological roots of the relevant statistical methods. (We comment more on this historical connection in §6.4, below.) These are perhaps the most general form of measured spatial data. A more precise descriptor might be *random field data*, corresponding to the stochastic phenomenon generating the data. Indeed, (6.6) can be formulated in terms of a more general theory of stochastic models or, for the spatial setting, in terms of the theory of random fields. Interested readers can find associated details in sources such as Cressie (1993, §2.3), Stein (1999, Ch. 2), or Adler (2002). We will employ some of these features in the constructions below.

By contrast, when the measurements are taken over a regular (or irregular) grid of locations, we often refer to the outcomes as *lattice data*. Lattice data occur in studies of environmental effects over defined regions; examples include ecological landscape pattern mapping, and satellite observation or other forms of *remote sensing* of the

Earth's surface. Models for spatial lattice data are perhaps closest in spirit and analogy to models for time series data seen in Chapter 5. In many cases, and where appropriate, they can be extremely effective in describing spatial variability. Cressie (1993, Ch. 6) and Schabenberger and Pierce (2002, §9.6) give good overviews.

For random field data, one often questions whether the measurements exhibit any dependencies that can be interpreted as a form of spatial correlation. In this section, we focus on quantification of such spatial dependencies, using measures of covariation and correlation similar in spirit to the temporal constructions seen in Chapter 5.

6.3.1 Spatial autocorrelation

Our fundamental principle for analyzing spatial dependency is similar to that employed in §5.3 for serial dependency: we assume that observations from locations spatially close to each other are more likely to be similar in value than those farther apart. The primary tools used to understand dependencies among the data are measures of autocorrelation. In §5.3, this was temporal or serial autocorrelation; here it is *spatial autocorrelation*.

To explore the autocorrelation evidenced in a set of spatial data, we will mimic the EDA strategy from Chapter 5. We start by defining the stochastic characteristics of the random process – i.e., the random field – represented by $Z(\mathbf{s})$.

Definition 6.2 A stochastic process $Z(\mathbf{s})$ on a d-dimensional space Ω_d is *mean stationary* if $E[Z(\mathbf{s})] = \mu(\mathbf{s})$ does not depend upon $\mathbf{s}$.

Definition 6.3 A stochastic process $Z(\mathbf{s})$ on a d-dimensional space Ω_d is *second-order stationary* if $\mathrm{Var}[Z(\mathbf{s})]$ exists for all $\mathbf{s} \in \Omega_d$ such that
(i) $Z(\mathbf{s})$ is mean stationary, and

(ii) $C(\mathbf{s}, \mathbf{t}) = \mathrm{Cov}[Z(\mathbf{s}), Z(\mathbf{t})]$ depends upon $\mathbf{s}$ and $\mathbf{t}$ only through the difference vector $\mathbf{h} = \mathbf{s} - \mathbf{t}$.

In Definition 6.2, mean stationarity implies that the mean response does not vary across locations in the region of interest. In Definition 6.3, the adjective *second-order* is used to indicate characteristics of the variance and covariances of $Z(\mathbf{s})$, that is, the 'second-order' expectations of $Z(\mathbf{s})$. Notice the similarities to a second-order stationary time series from Definition 5.2. In a similar sense, Ripley's K function in §6.2.3 measures certain second-order features of a spatial point pattern.

To illustrate these concepts, consider a simple example with the $n = 4$ vertices of the unit square: $\mathbf{s}_1 = (0, 0)$, $\mathbf{s}_2 = (1, 0)$, $\mathbf{s}_3 = (0, 1)$ and $\mathbf{s}_4 = (1, 1)$. If $Z(\mathbf{s})$ is second-order stationary, then $E[Z(\mathbf{s}_1)] = E[Z(\mathbf{s}_2)] = E[Z(\mathbf{s}_3)] = E[Z(\mathbf{s}_4)]$. Further, $C(\mathbf{s}_2, \mathbf{s}_1) = C(\mathbf{s}_4, \mathbf{s}_3)$ since $\mathbf{s}_2 - \mathbf{s}_1 = \mathbf{s}_4 - \mathbf{s}_3 = (1, 0)$, while $C(\mathbf{s}_3, \mathbf{s}_1) = C(\mathbf{s}_4, \mathbf{s}_2)$ since $\mathbf{s}_3 - \mathbf{s}_1 = \mathbf{s}_4 - \mathbf{s}_2 = (0, 1)$. But, $C(\mathbf{s}_2, \mathbf{s}_1)$ does not necessarily equal $C(\mathbf{s}_3, \mathbf{s}_1)$: $\mathbf{s}_2 - \mathbf{s}_1 = (1, 0) \neq (0, 1) = \mathbf{s}_3 - \mathbf{s}_1$ so, although the distance between the points is the same, they differ with respect to direction.

An immediate consequence of Definition 6.3 is that $\text{Var}[Z(\mathbf{s})]$ will not depend on $\mathbf{s}$, since, under (ii), $\text{Var}[Z(\mathbf{s})] = \text{Cov}[Z(\mathbf{s}), Z(\mathbf{s})]$ depends only upon the known vector $\mathbf{s} - \mathbf{s} = \mathbf{0}$. Some authors also define the less restrictive concept of *intrinsic stationarity*, where (ii) is written in terms of the difference $\text{Var}[Z(\mathbf{s}) - Z(\mathbf{t})]$ instead of $C(\mathbf{s}, \mathbf{t})$, and one requires only that $\text{Var}[Z(\mathbf{s}) - Z(\mathbf{t})]$ exist.

When $Z(\mathbf{s})$ satisfies Definition 6.3 we can simplify the notation for the covariance function by writing $C(\mathbf{h})$ instead of $C(\mathbf{s},\mathbf{t})$, since the covariance will then depend only on the distance and direction between $\mathbf{s}$ and $\mathbf{t}$. Also, the spatial correlation between $Z(\mathbf{s})$ and $Z(\mathbf{s} + \mathbf{h})$ is $\rho(\mathbf{h}) = \text{Cov}[Z(\mathbf{s}), Z(\mathbf{s} + \mathbf{h})]/\sqrt{\text{Var}[Z(\mathbf{s})]\text{Var}[Z(\mathbf{s} + \mathbf{h})]}$; this is analogous to the temporal autocorrelation function seen in §5.3.2. When $Z(\mathbf{s})$ is second-order station-ary, $\rho(\mathbf{h})$ simplifies: recognize that since a second-order stationary variance does not depend on $\mathbf{s}$, $\text{Var}[Z(\mathbf{s})] = \text{Var}[Z(\mathbf{s} + \mathbf{h})]$ for any $\mathbf{h}$. Thus $\sqrt{\text{Var}[Z(\mathbf{s})]\text{Var}[Z(\mathbf{s} + \mathbf{h})]} = \sqrt{\text{Var}^2[Z(\mathbf{s})]} = \text{Var}[Z(\mathbf{s})]$. But $\text{Var}[Z(\mathbf{s})]$ can also be expressed as $\text{Cov}[Z(\mathbf{s}), Z(\mathbf{s} + \mathbf{0})]$, where $\mathbf{0}$ is a column vector of zeros. Thus $\rho(\mathbf{h}) = \text{Cov}[Z(\mathbf{s}), Z(\mathbf{s} + \mathbf{h})]/\text{Cov}[Z(\mathbf{s}), Z(\mathbf{s} + \mathbf{0})]$, which under our notation for $C(\cdot)$ reduces to $\rho(\mathbf{h}) = C(\mathbf{h})/C(\mathbf{0})$.

Under second-order stationarity the covariance function $C(\mathbf{s}, \mathbf{t})$ depends only on the distance and direction between two points in Ω_d, and not on their actual loca-tions. For $d = 2$, this is tantamount to saying that $C(\mathbf{s},\mathbf{t})$ depends only on the length and angle of the line segment connecting the two points $\mathbf{s}$ and $\mathbf{t}$. In some cases, however, even the angle between the points is irrelevant. All that matters is the distance between them, $h = \sqrt{(x_s - x_t)^2 + (y_s - y_t)^2}$. This is the Euclidean distance in two dimensions, which, similar to our previous notation, we write as $\delta_{st} = \|\mathbf{s} - \mathbf{t}\|$; that is, $h = \delta_{st}$. Formalized, this leads to the following definition.

Definition 6.4 A stochastic process $Z(\mathbf{s})$ on a d-dimensional space Ω_d is *isotropic* if $\text{Var}[Z(\mathbf{s})]$ exists for all $\mathbf{s} \in \Omega_d$ such that

(i) $Z(\mathbf{s})$ is mean stationary, and

(ii) for $h = \delta_{st} = \|\mathbf{s} - \mathbf{t}\|$, $C(\mathbf{s}, \mathbf{t}) = \text{Cov}[Z(\mathbf{s}), Z(\mathbf{t})]$ depends upon $\mathbf{s}$ and $\mathbf{t}$ only through h.

An *anisotropic* process is a process $Z(\mathbf{s})$ that does not satisfy these properties.

If $Z(\mathbf{s})$ is isotropic, $\text{Var}[Z(\mathbf{s})]$ is again independent of $\mathbf{s}$, since under (ii), $\text{Var}[Z(\mathbf{s})] = \text{Cov}[Z(\mathbf{s}), Z(\mathbf{s})]$ will only depend upon the known distance $\delta_{ss} = 0$. We further sim-plify the notation for isotropic covariance and correlation functions by writing $C(h)$ and $\rho(h) = C(h)/C(0)$, respectively.

Unless otherwise indicated, we will throughout the rest of this chapter assume that the $Z(\mathbf{s}_i)$s are isotropic as per Definition 6.4. Testing for isotropy is discussed by Guan *et al.* (2004). Also, as was the case with stationary time series, any assumption that includes mean stationarity requires the user to ensure that no trend or drift exists in the data. Checks for this include data plots and other graphical diagnostics (Cressie, 1993, Ch. 4). If a spatial trend or drift pattern exists, some prewhitening or detrending will be necessary. We also assume that the variances are homogeneous. If preliminary diag-nostics or past experience suggests that this is not the case, adjustment via variance-stabilizing transformation is in order. More generally, heterogeneity and anisotropy can sometimes be corrected by rotating and scaling the coordinate axes. For more on analysis with anisotropic or otherwise non-stationary spatial data, see Zimmerman (1993), Guttorp and Sampson (1994), or Kitanidis (1997, Ch. 5).

6.3.2 Moran's *I* coefficient

An early measure of spatial autocorrelation was given by Moran (1950):

$$I(h) = \frac{N_h^{-1} \sum \sum_{(i,j) \in \mathbb{G}_h} \{Z(\mathbf{s}_i) - \overline{Z}\}\{Z(\mathbf{s}_j) - \overline{Z}\}}{n^{-1} \sum_{i=1}^{n} \{Z(\mathbf{s}_i) - \overline{Z}\}^2} \tag{6.7}$$

where $\mathbb{G}_h$ is the set of all pairs of indices whose corresponding points $(\mathbf{s}_i, \mathbf{s}_j)$ satisfy $\delta_{ij} = h$, N_h is the number of distinct pairs of indices (points) in $\mathbb{G}_h$, and $\overline{Z} = \sum_{i=1}^{n} Z(\mathbf{s}_i)/n$. The quantity in (6.7) is known as *Moran's I coefficient*. Readers should be careful not to confuse (6.7) with earlier quantities where the notation involves the symbol I, such as the indicator function from §A.2.1.

Many authors write (6.7) in the form

$$I(h) = \frac{\sum_{i=1}^{n} \sum_{j=1}^{n} \omega_{ij}(h) \{U_0(h)\}^{-1} \{Z(\mathbf{s}_i) - \overline{Z}\}\{Z(\mathbf{s}_j) - \overline{Z}\}}{n^{-1} \sum_{i=1}^{n} \{Z(\mathbf{s}_i) - \overline{Z}\}^2},$$

where $\omega_{ij}(h)$ is the (i, j)th element of a *weight matrix* such that $\omega_{ij}(h)$ equals one if the two data points $Z(\mathbf{s}_i)$ and $Z(\mathbf{s}_j)$ are separated by the lag distance h, and zero otherwise; also, we define $\omega_{ii}(h) = 0$ for all i at any h. (More advanced weighting schemes are possible for highly irregular neighborhood structures; see, for example, Schabenberger and Pierce, 2002, §9.6.1.) The quantity $U_0(h)$ is simply the sum of all the $\omega_{ij}(h)$ values: $U_0(h) = \sum_{i=1}^{n} \sum_{j=1}^{n} \omega_{ij}(h)$. As given here, the two formulations for $I(h)$ are identical; indeed, $U_0(h) = N_h$. For use in certain computer packages, however, users may be required to supply the weight matrix at each desired level of h rather than have the software perform the calculations.

The I coefficient is much more difficult to write than it is to describe: for a given lag distance $h > 0$, simply find all the points separated by that distance and calculate their average mean-corrected cross product (the numerator in (6.7)). Then, divide this by $(n - 1)/n$ times the sample variance of all the points in the sample. This latter quantity is not a function of h and need only be calculated once.

In practice, not every possible value of h can be represented among all the possible pairs of spatial locations (unless the spatial grid or field grows extremely dense), so one selects a discrete set of (ordered) lag distances $h_1 < h_2 < \ldots < h_B$ and performs the calculations over lag distance classes centered or otherwise constructed around each h_b, $b = 1, \ldots, B$. Even then, there may be cases where only a few points are available at a given h, making N_h too small for stable estimation of $I(h)$. This is quite common when h is large, and so we typically restrict calculation of $I(h)$ to small values of h. A useful recommendation – based on considerations from §6.3.6, below – is to set h_B to no higher than about half the maximum distance separating any two sample locations: $\frac{1}{2} \max_{i,j} \{\delta_{ij}\}$.

Note that when data are taken over a regular grid, the h_bs are typically self-defined. That is, a clear set of separable distances will be generated by the grid points. (To take a simple example, if spatial data are recorded at the vertices of a unit square, then h_b can take only two possible values: $h_1 = 1$ corresponding to points separated horizontally or vertically, and $h_2 = \sqrt{2}$ corresponding to points separated diagonally.)

For any set of spatial observations $Z(\mathbf{s}_1), Z(\mathbf{s}_2), \ldots, Z(\mathbf{s}_n)$, we have $-1 < I(h) < 1$. In the absence of any spatial autocorrelation, $I(h)$ has expected value $E[I(h)] = -1/(n-1)$, which for large n will be close to zero. If the $Z(\mathbf{s}_i)$s are normally distributed, the standard error of $I(h)$ under a null hypothesis of no spatial auto-correlation can be written as

$$se[I(h)] = \sqrt{\frac{n^2 U_1(h) - n U_2(h) + 3[U_0(h)]^2}{[U_0(h)]^2 (n^2 - 1)} - \left(\frac{1}{n-1}\right)^2}$$

(Cliff and Ord, 1981, §1.5.1), where

$$U_1(h) = \frac{1}{2} \sum_{i=1}^{n} \sum_{\substack{j=1 \\ j \neq i}}^{n} \{\omega_{ij}(h) + \omega_{ji}(h)\}^2$$

and

$$U_2(h) = \sum_{i=1}^{n} \left\{ \sum_{j=1}^{n} \omega_{ij}(h) + \sum_{j=1}^{n} \omega_{ji}(h) \right\}^2.$$

The I coefficient has strong connections to the estimated serial correlation coefficient in (5.10). Indeed, Moran's goal was to devise a statistic that measured a similar correlative effect, only now for spatially separated data. (Actually, he described a single statistic for testing spatial autocorrelation that did not depend on h. His construction has since been extended into (6.7) to show if and how the spatial autocorrelation varies with changing lag distance h.) A plot of $I(h)$ against h – called a *spatial correlogram* (Fortin *et al.*, 2002) – produces a graph not unlike the ACF plot from §5.3.2, and is used in similar fashion to identify lag distances at which strong correlation is evidenced in the spatial data: values of h past which $|I(h) - E[I(h)]| = |I(h) + (n-1)^{-1}|$ exceeds roughly $2se[I(h)]$ indicate lag distances for which the spatial autocorrelation may be important. (If the inferences are to be more formal, then over the B lag classes one should correct for multiplicity via a Bonferroni adjustment: use $z_{\alpha/2B}se[I(h)]$ for some prespecified level of α. Corrections are also appropriate in small samples; see Cliff and Ord, 1981, §2.5.1.)

Unfortunately, simple implementation of Moran's I coefficient via widely available software is problematic, at least as of the time of this writing. As with Ripley's K function, the Internet site at http://www.stat.lsu.edu/faculty/moser/spatial/spatial.html supplies SAS add-in routines that can perform the computations, but these require use of SAS's advanced, user-programmable IML component. Schabenberger and Pierce

(2002, §9.8.1) give a SAS macro, %MoranI, that can be useful. Analysts familiar with the more advanced R (or similarly, the S-Plus®) computer environment may appeal to the VR package by Venables and Ripley (2002), whose spatial sub-package can compute a quantity essentially similar to (6.7) via the *correlogram* function. The R package spdep can compute Moran's *I* directly, via its *moran* function. For both packages see http://www.r-project.org/. Also useful is the splancs R-library of spatial analysis tools (Rowlingson and Diggle, 1993). For Microsoft® Windows (PC) users, Sawada (1999) describes a Microsoft® Excel add-on, available at http://www.uottawa.ca/academic/arts/geographie/lpcweb/newlook/data_and_downloads/download/sawsoft/rooks.htm. Other programs for spatial data that include calculation of Moran's *I* are listed at the AI-GEOSTATS software site http://www.ai-geostats.org/software/index.htm. We should warn that the majority of our own experience is limited to the SAS add-ons mentioned above, and hence we neither recommend nor warn against use of these other routines. Users must proceed at their own risk.

6.3.3 Geary's *c* coefficient

An alternative measure of spatial autocorrelation similar in nature to Moran's *I* coefficient was proposed by Geary (1954):

$$c(h) = \frac{(2N_h)^{-1} \sum\sum_{(i,j) \in \mathbb{G}_h} \{Z(\mathbf{s}_i) - Z(\mathbf{s}_j)\}^2}{(n-1)^{-1} \sum_{i=1}^{n} \{Z(\mathbf{s}_i) - \overline{Z}\}^2}, \tag{6.8}$$

where, as in (6.7), $\mathbb{G}_h$ is the set of all pairs of indices whose corresponding points $(\mathbf{s}_i, \mathbf{s}_j)$ satisfy $\delta_{ij} = h$, N_h is the number of distinct pairs of indices (points) in $\mathbb{G}_h$, and $\overline{Z} = \sum_{i=1}^{n} Z(\mathbf{s}_i)/n$. The quantity in (6.8) is known as *Geary's c coefficient*. As with Moran's *I*, in practice one selects a set of lag distances $h_1 < h_2 < \ldots < h_B$ and performs the calculations over lag distance classes centered or otherwise constructed around each h_b, $b = 1, \ldots, B$.

Here again some authors write the numerator of (6.8) in terms of a *weight matrix* whose elements, $\omega_{ij}(h)$, equal one if the two data points $Z(\mathbf{s}_i)$ and $Z(\mathbf{s}_j)$ are separated by the lag distance h, and zero otherwise. If we define $\omega_{ii}(h) = 0$ for all i at any h, then this produces

$$c(h) = \frac{\frac{1}{2} \sum_{i=1}^{n} \sum_{j=1}^{n} \omega_{ij}(h) \{U_0(h)\}^{-1} \{Z(\mathbf{s}_i) - Z(\mathbf{s}_j)\}^2}{(n-1)^{-1} \sum_{i=1}^{n} \{Z(\mathbf{s}_i) - \overline{Z}\}^2},$$

where $U_0(h) = \sum_{i=1}^{n} \sum_{j=1}^{n} \omega_{ij}(h)$.

For any set of spatial observations $Z(\mathbf{s}_1), Z(\mathbf{s}_2), \ldots, Z(\mathbf{s}_n), c(h)$ is bounded between 0 and 2, such that, in the absence of any spatial autocorrelation, $E[c(h)] = 1$. Values near 0 indicate strong positive spatial autocorrelation, while values near 2 indicate strong negative spatial autocorrelation. If the $Z(\mathbf{s}_i)$s are normally distributed, the standard error of $c(h)$ under a null hypothesis of no spatial autocorrelation is

$$se[c(h)] = \sqrt{\frac{(n-1)\{2U_1(h) + U_2(h)\} - 4[U_0(h)]^2}{2[U_0(h)]^2(n+1)}}$$

(Cliff and Ord, 1981, §1.5.1), where $U_1(h)$ and $U_2(h)$ are as defined for Moran's I. Thus Geary's c coefficient can be used to identify lag distances past which the spatial autocorrelation is inconsequential, via, for example, plots of $c(h)$ vs. h. (Some authors also use the term *correlogram* to mean such a plot, although to avoid confusion we counsel against such use.) Values of h where $|c(h) - 1|$ exceeds roughly $2se[c(h)]$ may indicate important spatial autocorrelation(s). (If the inferences are to be more formal, then over the B lag classes one should correct for multiplicity via a Bonferroni adjustment: use $z_{\alpha/2B}se[c(h)]$ for some prespecified level of α. Corrections are also appropriate in small samples; see Cliff and Ord, 1981, §2.5.2.)

For computational acquisition of Geary's c, similar caveats apply as for the case with Moran's I. Many of the Internet sites mentioned in §6.3.2 also give Geary's c, although as noted there we can neither recommend nor warn against use of these routines. We will comment in §6.3.6 on certain SAS procedures that can calculate statistics very similar to (6.8), and how these can help quantify the autocorrelation in a set of spatial data.

6.3.4 The semivariogram

Both Moran's I and Geary's c coefficients use information in the data to quantify the covariation in response between two points, say $\mathbf{s}$ and $\mathbf{t}$, in the spatial region Ω_2. That is, they provide information on the covariance function $C(\mathbf{s}, \mathbf{t}) = \mathrm{Cov}[\varepsilon(\mathbf{s}), \varepsilon(\mathbf{t})]$, where $\varepsilon(\mathbf{s})$ is the additive error term from model (6.6). The distinction between them is small; $I(h)$ appears to perform slightly better in some cases (Cliff and Ord, 1981, §6.5.2; Walter, 1992), although in practice the two are often used in tandem to study spatial covariance. Another measure that we prefer is known as the *semivariogram* function: $\gamma(\mathbf{s}, \mathbf{t}) = \frac{1}{2}\mathrm{Var}[Z(\mathbf{s}) - Z(\mathbf{t})]$ for any $\mathbf{s}, \mathbf{t} \in \Omega_2$. (The *variogram* is the function formed by multiplying the semivariogram by 2.) The semivariogram and variogram are constructed to measure how variable the *difference* in response is between two spatially separated random variables as a function of the distance between them. Because they are defined in terms of such differences, the variogram and semivariogram can exist even when $\mathrm{Var}[Z(\mathbf{s})]$ does not. But, when $\mathrm{Var}[Z(\mathbf{s})]$ does exist we can relate the semivariogram to the covariance function via $\gamma(\mathbf{s}, \mathbf{t}) = \frac{1}{2}\{C(\mathbf{s}, \mathbf{s}) + C(\mathbf{t}, \mathbf{t})\} - C(\mathbf{s}, \mathbf{t})$; see Exercise 6.10.

These functions have been defined in various ways by various authors, going back at least as far as Kolmogorov's work in fluid mechanics (Kolmogorov, 1941). As a result, different fields of application often generate different names for them; see Cressie (2002a). The term *variogram* is due to Matheron (1962, §20), and this moniker appears to have gained greatest usage in the environmetric literature.

For random field data, characteristics of the stochastic process represented by $Z(\mathbf{s})$ are often described in terms of the semivariogram. Recall from Definition 6.3 that if $Z(\mathbf{s})$ is second-order stationary, $C(\mathbf{s}, \mathbf{t}) = \mathrm{Cov}[Z(\mathbf{s}), Z(\mathbf{t})]$ depends upon $\mathbf{s}$ and $\mathbf{t}$ only through the difference vector $\mathbf{h} = \mathbf{s} - \mathbf{t}$. The effect on the semivariogram is similar to that on the covariance function: we now write $\gamma(\mathbf{h})$ instead of $\gamma(\mathbf{s}, \mathbf{t})$, so that, for example, $\gamma(\mathbf{h}) = \frac{1}{2} \mathrm{Var}[Z(\mathbf{s} + \mathbf{h}) - Z(\mathbf{s})]$. This helps illustrate the consequent relationships $\gamma(\mathbf{h}) = \gamma(-\mathbf{h})$ for all $\mathbf{h}$, and $\gamma(\mathbf{0}) = 0$ (Exercise 6.11). (Recall that $\mathbf{0}$ is a column vector of zeros.) It can be shown (Exercise 6.12) that under second-order stationarity, $\gamma(\mathbf{h}) = C(\mathbf{0}) - C(\mathbf{h})$ for any difference vector $\mathbf{h}$.

The semivariogram is also related to the spatial correlation function of a second-order stationary process. In §6.3.1, we saw that the correlation between $Z(\mathbf{s})$ and $Z(\mathbf{s} + \mathbf{h})$ is $\rho(\mathbf{h}) = C(\mathbf{h})/C(\mathbf{0})$, while in Exercise 6.12 we find $\gamma(\mathbf{h}) = C(\mathbf{0}) - C(\mathbf{h})$. Thus while spatial correlation involves a ratio operation, the semivariogram involves a difference.

If the stochastic process is isotropic (Definition 6.4), the semivariogram depends on only the distance between $\mathbf{s}$ and $\mathbf{s} + \mathbf{h}$, say, $h = \| (\mathbf{s} + \mathbf{h}) - \mathbf{s} \| = \| \mathbf{h} \|$. As such, we simplify the notation further by writing $\gamma(\mathbf{h})$ as $\gamma(h)$. Figure 6.7 plots a representative version of an isotropic semivariogram. For such functions we can define a few specialized components, the terminology for which derives from the method's geostatistical roots. These are formalized in Definition 6.5.

Definition 6.5 If a stochastic process $Z(\mathbf{s})$ is isotropic, then

(i) the *nugget* is the value $\gamma_0 = 0$ such that $\lim_{h \to 0} \gamma(h) = \gamma_0$ (if the limit exists);

(ii) the *sill* is the value of $\lim_{h \to \infty} \gamma(h)$ (if the limit exists), that is, the upper asymptote of $\gamma(h)$; and

(iii) the *range* (if it exists) is the value of h at which $\gamma(h)$ first attains its sill.

In Fig. 6.7 we add some markers identifying the nugget, sill, and range. One can similarly plot the corresponding functions $C(h)$ and $\rho(h) = C(h)/C(0)$. The former is called the *covariogram* while the latter is the *correlogram*, since it represents a theoretical counterpart to Moran's $I(h)$ function. (Some authors use the term

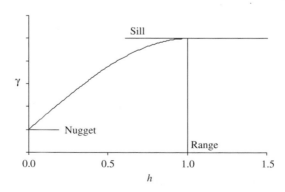

Figure 6.7 A representative version of an isotropic semivariogram

correlogram to mean a plot of $C(h)$; readers should be careful to avoid possible misinterpretations.)

Notice that some or all of the values in Definition 6.5 may not exist, depending on the structure of the underlying process $Z(\mathbf{s})$. For instance, if a semivariogram possesses a mathematical asymptote, $\lim_{h\to\infty} \gamma(h)$, then by definition that value is clearly the sill. If that semivariogram never actually achieves its limiting value, however, no range exists.

An alternative name for the range is the *scale* of the process, although we should caution readers that the two terms are not always used interchangeably in the literature. Nor, for that matter, may one analyst's use of the term 'range' be identical to another's. For example, suppose a sill exists as a true asymptote and an analyst sets the corresponding range (or scale) as the point at which the semivariogram comes so close to the sill as to be indistinguishable from a practical perspective. This is certainly reasonable; however, different analysts may interpret 'indistinguishable' in different ways and this adds ambiguity to the nomenclature. (Indeed, as noted above, the range here does not technically exist.) Our recommendation is to always be precisely aware of exactly what any given analyst means when he or she refers to the 'range' or to the 'scale' of a semivariogram.

There are some special interpretations of the nugget and sill that deserve clarification, in that they describe contrasting features of the underlying isotropic process. Consider first the nugget. In the broadest sense, $\gamma(\mathbf{0})$ is always zero since it is proportional to $\mathrm{Var}[Z(\mathbf{s}+\mathbf{0}) - Z(\mathbf{s})]$ (Exercise 6.11). In practice, however, one may find that as two spatial locations draw closer and closer together, there is still some covariation evidenced in $Z(\mathbf{s})$. Borrowing terminology from the mining engineers who popularized the technique, the composition or quality of two gold 'nuggets' can differ appreciably even when they are found at essentially the same spatial location (Matheron, 1962). This *nugget effect* can be quantified as $\lim_{h\to 0} \gamma(h) = \gamma_0$. Statisticians recognize this in part to be a problem of natural variation (sometimes called *measurement error*) in repeated sampling; i.e., if one were to take multiple measurements of a spatial process at the same location, the results would likely fluctuate about some underlying central value. This contribution to the nugget is the measurement-error variance, say, $v_{me} \geq 0$.

To explain the variation between two similarly located gold nuggets, however, we need to go further. Technically (and still assuming $d = 2$), if two nuggets are located at *exactly* the same $\mathbf{s}_i$, they should be the same nugget! In Matheron's mining-motivated terminology, however, there is a *micro scale* of spatial variation that lies below the level of detection/sampling most geologists would undertake. That is, as $h \to 0$ there is a certain point below which practical measurements cannot distinguish $h = 0$ from h close to 0. At and below that h, everything is, in effect, equi-located spatially, but the underlying microscale variation still induces variability from observation to observation. We denote this by $v_{ms} \geq 0$, and then recognize that the nugget represents a combination of the two sources of variability: $\gamma_0 = v_{me} + v_{ms}$. (Note, by the way, that either or both sources can be zero. In the latter case, we simply have a continuous (in mean square) random process at zero.) For more on the subtleties of the nugget effect, see Cressie (1993, §2.3.1) or Switzer (2002).

Regarding the sill, the interpretation becomes one of correlation at extreme (rather than microscale) separations. That is, suppose two points in Ω_2 separate farther and farther away from each other such that the spatial correlation between them drops to zero. Of course, this will depend on the true underlying features of the spatial random process, but it may be reasonable in many cases: sooner or later, one might expect the separation, h, between two points to be so large that the response seen in one is essentially unrelated to that seen in the other. If so, we would write $\lim_{h \to \infty} \rho(h) = \lim_{h \to \infty} C(h)/C(0) = 0$, which for $C(0) \neq 0$ implies $\lim_{h \to \infty} C(h) = 0$. But we know $\gamma(h) = C(0) - C(h)$, so if $\lim_{h \to \infty} C(h) = 0$, then $\lim_{h \to \infty} \gamma(h) = C(0)$. This says that in the limit, $\gamma(h)$ attains the constant value $C(0)$. This is just the definition of the sill: the constant level of the semivariogram as $h \to \infty$. Put another way, if the semivariogram of an isotropic, second-order stationary process reaches an asymptote (the sill) as $h \to \infty$, then there is evidence that the spatial correlation between any two points separated by a large distance is essentially zero. One might also ask: how large must h be before this lack of spatial correlation kicks in? The answer is of course the *range* from Definition 6.5(iii). In other words, the range corresponds to the distance beyond which spatial correlation appears inconsequential under an isotropic process.

In passing, we also note that if it exists, the sill may be interpreted as simply $\mathrm{Var}[Z(\mathbf{s})]$. That is, if $C(h) \to 0$ as $h \to \infty$, the sill is $C(0)$. But, as seen above, $C(0) = \mathrm{Cov}[Z(\mathbf{s}), Z(\mathbf{s} + \mathbf{0})] = \mathrm{Cov}[Z(\mathbf{s}), Z(\mathbf{s})] = \mathrm{Var}[Z(\mathbf{s})]$.

6.3.5 Semivariogram models

A number of parametric models exist for use as semivariograms with isotropic processes. We present a few of the more common ones here. These functions depend on possibly unknown parameters, say, $\theta_1, \theta_2, \ldots, \theta_L$, and to emphasize this we write them as $\gamma(h; \boldsymbol{\vartheta})$ for $\boldsymbol{\vartheta} = [\theta_1 \ \theta_2 \ \ldots \ \theta_L]^\mathrm{T}$. Note that all the functions are defined for use over $h \geq 0$.

The simplest semivariogram is linear in h:

$$\gamma(h; \boldsymbol{\vartheta}) = \begin{cases} 0 & \text{if } h = 0, \\ \theta_1 + \theta_2 h & \text{if } h > 0. \end{cases} \tag{6.9}$$

We can also express this simple linear model in terms of an indicator function (§A.2.1): $\gamma(h; \boldsymbol{\vartheta}) = (\theta_1 + \theta_2 h)I_{(0,\infty)}(h)$. A simple extension accepts an integer power for h: $\gamma_p(h; \boldsymbol{\vartheta}) = (\theta_1 + \theta_2 h^p)I_{(0,\infty)}(h)$. We assume $\theta_j \geq 0$ for $j = 1, 2$.

In (6.9), the nugget is θ_1. The sill and range are both undefined, since this semivariogram does not converge to an asymptote as $h \to \infty$.

More useful is a semivariogram model in what is called *spherical* form, given by

$$\gamma(h; \boldsymbol{\vartheta}) = \begin{cases} 0 & \text{if } h = 0, \\ \theta_1 + \theta_2 \left[\dfrac{3h}{2\theta_3} - \dfrac{1}{2}\left(\dfrac{h}{\theta_3}\right)^3 \right] & \text{if } 0 < h < \theta_3, \\ \theta_1 + \theta_2 & \text{if } h \geq \theta_3. \end{cases} \tag{6.10}$$

Here again, we assume all the parameters are non-negative: $\theta_j \geq 0$ for $j = 1, 2, 3$. The nugget under a spherical model is θ_1, the sill is $\theta_1 + \theta_2$, and the range is θ_3 (Exercise 6.14). This is, in fact, the form of model plotted in Fig. 6.7.

Another popular model is an exponential variant of the linear model in (6.9). Take

$$\gamma(h; \boldsymbol{\vartheta}) = \begin{cases} 0 & \text{if } h = 0, \\ \theta_1 + \theta_2[1 - \exp(-h/\theta_3)] & \text{if } h > 0. \end{cases} \qquad (6.11)$$

Once again, we assume $\theta_j \geq 0$ for $j = 1, 2, 3$. The nugget is θ_1 and the sill is $\theta_1 + \theta_2$ (Exercise 6.14). This is a case where the sill is never actually attained by the semivariogram function, but is instead a true asymptote. As a result, we leave the range undefined.

Notice that for analysts who wish to define the range as the point where $\gamma(h)$ attains some majority percentage of its sill, the exponential model presents a complication: such a range will always be a function of θ_3, but never exactly equal it. The best interpretation one can give θ_3 under this model is that it controls the rate at which $\gamma(h)$ approaches $\theta_1 + \theta_2$.

A modification of the exponential model that allows for inflections in the semivariogram as $h \to 0$ is the Gaussian model. Take

$$\gamma(h; \boldsymbol{\vartheta}) = \begin{cases} 0 & \text{if } h = 0, \\ \theta_1 + \theta_2[1 - \exp\{-(h/\theta_3)^2\}] & \text{if } h > 0. \end{cases} \qquad (6.12)$$

Here again, we assume $\theta_j \geq 0$ for $j = 1, 2, 3$. The nugget is θ_1 and the sill is $\theta_1 + \theta_2$ (Exercise 6.14). As in (6.11) the sill is never actually attained, hence the range is undefined. The θ_3 parameter has a similar interpretation as in (6.11).

6.3.6 The empirical semivariogram

In practice it may be the case that the semivariogram $\gamma(h)$ is unknown and must be estimated. We will see an important example of this in §6.4 when we discuss spatial prediction. Anticipating our needs, and for more general use, we consider here some introductory methods for semivariogram estimation.

For simplicity, we suppose that $Z(\mathbf{s})$ is intrinsically stationary (§6.3.1) with constant mean μ. Our goal is to estimate $\gamma(\mathbf{h}) = \frac{1}{2} \text{Var}[Z(\mathbf{s} + \mathbf{h}) - Z(\mathbf{s})]$ for any directional vector $\mathbf{h} \in \Omega_2$. A basic estimator based on the method of moments (§A.4.2) was suggested by Matheron (1963). Known as the *empirical semivariogram*, the estimator has the form

$$\hat{\gamma}(\mathbf{h}) = \frac{1}{2N_{\mathbf{h}}} \sum\sum_{(i,j) \in \mathbb{G}_{\mathbf{h}}} \{Z(\mathbf{s}_i) - Z(\mathbf{s}_j)\}^2, \qquad (6.13)$$

where, similar to the constructions in (6.7) and (6.8), $\mathbb{G}_{\mathbf{h}}$ is the set of all pairs of indices whose corresponding points $(\mathbf{s}_i, \mathbf{s}_j)$ satisfy $\mathbf{s}_i - \mathbf{s}_j = \mathbf{h}$, and $N_{\mathbf{h}}$ is the number of distinct pairs of indices (points) in $\mathbb{G}_{\mathbf{h}}$. As with the Moran and Geary coefficients, the empirical semivariogram is much more difficult to write than it is to describe: for a given value of $\mathbf{h}$, find all the points separated by that $\mathbf{h}$ and calculate half the average squared deviation between them. If we make the further assumption that $Z(\mathbf{s})$ is isotropic, then in (6.13) we can work with just the distance scalar $h = \|\mathbf{h}\|$ in place of $\mathbf{h}$. The aim then is to plot $\hat{\gamma}(h)$ vs. h and/or fit some known form for $\gamma(h)$ to the empirical semivariogram.

As with calculation of (6.7) or (6.8), not every possible value of h can be represented among all the possible pairs of locations (unless the spatial field under study grows extremely dense), so the empirical semivariogram must be limited to a discrete set of (ordered) distances $h_1 < h_2 < \ldots < h_B$. Here again, there may be cases where only a few points are available at a given h, making N_h too small to ensure stable estimation of $\gamma(h)$. This is particularly common when h is large, and so we place an upper bound on h_B: $h_B \leq \frac{1}{2}\max_{i,j}\{\delta_{ij}\}$ (Journel and Huijbregts, 1978). If the sample locations are defined over a regular grid, then the choice for the distance classes should be straightforward. Otherwise, one groups the observed distances δ_{ij} into a set of B classes between 0 and h_B. These are typically equi-spaced with, say, lag distances between their centroids taken as h_B/B. (The *centroid* of a region is the average of all the s_is in that region.) A standard rule of thumb is to choose B such that there are at least 30 distances in each class from which to compute $\hat{\gamma}(h)$. We also recommend B be at least 10; that is, set $B \geq 10$ such that $N_h \geq 30$.

Comparison of Geary's $c(h)$ in (6.8) with (6.13) under isotropy shows that the two are scaled versions of each other: $\hat{\gamma}(h) = S_Z^2 c(h)$, where $S_Z^2 = \sum_{i=1}^{n}\{Z(s_i) - \bar{Z}\}^2/(n-1)$ is the sample variance of all responses in the sample. This relationship allows us to use software programs that generate empirical semivariograms to find values of Geary's c. (In SAS, for example, the empirical semivariogram is available in PROC VARIOGRAM, as we illustrate in the next example.) Indeed, a plot of $c(h)$ vs. h provides essentially the same graphic information as a plot of $\hat{\gamma}(h)$ vs. h. Because of its usefulness for spatial prediction (§6.4), however, we emphasize the latter here.

Example 6.5 (Grass shrimp biomass) Porter *et al.* (1997) studied the effects of human development and urbanization on coastal ecosystems along the US Atlantic coast. One site of particular interest is Murrells Inlet, South Carolina, where encroaching urbanization endangers local salt-marsh estuaries. Among other outcomes, the investigators collected data on biomass of adult grass shrimp (*Palaemonetes pugio*). The shrimp comprise a large percentage of the faunal biomass in the marsh, and hence represent an important indicator for studying the ecosystem's health.

Sampled at $n = 58$ spatial locations in the estuary, data were collected as base-10 logarithms, $Z(s_i)$, of present biomass (grams per 25 m of stream reach) at each site s_i. A selection of the data appears in Table 6.2 (see http://www.wiley.com/go/environmental for the full data set). A three-dimensional scatterplot of the data (Exercise 6.16) shows some spatial variability in the $\log_{10}$-biomass, although no blatant spatial trend is evident (hence we apply no detrending measures).

Assuming the spatial process underlying the grass shrimp biomass is isotopic, we study the empirical semivariogram $\hat{\gamma}(h)$ for these data using (6.13). We start with determination of the set of h values. Figure 6.8 gives sample SAS code for this, using PROC VARIOGRAM. The process takes multiple steps, all available via SAS coding. First, we use PROC VARIOGRAM to calculate all possible distances between pairs of observations, $h_{ij} = \delta_{ij}$. Key features of the procedure call in Fig. 6.8 include (i) the `coordinates` statement that defines the horizontal (xc=) and vertical (yc=) coordinates of each s_i, and (ii) the `compute` statement, here with the `novariogram` option to cancel actual computation of the variogram. The procedure now only deposits the pairwise distances δ_{ij} into the output data set named by `outpair=`. The subsequent call to PROC UNIVARIATE in Fig. 6.8 calculates summary statistics for the $n(n-1)/2 = 1653$ pairwise distances δ_{ij} in the output data set `Findmax`. In particular,

Table 6.2 Selected base-10 logarithms of adult grass shrimp biomass in Murrells Inlet, SC. X-location and Y-location are in meters from an arbitrary origin

X-location	Y-location	$\log_{10}$-biomass
680070.500	3710087.00	1.853
680121.125	3710069.25	2.370
680114.438	3710861.00	0.758
⋮	⋮	⋮
685456.812	3716845.50	0.393
685472.938	3716865.25	0.646
685488.688	3716885.75	0.338

Source: Porter *et al.* (1997).

```
data ppugio;
  input xloc yloc zlogmas @@;
  datalines;
      680070.500 3710087.00 1.853
      680095.312 3710078.50 2.472
          ⋮          ⋮
      685182.438 3717120.50 1.925

* find all pairwise distances;
  proc variogram outpair=Findmax;
     var zlogmas;
     coordinates xc=xloc yc=yloc;
     compute novariogram;

* find max via UNIVARIATE;
  proc univariate data=Findmax;
     var distance;
```

Figure 6.8 Sample SAS program to determine summary statistics, including maximum separation distance, with grass shrimp biomass data

we wish to determine $\max_{i,j}\{\delta_{ij}\}$, which when divided by 2 yields h_B. From the resulting output (not shown) we find $\max_{i,j}\{\delta_{ij}\} = 8744.882$ m, so we take $h_B = 4372.441$ m.

Through some experimentation (using, for example, PROC VARIOGRAM with its `outdist=` option), we found that $B = 18$ produced distance classes whose N_h values were all at least 34. Hence using $B = 18$, we carry out another call to PROC VARIOGRAM, now to estimate the semivariogram using (6.13). The pertinent SAS commands are

```
proc variogram  data = ppugio  outvar = Gamma;
  var zlogmas;
  coordinates  xc = xloc  yc = yloc;
  compute  lagd = 242.9134  maxl = 18 ;
proc print;
  var  lag  distance  count  variog;
```

```
              The SAS System
   LAG      DISTANCE    COUNT      VARIOG
   -1          .          58         .
    0        33.38        50       0.13375
    1       282.90        40       0.20994
    2       479.35        63       0.44335
    3       772.16        60       0.40830
    4       971.49        84       0.34379
    5      1218.28        92       0.26436
    6      1459.12        71       0.55859
    7      1732.51        71       0.44898
    8      1929.25        46       0.66200
    9      2181.88        62       0.55099
   10      2452.57        88       0.40723
   11      2685.47        58       0.24229
   12      2901.36        53       0.46647
   13      3175.33        59       0.78379
   14      3369.29        62       0.43333
   15      3651.20        53       0.57914
   16      3903.66        50       0.34339
   17      4124.09        46       0.51669
   18      4396.77        34       0.49483
```

Figure 6.9 SAS output (edited) showing empirical semivariogram values for grass shrimp biomass data

Note the important difference between this new code and that in Fig. 6.8: the compute statement now calls for formal semivariogram calculation. The $\texttt{maxl} = 18$ option tells **PROC VARIOGRAM** to use $B = 18$ distance classes, while the $\texttt{lagd} = 242.9134$ option sets the distance between class centers to $h_{18}/18 = 4372.441/18 = 242.9134$. The resulting output (edited) appears in Fig. 6.9, where the column under LAG indexes the distance class ($\texttt{LAG} = -1$ corresponds to a preliminary computation **PROC VARIOGRAM** uses to build the semivariogram), DISTANCE is the average h for all the distances in that distance class (notice that these need not be half-multiples of h_B/B, even though the class is centered at such half-multiples), COUNT is N_h, and VARIOG is $\hat{\gamma}(h)$ from (6.13). (To find Geary's $c(h)$ over this same set of lag classes, simply divide VARIOG by $S_Z^2 = 0.5005$.)

Figure 6.10 plots the $\hat{\gamma}(h)$ values from Fig. 6.9 against distance, h. The graph shows a scattered semivariogram pattern, with a possible nugget effect evidenced by the limit of $\gamma_0 \approx 0.1$ as $h \to 0$. The semivariogram rises with h until a possible sill is reached near $\gamma \approx 0.5$. This corresponds to a possible range of between 400 and 700 m; that is, spatial correlation in the data appears to fade after observations are separated by approximately 400–700 m.

The empirical semivariogram in (6.13) can be regarded as a nonparametric estimator, since it makes no assumptions on the parametric form of $\gamma(h)$. (We discuss parametric estimation of γ in §6.4.4, below.) Nonetheless, it can suffer from instabilities when extreme outliers exist in the data; to remedy this, Cressie and Hawkins (1980) describe a robust version of (6.13):

$$\bar{\gamma}(\mathbf{h}) = \frac{1}{2} \frac{\left\{ N_{\mathbf{h}}^{-1} \sum \sum_{(i,j) \in G_{\mathbf{h}}} \sqrt{Z(\mathbf{s}_i) - Z(\mathbf{s}_j)} \right\}^4}{0.457 + \left(0.494 N_{\mathbf{h}}^{-1} \right)}. \tag{6.14}$$

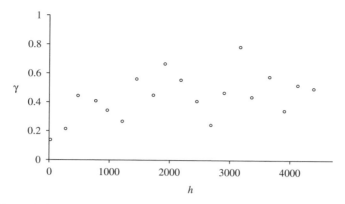

Figure 6.10 Empirical semivariogram for grass shrimp biomass data

Plotting or otherwise employing $\bar{\gamma}(\mathbf{h})$ instead of $\hat{\gamma}(\mathbf{h})$ may be appropriate when robust estimators are desired; see also Cressie (1993, §2.4.3) and Genton (1998).

To calculate the robust empirical semivariogram using **PROC VARIOGRAM**, include the `robust` option in the `compute` statement. The consequent values of (6.14) are stored in the variable `rvario` by the `outvar=` option.

Example 6.6 (Grass shrimp biomass, cont'd)   Returning to the grass shrimp biomass data from Example 6.5, suppose we wish to calculate a robust version of the empirical semivariogram. Sample code using **PROC VARIOGRAM** appears in Fig. 6.11 (output not shown). The resulting values of (6.14) are plotted in Fig. 6.12, along with the previous values found via (6.13).

In Fig. 6.12, we see that the robust empirical semivariogram (6.14) often agrees with its standard cousin (6.13), although for some distance classes the two quantities differ widely. This suggests that there may be selected outliers in the log-transformed data affecting the variogram estimates. The nugget drops to $\gamma_0 \approx 0.05$ as $h \to 0$, while the sill is now more difficult to gauge; a rough value of $\gamma \approx 0.4$ seems reasonable. The corresponding range appears to be between 400 and 500 m.   ☯

For more details on estimation of spatial covariance and the variogram, readers may consult sources such as Cressie (2002b), García-Soidán *et al.* (2004), and the references therein.

```
proc variogram data=ppugio outvar=Gamma;
   var zlogmas;
   coordinates xc=xloc yc=yloc;
   compute lagd=242.9134   maxl=18  robust ;

proc print;
   var lag distance count variog rvario;
```

Figure 6.11 Sample SAS code to find robust empirical semivariogram (6.14) with grass shrimp biomass data

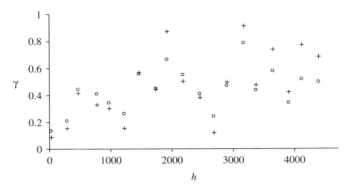

Figure 6.12 Robust (+) and standard (∘) empirical semivariograms for grass shrimp biomass data

6.4 Spatial prediction

In studies that generate spatial data, an important goal is prediction of a future measurement at a new location in the same spatial region. This is common when resources do not permit a full spatial census of the region. Examples include forestry and fishery surveys, geological/mining measurements in ore fields, or air pollution studies in large cities or towns.

To establish a methodology for performing spatial prediction, we continue with the notation of §6.3 and let $Z(s_1), Z(s_2), \ldots, Z(s_n)$ be measured, continuous observations over a sample of spatial sites s_i, $i = 1, \ldots, n$, in a predefined study region Ω_2. We retain the assumption that the data are generated under model (6.6) such that $E[\varepsilon(s_i)] = 0$ and $Cov[\varepsilon(s_i), \varepsilon(s_j)]$ can be described by a covariance function $C(s_i, s_j)$. Our goal is to use information in the $Z(s_i)$s to make predictions of some future or otherwise-unobserved outcome, $Z(s_0)$, at some known spatial location, s_0, in Ω_2. Note that this goal is interpolative rather than extrapolative: we typically wish to predict $Z(s)$ at a location s_0 that lies among the previously sampled locations. The warnings from §1.1 against extrapolating predictions in a linear model also apply to extrapolated predictions in a spatial model.

We formalize the concept of optimal spatial prediction in the following definition.

Definition 6.6 A predictor, $\hat{Z}(s_0)$, of some unobserved outcome $Z(s_0)$ is a *best linear unbiased predictor* if the following conditions hold:

(i) $\hat{Z}(s_0)$ takes the form

$$\kappa_0 + \sum_{i=1}^{n} \kappa_i Z(s_i) \tag{6.15}$$

for some set of known coefficients $\kappa_0, \kappa_1, \ldots, \kappa_n$.

(ii) $\hat{Z}(s_0)$ is unbiased in a stochastic sense, that is, $E[\hat{Z}(s_0) - Z(s_0)] = 0$.

(iii) $\text{Var}[\hat{Z}(s_0) - Z(s_0)]$ is a minimum among all estimators $\hat{Z}(s_0)$ satisfying (i) and (ii).

In Definition 6.6, condition (i) provides that $\hat{Z}(s_0)$ is a *linear* predictor, condition (ii) provides that it is *unbiased*, while condition (iii) provides that it is *best* in the sense of having minimum variance among linear unbiased predictors. Note that many authors replace condition (iii) with

(iii') $E[\{\hat{Z}(s_0) - Z(s_0)\}^2]$ is a minimum among all estimators $\hat{Z}(s_0)$ satisfying (i) and (ii).

We call $E[\{\hat{Z}(s_0) - Z(s_0)\}^2]$ the *mean squared prediction error*, and hence (iii') requires us to achieve minimum mean squared prediction error. The two conditions are equivalent, however, whenever condition (ii) holds; see Exercise 6.17.

The BLUP is similar to the BLUE mentioned in §5.6.2. It assures an investigator that the prediction being made at a particular s_0 lacks any stochastic bias and, within this context, is 'best' in the sense of attaining minimum variance (Robinson, 1991). By employing the linear form $\hat{Z}(s_0) = \kappa_0 + \sum_{i=1}^{n} \kappa_i Z(s_i)$ we also ensure that the data enter into the calculation in an easily assimilated fashion.

The issue of *prediction* is more intricate than that of simple estimation in a statistical sense. One possible goal of a spatial analysis would be to estimate $\mu(s)$, the wholly deterministic features of the spatial process under study (often called a *population-level analysis*). With prediction, however, we are asking the more complex question of what the actual value of an individual $Z(s)$ would be at some specific $s = s_0$. As a result, along with the deterministic contribution in $Z(s_0)$ due to $\mu(s_0)$ we must also incorporate the stochastic contribution due to $\varepsilon(s_0)$. A useful analogy exists here with the distinction between estimating the underlying mean response and predicting a new observation in the usual linear regression setting, as in Example 4.17.

Spatial prediction under such conditions has a colorful history. The standard term in modern parlance is *kriging*, due to Matheron (1962, Ch. 8). He intended the term to honor a pioneer of the method, D. G. Krige, who first described the concept for a geological mining application (Krige, 1951). As noted above, the geological connection has since stuck. (In correspondence with N.A. Cressie, however, Matheron noted that P. Carlier coined the French usage *krigeage* as early as the late 1950s. Indeed, Cressie, 1990, argues that along with the Soviet scientist L.S. Gandin, 1963, Matheron made perhaps the greatest contributions to spatial prediction for environmental data – what most now call kriging – in the mid-twenteeth century.) As a result, many authors refer broadly to the method as *geostatistics* and to spatial measurements arising within this context as *geostatistical data*, even if the observations have nothing to do with geology or mining. Cressie (1990) also notes that ideas behind minimum variance unbiased prediction with spatial data go back at least as far as Wold's work in the 1930s (Wold, 1938).

In the remaining sections, we give a simple introduction to the use of kriging for spatial prediction. We start with the simplest setting: known and constant mean response.

6.4.1 Simple kriging

Suppose we require in model (6.6) that $Z(\mathbf{s}_i)$ be mean stationary, as in Definition 6.2; that is,

$$Z(\mathbf{s}_i) = \mu + \varepsilon(\mathbf{s}_i), \qquad (6.16)$$

$i = 1, \ldots, n$. Suppose also that μ is *known*. This allows us to use μ in the selection of the κ_is. In particular, if we set $\kappa_0 = \mu(1 - \sum_{i=1}^{n} \kappa_i)$ in (6.15), we find that condition (ii) of Definition 6.6 is satisfied for *any* selection of the other κ_is; see Exercise 6.18.

To find the remaining κ_is, we turn to condition (iii). Our goal is to minimize $\mathrm{Var}[\hat{Z}(\mathbf{s}_0) - Z(\mathbf{s}_0)]$. From §A.1, we know this can be written as $\mathrm{Var}[\hat{Z}(\mathbf{s}_0) - Z(\mathbf{s}_0)] = \mathrm{Var}[\hat{Z}(\mathbf{s}_0)] + \mathrm{Var}[Z(\mathbf{s}_0)] - 2\mathrm{Cov}[\hat{Z}(\mathbf{s}_0), Z(\mathbf{s}_0)]$. For $\mathrm{Var}[\hat{Z}(\mathbf{s}_0)]$, recall from (6.15) that $\hat{Z}(\mathbf{s}_0)$ adds the constant κ_0 to a linear combination of random variables, $\sum_{i=1}^{n} \kappa_i Y_i$. But, also from §A.1, adding a constant to a random quantity does not affect its variance. Thus we have

$$\mathrm{Var}[\hat{Z}(\mathbf{s}_0)] = \mathrm{Var}\left[\kappa_0 + \sum_{i=1}^{n} \kappa_i Z(\mathbf{s}_i)\right] = \mathrm{Var}\left[\sum_{i=1}^{n} \kappa_i Z(\mathbf{s}_i)\right]$$

$$= \sum_{i=1}^{n} \kappa_i^2 \mathrm{Var}[Z(\mathbf{s}_i)] + 2\sum_{i=1}^{n-1}\sum_{j=i+1}^{n} \kappa_i \kappa_j \mathrm{Cov}[Z(\mathbf{s}_i), Z(\mathbf{s}_j)],$$

where the latter equality obtains from appeal to equation (A.6).

For the next term in $\mathrm{Var}[\hat{Z}(\mathbf{s}_0) - Z(\mathbf{s}_0)]$, $\mathrm{Var}[Z(\mathbf{s}_0)]$, we make no simplifications. Lastly, for $\mathrm{Cov}[\hat{Z}(\mathbf{s}_0), Z(\mathbf{s}_0)]$ we again apply (6.15) and write $\mathrm{Cov}[\hat{Z}(\mathbf{s}_0), Z(\mathbf{s}_0)] = \mathrm{Cov}[\kappa_0 + \sum_{i=1}^{n} \kappa_i Z(\mathbf{s}_i), Z(\mathbf{s}_0)]$. Using the definition of covariance, it can be shown that

$$\mathrm{Cov}\left[\kappa_0 + \sum_{i=1}^{n} \kappa_i Z(\mathbf{s}_i), Z(\mathbf{s}_0)\right] = \mathrm{Cov}\left[\sum_{i=1}^{n} \kappa_i Z(\mathbf{s}_i), Z(\mathbf{s}_0)\right] = \sum_{i=1}^{n} \kappa_i \mathrm{Cov}[Z(\mathbf{s}_i), Z(\mathbf{s}_0)]$$

(cf. Exercise 5.25). As a result, $\mathrm{Var}[\hat{Z}(\mathbf{s}_0) - Z(\mathbf{s}_0)]$ may be written as

$$\sum_{i=1}^{n} \kappa_i^2 \mathrm{Var}[Z(\mathbf{s}_i)] + 2\sum_{i=1}^{n-1}\sum_{j=i+1}^{n} \kappa_i \kappa_j \mathrm{Cov}[Z(\mathbf{s}_i), Z(\mathbf{s}_j)] + \mathrm{Var}[Z(\mathbf{s}_0)]$$

$$- 2\sum_{i=1}^{n} \kappa_i \mathrm{Cov}[Z(\mathbf{s}_i), Z(\mathbf{s}_0)].$$

To simplify this expression for $\mathrm{Var}[\hat{Z}(\mathbf{s}_0) - Z(\mathbf{s}_0)]$, denote $\mathrm{Cov}[Z(\mathbf{s}_i), Z(\mathbf{s}_0)]$ by σ_{i0} and $\mathrm{Cov}[Z(\mathbf{s}_i), Z(\mathbf{s}_j)]$ by σ_{ij}. To conform with this notation, we also take $\mathrm{Var}[Z(\mathbf{s}_i)]$ as σ_{ii}. Assuming that the σ_{ij} values are known, we can then write $\mathrm{Var}[\hat{Z}(\mathbf{s}_0) - Z(\mathbf{s}_0)]$ as

$$\mathrm{Var}[\hat{Z}(\mathbf{s}_0) - Z(\mathbf{s}_0)] = \mathrm{Var}[Z(\mathbf{s}_0)] + \sum_{i=1}^{n} \kappa_i^2 \sigma_{ii} + 2\sum_{i=1}^{n-1}\sum_{j=i+1}^{n} \kappa_i \kappa_j \sigma_{ij} - 2\sum_{i=1}^{n} \kappa_i \sigma_{i0}.$$

$$(6.17)$$

Thus we can restate our goal as follows: determine if constants $\kappa_1, \ldots, \kappa_n$ exist such that (6.17) is a minimum. Application of differential calculus (Exercise 6.19) leads to the system of n equations

$$-\sigma_{i0} + \sum_{j=1}^{n} \kappa_j \sigma_{ij} = 0 \qquad (i = 1, \ldots, n). \tag{6.18}$$

Thus to achieve minimum mean squared error of prediction, we solve (6.18) for each κ_i. Assuming the σ_{ij}s do not cause any numerical inconsistencies, this system has n equations and n unknowns and so can be solved in a straightforward fashion using the notation of matrix algebra (seen previously in our discussion of regression models in Chapter 1). Here, let $\boldsymbol{\kappa} = [\kappa_1 \ldots \kappa_n]^{\mathrm{T}}$ be the vector of kriging coefficients, $\mathbf{c}_0 = [\sigma_{10} \ldots \sigma_{n0}]^{\mathrm{T}}$ be the vector of (known) covariance terms $\mathrm{Cov}[Z(\mathbf{s}_i), Z(\mathbf{s}_0)]$, and collect the remaining covariance terms $\sigma_{ij} = \mathrm{Cov}[Z(\mathbf{s}_i), Z(\mathbf{s}_j)]$ into the variance–covariance matrix

$$\mathbf{V} = \begin{bmatrix} \sigma_{11} & \sigma_{12} & \cdots & \sigma_{1n} \\ \sigma_{12} & \sigma_{22} & \cdots & \sigma_{2n} \\ \vdots & \vdots & \ddots & \vdots \\ \sigma_{1n} & \sigma_{2n} & \cdots & \sigma_{nn} \end{bmatrix};$$

cf. §A.2.12. Recall that since $\mathrm{Cov}[Z(\mathbf{s}_i), Z(\mathbf{s}_j)] = \mathrm{Cov}[Z(\mathbf{s}_j), Z(\mathbf{s}_i)]$ for all $i \neq j$, we have $\sigma_{ij} = \sigma_{ji}$ and so $\mathbf{V}$ is a symmetric matrix. Using these vectors and matrices, it is possible to write the system in (6.18) as a matrix equation: $-\mathbf{c}_0 + \mathbf{V}\boldsymbol{\kappa} = \mathbf{0}$, where $\mathbf{0}$ is a column vector of zeros. This simplifies into the matrix equality $\mathbf{V}\boldsymbol{\kappa} = \mathbf{c}_0$. To solve for $\boldsymbol{\kappa}$, premultiply both components of this equation by $\mathbf{V}^{-1}$, where $\mathbf{V}^{-1}$ is the inverse of the matrix $\mathbf{V}$; see §A.4.3. This gives $\mathbf{V}^{-1}\mathbf{V}\boldsymbol{\kappa} = \mathbf{V}^{-1}\mathbf{c}_0$, or simply

$$\boldsymbol{\kappa} = \mathbf{V}^{-1}\mathbf{c}_0. \tag{6.19}$$

Due to the complexity of the matrix operations, computation of (6.19) typically proceeds by computer. We use the resulting values of $\kappa_1, \ldots, \kappa_n$ along with $\kappa_0 = \mu(1 - \sum_{i=1}^{n} \kappa_i)$ to construct the BLUP $\hat{Z}(\mathbf{s}_0) = \kappa_0 + \sum_{i=1}^{n} \kappa_i Z(\mathbf{s}_i)$. Matheron (1971) called (6.19) the solution to the *simple kriging* problem, ostensibly because the problem with known μ is relatively simple. This also corresponds to a solution discussed by Wold (1938).

Some comments:

(a) We have not made any strict distributional assumptions about the nature of the errors $\varepsilon(\mathbf{s}_i)$, other than that they have zero means and a known covariance matrix. Thus this BLUP analysis is distribution-free. It is not uncommon to further take the $\varepsilon(\mathbf{s}_i)$s as normally distributed, and we remark on this in §6.4.6.

(b) The minimum value of the mean squared prediction error attained under (6.19) is $E[\{\hat{Z}(\mathbf{s}_0) - Z(\mathbf{s}_0)\}^2] = \mathrm{Var}[Z(\mathbf{s}_0)] - \sum_{i=1}^{n} \sum_{j=1}^{n} \sigma_{i0} \sigma_{j0} \sigma^{(ij)}$, where $\sigma^{(ij)}$ is shorthand notation for the elements of the inverse matrix $\mathbf{V}^{-1}$. In matrix notation, this can be written more compactly as $E[\{\hat{Z}(\mathbf{s}_0) - Z(\mathbf{s}_0)\}^2] = \text{Var}[Z(\mathbf{s}_0)] - \mathbf{c}_0^T \mathbf{V}^{-1} \mathbf{c}_0$; see Exercise 6.20.

(c) Simple kriging can also be extended to the case where μ is allowed to vary with $\mathbf{s}$, that is, to model (6.6) with known spatial function $\mu(\mathbf{s})$. See Cressie (1993, pp. 109–110) for details.

6.4.2 Ordinary kriging

A practical concern regarding the use of simple kriging in §6.4.1 is that it requires known μ. For the more realistic case of unknown μ, we turn to what is called *ordinary kriging*. We continue to assume that (6.16) holds, but now allow μ to be unknown. Thus part of our effort must involve estimation of the unknown mean response.

We also assume that $Z(\mathbf{s})$ is a second-order stationary process, with known semivariogram $\gamma(\mathbf{h}) = \frac{1}{2}\text{Var}[Z(\mathbf{s}+\mathbf{h}) - Z(\mathbf{s})]$. We work with expressions of the form $\gamma(\mathbf{s}_i - \mathbf{s}_j)$, where $\mathbf{s}_i$ and $\mathbf{s}_j$ are the ith and jth locations at which observations of $Z(\mathbf{s})$ have been recorded ($i \neq j$).

It will be useful to constrain the BLUP slightly, by forcing $\kappa_0 = 0$. Thus we modify (6.15) into

$$\hat{Z}(\mathbf{s}_0) = \sum_{i=1}^{n} \kappa_i Z(\mathbf{s}_i) \tag{6.20}$$

for some set of known coefficients $\kappa_1, \ldots, \kappa_n$. Here again, our goal is to find a BLUP for $Z(\mathbf{s}_0)$ by determining expressions for the κ_is. We begin by exploring the restriction in Definition 6.6(ii) that the BLUP be stochastically unbiased: $E[\hat{Z}(\mathbf{s}_0) - Z(\mathbf{s}_0)] = 0$. Under (6.20), this is equivalent to $E[\sum_{i=1}^{n} \kappa_i Z(\mathbf{s}_i)] = E[Z(\mathbf{s}_0)]$, which simplifies to $\sum_{i=1}^{n} \kappa_i E[Z(\mathbf{s}_i)] = \mu$. But $E[Z(\mathbf{s}_i)]$ also equals μ, so we find $\sum_{i=1}^{n} \kappa_i \mu = \mu \sum_{i=1}^{n} \kappa_i = \mu$; that is, for $\mu \neq 0$,

$$\sum_{i=1}^{n} \kappa_i = 1. \tag{6.21}$$

Notice that any estimator satisfying (6.21) will by constraint be a weighted average of the $Z(\mathbf{s}_i)$s, that is, a weighted average of the observed spatial process at the sampled locations $\mathbf{s}_i$.

Our goal is to minimize the mean squared prediction error $E[\{\hat{Z}(\mathbf{s}_0) - Z(\mathbf{s}_0)\}^2] = E\{[\sum_{i=1}^{n} \kappa_i Z(\mathbf{s}_i) - Z(\mathbf{s}_0)]^2\}$, subject now to the constraint $\sum_{i=1}^{n} \kappa_i = 1$ from (6.21). This is known as a *constrained minimization* (or, more generally, a form of *constrained optimization*). Many approaches are available to perform constrained optimizations; kriging is usually undertaken via what is known as the method of Lagrange multipliers (Kotz *et al.*, 1983). (The approach requires multivariable calculus, and the derivation extends beyond the scope of our presentation. Advanced readers may explore its application to ordinary kriging in Cressie, 1993, §3.2, or Barnett,

2004, §11.2.) The resulting solution equations for the κ_is can be written in the (open) form

$$\gamma(s_0 - s_i) - \sum_{j=1}^{n} \kappa_j \gamma(s_i - s_j) + \frac{1 - \sum_{i=1}^{n} \sum_{j=1}^{n} \gamma^{(ij)} \gamma(s_0 - s_j)}{\sum_{i=1}^{n} \sum_{j=1}^{n} \gamma^{(ij)}} = 0 \qquad (6.22)$$

for all $i = 1, \ldots, n$. In (6.22), the quantities $\gamma^{(ij)}$ are the (i, j)th elements of the inverse of the matrix of observed semivariogram values $\gamma(s_i - s_j)$. We denote the original matrix as

$$\Gamma = \begin{bmatrix} \gamma(s_1 - s_1) & \gamma(s_1 - s_2) & \cdots & \gamma(s_1 - s_n) \\ \gamma(s_2 - s_1) & \gamma(s_2 - s_2) & \cdots & \gamma(s_2 - s_n) \\ \vdots & \vdots & \ddots & \vdots \\ \gamma(s_n - s_1) & \gamma(s_n - s_2) & \cdots & \gamma(s_n - s_n) \end{bmatrix} \qquad (6.23)$$

and take its inverse as Γ^{-1}.

Unfortunately, (6.22) does not give a closed-form expression for each κ_i. A closed-form matrix equation can be formed, however, by re-expressing the estimating equations via matrix quantities such as Γ. As in §6.4.1, let $\kappa = [\kappa_1 \ldots \kappa_n]^T$ be the vector of kriging coefficients, and now denote by γ_0 a special vector of semivariogram terms: $\gamma_0 = [\gamma(s_0 - s_1) \ldots \gamma(s_0 - s_n)]^T$. Then the kriging coefficients may be written in vector notation as

$$\kappa = \Gamma^{-1} \left\{ \gamma_0 - J \left(\frac{1 - \sum_{i=1}^{n} \sum_{j=1}^{n} \gamma^{(ij)} \gamma(s_0 - s_j)}{\sum_{i=1}^{n} \sum_{j=1}^{n} \gamma^{(ij)}} \right) \right\}, \qquad (6.24)$$

where J is a column vector made up exclusively of ones: $J = [1 \ 1 \ldots 1]^T$. As in equation (6.19), these kriging equations for the optimal κ_is require computer implementation.

Cressie (1993, §3.2) gives the corresponding minimized value of the mean squared prediction error as

$$\sigma^2 \{\hat{Z}(s_0)\} = 2 \sum_{i=1}^{n} \kappa_i \gamma(s_0 - s_i) - \sum_{i=1}^{n} \sum_{j=1}^{n} \kappa_i \kappa_j \gamma(s_i - s_j),$$

also called the *prediction variance* of $\hat{Z}(s_0)$. With this, an approximate (pointwise) $1 - \alpha$ prediction interval for $Z(s_0)$ is $\hat{Z}(s_0) \pm z_{\alpha/2} \sigma\{\hat{Z}(s_0)\}$, where $z_{\alpha/2}$ is the upper-$(\alpha/2)$ standard normal critical point.

6.4.3 Universal kriging

A natural extension to the mean-stationary kriging model in (6.16) is to allow μ to vary with s. That is, we return to the general model in (6.6) where both the deterministic term $\mu(\mathbf{s}_i)$ and the stochastic term $\varepsilon(\mathbf{s}_i)$ depend on the spatial location at which $Z(\mathbf{s}_i)$ is recorded. We assume the stochastic errors are second-order stationary with zero mean, $E[\varepsilon(\mathbf{s}_i)] = 0$, and known semivariogram $\gamma(\mathbf{h}) = \frac{1}{2}\text{Var}[Z(\mathbf{s} + \mathbf{h}) - Z(\mathbf{s})]$. As above, we work with expressions of the form $\gamma(\mathbf{s}_i - \mathbf{s}_j)$, where $\mathbf{s}_i$ and $\mathbf{s}_j$ are the ith and jth locations at which observations have been recorded ($i \neq j$). Where necessary, we collect these values into a matrix Γ as in (6.23).

If the deterministic effect on $\mu(\mathbf{s})$ is segmented into, say, P separate plateaus, each with a constant mean, then it might be natural to *stratify* the analysis into P sets of ordinary kriging calculations over each plateau. (This would not be uncommon in mining applications, where homogeneous rock strata often separate due to tectonic or other geological forces.) This effort might involve a variant of *local kriging* (Dillon, 1992), which is a way to restrict the spatial predictor to a specified region around each observation. Or one might be aware of other sources that separate responses in the plane into known sources of variation. It is natural then to consider some form of *block kriging* (Barnett, 2004, §11.2.3) to account for these sources.

More generally, however, we suppose that $\mu(\mathbf{s})$ is related to a set of known spatial predictor functions $u_1(\mathbf{s}_i), \ldots, u_p(\mathbf{s}_i)$ via the linear form

$$\mu(\mathbf{s}_i) = \sum_{j=1}^{P} \beta_j u_j(\mathbf{s}_i), \tag{6.25}$$

for each $\mathbf{s}_i$. For example, $u_j(\mathbf{s}_i)$ might be site elevation at $\mathbf{s}_i$ and $u_{j+1}(\mathbf{s}_i)$ might be soil acidity measured at $\mathbf{s}_i$, etc. Often, the linear model will contain an 'intercept' term, β_1, corresponding to $u_1(\mathbf{s}_i) = 1$ for all $i = 1, \ldots, n$. In what follows, we operate under this assumption and refer to β_1 as the intercept parameter.

As above, interest exists in predicting some future value of $Z(\mathbf{s})$ at the spatial location $\mathbf{s}_0$. This is known as *universal kriging*. We assume that $\mu(\mathbf{s}_0)$ also satisfies the linear model assumption: $\mu(\mathbf{s}_0) = \sum_{j=1}^{P} \beta_j u_j(\mathbf{s}_0)$. The BLUP, $\hat{Z}(\mathbf{s}_0)$, is based on the conditions given in Definition 6.6. To satisfy the unbiasedness requirement of Definition 6.6(ii), we set $E[\hat{Z}(\mathbf{s}_0) - Z(\mathbf{s}_0)] = 0$. One can show (Exercise 6.21) that under (6.25) this simplifies to $E[\hat{Z}(\mathbf{s}_0)] = \sum_{j=1}^{P} \beta_j u_j(\mathbf{s}_0)$.

Cressie (1993, §3.4) indicates that for a linear predictor $\hat{Z}(\mathbf{s}_0)$ of the form in (6.20), a necessary and sufficient condition for $E[\hat{Z}(\mathbf{s}_0)]$ to equal $\sum_{j=1}^{P} \beta_j u_j(\mathbf{s}_i)$ is

$$u_j(\mathbf{s}_0) = \sum_{i=1}^{n} \kappa_i u_j(\mathbf{s}_i), \tag{6.26}$$

for all $j = 1, \ldots, P$. This becomes our unbiasedness restriction on the κ_is. Under it, as per Definition 6.6(iii), our goal is to minimize the mean squared prediction error $E[\{\hat{Z}(\mathbf{s}_0) - Z(\mathbf{s}_0)\}^2]$. This is again a constrained minimization, here with $P \geq 1$ constraints via (6.26). The calculations again employ the method of Lagrange multipliers, only now the multiple constraints require us to apply it in a P-variate manner. The resulting equations for the κ_is can be written in matrix form, although the P-variate constraints increase the complexity over that in (6.24):

$$\kappa = \Gamma^{-1}\left\{\gamma_0 - U(U^T\Gamma^{-1}U)^{-1}(u_0 - U^T\Gamma^{-1}\gamma_0)\right\} \tag{6.27}$$

(Cressie, 1993, §3.4). In (6.27), U is a matrix whose (i, j)th element is $u_j(s_i)$, and u_0 is a column vector of the 'future' explanatory variates $[u_1(s_0)\; u_2(s_0)\ldots u_P(s_0)]^T$. The remaining terms are the same as those in §6.4.2. Once again, the complexity of these equations drives us to the computer for acquisition of the κ_is.

Cressie (1993, §3.4) also gives the corresponding mean squared prediction error as

$$\sigma^2\{\hat{Z}(s_0)\} = 2\sum_{i=1}^{n}\kappa_i\gamma(s_0 - s_i) - \sum_{i=1}^{n}\sum_{j=1}^{n}\kappa_i\kappa_j\gamma(s_i - s_j).$$

With this, an approximate (pointwise) $1 - \alpha$ prediction interval for $Z(s_0)$ is $\hat{Z}(s_0) \pm z_{\alpha/2}\sigma\{\hat{Z}(s_0)\}$, where $z_{\alpha/2}$ is the upper-$(\alpha/2)$ standard normal critical point.

6.4.4 Unknown γ

In practice, the semivariogram of the stochastic process $Z(s)$ is often unknown and hence must be estimated before it is used in the kriging equations. Even if the semivariogram could be assumed to take a specific form – e.g., as in §6.3.5 – the parameters of this functional form may still be unknown. To identify them and to estimate $\gamma(h)$ we turn to the empirical semivariogram from §6.3.6. We start by assuming an empirical semivariogram (6.13) has been fitted to spatial data, producing the values $\hat{\gamma}(h_b), b = 1,\ldots, B$. If desired, robust empirical semivariogram points, $\bar{\gamma}(h_b)$ from (6.14), may be used instead.

At its simplest level, such an effort would proceed using ordinary least squares (§A.4.1): given a postulated semivariogram function $\gamma(h; \vartheta)$, find $\hat{\vartheta}$ to minimize the objective quantity $D = \sum_{b=1}^{B}\{\hat{\gamma}(h_b) - \gamma(h_b; \hat{\vartheta})\}^2$. Use of ordinary least squares here does not incorporate the correlations between and heterogeneous variance of the values of $\hat{\gamma}(h_b)$, however, and hence can perform poorly. Cressie (2002b) suggests instead adjusting for this effect via weighted least squares (§A.4.1): find $\hat{\vartheta}$ to minimize the weighted objective quantity $D_w = \sum_{b=1}^{B} w_b\{\hat{\gamma}(h_b) - \gamma(h_b; \hat{\vartheta})\}^2$, where the weights w_b adjust for any differential covariation between estimated semivariogram points. Cressie recommends

$$w_b = \frac{N_{h_b}}{\gamma(h_b; \vartheta)^2}. \tag{6.28}$$

This can be performed using any nonlinear WLS program. For instance, we saw in Chapter 2 that PROC NLIN has such capabilities.

Example 6.7 (Grass shrimp biomass, cont'd) Consider again the grass shrimp biomass data from Example 6.5. After careful study, Porter *et al.* (1997) determined that the spherical form in (6.10) was a reasonable choice for modeling $\gamma(h)$ with these data. Using their guidance, we perform here a WLS fit of the spherical model to the

$\bar{\gamma}(h_b)$ values calculated in Example 6.6. (We should note that in some settings the spherical model can exhibit irregular behavior, particularly when an ML approach is used for the model fit; see Mardia and Watkins, 1989. The fit here is stable, however, and we continue with use of the spherical model.)

SAS code for performing the WLS fit appears in Fig. 6.13. The code begins with statements that find the robust semivariogram and then manipulate the SAS data set generated by the `outvar=` option to remove semivariogram values at non-informative lags. Notice that we rename the distance, count, and robust semivariogram variates as `h`, `nh`, and `gambar`, respectively.

The PROC NLIN code in Fig. 6.13 is written to fit a typical piecewise regression model, which is the essential nature of model (6.10) (See the discussion on piecewise models in §2.2.) For simplicity, we apply the derivative-free method in PROC NLIN (`method=dud`) for performing the nonlinear WLS fit. Three crucial features in Fig. 6.13 are (i) the parameter specifications, (ii) inclusion of a `_weight_` statement,

```
* WLS fit of spherical semivariogram;

   data ppugio;
   input xloc yloc z @@;
   datalines;
        680070.500 3710087.00 1.853
        680095.312 3710078.50 2.472
            ⋮            ⋮        ⋮
        685182.438 3717120.50 1.925

proc variogram data=ppugio outvar=Gamma;
   var z;
   coordinates xc=xloc yc=yloc;
   compute lagd=242.9134   maxl=18   robust;

data wlsfit;    set Gamma;
   h=distance;   nh=count;   gambar=rvario;
   if  h ne .   and   nh ne .   and   gambar ne . ;
   drop varname lag distance count average variog rvario covar;

   proc nlin method=dud data=wlsfit;
      parms t1 = 0.01 to 0.09 by .02
            t2 = 0.20 to 0.60 by .10
            t3 =  400 to 2400 by 200;
      if h<t3 then do;                     * cubic rise;
         model gambar = t1 + t2*((1.5*h/t3) - (.5*((h/t3)**3)));
         end;
      else do;                             * sill;
         model gambar = t1 + t2;
         end;
      bounds t1>=0 , t2>=0 , t3>=0 ;

      _weight_ = nh/(model.gambar**2);
```

Figure 6.13 Sample SAS code for weighted least squares fit of spherical model to robust empirical semivariogram with grass shrimp biomass data: `t1` indicates θ_1 parameter, `t2` indicates θ_2 parameter, `t3` indicates θ_3 parameter

and (iii) parameter restrictions to perform the WLS fit. In the former case, the `parms` statement is used to define the nonlinear model parameters and give starting values for the nonlinear WLS iterations. We selected the starting values via visual inspection of the plot in Fig. 6.12. The nugget was near $\theta_1 \approx 0.05$, while the sill was near $\theta_1 + \theta_2 \approx 0.4$, hence $\theta_2 \approx 0.35$. The range appeared to be near $\theta_3 \approx 450$. To allow for some inaccuracy in these visual estimates, we instructed PROC NLIN to begin over a grid of possible parameters: $0.01 \leq \theta_1 \leq 0.09$, $0.2 \leq \theta_2 \leq 0.6$, and $400 \leq \theta_3 \leq 2400$. For the `_weight_` statement, we used weights given by (6.28). Since the weights include values of $\gamma(h_b; \vartheta)$, they can change with each iteration. To incorporate this, we used the program-level variable `model.gamhat` to instruct SAS to use the current model value of $\gamma(h; \vartheta)$ as part of the weight. And, lastly, to incorporate the constraint that $\theta_j \geq 0$ for $j = 1, 2, 3$, we used a `bounds` statement.

The resulting output (edited to give only the WLS search and, in the last row, the final parameter estimates) appears in Fig. 6.14. A number of features are of interest. First, PROC NLIN reports on the grid search for initial values. The combination of $\theta_1 = 0.09$, $\theta_2 = 0.5$, and $\theta_3 = 2000$ corresponds to the lowest D_w on the grid, hence PROC NLIN chose these for its starting values. The output follows with the results of the nonlinear WLS fit, where we see convergence does occur. (As noted above, this

```
                          The SAS System
                         The NLIN Procedure

                            Grid Search
        t1              t2              t3       Weighted SS
      0.0100          0.2000          400.0         2920.1
      0.0300          0.2000          400.0         2155.1
        ⋮               ⋮               ⋮             ⋮
      0.0700          0.5000         2000.0          180.8
      0.0900          0.5000         2000.0          171.8
        ⋮               ⋮               ⋮             ⋮
      0.0900          0.6000         2400.0          172.8

                        DUD Initialization
      DUD        t1              t2         t3       Weighted SS
       -4      0.0900          0.5000      2000.0        171.8
       -3      0.0990          0.5000      2000.0        170.5
       -2      0.0900          0.5500      2000.0        169.7
       -1      0.0900          0.5000      2200.0        174.5

                         Iterative Phase
      DUD        t1              t2         t3       Weighted SS
        0      0.0900          0.5500      2000.0        169.7
        1      0.0903          0.5375      2034.8        168.9
        2      0.0904          0.5348      2042.2        168.9
        ⋮        ⋮               ⋮           ⋮             ⋮
        9      0.0904          0.5347      2049.0        168.9
       10      0.0904          0.5347      2049.1        168.9
NOTE:  Convergence criterion met.
```

Figure 6.14 SAS output (edited) from WLS fit of spherical semivariogram to grass shrimp biomass data: `t1` indicates θ_1 parameter, `t2` indicates θ_2 parameter, `t3` indicates θ_3 parameter

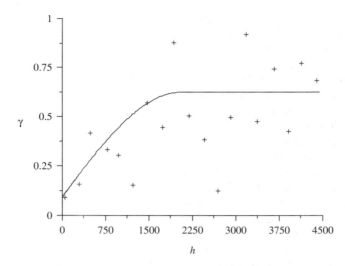

Figure 6.15 WLS spherical semivariogram from (6.10) and robust empirical points (+) for grass shrimp biomass data

result is not always guaranteed when performing a nonlinear WLS fit, especially with the spherical model. One may have to modify the starting values or possibly turn to a different fitting algorithm to achieve convergence. Moving to an ordinary LS fit by removing the _weight_ statement might also be necessary, if not desirable.) The final estimates for the spherical semivariogram parameters are $\hat{\theta}_1 = 0.0904$ (the estimated nugget), $\hat{\theta}_2 = 0.5347$, and $\hat{\theta}_3 = 2049.1$. Notice that this gives an estimated sill of $\hat{\theta}_1 + \hat{\theta}_2 = 0.6251$.

Plotted as a single function, the fitted semivariogram appears in Fig. 6.15. The graph suggests that spatial correlation in grass shrimp biomass does appear to exist at small scales, but as distances expand the correlation lessens. After separation distances of approximately $\hat{\theta}_3 = 2049.1$ m spatial correlation is essentially undetectable, relative to natural population variability.   ☉

6.4.5 Two-dimensional spatial prediction

Given a specific form for the semivariogram $\gamma(\cdot)$, we use it in the ordinary kriging equations (6.20) and (6.24) or in the universal kriging equations (6.20) and (6.27) for predicting unobserved outcomes of the spatial process $Z(\mathbf{s}_0)$. As these equations illustrate, this effort generally requires computer calculation. In SAS, the pertinent procedure for performing ordinary kriging when $d = 2$ is PROC KRIGE2D. If the likelihood for the spatial data is normal, then one can also use PROC MIXED to derive semivariogram estimates and spatial predictions. We discuss this in more detail in §6.4.6.

Kriging requires an input data set in (x, y, Z) coordinates, specification of a known form for the semivariogram, and a set of grid coordinates $\{(x_0, y_0): \mathbf{s}_0 = (x_0, y_0)\}$ over which to construct the kriged values $\hat{Z}(s_0)$. If the parameters of the semivariogram model are unknown, as is common, one uses estimated values from §6.4.4. In this latter case, we refer to $\hat{Z}(s_0)$ as the *empirical best linear unbiased predictor (EBLUP)*.

Example 6.8 (Grass shrimp biomass, cont'd) To illustrate EBLUP construction, we again return to the grass shrimp biomass data from Example 6.5. Suppose we wish to visualize the spatial biomass outcome across the salt-marsh estuary. As in Example 6.7, we assume that the underlying semivariogram is of the spherical form (6.10). For the spherical model parameters we use the estimates determined from the WLS fit to the robust empirical semivariogram: $\hat{\theta}_1 = 0.0904$, $\hat{\theta}_2 = 0.5347$, and $\hat{\theta}_3 = 2049.1$. Figure 6.16 displays sample SAS code to perform the fit using PROC KRIGE2D. Notice in the figure that the PROC KRIGE2D `model` statement identifies the θ_2 parameter as the *scale*. Do not confuse this with θ_3; the final parameter is still called the *range*. (This distinction in nomenclatures is important. Users should be acutely aware of the difference when constructing their own PROC KRIGE2D code.)

For purposes of the kriging output, PROC KRIGE2D requires the user to provide a group of values over which to calculate the EBLUPs, via the `grid` statement. This can be as simple as a single (x, y) point, a rectangular grid of points in $\mathbb{R}^2$, or even a preassigned SAS data set of irregularly spaced points. For this example, we chose a highly dense grid of points defined by the roughly elliptical region containing the original data locations, using the `griddata=` option. The

```
      data ppugio;
      input xloc yloc z @@;
      datalines;
          680070.500   3710087.00   1.853
          680095.312   3710078.50   2.472
             :            :           :
          685182.438   3717120.50   1.925

  *  first define external griddata set;
     data gridxy;
        input x y @@;
        datalines;
        680000.0   3710060.0        680000.0   3710095.5
           :          :                :          :
        685500.0   3717124.5        685500.0   3717160.0

  *  2D kriging & spatial prediction;
     proc krige2d data=ppugio  outest=zhat;
        predict var=z;

  *  in model statement, "nugget" is theta1,
                        "scale"  is theta2, "range" is theta3;
     model nugget = 0.0904
           scale  = 0.5347
           range  = 2049.1
          form = spherical;
     coord xc=xloc yc=yloc;
     grid griddata=gridxy xc=x yc=y;
```

Figure 6.16 Sample SAS code for kriging EBLUPs with grass shrimp biomass data

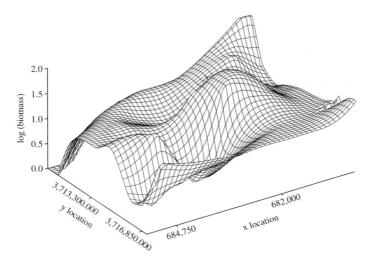

Figure 6.17 Kriging EBLUPs for grass shrimp biomass data

corresponding values (not shown) of $\hat{Z}(s_0)$ are stored in an output data set via the `outest=` option. PROC KRIGE2D gives them the variable name `estimate`; it also calculates approximate standard errors of prediction, $se[\hat{Z}(s_0)]$, under the variable name `stderr`. In both cases, the variables can be ported to a three-dimensional graphics plotter for display and study.

Note that the standard errors of prediction are *unadjusted* for any estimation performed to identify the semivariogram parameters. Thus if these parameters have been estimated from the data, it is likely that the standard errors underestimate the true variability existent in the EBLUPs (Wang and Wall, 2003).

The output EBLUPs, $\hat{Z}(s_0)$, from the code in Fig. 6.16 are plotted in Fig. 6.17. (The corresponding standard errors of prediction are computed as part of Exercise 6.16.) The EBLUP plot shows clear spatial variation: sharp increases and decreases appear in predicted log-biomass at the edges of the sampled region. How these locations compare to the sites of nearby urbanization could be an issue of further interest for environmental scientists studying the estuary. ✪

6.4.6 Kriging under a normal likelihood

The BLUP in (6.20) for both ordinary and universal kriging may be regarded as a distribution-free quantity, since no formal distribution has been assumed for the stochastic error terms $\varepsilon(s_i)$. In some settings, however, we may wish to specify a distribution for the errors in order to predict $Z(s_0)$. Most common for continuous measurements under (6.6) or (6.16) is the assumption of a normal distribution: $\varepsilon(s_i) \sim N(0, \sigma^2)$. Where necessary, one can apply a logarithmic transformation or other function to stabilize any variance heterogeneity and/or transform the data into a more approximately normal form; cf. the similar discussion for linear regression towards the end of §1.1.

As part of any distributional specification, it is also natural to assign to $\text{Cov}[\varepsilon(\mathbf{s}_i), \varepsilon(\mathbf{s}_j)]$ a functional form dependent upon $\mathbf{s}_i$ and $\mathbf{s}_j$ $(i \neq j)$. For isotropic models, $\text{Cov}[\varepsilon(\mathbf{s}_i), \varepsilon(\mathbf{s}_j)]$ is defined via the function $C(h_{ij})$, where $h_{ij} = \delta_{ij}$ is the Euclidean distance between the points $\mathbf{s}_i$ and $\mathbf{s}_j$. If the semivariogram is known, then $\text{Cov}[\varepsilon(\mathbf{s}_i), \varepsilon(\mathbf{s}_j)] = C(h_{ij})$ is further defined from $\gamma(h) = C(0) - C(h)$, where $C(0)$ is the sill of the model. For example, the exponential semivariogram $\gamma(h) = \theta_1 + \theta_2[1 - \exp(-h/\theta_3)]$ corresponds to $C(h) = \theta_2 \exp\{-h/\theta_3\}$ (Exercise 6.24).

Under normal errors, the unknown model parameters may be estimated by maximizing the normal likelihood. For spatial applications, residual maximum likelihood is often preferred since it can exhibit lower estimation bias than full maximum likelihood (Mardia and Marshall, 1984). Recall from §5.6.2 that the REML method maximizes a log-likelihood based on residuals found by adjusting or 'filtering' the contributions of terms in the model that do not affect the covariance structure. In SAS, REML estimation for kriging may be performed via PROC MIXED (Littell *et al.*, 1996). The procedure can estimate the unknown parameters from a pre-specified variogram or covariance model, and can then be manipulated to output spatial EBLUPs for any number of future spatial locations $\mathbf{s}_0$.

If the normal distribution assumption is correct, then approximate standard errors, confidence intervals, likelihood ratio tests, and other useful inferences may be derived from the REML estimates. For example, PROC MIXED produces standard errors and approximate Wald tests for the θ-parameters via use of the `covtest` option in the procedure call; approximate $1 - \alpha$ confidence intervals are provided by the `cl` option. LR statistics may be derived by fitting two models, the first with all the model parameters, and the second with one or more of these missing (see §A.5.3). The PROC MIXED output will give the corresponding REML log-likelihoods (as `-2REML_LL`); the LR test statistic is then the difference between them. Compare this in large samples to a χ^2 distribution with df equal to the number of parameters removed in the second fit.

Example 6.9 (TCE groundwater concentrations) Kitanidis (1997, Table 2.2) presents data on concentrations (in ppb) of the industrial chemical trichloroethylene (TCE) in groundwater of a fine-sand surface aquifer. The data are similar to the TCE groundwater contamination study from Example 3.5, except in this new study the actual TCE concentrations were recorded. As discussed in that earlier example, the possible human carcinogenicity of TCE makes predicting groundwater concentrations of this chemical an important problem. A selection of the data appears in Table 6.3 (see http://www.wiley.com/go/environmental for the complete data set).

It is common for concentration data to skew right, so we operate with the natural logarithms of the original observations. We assign a normal model to the error terms for the log-concentrations and consider application of ordinary kriging via REML. Assuming isotropy, we start by calculating an empirical semivariogram from (6.13). For the distance classes, we set $B = 10$ and find using the methods of §6.3.6 that the maximum distance among the locations is 371.652 ft. This leads to $h_{10} = 185.826$. Sample SAS code to find the semivariogram values appears in Fig. 6.18. Output (edited) containing a simple PROC PLOT graphic of the semivariogram appears in Fig. 6.19.

From the plot in Fig. 6.19, an exponential semivariogram (6.11) may be a reasonable choice for $\gamma(h)$ with these data. We can also use the plot to identify some initial

Table 6.3 Selected concentrations (in ppb) of trichloroethylene (TCE) in groundwater of a fine-sand surface aquifer. X-location and Y-location are in feet from a source location

X-location	Y-location	Concentration
0	−45	10.0
0	−50	10.0
0	−55	64.0
⋮	⋮	⋮
370	−65	7.0
370	−70	2.5
370	−75	2.5

Source: Kitanidis (1997).

```
data tcegw;
input xloc yloc tce @@;
z = log(tce);
datalines;
  0   -45   10.0          0  -50   10.0          0  -55   64.0
      :                       :                      :
370   -65   7.0         370  -70   2.5         370  -75   2.5

proc variogram data=tcegw    outvar=gamma;
   var z;
   coordinates xc=xloc yc=yloc;
      compute lagd=18.5826   maxl=10;

proc print;
   var lag distance count variog;

data wlsfit;   set gamma;
    h=distance;   nh=count;   gamhat=vario;
    if  h ne .   and  nh ne .   and   gamhat ne . ;
drop varname lag distance count average variog covar;

proc plot;
    plot  gamhat*h;
```

Figure 6.18 Sample SAS code for empirical semivariogram estimation with TCE groundwater concentration data

estimates for the REML fit. The nugget appears to be at or near zero, which is not unreasonable with these groundwater concentration data. Of course, we cannot rule out the case $\theta_1 > 0$, and so we consider a tight span near zero, say, $0.001 \le \theta_1 \le 0.009$. The sill is roughly $\theta_1 + \theta_2 \approx 8$, but since $\theta_1 \approx 0$ we simply set $\theta_2 \approx 8$. Lastly, after about 20 ft the semivariogram appears to level off, so start with $\theta_3 \approx 20$. Sample SAS code for invoking **PROC MIXED** and fitting an exponential semivariogram model to the log-transformed data using these initial estimates appears in Fig. 6.20.

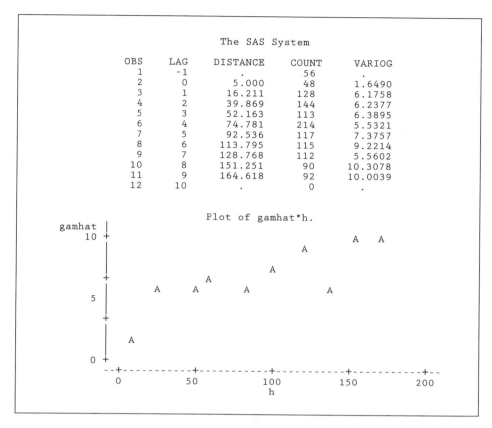

The SAS System

OBS	LAG	DISTANCE	COUNT	VARIOG
1	-1	.	56	.
2	0	5.000	48	1.6490
3	1	16.211	128	6.1758
4	2	39.869	144	6.2377
5	3	52.163	113	6.3895
6	4	74.781	214	5.5321
7	5	92.536	117	7.3757
8	6	113.795	115	9.2214
9	7	128.768	112	5.5602
10	8	151.251	90	10.3078
11	9	164.618	92	10.0039
12	10	.	0	.

Figure 6.19 SAS output and plot (edited) of empirical semivariogram for TCE groundwater concentration data

In the PROC MIXED code from Fig. 6.20, a number of features are critical. First, in the procedure call we include the convh= option, which relaxes slightly PROC MIXED's convergence tolerance of less than 10^{-8} for stopping the iterations. (Here, we have set it to 10^{-5}.) It sometimes occurs that complex semivariogram models push the iterations to PROC MIXED's maximum without converging or take an exceptionally long time to converge under the stricter tolerance (Littell *et al.*, 1996, §9.5). The relaxed tolerance helps avoid this concern. This modification is not always necessary, however, and in practice analysts may have to experiment with the convh= option to achieve an acceptable model fit.

Next, the parms statement is used to provide initial values (or a grid of initial values) from which the iterations may proceed. It is similar in style to its counterpart under PROC NLIN, but with slightly different syntax: parentheses bracket the initial parameter values. θ-parameter order in the parms statement is determined by the order of the variance–covariance parameters embedded in the repeated statement (see below). For most spatial models that PROC MIXED supports, this order is $(\theta_2)(\theta_3)(\theta_1)$ in the notation from §6.3.5. Here, we specify an initial grid based on the corresponding values mentioned above.

```
data tcegw;
input xloc yloc tce @@;
z = log(tce);
datalines;
   0  -45   10.0          0  -50   10.0          0  -55   64.0
        :                     :                     :
 370  -65    7.0        370  -70    2.5        370  -75    2.5

   proc mixed data=tcegw convh=1e-05   covtest   cl;
 * in parms statement,  first bracket is theta2;
 *                      second bracket is theta3;
 *                       third bracket is theta1;
   model z = ;
     parms (8) (20) (0.001 to 0.009 by .002);
   repeated / subject=intercept   local   type=sp(exp)(xloc yloc);
```

Figure 6.20   Sample SAS code for REML semivariogram estimation with TCE groundwater concentration data

   Lastly, the `repeated` statement invokes the spatial correlation model. The `subject=intercept` option instructs **PROC MIXED** to view all observations as potentially correlated. (Block correlation can also be modeled, via use of a classifying block variable in place of `intercept`.) The `local` option is required if a nugget parameter is to be included in the semivariogram. The default in PROC MIXED is a no-nugget model. Specifying the option `type=sp(exp)(xloc yloc)` sets the spatial covariance structure: `sp(exp)` calls for an exponential semivariogram, while `(xloc yloc)` indicates to **PROC MIXED** the x- and y-coordinates of the points s_i.

   The REML fit to these data produces the point estimators $\hat{\theta}_1 = 0.0002$, $\hat{\theta}_2 = 9.204$, and $\hat{\theta}_3 = 28.694$ (output not shown). As expected, the estimated nugget is very close to zero. Indeed, to test $H_0: \theta_1 = 0$ we can perform a REML fit using a no-nugget (two-parameter) exponential model and compare this to the full three-parameter model. The only required changes to the SAS code in Fig. 6.20 are (i) removal of the third bracket of initial values in the `parms` statement, and (ii) removal of the `local` option in the `repeated` statement. (Since the reduced model fit is likely to be more stable, we also remove the `convh=1e-05` limitation on the convergence criterion.) Doing so leads to the REML output (edited) in Fig. 6.21. To assess the reduced model under H_0, we appeal to an approximate LR test: the summary fit statistic `-2 Res Log Like` under H_0 is given in Fig. 6.21 as $-2\ell(0, \theta_2, \theta_3) = 220.4537$, while from the full-model fit we can find $-2\ell(\theta_1, \theta_2, \theta_3) = 220.4553$ (output not shown). Under $H_0: \theta_1 = 0$, the difference between them is approximately χ^2 with df $= 1$. Here this is $\chi^2_{calc} = 0.0016$, with P-value $P[\chi^2(1) \geq 0.0016]$ well above 0.20 from Table B.3. Thus we conclude that θ_1 is not significantly different from zero.

   Under the reduced/no-nugget model, the REML fit in Fig. 6.21 gives the sill as $\hat{\theta}_2 = 9.21$ and the range as $\hat{\theta}_3 = 28.70$ ft. (For models not containing a nugget, PROC MIXED gives $\hat{\theta}_2$ under `SP(EXP)` and $\hat{\theta}_3$ under `Residual`. For models with a non-zero nugget parameter, **PROC MIXED** gives $\hat{\theta}_2$ under `Variance`, $\hat{\theta}_3$ under `SP(EXP)`, and the estimated nugget $\hat{\theta}_1$ under `Residual`.) One conclusion to reach from these results is that under the normal assumption the spatial autocorrelation

```
                          The SAS System
                        The MIXED Procedure

                         Parameter Search
      CovP1       CovP2   Variance    Res Log Like   -2 Res Log Like
      8.0000     20.0000   5.0358       -115.8948          231.7896

                    REML Estimation Iteration History
         Iteration    Evaluations     -2 Res Log Like        Criterion
             1             2           224.80465217         0.03724199
             2             1           221.87904187         0.01426306
             3             1           220.80473376         0.00418307
             4             1           220.50349882         0.00070843
             5             1           220.45591460         0.00003523
             6             1           220.45373468         0.00000012
             7             1           220.45372760         0.00000000
    Convergence criteria met.

                    Covariance Parameter Estimates
                                          Standard        Z
       Cov Parm   Subject    Estimate       Error      Value      Pr Z
       SP(EXP)    Intercept   28.6981      14.8692      1.93     0.0268
       Residual                9.2055       4.1023      2.24     0.0124

       Cov Parm   Subject     Alpha        Lower       Upper
       SP(EXP)    Intercept    0.05        12.8009     112.16
       Residual                0.05         4.5035      28.2074

                        Fit Statistics
              -2 Res Log Likelihood                220.5
              AIC (smaller is better)              224.5
```

Figure 6.21 SAS output (edited) of reduced-model/no-nugget semivariogram REML fit to TCE groundwater concentration data

between TCE concentrations drops to near zero for locations separated by at least 28.7 ft, with 95% limits given by $12.8 < \theta_3 < 112.2$ ft.

Given a REML estimate for the semivariogram, we can perform ordinary kriging/prediction of the spatial surface via EBLUPs. In PROC MIXED, the effort is straightforward: call for predicted values via the `predicted` option in the model statement. This produces EBLUPs for any combination of spatial coordinates in the input SAS data set. (To generate EBLUPs at spatial locations where no observation was recorded, simply enter a missing observation at those coordinates in the input SAS data set. SAS notation for a missing value is a period: '.') The output also includes approximate standard errors for each EBLUP and an approximate $1 - \alpha$ prediction interval.

Example 6.10 (TCE groundwater concentrations, cont'd) Returning to the TCE groundwater contamination data from Example 6.9, suppose that three new wells are to be drilled at the following spatial locations: $s_0 = (50, -50), (100, -50)$, and $(170, -50)$. It may be of interest to determine if and to what extent there is TCE contamination at the new sites. Thus we wish to find EBLUPs, $\hat{Z}(s_0)$, for the TCE concentration at these three locations. Towards this end, Fig. 6.22 displays sample SAS ver. 8 code for fitting a no-nugget exponential semivariogram model and

```
data tcepred;
input xloc yloc tce @@;
z = log(tce);
datalines;
  50   -50     .          100  -50     .          170  -50     .
   0   -45   10.0           0  -50   10.0           0  -55   64.0
      :                        :                        :
 370   -65    7.0          370  -70    2.5          370  -75    2.5

proc mixed ;
model z = / outpred=zpred ;
parms (8) (20) ;
repeated / subject=intercept type=sp(exp)(xloc yloc);

proc print data=zpred;
var xloc yloc z Pred  Lower Upper;
```

Figure 6.22 Sample SAS code for no-nugget REML EBLUP with TCE groundwater concentration data

calculating the EBLUPs. Notice, in particular, use of missing observations (as periods, '.') in the data step to identify the new sites for spatial prediction, and of the outpred= option in the model statement to define the output data set containing the predicted values. Figure 6.23 shows the pertinent EBLUPs from the PROC MIXED output (edited).

We find the EBLUPs at the new locations are $\hat{Z}(50, -50) = 4.145$, $\hat{Z}(100, -50) = 5.095$, and $\hat{Z}(170, -50) = 5.546$. Pointwise (unadjusted) 95% prediction intervals at each s_0 are also given as $-0.909 < Z(50, -50) < 9.199, 1.171 < Z(100, -50) < 9.018$, and $0.717 < Z(170, -50) < 10.375$. Thus, for example, at $s_0 = (100, -50)$ we predict with approximately 95% confidence that the TCE concentration on the original scale is between $e^{1.171} = 3.23$ and $e^{9.018} = 8250.26$ ppb. ✪

Universal kriging via REML is also possible using PROC MIXED. Continuing with the normal assumption on $\varepsilon(s_i)$, suppose now that the linear model in (6.25) holds at each spatial location s_i. Assume that the mean contains an intercept term, β_1. Our goal remains that of predicting some future value of Z at the spatial location s_0, but now using REML estimates under a normal sampling assumption. The next example illustrates use of the methodology.

```
                       The SAS System

                      Predicted Values
  Obs  xloc   yloc    z      Pred       Lower      Upper
   1     50   -50      .    4.14505   -0.90899    9.1991
   2    100   -50      .    5.09454    1.17115    9.0179
   3    170   -50      .    5.54612    0.71714   10.3751
```

Figure 6.23 SAS output (edited) of no-nugget REML EBLUPs for TCE groundwater concentration data

Example 6.11 (Reventazon rainfall) Gaudard *et al.* (1999) report data on mean annual rainfall (in m) for the Reventazon River basin in Costa Rica, as a function of site elevation (in km). Of interest are description and prediction of the spatial pattern the rainfall exhibits. Since the elevation of a location may impact its precipitation rate, we include elevation in the analysis to adjust any statements made on the rainfall amounts. A selection of the data appears in Table 6.4 (see http://www.wiley.com/go/environmental for the complete data set; lower elevations in the data correspond to weather reporting stations near the river basin, while higher elevations are stations up the basin slope).

We consider here a universal kriging analysis including the known predictor variables $u_1(s_i) = 1$ and $u_2(s_i) = $ elevation. Our first step is to fit a simple linear model to the rainfall data using an ordinary least squares regression on $u_2(s_i)$. From this fit we find the residuals and, assuming isotropy, use them to calculate an empirical semivariogram via (6.13); that is, we replace the $Z(s_i)$ terms in (6.13) with the residuals from the LS fit. (In effect, this 'detrends' the outcome variable to remove any non-stationarity induced by the elevation predictor.) A plot of the result (Exercise 6.27) shows that a Gaussian semivariogram (6.12) may be appropriate for the fit here.

Assuming a Gaussian form for $\gamma(h)$, we turn to PROC MIXED and estimate the corresponding semivariogram via REML under a normal-error model. Sample SAS code for these operations appears in Fig. 6.24. (The initial estimates for the semivariogram are based on the empirical semivariogram calculated above.) Notice inclusion of elevation as a predictor variable in the `model` statement.

The REML output (edited) from the code in Fig. 6.24 appears in Fig. 6.25. There, we see that PROC MIXED converges comfortably, giving final REML estimators $\hat{\theta}_1 = 0.257$ (the estimated nugget), $\hat{\theta}_2 = 2.468$, and $\hat{\theta}_3 = 12.109$. Notice that this gives an estimated (asymptotic) sill of $\hat{\theta}_1 + \hat{\theta}_2 = 2.725$. (Recall that for models containing a nugget, PROC MIXED gives $\hat{\theta}_2$ under `Variance`, the estimated range $\hat{\theta}_3$ under `SP(GAU)`, and the estimated nugget $\hat{\theta}_1$ under `Residual`.)

Table 6.4 Mean annual rainfall (m) and elevation (km) for locations in the Reventazon River basin, Costa Rica. X-location and Y-location are in kilometers east and north, respectively, from a fixed origin

X-location	Y-location	Elevation	Rainfall
540.10	197.93	1.830	1.790
545.58	197.93	1.410	1.920
545.58	205.33	1.440	1.367
⋮	⋮	⋮	⋮
569.30	190.52	1.360	6.706
576.60	207.19	0.602	2.610
578.43	203.48	0.620	2.464

Source: Gaudard *et al.* (1999).

```
data rain;
input  xloc  yloc  elev  zrain @@;
datalines;
545.58  205.33  1.440  1.367    576.60  207.19  0.602  2.610
                      :
552.88  218.30  3.400  2.081    562.00  186.81  1.921  6.124

  proc mixed ;
* in parms statement, first bracket is theta2;
*                     second bracket is theta3;
*                     third bracket is theta1;
  model zrain = elev ;
  parms (.5) (3.5 to 4.5 by .5) (10 to 15 by 2.5);
  repeated / subject=intercept local type=sp(gau)(xloc yloc);
```

Figure 6.24   Sample SAS code for REML semivariogram estimation with Reventazon rainfall data

```
                        The SAS System

                      Parameter Search
  CovP1    CovP2     CovP3 Variance  Res LogLike  -2 Res LogLike
  0.5000   3.5000   10.0000   3.1529    -67.8372      135.6744
  0.5000   3.5000   12.5000   3.2122    -67.9905      135.9810
  0.5000   3.5000   15.0000   3.2531    -68.0959      136.1918
  0.5000   4.0000   10.0000   3.1192    -67.6552      135.3104
  0.5000   4.0000   12.5000   3.1840    -67.8417      135.6834
  0.5000   4.0000   15.0000   3.2290    -67.9700      135.9401
  0.5000   4.5000   10.0000   3.0864    -67.4751      134.9503
  0.5000   4.5000   12.5000   3.1565    -67.6944      135.3887
  0.5000   4.5000   15.0000   3.2053    -67.8453      135.6906

                     Iteration History
  Iteration    Evaluations   -2 Res Log Like        Criterion
          1            2       122.90945669       2.48057791
          2            1       113.31636911       0.81471703
          :            :             :                 :
         11            1        89.52784184       0.00002029
         12            1        89.52773848       0.00000927
Convergence criteria met.

               Covariance Parameter Estimates
                                   Standard         Z
  Cov Parm    Subject    Estimate    Error    Value      Pr Z
  Variance    Intercept    2.4675   1.1820    2.09     0.0184
  SP(GAU)     Intercept   12.1088   2.1566    5.61     <.0001
  Residual                 0.2574   0.09237   2.79     0.0027

               Type 3 Tests of Fixed Effects
          Effect    Num DF   Den DF   F Value    Pr > F
          elev         1       32       0.32     0.5755
```

Figure 6.25   SAS output (edited) for REML semivariogram with Reventazon rainfall data

For purposes of spatial prediction, we can employ the REML fit to predict the mean annual rainfall at some new site s_0. For example, suppose there is a location up from the river basin floor at coordinates $s_0 = (555.0, 215.0)$, with an associated elevation of $u_2(s_0) = 2.5$ km. To predict $Z(s_0)$, and also to find an approximate 95% prediction interval, we modify the SAS code in Fig. 6.24 by adding the 'observation'

```
555.00  215.00  2.500  .
```

to the SAS data set (notice the missing value, denoted as the trailing period), then modifying the `model` statement in the SAS ver. 8 code

```
model zrain = elev / outpred = zpred;
```

and finally calling for a printout of the values in the SAS data set `zpred`. The corresponding output (not shown) produces an EBLUP of $\hat{Z}(s_0) = 2.321$ m, with approximate 95% prediction limits given by $1.095 < Z(s_0) < 3.547$ m. Our prediction at this higher location appears to involve slightly lower rainfall amounts than those seen closer to the basin floor – which is not surprising – although the approximate 95% confidence bounds give a fairly wide spread for the predicted rainfall. Gaudard et al. (1999) give a further analysis of these complex spatial data. ✪

Although we have framed the use of REML methods here in terms of a normal likelihood, REML estimating equations can be applied successfully to non-normal data (Heyde, 1994). Indeed, when employed for spatial prediction, Cressie and Lahiri (1996) showed that REML estimates actually solve a set of unbiased estimating equations, and so may prove useful in a number of more general settings. See Zimmerman (1989), Cressie (1993, §2.6; 2002b), and the references therein for more on REML estimation for spatial prediction.

Exercises

6.1. Return to the redwood seedlings data in Example 6.1.

(a) Verify under the 3×3 configuration that both regularity conditions listed in §6.2.1 are met for the χ^2 approximation to hold.

(b) Construct a table of Pearson residuals corresponding to the 3×3 configuration used in the example. (If you have access to sufficient computer resources, find the standardized Pearson residuals as well.) Does any obvious pattern of departure from CSR emerge?

(c) Modify Fig. 6.2 by moving to a 4×4 grid. Find the resulting 4×4 table of counts and apply X^2_{calc} from (6.1) to test for CSR. What do you conclude at $\alpha = 0.10$? (Diggle, 2003)

(d) Were both regularity conditions met for the χ^2 approximation under the 4×4 configuration in part (c)?

6.2. Show that the Pearson statistic in (6.1) can also be written as $X_{calc}^2 = (\sum_{j=1}^J \sum_{i=1}^I Y_{ij}^2/\bar{Y}) - n$. [*Hint*: we know that $\sum_{j=1}^J \sum_{i=1}^I Y_{ij} = n$.]

6.3. Dixon (2002b), following Jones *et al.* (1994), describes a study on the spatial pattern of $n = 98$ ash trees (genus *Fraxinus*) in a swamp hardwood forest along the US Atlantic coast. The region under study is a 200 m × 50 m rectangle in the central portion of the forest. The paired (x, y) locations are available online at http://www.wiley.com/go/environmental (the data are in two columns: X-location | Y-location); a sample follows:

X-location	Y-location
2.9	177.2
3.1	3.4
4.8	6.9
⋮	⋮
49.0	39.2
49.1	192.7
49.8	3.3

To assess whether the trees' spatial distribution deviates from CSR, perform the following.

(a) Plot the spatial coordinates over an appropriately shaped rectangle. Do you see any obvious spatial pattern?

(b) Overlay a contiguous, equal-area, 4 × 2 grid on the study region and count the number of trees in each quadrat. Construct from this a 4 × 2 table of counts and use X_{calc}^2 from (6.1) to assess the CSR null hypothesis. What do you conclude at $\alpha = 0.05$?

(c) Are both regularity conditions met for the χ^2 approximation on X_{calc}^2 to hold?

6.4. Geologists study surface features of planets to understand how their environments shape and are shaped by geological processes. On Venus, for example, impact cratering appears to be highly complex (Schaber *et al.*, 1998); consider the $n = 20$ heavily cratered ('type f2') impact locations on the planet, given by paired longitude and latitude coordinates as follows:

°Longitude	°Latitude	°Longitude	°Latitude	°Longitude	°Latitude
337.700	86.800	85.000	81.700	337.474	72.448
6.416	56.962	280.400	38.200	283.800	26.100
204.568	23.626	240.216	17.888	215.340	17.237
288.784	9.733	97.388	6.233	90.000	2.300
286.289	−4.162	132.750	−6.673	277.959	−8.981
174.747	−11.678	308.862	−12.437	188.099	−20.722
258.838	−23.264	132.830	−27.000		

(a) Plot the 'f2' impact locations as a two-dimensional representation. Do the locations appear to cluster in selected small groups?

(b) Overlay a contiguous, equal-area, 2×3 grid on the study region and count the number of craters in each quadrat. Construct from this a 2×3 table of counts and use X^2_{calc} from (6.1) to assess the CSR null hypothesis. (Are both regularity conditions met for the χ^2 approximation on X^2_{calc} to hold?) What do you conclude at $\alpha = 0.01$?

(c) Although edge effects are not an issue with these data (why not?), we can also apply the nearest-neighbor methods from §6.2.2 to study CSR. Venus has a circumference of approximately 38 026 km. Assume the planet is a sphere (which is a slight simplification) and translate the location data into kilometers. Calculate the individual nearest-neighbor distances, W_i, and from these the test statistic in (6.2). Is there significant deviation from CSR at $\alpha = 0.01$?

(d) Notice that a number of the nearest-neighbor distances in part (c) are *reflexive*; i.e., if s_i's nearest neighbor is s_j, s_j's nearest neighbor is s_i ($i \neq j$). But this is not always the case. Explain.

6.5. Return to the ash hardwood data in Exercise 6.3 and test for CSR via the edge effect-adjusted Wald statistic from (6.2). Do the inferences differ?

6.6. Return to the following data sets. Update your test for CSR by including a buffer to help correct further for edge effects. (Give your buffer a constant length as prescribed below.) Construct a new table of counts and apply X^2_{calc} from (6.1) to test for CSR. Do your inferences differ from those seen previously?

(a) The redwood seedlings data in Example 6.1. Give your buffer a constant length of 2 m from all edges of the 23 m square.

(b) The ash hardwood data in Exercise 6.3. Buffer the 200 m-long vertical edges of the region by 5 m left and right, and the 50 m-long horizontal edges of the region by 20 m top and bottom. The result should be a buffered rectangular region of dimension 160 m $\times$ 40 m.

6.7. Let $L(h) = \sqrt{K(h)/\pi}$. Verify that if $K(h) = \pi h^2$, then $L(h) - h = 0$.

6.8. Return to the ash hardwood data in Exercise 6.3.

(a) Estimate the $K(h)$ and $L(h) - h$ functions for these data. Plot your results over an appropriate range for h. Does the graphic confirm your previous results with these data?

(b) Recall that in Exercise 6.6(b) concern was raised over possible edge effects with these data. Calculate the edge-adjusted estimator of $K(h)$ from (6.5) on the original data. Do the K-function and L-function patterns change drastically?

6.9. Return to the K-function analysis of the redwood seedlings data in Example 6.4. Recall that in the example concern was raised over possible edge effects. Use the edge-buffered data from Exercise 6.6(a) and repeat the K- and L-plot analyses. Do you see any substantive change from the results in Example 6.4?

6.10. Show that when $\text{Var}[Z(\mathbf{s})]$ and $C(\mathbf{s},\mathbf{t})$ exist, the semivariogram is $\gamma(\mathbf{s},\mathbf{t}) = \frac{1}{2}\{C(\mathbf{s},\mathbf{s}) + C(\mathbf{t},\mathbf{t})\} - C(\mathbf{s},\mathbf{t})$.

6.11. If $Z(\mathbf{s})$ is second-order stationary as per Definition 6.3, establish the following:

(a) $\gamma(\mathbf{h}) = \gamma(-\mathbf{h})$ for all $\mathbf{h}$.

(b) $\gamma(\mathbf{0}) = 0$.
[*Hint*: use the fact that the semivariogram can be written as $\gamma(\mathbf{h}) = \frac{1}{2}E[\{Z(\mathbf{s}+\mathbf{h}) - Z(\mathbf{s})\}^2]$.]

(c) Show why $\gamma(\mathbf{h})$ can be written as $\frac{1}{2}E[\{Z(\mathbf{s}+\mathbf{h}) - Z(\mathbf{s})\}^2]$.

6.12. Show that if $Z(\mathbf{s})$ is second-order stationary as per Definition 6.3, then $\gamma(\mathbf{h}) = C(\mathbf{0}) - C(\mathbf{h})$, for any difference vector $\mathbf{h}$. [*Hint*: use the result from Exercise 6.10.]

6.13. Plot the following models. In each case, identify the nugget, the sill and, where appropriate, the range.

(a) The spherical model from (6.10) for $\theta_1 = \theta_2 = \theta_3 = 1$. Graph your plot over $0 \le h < 1.5$.

(b) The spherical model from (6.10) for $\theta_1 = 1, \theta_2 = 2, \theta_3 = 0.75$. Graph your plot over $0 \le h < 1.5$.

(c) The exponential model from (6.11) for $\theta_1 = \theta_2 = \theta_3 = 1$. Graph your plot over $0 \le h < 5$.

(d) The exponential model from (6.11) for $\theta_1 = \theta_2 = 2, \theta_3 = 0.5$. Graph your plot over $0 \le h < 5$.

(e) The Gaussian model from (6.12) for $\theta_1 = 1, \theta_2 = 3, \theta_3 = 2$. Graph your plot over $0 \le h < 5$.

6.14. Show that for the following models, the nugget is θ_1 and the sill is $\theta_1 + \theta_2$, where $\theta_j \ge 0$ for $j = 1, 2$:

(a) The spherical model from (6.10). Also show that the range is θ_3.

(b) The exponential model from (6.11).

(c) The Gaussian model from (6.12).

6.15. Cressie (1993, §8.2.1) gives data on the locations of $n = 584$ longleaf pine trees (*Pinus palustris*) from a central section of an old-growth forest in the southeastern US. The study area is a square of sides 200 m. Also measured were the diameters at breast height (in cm) of the trees at each spatial location. The data are available online at http://www.wiley.com/go/environmental (the data are in three columns: X-location | Y-location | diameter); a sample follows:

X-location	Y-location	Diameter
0.0	177.5	50.8
0.4	175.2	50.9
0.9	100.0	41.4
⋮	⋮	⋮
199.3	10.0	53.5
199.5	179.4	9.7
200.0	8.8	32.9

Assuming isotropy, construct an empirical semivariogram for these data, using the estimator in (6.13). The size of this data set is sufficient to allow for a large number of distance classes: use $B = 30$. Plot the empirical semivariogram values and determine if they appear to match any of the functional forms in §6.3.5.

6.16. Return to the grass shrimp biomass data from Example 6.5. Download the full data set from http://www.wiley.com/go/environmental.

(a) Construct a three-dimensional scatter plot of the original data. Comment on the spatial pattern. Compare your plot with the EBLUP plot in Fig. 6.17.

(b) Find the standard errors of prediction for the EBLUPS, $\hat{Z}(s_0)$, calculated in Example 6.8. Plot them in three dimensions over the same spatial coordinates and comment on the graph.

6.17. Show that in Definition 6.6, the two conditions
(iii) $\text{Var}[\hat{Z}(s_0) - Z(s_0)]$ is a minimum among all estimators $\hat{Z}(s_0)$ satisfying (i) and (ii), and

(iii') $E[\{\hat{Z}(s_0) - Z(s_0)\}^2]$ is a minimum among all estimators $\hat{Z}(s_0)$ satisfying (i) and (ii),
are equivalent when $E[\hat{Z}(s_0) - Z(s_0)] = 0$.

6.18. For the case of simple kriging – i.e., known and constant μ in (6.16) – show that by using $\kappa_0 = \mu(1 - \sum_{i=1}^{n} \kappa_i)$ in (6.15) we force $E[\hat{Z}(s_0) - Z(s_0)]$ to be zero for any choice of (real-valued) κ_is.

6.19. Show that by selecting constants $\kappa_1, \ldots, \kappa_n$ satisfying (6.18), we minimize the mean squared error of prediction in (6.17). Use the following steps:

(a) Write the objective function in (6.17) as $V(\kappa_1, \ldots, \kappa_n) = \text{Var}[Z(s_0)] + \sum_{i=1}^{n} \kappa_i^2 \sigma_{ii} + 2 \sum_{i=1}^{n-1} \sum_{j=i+1}^{n} \kappa_i \kappa_j \sigma_{ij} - 2 \sum_{i=1}^{n} \kappa_i \sigma_{i0}$ and apply differential calculus. That is, find the derivative of $V(\kappa_1, \ldots, \kappa_n)$ with respect to each κ_i. [*Hint*: the first derivative with respect to κ_i of $\sum_{i=1}^{n} \kappa_i^2 \sigma_{ii}$ is $\sigma_{ij} \partial \kappa_i^2 / \partial \kappa_i$, while the first derivative with respect to κ_i of $2 \sum_{i=1}^{n-1} \sum_{j=i+1}^{n} \kappa_i \kappa_j \sigma_{ij}$ is $2 \sum_{\substack{j=1 \\ j \neq i}}^{n} \kappa_j \sigma_{ij}$.]

(b) Take the result from part (a) and set it equal to zero. Show that this is equivalent to $\kappa_i \sigma_{ii} - \sigma_{i0} + \sum_{j \neq i} \kappa_j \sigma_{ij} = 0$.

(c) Simplify the result in part (b) to match (6.18).

6.20. (a) Use the matrix notation for $\boldsymbol{\kappa} = [\kappa_1 \ldots \kappa_n]^T$, $\mathbf{c}_0 = [\sigma_{10} \ldots \sigma_{n0}]^T$, and $\mathbf{V}$ from §6.4.1 to show that $\mathrm{Var}[\hat{Z}(\mathbf{s}_0) - Z(\mathbf{s}_0)]$ in (6.17) can be written in matrix form as $\mathrm{Var}[Z(\mathbf{s}_0)] + \boldsymbol{\kappa}^T \mathbf{V} \boldsymbol{\kappa} - 2\mathbf{c}_0^T \boldsymbol{\kappa}$.

(b) Show that the minimum value of the mean squared prediction error attained under (6.19) is $E[\{\hat{Z}(\mathbf{s}_0) - Z(\mathbf{s}_0)\}^2] = \mathrm{Var}[Z(\mathbf{s}_0)] - \mathbf{c}_0^T \mathbf{V}^{-1} \mathbf{c}_0$. [*Hint*: use the results from part (a) and Exercise 6.17.]

6.21. Show that under the model assumptions in equations (6.6) and (6.25), the relationship $E[\hat{Z}(\mathbf{s}_0) - Z(\mathbf{s}_0)] = 0$ is equivalent to $E[\hat{Z}(\mathbf{s}_0)] = \sum_{j=1}^{P} \beta_j u_j(\mathbf{s}_0)$, where $\hat{Z}(\mathbf{s}_0) = \sum_{i=1}^{n} \kappa_i Z(\mathbf{s}_i)$ is based on equation (6.20).

6.22. Return to the longleaf pine data and, in particular, to the empirical semivariogram you calculated in Exercise 6.15. Estimate the semivariogram by weighted least squares, using the form you specified in that exercise. Include an overlay plot of the estimated semivariogram and the empirical points. Discuss your findings.

6.23. Ecker and Heltsche (1994) describe a study of aquatic species abundance in the Atlantic ocean off the US and Canadian coasts. Measured was the total number of scallops caught in a standardized scallop dredge over a fixed 15 min. tow period. The data are available online at http://www.wiley.com/go/environmental (the data are in three columns: °Latitude | °Longitude | Species count); a sample follows:

°Latitude	°Longitude	Species count
38.60000	73.38333	1
38.65000	73.25000	0
38.65000	73.53333	942
⋮	⋮	⋮
40.86667	71.85000	23
40.90000	71.95000	1
40.91667	71.76667	297

Since the data are counts, apply a Freeman–Tukey variance stabilizing transform (Freeman and Tukey, 1950): $Z = \sqrt{\text{count}} + \sqrt{\text{count} + 1}$.

(a) Plot the transformed data in a three-dimensional fashion to help visualize the observed random field.

(b) Conduct a semivariogram analysis by assuming isotropy and calculating the empirical semivariogram using the estimator in (6.13). Use $B = 20$ distance classes.

(c) Plot the values from part (b) and determine if they appear to match any of the functional forms in §6.3.5. If so, estimate the semivariogram using weighted least squares. Discuss your findings.

6.24. Show that the following semivariogram models correspond to the specified covariance functions under isotropy.

(a) Exponential model (6.11): over $h > 0$, $\gamma(h) = \theta_1 + \theta_2[1 - \exp(-h/\theta_3)]$ corresponds to $C(h) = \theta_2 \exp\{-h/\theta_3\}$.

(b) Gaussian model based on (6.12): over $h > 0$, $\gamma(h) = \theta_1 + \theta_2[1 - \exp(-h^2/\theta_3^2)]$ corresponds to $C(h) = \theta_2 \exp\{-h^2/\theta_3^2\}$.

6.25. Extend your analysis of the longleaf pine data in Exercise 6.22 by finding EBLUPs of the tree diameters, $\hat{Z}(s_0)$, via ordinary kriging. Use a square 200 m × 200 m grid for the spatial locations. Also find the standard errors of prediction $se[\hat{Z}(s_0)]$. Construct surface plots of both quantities. Discuss your findings.

6.26. Return to the scallop abundance data from Exercise 6.23. Assume a normal distribution for the Freeman–Tukey transformed data, with an isotropic, spherical semivariogram for the spatial errors. Perform a REML fit and use your results to complete the following:

(a) Estimate the semivariogram and plot the resulting function.

(b) Find EBLUPs of the transformed abundances via ordinary kriging. Also find the standard errors of prediction. Construct surface plots of both quantities over the original spatial locations.

(c) Suppose we wish to predict the scallop abundance at a new location: latitude 39.5°, longitude 72.75°. Find the EBLUP and calculate an approximate 95% prediction interval for the abundance (number of scallops) at this new location.

6.27. Download the complete Reventazon rainfall data set from http://www.wiley.com/go/environmental and verify the various indications given in Example 6.11. Specifically:

(a) Construct an empirical semivariogram to identify a possible parametric model for use in the universal kriging analysis. Do this by analyzing the location data to find a value for h_B. Operate here with $B = 10$ to compute (6.13). For your detrended outcome variable, use the residuals from a simple linear regression fit of the rainfall rates on u_2 = elevation. Plot the empirical semivariogram; does it appear Gaussian?

(b) Use the plot from part (a) to identify possible initial estimates for the Gaussian semivariogram (6.12) to be fitted via REML. Compare these to the values used in Fig. 6.24.

6.28. In the German state of Niedersachen (Lower Saxony) fossil fuel consumption has led to emissions and wet deposition of highly acidic rainfall ('acid rain'). Berke (1999) gives data on the pH of rainfall at 37 of Niedersachen's mainland weather stations. Also measured was the distance of each station from the

North Sea coast. (A spatial trend may exist where stations close to heavy commercial activity near the coast receive higher levels of acid precipitation.) Distances were taken as km from the separate weather station on the North Sea island of Norderney (longitude 7.15 °N, latitude 53.7 °E) just off the coast in the northwest corner of the state. The data are:

Location	°E Longitude	°N Latitude	pH	Distance (km)
Ahlhorn	8.233	52.900	6.77	14.975
Bad Bentheim	7.167	52.317	5.93	15.384
Bad Rothenfelde	8.167	52.117	5.03	20.924
Berkhof	9.733	52.600	4.97	31.222
Brockel	9.517	53.083	5.33	27.196
Bückeberge	9.400	52.067	4.93	30.917
Buer-Osterwalde	8.400	52.250	4.57	21.288
Damme-Harlinghausen	8.200	52.517	6.00	17.592
Diekholzen-Petze	9.933	52.100	5.10	35.700
Dören-Berg I (hillside)	8.054	52.183	4.53	19.634
Dören-Berg I (hilltop)	8.090	52.183	4.47	19.842
Drangstedt	8.767	53.617	5.70	18.001
Edewecht	7.983	53.133	6.43	11.206
Emden-Knock	7.217	53.367	6.27	3.780
Friedeburg	7.833	53.450	5.97	8.091
Hann.-Münden-Hemeln	9.683	51.417	4.63	37.925
Holzminden-Schießhaus	9.582	51.898	4.43	33.659
Hude-Hasbruch	8.475	53.092	5.73	16.213
Knesebeck	10.700	52.683	4.63	41.063
Königslutter	10.817	52.250	5.63	43.846
Kreuz-Berg	7.717	52.633	5.77	13.431
Lingen-Baccum	7.317	52.517	6.13	13.289
Neuenhaus-Itterbeck	6.967	52.500	6.50	13.498
Northeim-Westerhof	10.000	51.700	4.67	38.717
Osnabrück (city)	8.050	52.267	6.00	18.820
Osterholz-Scharmbeck	8.900	52.917	5.90	21.321
Ovelgönne-Colmar	8.367	53.342	5.83	14.104
Riefensbeek	10.400	51.767	4.60	42.051
Rütenbrock	7.100	52.850	6.60	9.468
Scharnebeck	10.517	53.300	5.03	37.701
Schwaförden	8.833	52.733	5.13	21.586
Seesen-Hohestein	10.183	51.900	4.67	39.222
Siemen	11.233	53.050	4.63	45.978
Stade-Mulsum	9.392	53.558	5.87	24.977
Stuvenwald	9.817	53.367	4.83	29.884
Trillen-Berg	7.850	52.500	6.07	15.448
Unterlüß	10.283	52.833	4.87	36.151

(a) Construct an empirical semivariogram to identify a possible parametric model for use in a universal kriging analysis. Start by analyzing the location data to find a value for h_B. Set $B = 10$ to compute (6.13). Remove the possible spatial trend by operating with the residuals from a linear regression fit of pH on $u_2 =$ distance and $u_3 =$ distance^2. Plot the empirical semivariogram.

(b) Use the plot from part (a) to identify possible initial estimates for an exponential semivariogram (6.11) to be fitted via REML. Perform the fit via universal kriging and include the explanatory variables $u_2 =$ distance and $u_3 =$ distance^2. Compare the fitted semivariogram to the empirical values in part (a).

(c) Niedersachen includes the major city of Hannover (longitude 9.667 °N, latitude 52.4 °E; distance 31.498 km) and completely surrounds the city-state of Bremen (longitude 8.81 °N, latitude 53.08 °E; distance 19.705 km), neither of which had a reporting station within the city limits. From your universal kriging fit in part (b), predict the pH precipitation level for each city. Also include (pointwise) 95% prediction intervals.

7

Combining environmental information

An important area of environmental science involves the combination of information from diverse sources relating to a similar endpoint. A common rubric for combining the results of independent studies is to apply a *meta-analysis*. The term suggests a move past an analysis of standalone data or of a single analysis of pooled, multi-sourced data, to one incorporating and synthesizing information from many associated sources. It was first coined by Glass (1976) in an application combining results in multiple social science studies and is now quite common in many social and biomedical applications.

The typical goal of a meta-analysis is to consolidate outcomes of independent studies, reanalyze the possibly disparate results within the context of their common endpoints, increase the sensitivity of the analysis to detect the presence of environmental effects, and provide a quantitative analysis of the phenomenon of interest based on the combined data. The result is often a pooled estimate of the overall effect. For example, when assessing environmental risks of some chemical hazard – as we discussed in Chapter 4 – it is increasingly difficult for a single, large, well-designed ecological or toxicological evaluation to assess definitively the risk(s) of exposure to the chemical. Rather, many small studies may be carried out under different conditions, on different populations, using different exposure regimens or different chemical metabolites, etc. In some cases, the effects of interest are small and therefore hard to detect with limited sample sizes; or, data on too many multiple endpoints may mask or divert attention from limited, highly localized effects. In these instances, quantitative strategies that can *synthesize* the independent information into a single, well-understood inference will be of great value.

Such strategies can be highly varied, however, and will depend strongly on the environmental endpoint under study, the underlying models assumed for the

Analyzing Environmental Data   W. W. Piegorsch and A. J. Bailer
© 2005 John Wiley & Sons, Ltd   ISBN: 0-470-84836-7 (HB)

data, and the goal/questions of the investigators. In all cases, one must assume that the different studies are considering equivalent endpoints, and that data derived from them will provide essentially similar information when consolidated with similar study conditions. Statistically, this is known as *homogeneity* or *exchangeability* among studies. More formally, the following assumptions should be satisfied:

1. All studies/investigations meet basic scientific standards of quality (proper data reporting/collecting, random sampling, avoidance of bias, appropriate ethical considerations, fulfilling quality assurance/QA or quality control/QC guidelines, etc.).

2. All studies provide results on the same endpoint.

3. All studies operate under (essentially) the same conditions.

4. The underlying effect is a fixed effect as defined in §1.4; that is, it is non-stochastic and homogeneous across all studies. (This assumption relates to the exchangeability feature mentioned above.)

In practice, violations of some of these assumptions may be overcome by modifying the statistical model: for example, differences in sampling protocols among different studies – violating Assumption 3 – may be incorporated via some form of weighting to de-emphasize the contribution of lower-quality studies; or violation of the fixed-effects feature in Assumption 4 may be overcome by employing some form of random-effects model. Throughout the data combination effort, however, the goal is to retain interpretability in the resulting statistical inferences. Adherence to the four main assumptions above will help to achieve this.

We also make the implicit assumption that results from all relevant studies in the scientific literature are available and accessible for the meta-analysis. Failure to meet this assumption is often called the *file drawer problem*; it is a form of publication bias and, if present, can undesirably affect the analysis. (We discuss ways to identify publication bias in §7.3.3, below.)

7.1 Combining *P*-values

Perhaps the best-known and simplest approach to combining information collects together *P*-values from individual, independent studies of the same null hypothesis, H_0, and aggregates the associated statistical inferences into a single, combined *P*-value. R. A. Fisher gave a basic meta-analytic technique towards this end; a readable exposition is given in Fisher (1948). Suppose we observe $K \geq 1$ studies of the same H_0, from which we find the associated *P*-values P_k, $k = 1, \ldots, K$. To combine these values, Fisher recognized that each *P*-value is a random variable, and that under H_0 these are distributed as independently uniform (§A.2.7) on the interval $0 < P_k < 1$. He then showed that the transformed quantities $-2 \log(P_k)$ are each distributed as χ^2 with 2 df. Since a sum of K independent $\chi^2(2)$ random variables is itself χ^2 with $2K$ df (§A.2.8), one can combine the independent P-values together into an aggregate statistic:

$$X^2_{\text{calc}} = -2 \sum_{k=1}^{K} \log(P_k), \tag{7.1}$$

and refer this to $\chi^2(2K)$.

Fisher's method often is called the *inverse χ^2 method*, since it inverts the P-values to construct the combined test statistic in (7.1). The resulting combined P-value is $P[\chi^2(2K) \geq X^2_{\text{calc}}]$, calculable by manipulating the SAS function `probchi`; see Fig. A.5. Alternatively, Piegorsch (2002) gives tables of P-values for χ^2 or t-distributions. Report combined significance when the combined P-value drops below a prespecified significance level, α.

Example 7.1 (Herbivore pine cone usage) Christensen and Whitham (1993) compare the use and disposition of pinyon pine (*Pinus edulis*) cones by animal herbivores in the US state of Arizona. Data were summarized as the correlation in cone usage between insects and mammals at a number of independent sites over a three-year period. Of interest was whether the two animal groups used pine cones for different purposes, which translates here to testing for a significant negative correlation between groups. These led to the following $K = 14$ P-values: $P_1 = 0.046$, $P_2 = 0.072, P_3 = 0.0005, P_4 = 0.999, P_5 = 0.198, P_6 = 0.944, P_7 = 0.0015, P_8 = 0.0075, P_9 = 0.0625, P_{10} = 0.003, P_{11} = 0.0014, P_{12} = 0.0005, P_{13} = 0.238$, and $P_{14} = 0.988$.

At $\alpha = 0.01$ only six of the 14 P-values are separately significant. (P_4 is significant if testing for a positive correlation, but positive correlations were not considered pertinent for the study at hand.) Applying the inverse χ^2 method gives, however, $X^2_{\text{calc}} = -2\{\log(0.046) + \log(0.072) + \cdots + \log(0.238) + \log(0.988)\} = -2(-50.585) = 101.17$. The combined P-value is $P[\chi^2(28) \geq 101.17] < 0.0001$, which is clearly significant at $\alpha = 0.01$. One concludes that, aggregated over all 14 cases, a significant negative correlation exists in cone usage between insects and mammals. ☙

One can construct alternative methods for combining P-values, useful, for example, when χ^2 tables are not immediately available. For example, from a set of independent P-values, $P_1, P_2, \ldots, P_K$, each testing the same H_0, the quantities $\Phi^{-1}(P_k)$ can be shown to be distributed as i.i.d. $N(0,1)$, where $\Phi^{-1}(P)$ is the inverse of the standard normal c.d.f. (This is related to a famous result on c.d.f.s known as the *probability integral transform*; see Hogg and Tanis, 2001, §4.5.) Recall from §A.2.10 that this inverse c.d.f. also may be characterized in terms of standard normal critical points: since the area to the right of the point $\Phi^{-1}(P)$ is $1 - P$, we denote this as $\Phi^{-1}(P) = z_{1-P}$. In SAS, the inverse c.d.f./critical points can be calculated via the `probit` function; see Fig. A.2.

In §A.2.10, we also noted that sums of independent normals are again normal. Applied here, we can sum the individual terms $\Phi^{-1}(P_k)$ to find $\sum_{k=1}^{K} \Phi^{-1}(P_k) \sim N(0, K)$. Dividing by the standard deviation, $\sqrt{K}$, yields yet another standard normal random variable. Thus, the quantity

$$Z_{calc} = \frac{1}{\sqrt{K}} \sum_{k=1}^{K} \Phi^{-1}(P_k), \tag{7.2}$$

can be used to make combined inferences on H_0. Due to Stouffer *et al.* (1949), this is known as the *inverse normal method*, or also *Stouffer's method*.

To combine the *P*-values into a single aggregate *P*-value, calculate from Z_{calc} the lower tail quantity $P_+ = \Phi(Z_{calc})$. This can be computed using the SAS function `probnorm` – see Fig. A.1 – or by referring to Table B.1 and appealing to the symmetry of the standard normal about $z = 0$. As with the inverse χ^2 method, report combined significance if $P_+ < \alpha$.

In some instances, the different *P*-values may be derived from studies of varying precision, for example, if the number of observations differs across studies. We then weight the P_ks differentially, using a set of preassigned weights $w_k > 0$, $k = 1, \ldots, K$. For example, if the individual study sample sizes are n_k, use $w_k \propto n_k$. To modify the inverse normal method to account for differential weighting, replace (7.2) with

$$Z_{calc} = \frac{\sum_{k=1}^{K} w_k \Phi^{-1}(P_k)}{\sqrt{\sum_{k=1}^{K} w_k^2}} \tag{7.3}$$

and continue to reject if $P_+ = \Phi(Z_{calc})$ drops below the desired α-level (Liptak, 1958). This weighted version of the inverse normal method is sometimes called the *Liptak–Stouffer method*. With proper modification, it can also be employed to combine *P*-values from dependent or correlated studies; see, for example, Hartung (1999).

Historically, the first successful approach to combining *P*-values involved neither of these two methods. Rather, it employed the ordered *P*-values, denoted as $P_{[1]} \leq P_{[2]} \leq \cdots \leq P_{[K]}$. Originally proposed by Tippett (1931), the method operated by taking the smallest *P*-value, $P_{[1]}$, and rejecting H_0 from the combined data if $P_{[1]} < 1 - (1 - \alpha)^{1/K}$. Wilkinson (1951) gave an extension of the method using the *L*th ordered *P*-value: reject H_0 from the combined data if $P_{[L]} < C_{\alpha, K, L}$, where $C_{\alpha, K, L}$ is a critical point found using specialized tables (Hedges and Olkin, 1985, Table 3.2). Along with α, the value of $L \leq K$ must be specified in advance.

Wilkinson's extension is more resilient to possible outlying effects than Tippett's method, since it does not rest on the single lowest *P*-value. On the other hand, since it relies upon specialized tables it is also far less useful than the other methods discussed above. Hedges and Olkin (1985) note some other concerns with the Wilkinson extension.

Example 7.2 (Herbivore pine cone usage, cont'd) Returning to the results on herbivore pine cone usage from Example 7.1, we can also apply the inverse normal method of combining *P*-values. This requires preliminary calculation of the inverse normal c.d.f. values (found, for example, using the `probit` function in SAS): $\Phi^{-1}(0.046) = -1.685$, $\Phi^{-1}(0.072) = -1.461, \ldots, \Phi^{-1}(0.238) = -0.713$, and $\Phi^{-1}(0.988) = 2.257$. From these, we find $Z_{calc} = (-1.685 - 1.461 - \cdots - 0.713 + 2.257)/\sqrt{14} = -4.550$, leading to a combined *P*-value of $P_+ = \Phi(-4.550) < 0.0001$. Again, the result is strongly significant.

Alternatively, applying Tippet's method we find $P_{[1]} = 0.0005$ among the $K = 14$ ordered *P*-values. At $\alpha = 0.01$, this is compared to the quantity

$1 - (0.99)^{0.071} = 1 - 0.9993 = 0.0007$. Since $P_{[1]}$ does lie below α, we once again find the combined result to be significant. ✪

Fisher's observation that a P-value under H_0 is uniformly distributed can motivate a variety of statistical manipulations to combine independent P-values. The few described above represent only the more traditional approaches. For a discussion of some others, see Hedges and Olkin (1985).

7.2 Effect size estimation

While useful and simple, the approach of combining P-values does have its drawbacks. By their very nature, P-values are summary measures – sometimes crudely so – and by relying solely upon them one may overlook or fail to emphasize relevant scientific differences among the various independent studies (Gaver *et al.*, 1992). This is typically a problem only when the data combination is disturbed by some complicating factor, but it is still an important concern.

To compensate for potential loss of information, one can calculate directly the size of the effect detected by a significant P-value. Suppose we have a simple two-group experiment where the effect of an external environmental stimulus is to be compared with an appropriate control group. Often, the simplest way to index the effect of the stimulus is to take the difference in observed mean responses between the two groups. For combining information over two such independent studies, it is necessary to standardize the difference in means by scaling inversely to its standard deviation. Championed by Cohen (1969), this is known as a *standardized mean difference*, which for use in meta-analysis is often called an *effect size*. The goal is to represent events on a common metric, in order to quantify departures from a null hypothesis of no difference in mean response between the control and exposure groups.

Formally, consider a series of two-sample studies. Model each observation Y_{gikr} as the sum of a group mean μ_g and an experimental error term ε_{gikr}:

$$Y_{gikr} = \mu_g + \varepsilon_{gikr}, \tag{7.4}$$

where $g = C$ (control group), T (target group); $i = 1, \ldots, m$ strata (if present); $k = 1, \ldots, K_i$ studies per stratum, and $r = 1, \ldots, N_{gik}$ replicates per study. (The index i is used if some stratification variable is to be included in the experimental design; if not present, then i can be ignored or dropped.) The μ_gs are viewed as unknown constants. This statistical model assumes that the additive error terms are normally distributed: $\varepsilon_{gikr} \sim$ indep. $N(0, \sigma_{ik}^2)$. Notice that the population variances are allowed to differ across strata and/or studies, but not between groups.

The goal is to combine effect size information over the K_i studies in each stratum. Under this model, effect size is measured via the standardized mean difference

$$d_{ik} = \frac{\varphi_{ik}(\overline{Y}_{Tik} - \overline{Y}_{Cik})}{s_{ik}}, \tag{7.5}$$

where $\overline{Y}_{gik}$ is the sample mean of the N_{gik} observations in the (i, k)th study ($g = C, T$), s_{ik} is the square root of the pooled variance estimator

$$s_{ik}^2 = \frac{(N_{Tik} - 1)s_{Tik}^2 + (N_{Cik} - 1)s_{Cik}^2}{N_{Tik} + N_{Cik} - 2}$$

using the corresponding per-study sample variances s_{gik}^2, and φ_{ik} is an adjustment factor to correct for bias in small samples:

$$\varphi_{ik} = 1 - \frac{3}{4(N_{Tik} + N_{Cik} - 2) - 1}.$$

For instance, if $N_{Tik} = N_{Cik} = 10$, then $\varphi_{ik} = 1 - \frac{3}{71} = 0.958$. Notice that $\varphi_{ik} \to 1$ as $N_{gik} \to \infty$.

Within each stratum, we combine the individual effect sizes in (7.5) over the K_i independent studies. To do so, we find variance estimators for the effect sizes, $\mathrm{Var}[d_{ik}]$, and weight the d_{ik} inversely to these variances: the weights are $w_{ik} = 1/\mathrm{Var}[d_{ik}]$. (As we mentioned in §1.1, inverse-variance weighting is a common way to correct for heterogeneous precision when consolidating statistical information.) §7.3.1 discusses in more detail its use for combining environmental information.) As might be expected, the variance estimators, $\mathrm{Var}[d_{ik}]$, are quite complex. Here, we employ a large-sample approximation that operates well when the samples sizes, N_{gik}, are roughly equal and are at least 10 for all g, i, and k (Hedges and Olkin, 1985):

$$\mathrm{Var}[d_{ik}] \approx \frac{N_{Tik} + N_{Cik}}{N_{Tik} N_{Cik}} + \frac{d_{ik}^2}{2(N_{Tik} + N_{Cik})}. \tag{7.6}$$

With these, the weighted averages are

$$\overline{d}_{i+} = \frac{\sum_{k=1}^{K_i} w_{ik} d_{ik}}{\sum_{k=1}^{K_i} w_{ik}}. \tag{7.7}$$

Standard practice views a combined effect size as minimal (or 'none') if in absolute value it is near zero, as 'small' if it is near $d = 0.2$, as 'medium' if it is near $d = 0.5$, as 'large' if it is near $d = 0.8$, and as 'very large' if it exceeds 1.0. To assess this statistically, we find the standard error of $\overline{d}_{i+}$, given by

$$se[\overline{d}_{i+}] = \frac{1}{\sqrt{\sum_{k=1}^{K_i} w_{ik}}},$$

and build an approximate $1 - \alpha$ confidence interval on the true effect size. Simplest is the Wald interval $\overline{d}_{i+} \pm z_{\alpha/2} se[\overline{d}_{i+}]$.

Example 7.3 (Heavy metal toxicity) In a study of how heavy metals can collect in human tissues, Ashraf and Jaffar (1997) reported on metal concentrations in scalp hair of males exposed to industrial plant pollution in Pakistan. For purposes of contrast and control, scalp hair concentrations in males were compared to an unexposed urban population. Of interest was whether exposed individuals exhibit increased metal concentrations in their scalp hair, and, if so, how this could be quantified via effect size calculations.

The study was stratified over different essential metals and other elements (copper, iron, calcium, etc.). For simplicity, we consider a single stratum with the response taken as manganese concentration in the scalp hair. Thus $m = 1$ and we drop use of the i subscript.

Based on Ashraf and Jaffar's presentation, we give in Table 7.1 summary data over $K = 6$ cohorts. The per-cohort sample sizes are given in the columns under N_T and N_C. Note that these are all large enough to validate use of the large-sample approximation for $\bar{d}_{i+}$. The table also includes results from intermediate calculations of the effect sizes.

In Table 7.1 the per-cohort effect sizes, d_k, are given in the penultimate column. For example, consider the first effect size, d_1. The small-sample correction is

$$\varphi_1 = 1 - \frac{3}{4(16 + 18 - 2) - 1} = 1 - 0.024 = 0.976,$$

while the pooled sample variance is

$$s_1^2 = \frac{(18 - 1)2.43 + (16 - 1)5.57}{16 + 18 - 2} = \frac{124.86}{32} = 3.902.$$

These lead to the result in Table 7.1:

$$d_1 = 0.976 \frac{4.63 - 3.50}{\sqrt{3.902}} = 0.558.$$

The other effect sizes follow similarly. As can be seen in the table, for all cohorts the exposure effect leads to increased manganese concentrations in the T group relative to the C group (all d_ks are positive), ranging from an increase of 0.558 standard deviations for cohort 1 to an increase of over one standard deviation for cohort 5.

Table 7.1 Summary data and intermediate effect size statistics for manganese concentrations in human scalp hair; T group is exposure to industrial emissions, C group is unexposed control

Cohort (k)	N_T	$\bar{Y}_T$	s_T^2	N_C	$\bar{Y}_C$	s_C^2	φ_k	d_k	w_k
1	16	4.63	5.57	18	3.50	2.43	0.976	0.558	8.154
2	14	6.66	12.74	17	3.70	14.06	0.974	0.785	7.133
3	23	6.46	8.24	22	4.50	4.41	0.982	0.763	10.482
4	26	5.82	9.73	19	4.00	12.25	0.982	0.544	10.595
5	24	9.95	18.58	12	5.56	9.99	0.978	1.080	7.082
6	18	6.30	5.76	18	4.85	4.54	0.978	0.625	8.581

To determine a combined effect size, we begin with the inverse-variance weights, w_k, using the variance terms from (7.6); for example, at $k = 1$,

$$\text{Var}[d_1] = \frac{16 + 18}{288} + \frac{0.588^2}{2(16 + 18)} = \frac{34}{288} + \frac{0.346}{68} = 0.123.$$

These lead to weights $w_k = 1/\text{Var}[d_k]$, which are listed in the last column of Table 1. The summary data can be displayed in what is known as a *forest plot* (Sutton *et al.*, 2000, §10.3.1), which shows the estimated effect sizes along with corresponding, approximate, 95% confidence intervals; see Fig. 7.1. (One can also apply a variant of the forest plot, and display the contribution of each study by representing the point estimate as a variably sized box proportional to w_k.) Sample SAS code to produce a rougher version of the forest plot for these data is given in Fig. 7.2 (output not shown).

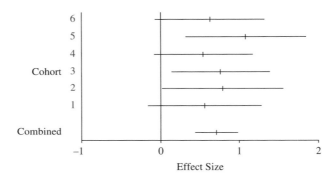

Figure 7.1 Forest plot of effect sizes from heavy metal toxicity data. Vertical line at $d = 0$ indicates no effect. Combined result is based on $\bar{d}_+ \pm z_{0.025}se[\bar{d}_+]$

```
data metal;
  input grpid $ 1-8   upper 11-16   lower 19-25   d 28-32;
  datalines;
  Group 1    1.2713   -0.1553   0.558
  Group 2    1.5507    0.0192   0.785
  Group 3    1.3858    0.1401   0.763
  Group 4    1.1635   -0.0755   0.544
  Group 5    1.8436    0.3163   1.080
  Group 6    1.3187   -0.0687   0.625
  Combined   0.9814    0.4380   0.710

  proc timeplot;
  plot lower="<" d="+" upper=">" / overlay
                       hiloc    ref=0    refchar="|";
  id grpid;
```

Figure 7.2 Sample SAS program for forest plot with heavy metal toxicity data

From these values, one has

$$\bar{d}_+ = \frac{\sum_{k=1}^{6} w_k d_k}{\sum_{k=1}^{6} w_k} = \frac{(8.154 \times 0.558) + \cdots + (7.133 \times 0.625)}{8.154 + \cdots + 7.133} = 0.710.$$

For a 95% confidence interval, calculate

$$s^2[\bar{d}_+] = \frac{1}{\sum_{k=1}^{6} w_k} = \frac{1}{8.154 + \cdots + 7.133} = 0.019,$$

with which we find $\bar{d}_+ \pm z_{0.025} se[\bar{d}_+] = 0.71 \pm 1.96\sqrt{0.019} = 0.71 \pm 0.27$.

The combined estimate of effect is also graphed in the forest plot of Fig. 7.1. The figure clearly indicates the advantage of the data combination: the approximate 95% confidence interval on the overall effect is reduced in length and more separated from zero compared to those of the individual cohorts.

Overall, a 'large' combined effect size is indicated with these data: on average, an exposed male has scalp hair manganese concentrations between 0.44 and 0.98 standard deviations larger than an unexposed male. There appears to be a strong effect driving increased manganese concentrations in Pakistani males exposed to the plant pollution. ✿

One can also assess homogeneity of studies with respect to the effect of interest *within* each stratum (i.e., for fixed i). A measure for within-stratum homogeneity is $Q_{Wi} = \sum_{k=1}^{K_i} w_{ik}(d_{ik} - \bar{d}_{i+})^2$ (Cochran, 1937) where as above, $w_{ik} = 1/\text{Var}[d_{ik}]$. For computing purposes we use the equivalent form

$$Q_{Wi} = \sum_{k=1}^{K_i} w_{ik} d_{ik}^2 - \frac{\left(\sum_{k=1}^{K_i} w_{ik} d_{ik}\right)^2}{\sum_{k=1}^{K_i} w_{ik}}; \tag{7.8}$$

see Exercise 7.5. Under the null hypothesis of homogeneity across studies within the ith stratum, $Q_{Wi} \sim \chi^2(K_i - 1)$. Here again, this is a large-sample approximation that operates well when the samples sizes, N_{gik}, are roughly equal and are at least 10 for all g, i, and k. For smaller sample sizes the test can lose sensitivity to detect departures from homogeneity, and caution is advised.

Reject within-stratum homogeneity in the ith stratum if $P[\chi^2(K_i - 1) \geq Q_{Wi}]$ is smaller than some predetermined α. Correct for multiplicity over the m strata via application of Bonferroni's adjustment (§A.5.4). If study homogeneity is rejected in any stratum, combination of the effect sizes via (7.7) is contraindicated since the d_{ik}s may no longer estimate a homogeneous quantity. Hardy and Thompson (1998) give additional details on use of Q_{Wi} and other tools for assessing homogeneity within a stratum.

Example 7.4 (Heavy metal toxicity, cont'd) Returning to the heavy metal toxicity data in Example 7.3, we apply the homogeneity test statistic in (7.8). Recall that there is only one stratum, so we disregard the i subscript.

From Table 7.1, $\sum_{k=1}^{6} w_k d_k^2 = 27.78$ so, using (7.8), we find $Q_W = 27.78 - (36.92^2/52.03) = 1.58$. Under the null hypothesis of homogeneity among sites, Q_W

is distributed as $\chi^2(6-1)$. The corresponding P-value for testing homogeneity is $P[\chi^2(5) \geq 1.58] = 0.90$. At $\alpha = 0.05$ this is insignificant, so we conclude that no significant heterogeneity exists for this endpoint among the different cohorts. The calculation of a combined effect size in Example 7.3 was indeed valid.   ☯

Example 7.5 (Pollutant site remediation)   Ralph and Petras (1997) report data on DNA damage in green frog (*Rana clamitans*) tadpoles near sites undergoing pollution restoration in southern Ontario. Data were taken in 1994, serving here as the control (C) group, and also one year later in 1995, serving as the target (T) group. Of interest is whether resident tadpoles exhibit decreased DNA damage between the two groups. The endpoint measured per animal is the average length-to-width ratio in DNA fragments from 25 of the tadpole's peripheral blood erythrocytes. Higher ratios indicate greater DNA fragmentation, due ostensibly to pollutant exposure. Thus, a negative effect size indicates improved conditions. The summary data appear in Table 7.2. The table also includes results from the intermediate statistics used in calculating the effect sizes.

Only one stratum is involved with these data, so $m = 1$ and we drop the i subscript. The sample sizes in Table 7.2 are large enough to justify use of the large-sample approximation, although the imbalances between them may lead to slight decreases in sensitivity with the Q_W test.

To calculate the homogeneity test statistic, start with $\sum_{k=1}^{3} w_k d_k = -15.985$, $\sum_{k=1}^{3} w_k = 23.940$, and $\sum_{k=1}^{3} w_k d_k^2 = 19.675$. From these, (7.8) gives $Q_W = 19.675 -[(-15.985)^2/23.940] = 9.002$. With $3 - 1 = 2$ df, the corresponding P-value for testing homogeneity is $P[\chi^2(2) \geq 9.002] = 0.011$. Hence at $\alpha = 0.05$, we conclude that significant heterogeneity exists among pollutant sites. (Lack of sensitivity due to sample-size imbalance did not appear to greatly affect our ability to identify the heterogeneity here.)

This result is consistent with an examination of the site-specific effects. At site 1, the T group exhibits slightly elevated average length-to-width ratios. By contrast, however, the other two sites exhibit moderate to strong decreases in average length-to-width ratios. Given such a change in the direction of the effect, it would be surprising not to reject homogeneity. With these data, it is inadvisable to aggregate the per-site effect sizes into a simple combined estimator.   ☯

If no significant within-stratum heterogeneity is evidenced, it is then possible to assess *between*-stratum heterogeneity (Gurevitch and Hedges, 1993). First, sum the

Table 7.2   Summary statistics for DNA damage in tadpoles: T group is 1995, C group is 1994

Site (k)	N_T	$\overline{Y}_T$	s_T^2	N_C	$\overline{Y}_C$	s_C^2	φ_k	d_k	w_k
1	46	2.129	0.129	12	2.114	0.057	0.987	0.043	9.516
2	36	1.958	0.166	11	2.525	0.128	0.983	-1.405	7.160
3	12	2.087	0.140	23	2.371	0.082	0.977	-0.872	7.264

Source: Ralph and Petras (1997).

individual within-stratum measures into $Q_{W+} = \sum_{i=1}^{m} Q_{Wi}$, and then find the between-stratum measure $Q_B = \sum_{i=1}^{m} \sum_{k=1}^{K_i} w_{ik}(\bar{d}_{i+} - \bar{d}_{++})^2$, where

$$\bar{d}_{++} = \frac{\sum_{i=1}^{m} \sum_{k=1}^{K_i} w_{ik} d_{ik}}{\sum_{i=1}^{m} \sum_{k=1}^{K_i} w_{ik}}. \tag{7.9}$$

Under a null hypothesis of homogeneity across strata, $Q_B \sim \chi^2(m-1)$. Reject the homogeneity hypothesis when the approximate P-value $P[\chi^2(m-1) \geq Q_B]$ drops below some prespecified α.

Q_B is perhaps best computed by recognizing the similarity to partitioned sums of squares in the analysis of variance (Neter *et al.*, 1996, §2.7): the total heterogeneity is measured by $Q_T = \sum_{i=1}^{m} \sum_{k=1}^{K_i} w_{ik}(d_{ik} - \bar{d}_{++})^2 = Q_B + Q_{W+}$, and this may be computed as

$$Q_T = \sum_{i=1}^{m} \sum_{k=1}^{K_i} w_{ik} d_{ik}^2 - \frac{\left(\sum_{i=1}^{m} \sum_{k=1}^{K_i} w_{ik} d_{ik}\right)^2}{\sum_{i=1}^{m} \sum_{k=1}^{K_i} w_{ik}}.$$

Then, find Q_B by subtraction: $Q_B = Q_T - Q_{W+}$.

If between-strata homogeneity is valid, we can pool the effect sizes into the single estimator $\bar{d}_{++}$ from (7.9) An approximate $1 - \alpha$ Wald interval for the combined effect size is then $\bar{d}_{++} \pm z_{\alpha/2} se[\bar{d}_{++}]$, where

$$se[\bar{d}_{++}] = \frac{1}{\sqrt{\sum_{i=1}^{m} \sum_{k=1}^{K_i} w_{ik}}}$$

is the large-sample standard error of $\bar{d}_{++}$.

7.3 Meta-analysis

7.3.1 Inverse-variance weighting

Taken broadly, an 'effect size' in an environmental study can be any valid quantification of the effect, change, or impact under study (Umbach, 2002), not just the difference in sample means employed in §7.2. Thus, for example, we might use estimated coefficients from a regression analysis (Exercise 7.10), potencies (Exercise 7.11), or correlation coefficients (Exercise 7.12).

For any measure of effect, the approach used in (7.7) to produce a combined estimator via inverse-variance weighting remains applicable. Suppose we are interested in an effect measured by some unknown parameter ξ, with estimators $\hat{\xi}_k$ found from a series of independent, homogeneous studies, $k = 1, \ldots, K$. Assume these estimators have variances $Var[\hat{\xi}_k]$, and define the inverse-variance weights as $w_k = 1/Var[\hat{\xi}_k]$. A combined estimator of ξ is then

$$\bar{\xi} = \frac{\sum_{k=1}^{K} w_k \hat{\xi}_k}{\sum_{k=1}^{K} w_k} \tag{7.10}$$

with standard error (Exercise 7.13)

$$se[\bar{\xi}] = \frac{1}{\sqrt{\displaystyle\sum_{k=1}^{K} w_k}}. \tag{7.11}$$

If the $\hat{\xi}_k$s are approximately normal, then an approximate $1 - \alpha$ Wald interval on the common value of ξ is $\bar{\xi} \pm z_{\alpha/2} se[\bar{\xi}]$. Indeed, even if the $\hat{\xi}_k$s are not close to normal, for large K the averaging effect in (7.10) may be strong enough to allow for approximate normality of $\bar{\xi}$, and hence the Wald interval may still be approximately valid. For cases where approximate normality is difficult to achieve, use of the bootstrap method from §A.5.2 can be useful in constructing confidence limits on ξ.

Inverse-variance weighting is a common technique for combining independent, homogeneous information into a single summary measure, as described in early reports by Birge (1932) and Cochran (1937); see Gaver *et al.* (1992) for a concise overview.

In some cases, the large-sample distribution of $\log\{\hat{\xi}_k\}$ is more closely normal than that of $\hat{\xi}_k$. In this case we find instead $\mathrm{Var}[\log\{\hat{\xi}_k\}]$ and take $w_k = \mathrm{Var}[\log\{\hat{\xi}_k\}]^{-1}$ for use in the data combination. That is, calculate

$$\overline{\log(\xi)} = \frac{\sum_{k=1}^{K} w_k \log\{\hat{\xi}_k\}}{\sum_{k=1}^{K} w_k}, \tag{7.12}$$

with $se[\overline{\log(\xi)}] = 1/\sqrt{\sum_{k=1}^{K} w_k}$. A $1 - \alpha$ Wald interval on ξ is then

$$\exp\left\{ \overline{\log(\xi)} \pm \frac{z_{\alpha/2}}{\sqrt{\sum_{k=1}^{K} w_k}} \right\}. \tag{7.13}$$

When $\mathrm{Var}[\log\{\hat{\xi}_k\}]$ is difficult to find but $\mathrm{Var}[\hat{\xi}_k]$ is simple, a delta-method approximation (§A.6) for $\mathrm{Var}[\log\{\hat{\xi}_k\}]$ is $\mathrm{Var}[\hat{\xi}_k]/(\hat{\xi}_k)^2$; cf. Example A.1. From this, the w_ks can be quickly determined.

Example 7.6 (Environmental tobacco smoke)   A major concern in environmental epidemiology is whether tobacco smoke exhaled or ambient from a lit cigarette detrimentally affects non-smokers. Known formally as *environmental tobacco smoke* (*ETS*) and also colloquially as 'passive' smoke, ETS exposure may be an important determinant for lung cancer. A variety of studies have combined dose-response data over multiple sources to investigate this potential environmental risk factor. Givens *et al.* (1997) review the issue and the literature.

To study ETS cancer risk in greater detail, the US Environmental Protection Agency produced a report (US EPA, 1992) that included a meta-analysis of previous ETS studies. At that time, only a few studies of sufficiently high quality were available for examination. The outcome measured in these studies was the relative increase in risk for lung cancer mortality in non-smoking women whose partners smoked (ETS group) over that for non-exposed, non-smoking women (controls). This was measured as the *relative risk* of exposure death to unexposed death

$$RR = \frac{\text{Pr\{death due to lung cancer, given ETS exposure\}}}{\text{Pr\{death due to lung cancer, given no ETS exposure\}}}$$

(Tweedie, 2002). If RR equals 1.0, no increased risk is evidenced. When the ratio is significantly greater (less) than 1.0, we infer instead a significant increase (decrease) in risk. Thus in terms of the notation in this section, $\xi = RR$.

For example, in the $K = 11$ studies the EPA collected from the US, the relative risks for women after ETS exposure are summarized in Table 7.3, taken for use here from Piegorsch and Cox (1996). Since the natural logarithm of $\hat{R}R$ here tends to be more stable and symmetric in distribution than the original ratio (Woolf, 1955), Table 7.3 gives the $\log\{\hat{R}R_k\}$ values. It also reports the inverse-variance weights, $w_k = 1/\text{Var}[\log\{\hat{R}R_k\}]$. Note then that $\log\{RR\} = 0$ is identical to $RR = 1$, and that either case indicates no additional lung cancer mortality risk due to ETS exposure.

The results in Table 7.3 show a range of results: from $\hat{R}R_2 = e^{-0.386} = 0.68$ (i.e., an estimated 32% drop in lung cancer mortality), to $\hat{R}R_3 = e^{0.698} = 2.01$ (i.e., an estimated 101% increase in lung cancer mortality), giving contradictory information. In fact, only one study showed an increase in relative risk that significantly exceeded 1: $\hat{R}R_5 = 1.28$, with an individual Wald statistic for testing $H_0: \log(RR) = 0$ given by $z_5 = \log\{\hat{R}R_5\}/se[\log\{\hat{R}R_5\}] = 1.845$; the corresponding one-sided P-value is 0.033. (Individual Wald statistics for all 11 studies are presented in Table 7.3.) For the other ten US studies, it was unclear if the lack of significant departure from $RR = 1.0$ was due to a truly equal risk between exposed and unexposed groups, or whether it was simply a case of not having enough data to detect a small but important environmental

Table 7.3 Relative risk (RR) statistics and inverse-variance weights for 11 individual US studies of lung cancer risk after ETS exposure

Study number (k)	$\hat{R}R_k$	$\log\{\hat{R}R_k\}$	$w_k = \dfrac{1}{\text{Var}[\log\{\hat{R}R_k\}]}$	$z_k = \dfrac{\log\{\hat{R}R_k\}}{se[\log\{\hat{R}R_k\}]}$
1	1.501	0.406	2.071	0.584
2	0.680	−0.386	4.921	−0.856
3	2.010	0.698	1.877	0.956
4	1.891	0.637	4.318	1.324
5	1.280	0.247	55.787	1.845
6	1.270	0.239	23.645	1.162
7	1.160	0.148	37.776	0.910
8	2.000	0.693	3.379	1.274
9	0.790	−0.236	16.457	−0.957
10	0.730	−0.315	2.858	−0.533
11	1.320	0.278	4.214	0.571

Source: Piegorsch and Cox (1996).

effect. This is a classic instance where combination of the information over all 11 studies might lead to a clearer picture of ETS's impact on cancer mortality.

To perform the calculations, we operate with the values in Table 7.3 as $\log\{\hat{\xi}_k\} = \log\{\hat{RR}_k\}$. From them, we find $\sum_{k=1}^{11} w_k \log\{\hat{\xi}_k\} = 26.768$ and $\sum_{k=1}^{11} w_k = 157.303$. Then (7.12) yields

$$\overline{\log(\xi)} = \frac{26.768}{157.303} = 0.170.$$

That is, a combined point estimate of the relative risk is $e^{0.170} = 1.186$, representing an 18.6% increase in risk of female lung cancer mortality after exposure to ETS. We also find $se[\overline{\log(\xi)}] = 1/\sqrt{157.303} = 0.080$, so that (7.13) gives an approximate 95% confidence interval on $\xi = RR$ as $\exp\{0.170 \pm (1.96 \times 0.080)\}$; i.e., $1.014 < RR < 1.386$. Thus from the combined analysis we find that an increase of between 1.4% and 38.6% in lung cancer mortality is possible for US non-smoking women exposed to ETS. We explore similar results with some other regions from the EPA data in Exercise 7.9. ☯

7.3.2 Fixed-effects and random-effects models

At the beginning of this chapter, we specified in Assumption 4 that the effect under study should be non-stochastic and homogeneous across all studies. This models the effect or phenomenon as a fixed constant; for example, μ_g in equation (7.4). In §1.4 and §5.6 we called this a *fixed effect*, in order to distinguish it from a *random effect* where some additional variation exists among levels of the effect factor. For instance, when estimating effects sizes as in §7.2 it is crucial that the underlying effect size remain constant across the various studies. If not, the calculated values will no longer estimate the same quantity, confusing interpretation of the combined effect size.

To clarify, assume for simplicity that we are studying only a single stratum, so that $m = 1$ in the notation of §7.2. As such, we drop the subscript i in (7.4), producing $Y_{gkr} = \mu_g + \varepsilon_{gkr}$ for $g = C$ (control group), T (target group); $k = 1, \ldots, K$ studies, and $r = 1, \ldots, N_{gk}$ replicates per study. We continue to assume that $\varepsilon_{gkr} \sim$ indep. $N(0, \sigma_k^2)$, that is, that error variances are homogeneous across treatments within a study but may differ across studies. Extensions to multiple groups are also possible (Normand, 1999). Under this model, the underlying parameter being estimated by d_k in (7.5) can be written as

$$\delta_k \propto \frac{\mu_{Tk} - \mu_{Ck}}{\sigma_k}$$

for each $k = 1, \ldots, K$. Homogeneity corresponds to the requirement that all the δ_ks remain constant: $\delta_1 = \delta_2 = \cdots = \delta_K$. If so, we say the effect sizes represent the fixed effects of the target (vs. the control), and we can proceed as described in §7.2 or §7.3.1.

If, however, each study represents a different effect, or if they comprise a random sample of all possible studies of the environmental phenomenon, homogeneity may be violated and the δ_ks could differ. A possible model to describe this adds a second stage, or *hierarchy*, to the definition of δ_k:

$$\delta_k = \Delta + \gamma_k, \tag{7.14}$$

where γ_k, when added to the common, population-averaged effect, Δ, acts as an additional error term that induces a source of heterogeneity among the δ_ks.

The customary specification for γ_k mimics the normal-error assumption at the first stage, such that $\gamma_k \sim$ indep. $N(0, \tau^2)$ for all $k = 1, \ldots, K$. We also assume the γ_ks are mutually independent of the ε_{gkr}s. Notice that this allows the δ_ks to share the same expected value $\mathrm{E}[\delta_1] = \mathrm{E}[\delta_1] = \cdots = \mathrm{E}[\delta_K] = \Delta$. Given the strong similarity between this hierarchical specification and the more general random-effects model in §1.4, we call this a *random-effects model* for the effect sizes. When the hierarchical variance, τ^2, is zero, the model collapses to the simple fixed-effects form.

The random-effects model is applicable when the Q_W test rejects homogeneity of the δ_ks. In this case, the goal becomes one of estimating the underlying mean, Δ, in (7.14). The result is again a form of inverse-variance weighted average of the d_ks:

$$\overline{\Delta} = \frac{\sum_{k=1}^{K} \omega_k d_k}{\sum_{k=1}^{K} \omega_k}, \tag{7.15}$$

where

$$\omega_k = \frac{1}{\mathrm{Var}[d_k] + \tau^2},$$

and $\mathrm{Var}[d_k]$ mimics (7.6):

$$\mathrm{Var}[d_k] = \frac{N_{Tk} + N_{Ck}}{N_{Tk} N_{Ck}} + \frac{d_k^2}{2(N_{Tk} + N_{Ck})}. \tag{7.16}$$

In practice, the hierarchical variance, τ^2, is typically unknown and must be estimated. A simple estimator based on the method of moments (§A.4.2) is

$$\hat{\tau}^2 = \frac{Q_W - (K-1)}{\sum_{k=1}^{K} w_k - \left\{ \sum_{k=1}^{K} w_k^2 / \sum_{k=1}^{K} w_k \right\}} \tag{7.17}$$

(DerSimonian and Laird, 1986), where Q_W is the value of the homogeneity test statistic from (7.8) and $w_k = 1/\mathrm{Var}[d_k]$ as above. Since we assume $\tau^2 \geq 0$, we truncate (7.17) if it drops below 0. (In effect, this then collapses to a fixed-effects calculation.) To complete the analysis, use $\hat{\tau}^2$ in $\omega_k = 1/(\mathrm{Var}[d_k] + \hat{\tau}^2)$, and estimate the mean effect size via (7.15). For the standard error of $\overline{\Delta}$, mimic (7.11) and take $se[\overline{\Delta}] = 1/\sqrt{\sum_{k=1}^{K} \omega_k}$. Note that we use the values of ω_k and *not* w_k in $se[\overline{\Delta}]$; indeed, $\omega_k \leq w_k$ since the latter values do not include study-to-study variability as a component of overall response variability. An approximate $1 - \alpha$ Wald interval for Δ is then $\overline{\Delta} \pm z_{\alpha/2} se[\overline{\Delta}]$.

Example 7.7 (Pollutant site remediation, cont'd) Returning to the meta-analysis on DNA damage in *R. clamitans* tadpoles from Example 7.5, recall that the test for homogeneity yielded $Q_W = 9.002$ for these data ($P = 0.011$). We concluded that significant heterogeneity exists among the $K = 3$ sites in this study. One possible approach to modeling this heterogeneity is to assume that the sites themselves exert a component of variability to the observations, and view this as a form of random effect. Thus, suppose model (7.14) holds, and consider estimating the mean effect size Δ.

To find $\overline{\Delta}$, we begin by estimating the hierarchical variance parameter τ^2. Recall from Table 7.2 that $\sum_{k=1}^{3} w_k = 23.940$, while $\sum_{k=1}^{3} w_k^2 = 194.586$. Applying these in (7.17) produces

$$\hat{\tau}^2 = \frac{9.002 - 2}{23.94 - (194.586/23.94)} = 0.443.$$

We use this in the random-effects weights $\omega_k = 1/(\hat{\tau}^2 + \text{Var}[d_k])$, where $\text{Var}[d_k] = 1/w_k$ from Table 7.2. This yields $\omega_1 = 1.825$, $\omega_2 = 1.717$, and $\omega_3 = 1.723$, so that $\sum_{k=1}^{3} \omega_k = 5.265$ and $\sum_{k=1}^{3} \omega_k d_k = -3.836$. The combined estimator of mean effect size becomes $\overline{\Delta} = -0.729$. (The result is negative, suggesting greater pollutant damage in the control population.)

The standard error is $se[\overline{\Delta}] = 1/\sqrt{5.265} = 0.436$, yielding an approximate 90% confidence interval for Δ of $-0.729 \pm (1.645 \times 0.436)$ or simply -0.729 ± 0.717. A 'large' combined effect size is indicated, although the interval's width is broad enough to suggest that the effect may be somewhat less extensive than indicated by the point estimate alone. Nonetheless, this still represents an important effect between the target and control populations. ✇

For more advanced estimation of τ^2 computer iteration is necessary, and this can require substantial computational resources. If such resources are available, an approximate residual maximum likelihood (see §5.6.2) solution can be achieved by solving the REML estimating equations

$$\overline{\Delta} = \sum_{k=1}^{K} \left\{ \frac{d_k}{\hat{\tau}^2 + \text{Var}[d_k]} \right\} \bigg/ \left(\sum_{k=1}^{K} \left\{ \frac{1}{\hat{\tau}^2 + \text{Var}[d_k]} \right\} \right) \tag{7.18}$$

and

$$\hat{\tau}^2 = \sum_{k=1}^{K} \left\{ \frac{\frac{K(d_k - \overline{\Delta})^2}{K - 1} - \text{Var}[d_k]}{(\hat{\tau}^2 + \text{Var}[d_k])^2} \right\} \bigg/ \left(\sum_{k=1}^{K} \left\{ \frac{1}{\hat{\tau}^2 + \text{Var}[d_k]} \right\}^2 \right) \tag{7.19}$$

(Normand, 1999), where $\text{Var}[d_k]$ is given by (7.16). Solutions to the two equations cannot be written in closed form, unfortunately, so computer iteration is required. We recommend using (7.17) for an initial value of $\hat{\tau}^2$ in (7.18) and then iterating between (7.18) and (7.19) until convergence is achieved. Note that some simplification in these formulas is available by writing $\omega_k(\hat{\tau}) = 1/(\hat{\tau}^2 + \text{Var}[d_k])$. Once the solution is attained, use $se[\overline{\Delta}] = 1/\sqrt{\sum_{k=1}^{K} \omega_k(\hat{\tau})}$ for building confidence intervals or making other inferences on Δ. Normand (1999) illustrates ways to manipulate

PROC MIXED into performing some of these REML operations, while Gaver *et al.* (1992) discuss further uses of $\hat{\tau}^2$ and gives other considerations for applying random-effects and fixed-effects models when combining environmental information.

7.3.3 Publication bias

As a tool, meta-analysis attempts to consolidate multiple, independent sources of information into a single, summary inference. Particularly in the social and biomedical sciences, this can involve extensive surveys of published results from the scientific literature. This is less common in the environmental sciences; however, when a literature survey is used there are certain considerations the analyst must keep in mind. These include the four assumptions listed at the beginning of this chapter, along with a fifth realization: modern editors and/or authors may be reticent to publish negative results that do not advance the field. In effect, there is a natural bias against publication of insignificant results, and simple surveys of the published literature may fail to identify studies that themselves failed to identify an effect. This is a form of *selection bias*; specific terms for it in meta-anlaysis include *publication bias* or the *file drawer problem*.

If present, publication bias can act to skew upwards any combined effect sizes or other meta-analyses of an environmental phenomenon. A simple diagnostic for it is due to Light and Pillemer (1984). Known as a *funnel plot*, the concept is straightforward: graph the sample size of each study included in a meta-analysis against its corresponding effect size d_k (or other measure of interest, such as $\log\{\hat{RR}_k\}$). Where possible, we recommend a slightly more advanced technique: replace sample size on the vertical axis with an inverse measure of variation; for example, plot $1/\sqrt{\mathrm{Var}[d_k]}$ vs. d_k, or $1/\sqrt{\mathrm{Var}[\log(\hat{RR}_k)]}$ vs. $\log(\hat{RR}_k)$, etc. Inverse measures of variation are indicators of the precision in a study. Large variation translates into low precision, while small variation translates into high precision. Since we expect variation to decrease as sample size increases, the plot should resemble an inverted funnel with a wide base and a very narrow top. If, however, there is an active publication bias, large portions of the funnel will appear empty.

The funnel plot is most effective when the number of studies is at least 15 or 20. Even then, it may be difficult to draw out very subtle patterns of publication bias or other selective effects. Indeed, to retain interpretability, examination of publication bias and construction of funnel plots should conform to the same restrictions (and reservations) we mention at the beginning of the chapter.

For summary purposes, Begg and Mazumdar (1994) give a test statistic based on rank correlations that quantifies the graphical features of a funnel plot; also see Duval and Tweedie (2000). Perhaps the best advice to give for avoiding publication bias is, however, always to conduct an objective and thorough search of the literature when conducting a meta-analysis.

Example 7.8 (Environmental tobacco smoke, cont'd) For the ETS analysis discussed in Example 7.6 publication bias was thought to be a concern, since some cancer researchers may have been reticent to publish results on the effects of 'passive' smoking that did not indicate a link to lung cancer. One way to assess this with the data collected by the EPA is to construct a funnel plot. Here, we plot the inverse measure of variation $1/\sqrt{\text{Var}[\log(\hat{RR}_k)]}$ vs. the corresponding summary effect measure $\log(\hat{RR}_k)$.

A funnel plot of the $K = 11$ studies in Table 7.3 may not be helpful, due to the small value of K for just these US studies (however, see Exercise 7.16). Instead, we will pool all $K = 30$ studies given in the US EPA (1992) report. The data for such, taken from Piegorsch and Cox (1996) and unadjusted for region of origin, appear in Table 7.4.

Table 7.4 Log relative risks and reciprocal standard errors for 30 studies of lung cancer risk after ETS exposure

$\log\{\hat{RR}_k\}$	$1/\sqrt{\text{Var}[\log(\hat{RR}_k)]}$
0.406	1.439
−0.386	2.218
0.698	1.370
0.637	2.078
0.247	7.469
0.239	4.863
0.148	6.146
0.693	1.838
−0.236	4.057
−0.315	1.691
0.278	2.053
0.652	3.132
0.732	3.572
−0.301	3.607
0.432	3.623
0.495	5.461
0.920	3.153
0.405	3.590
0.315	5.476
0.936	1.582
0.068	3.784
0.451	5.114
0.678	0.931
0.010	2.164
0.157	3.601
0.182	2.491
0.174	5.292
0.770	2.849
−0.261	2.097
−0.248	7.810

Source: Piegorsch and Cox (1996).

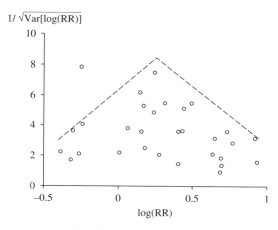

Figure 7.3 Funnel plot of $K = 30$ studies on ETS and lung cancer mortality. Horizontal axis is estimated log-relative risk; vertical axis is reciprocal standard error. Dashed lines indicate 'funnel' effect

The funnel plot of the data from Table 7.4 appears in Fig. 7.3. The result shows a reasonable funnel: a broad base of points slowly narrowing as the points rise, with perhaps an unusual outlier in the upper left corner. Publication bias usually acts to remove insignificant studies, so if present we expect it to affect the funnel plot somewhere near its base; see, for example, Tweedie (2002, Fig. 1). Thus the funnel in Fig. 7.3 does not imply any bias of this form. As for the outlier, this particular point has $\log(\hat{RR}_k) = -0.248$ and $1/\sqrt{\text{Var}[\log(\hat{RR}_k)]} = 7.810$. The latter value is an unusually large measure of precision. It is from a series of ETS studies reported in mainland China (see Exercise 7.9), and may be unusual for some reason – perhaps a signal of study-to-study heterogeneity – other than publication bias. ✪

For more on publication bias in some environmental settings, see Givens *et al.* (1997) or Priyadarshi *et al.* (2001). Hedges (1992) and Thornton and Lee (2000) give useful overall reviews.

7.4 Historical control information

The basic paradigm we employed for constructing the effect size measures in §7.2 involved a simple two-group comparison with data collected on a control group of subjects and on an independent target group of subjects. This is, in fact, a common problem in many environmental studies (Piegorsch and Bailer, 1997, Ch. 4). In some cases, the studies or experiments may involve a number of different target groups, but still use the same protocol for collection of the control-level information. The consequent, ongoing collection of 'historical' control data represents a large base of information which could be combined for statistical purposes. By incorporating the historical data on control responses into the analysis, we may be able to improve sensitivity to detect a significant effect of the target, treatment, exposure, etc., over the controls.

In this section, we discuss some basic methods for combining historical control data for target-vs.-control comparisons. To match the traditional focus in the literature, we emphasize differences between the two groups found via two-sample hypothesis tests. We begin by reviewing some fundamental guidelines and protocol similarities that must be met in order for the historical data to be valid for use with the current experiment or investigation under study.

7.4.1   Guidelines for using historical data

Historical information on environmental phenomena can be easily misapplied, and care must be taken when combining historical data into any statistical analysis. Nomenclature conventions and diagnostic criteria must be calibrated carefully, so as to allow meaningful comparisons among similar effects. Also, any outside sources of variability – such as laboratory differences or changes in susceptibility of a strain or species over time – must be isolated and controlled. Based on suggestions by Pocock (1976), we recommend adherence to the following four criteria when forming a database of historical outcomes, and when employing these data in a statistical analysis. (Although similar in spirit to the general assumptions for data combination we gave at the beginning of the chapter, the criteria here are somewhat more specialized for the target-vs.-control problem.)

1. The historical data must be gathered by the same research organization, and preferably the same research team, that conducts or oversees the current study.

2. The study protocol must remain fixed throughout the period covering the historical and current studies; this includes the method of scoring outcomes of interest.

3. The historical and concurrent control groups must be comparable with regard to age, sex, and other explanatory factors. Where doubt exists as to such comparability, the burden of proof rests upon the research team performing the study.

4. There must be no detectable systematic differences in response between the various control groups.

In practice, selected components of these criteria may be difficult to achieve. For example, if a study has a multi-location/multi-center design, it is difficult for the same research team to conduct each experiment. In such situations, the coordinating investigator must meet Criterion 1 to the best extent possible. Also, if a study protocol has been improved to incorporate technological improvements over time, differences between the old and the improved protocol may invalidate Criterion 2.

These criteria are intended to prevent improper or cavalier incorporation of historical data in environmental decision-making. The analyst must acknowledge at all times the basic premise of any target-vs.-control comparison: a properly constructed, concurrent control represents the most appropriate group to which any target may be compared. Any additional historical information must match the concurrent control with respect to this comparability; cf. Criterion 3. If comparability is compromised, inclusion of historical data is contraindicated. (In some cases, however, it may be possible to include some form of covariate adjustment in the statistical model to correct for poor comparability; for example, adjusting observed responses for age if the subjects have different ages or life spans.) In effect, comparability is a form of statistical exchangeability, as discussed at the beginning of the chapter.

7.4.2 Target-vs.-control hypothesis testing

Formally, suppose the current data are taken in two independent samples, $Y_{ij}, i = 0$ (control), 1 (target); $j = 1, \ldots, n_i$. To incorporate historical control data in the analysis, we mimic a construction due to Pocock (1976) and discussed by Margolin and Risko (1984): assume that Y_{ij} may be modeled as an extension of the one-factor ANOVA model from (1.8),

$$Y_{ij} = \theta + \alpha + i\beta + \varepsilon_{ij}, \tag{7.20}$$

where heterogeneous variances are accepted in the error terms, $\varepsilon_{ij} \sim$ indep. $N(0, \sigma_i^2)$, $i = 0, 1; j = 1, \ldots, n_i$. In addition, a random-effect assumption is made on α to model experiment-to-experiment variability: $\alpha \sim N(0, \tau^2/2)$, independent of ε_{ij}. We incorporate a similar parameter in the historical control assumption, below. (Division by 2 in Var[α] is done for convenience; it leads to some simplifications in the statistical formulas.) θ corresponds to the mean control response. β is the unknown, fixed-effect, treatment parameter corresponding to the difference in mean response between the two groups (Exercise 7.18): $E[Y_{1j}] - E[Y_{0j}] = \beta$. By testing $H_0: \beta = 0$ vs. the one-sided alternative $H_a: \beta > 0$, we assess whether the target group mean response significantly exceeds that in the control group. (Two-sided tests are also possible by moving to $H_a: \beta \neq 0$.)

The independent historical control data are incorporated as a single average value, $\overline{Y}_h$, based on n_h observations. We model this to shadow (7.20):

$$\overline{Y}_h = \theta + \delta + \gamma_h, \tag{7.21}$$

where the control error term is $\gamma_h \sim N(0, \sigma_h^2/n_h)$, independent of the control random effect $\delta \sim N(0, \tau^2/2)$. Some authors refer to this as a *superpopulation model*, emphasizing that the historical controls are drawn from some overarching control population independent of and encompassing the concurrent control. Notice that the experiment-to-experiment variability, as captured by $\tau^2/2$, is modeled identically in both (7.20) and (7.21).

In the model assumptions for (7.20) and (7.21), the variance parameters are assumed known. For σ_h^2 and τ^2 this is not unreasonable, since we expect a large database of control information to be available. (That is, if variance estimates are drawn from this database, the associated df should be large enough to approximate known information.) For the current experiment, the variances σ_0^2 and σ_1^2 also are assumed known. The latter assumption is made primarily for mathematical simplicity. If inappropriate, sample variances can be substituted for the unknown σ_i^2-values, changing the approximate reference distribution from standard normal to t.

It can be shown that the maximum likelihood estimate for β under this model is a linear combination of sample means:

$$b = \overline{Y}_1 - \hat{\alpha}\overline{Y}_0 - (1 - \hat{\alpha})\overline{Y}_h, \tag{7.22}$$

where $\overline{Y}_0$ and $\overline{Y}_1$ are the sample control and treatment means, respectively, from the current data, $\overline{Y}_h$ is the historical control mean, and $\hat{\alpha}$ is an estimate of α based on the known variance components σ_0^2, σ_h^2, and τ^2:

$$\hat{\alpha} = \frac{(\sigma_h^2/n_h) + \tau^2}{(\sigma_0^2/n_0) + (\sigma_h^2/n_h) + \tau^2}.$$

In (7.22), the lagging terms $\hat{\alpha}\overline{Y}_0 + (1 - \hat{\alpha})\overline{Y}_h$ represent a weighted estimate of control responses using both the concurrent control and the historical control information. If historical controls are highly variable relative to the concurrent controls, σ_h^2/n_h or τ^2 (or both) will be large relative to σ_0^2/n_0, so that $\hat{\alpha} \approx 1$. As a result, the historical control information will be weighted lower than the concurrent control information. If the reverse is true and $\hat{\alpha} \approx 0$, the historical controls will be given larger weight.

For simplicity, let $\Omega = (\sigma_1^2/n_1)/(\sigma_0^2/n_0)$. Then, the standard error of b under $H_0: \beta = 0$ was given by Margolin and Risko (1984) as

$$se[b] = \left(\frac{n_0}{\sigma_0^2(1 + \Omega)} + \frac{1}{\left\{ \Omega(1 + \Omega)\sigma_0^2/n_0 \right\} + \left\{ (1 + \Omega)^2\sigma_h^2/n_h \right\} + (1 + \Omega + \Omega^2)\tau^2} \right)^{-1/2}.$$

In large samples, reject H_0 in favor of H_a when the Wald statistic $W_{calc} = b/se[b]$ exceeds z_α.

We should warn that in practice these calculations may produce only limited increases in statistical power to detect $\beta > 0$. For a similar hierarchical model, Hoel (1983) found that power increases were moderate, only achieving substantial improvements when σ_h^2 was much smaller than σ_0^2 and σ_1^2.

Another important use of historical control information occurs for data in the form of proportions. In this case, it is possible to construct a model similar to that used above, but with slightly better sensitivity. We begin by assuming $Y_i \sim$ indep. $\text{Bin}(N_i, \pi_i), i = 0, 1$. Our goal is to test $H_0: \pi_0 = \pi_1$ vs. $H_a: \pi_0 < \pi_1$. (Again, two-sided alternatives are also possible, with appropriate modifications.)

Suppose a set of M_h independent historical control proportions are available, say, $Y_{hj}/N_{hj}, j = 1, \ldots, M_h$. We incorporate these into the model by assuming $Y_{hj} \sim$ indep. $\text{Bin}(N_{hj}, \pi_0)$. These are pooled into a single historical proportion by dividing the sum of all the historical counts by the sum of the historical sample sizes: $Y_{h+}/N_{h+} = \sum_{j=1}^{M_h} Y_{hj}/\sum_{j=1}^{M_h} N_{hj}$. To account for the additional information in the historical data, we assume further that the control response probability π_0 is itself random. To model this over $0 < \pi_0 < 1$, we use a distribution specific to the unit interval known as the *beta distribution*. This assigns to π_0 the p.d.f.

$$f(\pi_0) = \frac{1}{\Gamma(u)\Gamma(v)}\Gamma(u + v)\pi_0^{u-1}(1 - \pi_0)^{v-1}I_{(0,1)}(\pi_0);$$

recall that $\Gamma(u)$ is the gamma function from (A.10). The beta distribution assumption used here for π_0 induces a beta-binomial distribution (§A.2.2) on the marginal distribution of Y_{h+} (Hoel, 1983). The marginal mean and variance are $E[Y_{h+}] = N_{h+}u/(u + v)$ and $\text{Var}[Y_{h+}] = N_{h+}uv(u + v + N_{h+})/\{(u + v)^2(u + v + 1)\}$. It will be convenient to reparameterize u and v into $\mu = u/(u + v)$ and $\varphi = 1/(u + v)$, so that $E[Y_{h+}] = N_{h+}\mu$ and $\text{Var}[Y_{h+}] = N_{h+}\mu(1 - \mu)(1 + N_{h+}\varphi)/(1 + \varphi)$; cf. equation (A.8).

We use Y_{h+} and N_{h+} to estimate the hierarchical beta parameters, u and v. For simplicity, we appeal to the method of moments (§A.4.2), although to do so we require additional information on variability in the control population. This is usually in the form of a sample variance, $S^2 = \sum_{j=1}^{M_h}\{(Y_{hj}/N_{hj}) - (Y_{h+}/N_{h+})\}^2/(M_h - 1)$, or in some cases a range, of the historical proportions. In the latter case, a useful approximation is $S^2 \approx (\text{Range})^2/26.5$, where 'Range' is the observed range of the proportion outcomes. This relationship comes about by assuming that the N_{hj}s are large enough to approximate the historical proportions via a normal distribution. Then, one appeals to basic properties of the normal: if $X \sim N(\mu, \sigma^2)$ we expect it to lie between $\pm z_{0.995}\sigma = \pm 2.576\sigma$ of μ 99% of the time, that is, $P[\mu - z_{0.995}\sigma < X < \mu + z_{0.995}\sigma] = 0.995 - 0.005 = 0.99$. The corresponding range of probable data values is then Range $\approx (\mu + 2.576\sigma) - (\mu - 2.576\sigma) = 2(2.576\sigma) = 5.152\sigma$. Solving for σ^2 yields $\sigma^2 \approx (\text{Range}/5.152)^2 \approx (\text{Range})^2/26.5$.

To apply the method of moments, we equate the first moment of Y_{h+}/N_{h+} to the historical proportion. This gives $\mu = Y_{h+}/N_{h+}$. Next, equating the variance of Y_{h+}/N_{h+} to the sample variance produces

$$\varphi = \frac{S^2 - \{\mu(1 - \mu)/N_{h+}\}}{\mu(1 - \mu) - S^2};\qquad (7.23)$$

see Exercise 7.19. Transforming back to u and v via the inverse relationship $u = \mu/\varphi$, $v = (1 - \mu)/\varphi$ gives

$$u = \mu N_{h+}\frac{\mu(1 - \mu) - S^2}{N_{h+}S^2 - \mu(1 - \mu)}$$

and

$$v = N_{h+}(1 - \mu)\frac{\mu(1 - \mu) - S^2}{N_{h+}S^2 - \mu(1 - \mu)},$$

where $\mu = Y_{h+}/N_{h+}$ from above. If the estimated values of u and/or v drop below their lower bounds of zero, simply truncate them there.

To test H_0 vs. H_a, a statistic incorporating u and v that measures departure from H_0 is

$$T_{\text{calc}} = \frac{(N_0 + u + v)Y_1 - N_1(Y_0 + u)}{u + v + N_+},$$

where $N_+ = N_0 + N_1$ (Hoel and Yanagawa, 1986; Krewski et al., 1988). Unfortunately, the small-sample distribution of T_{calc} can be highly skewed and is difficult to approximate. If computer resources are available, a Monte Carlo approach as described in §4.4.2 or the similar bootstrap method from §A.5.2 may be useful. Otherwise, to overcome the distributional difficulties Hoel and Yanagawa (1986) recommend performing the analysis conditional on the concurrent control value, Y_0, similar to the exact (conditional) logistic regression analysis mentioned in §3.3.2. Hoel and Yanagawa note that for testing H_0 vs. H_a, the conditional analysis is locally most powerful against alternative configurations close to H_0. The conditional standard error of T_{calc} is

$$se[T_{calc}|Y_0] = \sqrt{\frac{N_0 + u + v}{N_0 + u + v + 1} p_0(1 - p_0)\left\{N_1 - \frac{N_1^2}{u + v + N_+}\right\}},$$

where p_0 is an updated estimator of π_0 based on the historical control information: $p_0 = (Y_0 + u)/(N_0 + u + v)$. In large samples, one builds the Wald statistic $W_{calc} = T_{calc}/se[T_{calc}|Y_0]$ and rejects H_0 in favor of H_a when $W_{calc} > z_\alpha$. The corresponding one-sided P-value may be found from Table B.1. In small samples, Hoel and Yanagawa (1986) give an exact P-value that can be calculated using probabilities from the corresponding conditional beta-binomial distribution.

Example 7.9 (Environmental carcinogenesis of urethane) Ethyl carbamate (also called urethane) is a chemical intermediate used in the production of outdoor pesticides. To test its potential for inducing mammalian carcinogenicity, the US National Toxicology Program (NTP) exposed $N_1 = 47$ female B6C3F$_1$ mice to 10 ppm urethane in their drinking water over a two-year period (National Toxicology Program, 2004). At the end of the study, the mice were examined histopathologically for evidence of tumor induction. The study found that $Y_1 = 11$ of the 47 mice (23.4%) exhibited hepatocellular tumors. In a concurrent control group of $N_0 = 48$ mice, only $Y_0 = 5$ such tumors (10.4%) were observed. Of interest is whether the rate in the exposed animals indicates a significant increase in carcinogenesis due to the chemical.

To assess this question, we assume a binomial model for the observed proportions: $Y_i \sim$ indep. Bin(N_i, π_i), $i = 0, 1$. We wish to test $H_0: \pi_0 = \pi_1$ vs. $H_a: \pi_0 < \pi_1$. Studied in isolation, such an analysis is typically conducted using what is known as *Fisher's exact test*, which is similar to the exact tests discussed in §3.3.2. The method apportions the data into a 2×2 contingency table (cf. Example 3.10), comparing the frequencies of success (here animals with tumors) and failure (no tumors) between the exposed and control groups. Table 7.5 displays the urethane data in such a classification; note the inclusion of column and row totals.

Table 7.5 Liver tumor induction in female B6C3F$_1$ mice

	Control	Exposed	Total
Animals with tumor	5	11	16
Animals without tumor	43	36	79
Total	48	47	95

Source: National Toxicology Program (2004, Table 6).

Fisher (1935) argued that by computing all possible tabular configurations of the 2×2 data under the condition that the row and column marginal totals are held fixed, the corresponding conditional probability of recovering a tabular configuration as extreme as or more extreme than that actually observed represents a P-value for testing H_0 against H_a. If $P < \alpha$, reject H_0.

To implement Fisher's exact test, a useful site on the World Wide Web is http://www.matforsk.no/ola/fisher.htm. SAS users can employ PROC FREQ; Fig. 7.4 gives sample PROC FREQ code for the data in Table 7.5. In the figure, we use the SAS variable `tumor` to classify whether liver tumors were observed (`tumor=0`) or not (`tumor=1`), and the SAS variable `trt` to classify the group status: control (`trt=0`) or exposed (`trt=1`). Figure 7.5 displays the corresponding output (edited for presentation). From it, Fisher's exact test gives the P-value for increasing (`Left-sided`) cancer induction due to urethane exposure as $P = 0.078$. At $\alpha = 0.05$, no significant increase is suggested.

Can incorporation of historical data improve the statistical sensitivity here? To assess this, consider the NTP's reported historical control rates in these female mice:

```
data urethane;
input tumor trt ycount @@;
datalines;
0 0 5     0 1 11     1 0 43     1 1 36

proc freq;
weight ycount;
tables tumor*trt/exact;
```

Figure 7.4 Sample SAS program for Fisher's exact test with urethane carcinogenesis data

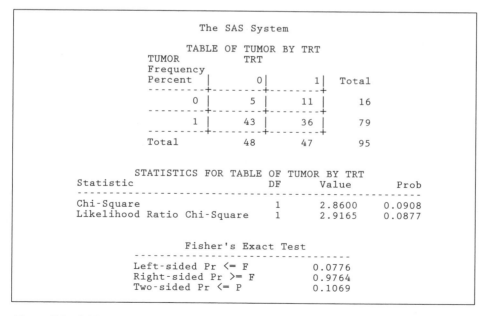

Figure 7.5 SAS output (edited) for Fisher's exact test with urethane carcinogenesis data

$Y_{h+} = 29$ control animals out of $N_{h+} = 515$ exhibited hepatocellular tumors in comparable two-year NTP studies, with an historical range of 11%. We can use these values to derive estimates of the hierarchical parameters u and v. First, we use the reported range to approximate the historical variance as $S^2 \approx 0.11^2/26.5 = 4.566 \times 10^{-4}$. We also take $\mu = \frac{29}{515} = 0.056$. (Compare this to the concurrent control response of $\frac{5}{48} = 0.104$.) Via (7.23), this leads to $\varphi = 0.0067$. Next, we apply the MOM equations to find $u = 8.3618$ and $v = 140.1315$, producing the associated test statistic

$$T_{calc} = \frac{(48 + 8.3618 + 140.1315)11 - 47(5 + 8.3618)}{8.3618 + 140.1315 + 95} = \frac{1533.422}{243.493} = 6.298.$$

The conditional standard error is

$$se[T_{calc}|Y_0] = \sqrt{\frac{196.4933}{197.4933} 0.068 \times 0.932 \left\{ 47 - \frac{2209}{243.4933} \right\}} = 1.546.$$

From these, the Wald statistic is $W_{calc} = 6.298/1.546 = 4.07$. From Table B.1, we see the approximate one-sided P-value is well below 0.0001. A significant increase in hepatocellular carcinogenesis is suggested, after incorporation of the historical information.

We end this example with an important caveat: since the animals' exposure to the chemical was taken over an extended period in this study, intercurrent mortality can exist between the exposed and unexposed animals. If not taken into account, this can affect the statistical inferences (Bailer and Portier, 1988). Here, there was no significant difference in survival between the control group and the exposure group (National Toxicology Program, 2004, Table 2), so survival adjustment was not critical. ✪

Exercises

7.1. Exposure to hazardous chemical agents can induce cancer in a variety of ways. For example, if an agent *promotes* carcinogenesis the form of the dose response often appears more curvilinear as exposure increases. That is, if x quantifies the exposure or dose a common model for the dose response is the simple linear function, $\beta_0 + \beta_1(x - \bar{x})$. If the promoter induces a super-linear response, however, it may be better explained via inclusion of a quadratic dose term: $\beta_0 + \beta_1(x - \bar{x}) + \beta_2(x - \bar{x})^2$. When $\beta_2 \neq 0$, simple linearity is insufficient to explain the dose response. To assess this possibility, Kitchin *et al.* (1994) reported on promotion of skin cancers by the chemical 12-*O*-tetradecanoylphorbol-13-acetate (TPA) in Swiss CD-1 mice. Five separate studies gave P-values for testing $H_0: \beta_2 = 0$ as $P_1 = 0.0048$, $P_2 = 0.0032$, $P_3 = 0.86$, $P_4 = 0.14$, and $P_5 = 0.27$.

(a) Apply the Fisher inverse χ^2 method of combining P-values. Is the aggregate inference significant at $\alpha = 0.01$?

(b) Apply the inverse normal method of combining P-values. Is the aggregate inference significant at $\alpha = 0.01$?

(c) Apply Tippet's method of combining P-values. Is the aggregate inference significant at $\alpha = 0.01$?

(d) How do your results compare across the three methods? What is your assessment of the carcinogenic promotion capability of TPA?

7.2. In Exercise 3.10 we described a study by Burkhart *et al.* (1998) on embryonic deformities in pond frogs (*Xenopus laevis*) taken from Minnesota wetlands in the summer of 1997. In fact, those data were part of a larger set of $K = 3$ investigations, each questioning whether the probability of malformation in frogs increased over time. This led to three independent P-values for assessing the malformation rate: $P_1 = 0.108$, $P_2 = 0.039$, and $P_3 = 0.090$.

(a) Apply the Fisher inverse χ^2 method of combining P-values. Is the aggregate inference significant at $\alpha = 0.01$?

(b) Apply the inverse normal method of combining P-values. Is the aggregate inference significant at $\alpha = 0.01$?

(c) Apply Tippet's method of combining P-values. Is the aggregate inference significant at $\alpha = 0.01$?

(d) How do your results compare across the three methods? What is your assessment of the malformation rate?

7.3. The studies in Exercise 7.2 used different sample sizes (numbers of frogs) to produce the three P-values. These were $n_1 = 370, n_2 = 371$, and $n_3 = 759$. Return to the individual P-values and apply the weighted inverse normal method in (7.3) to assess the aggregate evidence in favor of a malformation effect over time. Operate at $\alpha = 0.01$. Comment on any differences in the inference from Exercise 7.2(b).

7.4. The pine cone usage studies in Example 7.1 used different sample sizes (numbers of trees) to produce the $K = 14$ P-values. These were $n_1 = 33, n_2 = 20, n_3 = 12$, $n_4 = 9, n_5 = 11, n_6 = 11, n_7 = 11, n_8 = 15, n_9 = 10, n_{10} = 10, n_{11} = 10, n_{12} = 13$, $n_{13} = 8$, and $n_{14} = 8$. Return to the individual P-values and apply the weighted inverse normal method in (7.3) to assess the aggregate evidence for a usage correlation. Operate at $\alpha = 0.01$. Comment on any differences in the inference from Example 7.2.

7.5. Verify computing formula (7.8) for the Q_{Wi} homogeneity statistic; that is, for fixed i show

$$\sum_{k=1}^{K_i} w_{ik}(d_{ik} - \bar{d}_{i+})^2 = \sum_{k=1}^{K_i} w_{ik}d_{ik}^2 - \frac{\left(\sum_{k=1}^{K_i} w_{ik}d_{ik}\right)^2}{\sum_{k=1}^{K_i} w_{ik}},$$

where $\bar{d}_{i+}$ is given in (7.7).

7.6. DeLeve (1996) gave data on genetic damage (as rates of sister chromatid exchange) in cultured human skin cells after *in vitro* exposure to dinitrochlorobenzene.

A series of multiple experiments over three different types ($i = 1, 2, 3$) of skin cells were reported, each with a separate control (C) group. Each target (T) group represented exposure to $5\,\mu M$ of dinitrochlorobenzene. Three different experiments ($k = 1, 2, 3$) for each cell type provided the following summary information:

Type (i)	Expt. (k)	N_T	$\overline{Y}_T$	s_T^2	N_C	$\overline{Y}_C$	s_C^2
1	1	20	12.86	0.86	20	9.38	0.10
	2	20	11.11	1.14	20	7.46	0.36
	3	20	13.86	0.77	20	8.92	0.59
2	1	20	11.16	2.37	20	9.55	1.38
	2	20	14.10	0.72	20	9.26	0.10
	3	20	10.89	2.16	20	7.69	1.11
3	1	20	10.86	0.86	20	6.32	0.27
	2	20	11.24	0.41	20	7.65	0.76
	3	20	11.22	0.83	20	7.55	0.97

(a) Find the effect sizes d_{ik} using (7.5), and also calculate the variances $\mathrm{Var}[d_{ik}]$ and weights $w_{ik} = 1/\mathrm{Var}[d_{ik}]$ for $i = 1, 2, 3$; $k = 1, 2, 3$.

(b) Use the Q_{Wi} statistic to assess if the three experiments appear homogeneous within types, $i = 1, 2, 3$. Set $\alpha = 0.10$.

(c) If your analysis in part (b) shows no significant departure from homogeneity in any type, calculate a 90% confidence interval for the combined effect size for that type.

7.7. Schlekat *et al.* (1995) compared polluted sediment in a marine estuary in the US state of Connecticut with ostensibly pristine sediment from nearby Long Island Sound. The measure of contamination was survival rates in each estuary of the estuarine amphipod *Leptocherirus plumulosus*. We view the polluted estuary as the target (T) site and the pristine estuary as the control (C) site. Analyzed by $K = 7$ different laboratories, the summary survival rate data are:

Lab. (k)	N_T	$\overline{Y}_T$	s_T	N_C	$\overline{Y}_C$	s_C	φ_k	d_k	w_k
1	5	60.0	4.2	5	91.3	4.8	0.903	−6.269	0.423
2	5	62.0	11.0	5	91.0	8.9	0.903	−2.618	1.346
3	5	34.0	15.2	5	88.0	8.4	0.903	−3.972	0.841
4	5	48.0	23.9	5	92.0	7.6	0.903	−2.241	1.536
5	5	20.0	9.4	5	86.0	10.2	0.903	−6.078	0.445
6	5	76.0	10.2	5	95.0	6.1	0.903	−2.042	1.643
7	5	78.0	13.0	5	99.0	2.2	0.903	−2.034	1.648

Use the Q_W statistic to show that the effect size across laboratories is not homogeneous at $\alpha = 0.10$. Are there any concerns you should have with this analysis?

7.8. In Example 7.5, one can *partition* the homogeneity calculations on tadpole damage effect sizes into separate components, in order to determine which of the separate sites may have contributed to the departure from homogeneity. To do so, recognize that the homogeneity χ^2 statistic, Q_W, is

$$\sum_{k=1}^{3} w_k (d_k - \bar{d}_+)^2 = w_1 (d_1 - \bar{d}_+)^2 + w_2 (d_2 - \bar{d}_+)^2 + w_3 (d_3 - \bar{d}_+)^2$$

and so each separate term $w_k (d_k - \bar{d}_+)^2$ represents a contribution to the departure from homogeneity. Find each of the three terms for these data, and identify which are large relative to the other(s). (Descriptively, the larger terms are contributing in a greater sense to the departure from homogeneity.)

7.9. The data from Example 7.8 in Table 7.4 give relative risks from the US EPA study on ETS from geographic regions across the globe. These can, in fact, be stratified by region of origin. We give below a sampling. For each region, calculate the combined relative risk estimate. Also find 90% confidence intervals on the relative risk.

(a) Region: Western Europe ($K = 4$).

Study number (k)	1	2	3	4
$\log\{\hat{RR}_k\}$	0.678	0.010	0.157	0.182
$w_k = 1/\text{Var}[\log\{\hat{RR}_k\}]$	0.87	4.68	12.97	6.21

(b) Region: Japan ($K = 5$).

Study number (k)	1	2	3	4	5
$\log\{\hat{RR}_k\}$	0.405	0.315	0.936	0.068	0.451
$w_k = 1/\text{Var}[\log\{\hat{RR}_k\}]$	12.89	29.98	2.50	14.32	26.16

(c) Region: China, including Hong Kong ($K = 8$).

Study number (k)	1	2	3	4	5
$\log\{\hat{RR}_k\}$	0.174	0.770	-0.261	-0.248	-0.301
$w_k = 1/\text{Var}[\log\{\hat{RR}_k\}]$	28.00	8.12	4.40	60.99	13.01

Study number (k)	6	7	8
$\log\{\hat{RR}_k\}$	0.432	0.495	0.920
$w_k = 1/\text{Var}[\log\{\hat{RR}_k\}]$	13.12	29.83	9.94

7.10. Link and Sauer (1996) give results on trends in population decline/increase of North American breeding birds during the period 1966–1993. Data on $K = 27$ similar grassland nesting species were summarized as trend coefficients, b_k, from a regression of population on time, along with their inverse variances. Ordered from smallest to largest, these were:

Species number (k)	1	2	3	4	5	6	7
b_k	−5.12	−4.65	−4.04	−3.36	−3.61	−2.51	−2.25
$w_k = 1/\mathrm{Var}[b_k]$	0.046	0.263	1.602	0.518	1.132	1.352	0.756
Species number (k)	8	9	10	11	12	13	14
b_k	−2.23	−1.69	−1.56	−1.53	−1.43	−1.26	−0.72
$w_k = 1/\mathrm{Var}[b_k]$	7.305	0.323	5.408	10.406	0.422	2.778	10.406
Species number (k)	15	16	17	18	19	20	21
b_k	−0.57	−0.56	−0.55	−0.38	−0.36	−0.25	0.77
$w_k = 1/\mathrm{Var}[b_k]$	2.973	9.183	13.717	9.183	7.716	4.726	0.980
Species number (k)	22	23	24	25	26	27	
b_k	0.92	1.72	2.29	2.34	2.64	7.53	
$w_k = 1/\mathrm{Var}[b_k]$	0.706	1.452	0.253	1.108	1.731	0.157	

Use (7.10) to calculate a combined trend estimate, $\bar{b}$. Also find a 95% confidence interval on the trend coefficient.

7.11. As mentioned in §7.3.1, effect sizes can take on a variety of quantitative forms. For example, one might view a potency estimator (§4.1) as an effect size, and wish to combine potencies across disparate studies of the same environmental agent. Along these lines, Simmons *et al.* (2003) gave data on potencies of the monoazo dye CI solvent yellow, assayed for its mutagenic potential in the bacterium *Salmonella typhimurium*. Those authors used a form of slope estimator from an exponential dose-response model (Leroux and Krewski, 1993) to quantify the mutagenic potency. Over $K = 4$ strains of the bacterium, the following potency estimates and standard errors were calculated:

Strain (k)	1	2	3	4
Potency, $\hat{\xi}_k$	6.43	0.291	13.60	5.83
$se[\hat{\xi}_k]$	2.09	0.327	9.58	1.49

Compute a combined estimator of mutagenic potency from these values, using (7.10). Also find the associated standard error from (7.11), and use this to build an approximate 90% confidence interval for the underlying potency, ξ.

7.12. Combining correlation coefficients is common in many application areas (Hedges and Olkin, 1985, Ch. 11; Hunter and Schmidt, 2004, Ch. 5). For example, return to the pine cone usage studies in Example 7.1. The estimated correlation coefficients (and their sample sizes, n_k) from that study were:

Study number (k)	1	2	3	4	5	6	7
r_k	-0.367	-0.431	-0.945	0.606	-0.191	0.476	-0.682
n_k	33	20	12	9	11	11	11
Study number (k)	8	9	10	11	12	13	14
r_k	-0.593	-0.444	-0.829	-0.646	-0.894	-0.298	0.411
n_k	15	10	10	10	13	8	8

(a) To construct a combined correlation coefficient for these data, first transform them into Fisher Z-variates (Fisher, 1921), $Z_k = \frac{1}{2}\log\{(1+r_k)/(1-r_k)\}$. These transformed variates have variances $\mathrm{Var}[Z_k] \approx 1/(n_k - 3)$, so to build the combined estimator $\overline{Z}$, apply (7.10) with $w_k = n_k - 3$. Also find the standard error $se[\overline{Z}]$ and from this calculate an approximate 95% confidence interval on the combined Z-variate.

(b) Complete the calculations in part (a) by returning to the original correlative scale: apply the inverse Fisher Z-transform $r = (e^{2Z} - 1)/(e^{2Z} + 1)$ to the combined point estimate and 95% limits you computed above.

7.13. Verify that the standard error of $\overline{\xi}$ from (7.10) is given by (7.11). [*Hint*: assume that the $\hat{\xi}_k$s are independent random variables with variances $\mathrm{Var}(\hat{\xi}_k) = 1/w_k$ and show that $\mathrm{Var}(\overline{\xi})$ is the square of (7.11); equation (A.6) may prove useful here.]

7.14. Return to the data on survival in *L. plumulosus* between $K = 7$ different laboratories in Exercise 7.7. Recall that the test for homogeneity was significant with these data, implying the existence of significant between-laboratory heterogeneity. To estimate an effect size in this case, consider a random-effects model for the data as in (7.14).

(a) Use your results from Exercise 7.7 to estimate the hierarchical variance parameter τ^2.

(b) Using the value found in part (a), find the random-effects weights ω_k and from these estimate the effect size parameter Δ via (7.15).

(c) Calculate the standard error $se[\overline{\Delta}]$ and use this to construct a 95% Wald interval for Δ.

7.15. If you have access to sufficient computational resources, return to the following data and solve the REML estimating equations for Δ and τ^2 in (7.18) and (7.19), respectively. Also find $se[\overline{\Delta}]$ and use your results to construct a 95% confidence interval for Δ. Compare your results with those achieved previously.

(a) The pollutant site remediation data in Example 7.7.

(b) The *L. plumulosus* survival rate data in Exercise 7.14.

7.16. Return to the funnel plot in Example 7.8 and plot only the $K = 11$ US studies from Table 7.3. How does this smaller plot compare with that in Fig. 7.3?

7.17. Gurevitch and Hedges (1993) report a meta-analysis on the effects of species competition on an organism's biomass. They studied the effect of competition on each species, compared with a control group of the same species. The full meta-analysis included $K = 43$ different studies, producing the following effect sizes d_k and reciprocal variances w_k:

d_k	1.218	1.288	1.818	2.223	0.767	0.524	0.163
$w_k = 1/\mathrm{Var}[d_k]$	2.953	2.011	7.077	6.182	9.315	4.834	4.983
d_k	0.449	2.218	0.571	−0.326	0.735	0.869	0.463
$w_k = 1/\mathrm{Var}[d_k]$	4.877	0.619	0.961	0.987	1.873	1.827	1.948
d_k	0.432	0.728	0.629	0.501	−0.367	0.820	−1.956
$w_k = 1/\mathrm{Var}[d_k]$	1.954	1.876	1.906	1.939	1.967	1.845	1.353
d_k	−0.399	0.036	0.729	0.565	1.533	2.014	1.880
$w_k = 1/\mathrm{Var}[d_k]$	2.451	3.499	3.282	2.885	1.932	2.322	2.081
d_k	1.181	1.400	1.089	1.091	0.000	−0.917	1.214
$w_k = 1/\mathrm{Var}[d_k]$	1.277	1.205	1.306	1.306	2.500	2.262	1.689
d_k	1.694	1.273	1.303	1.519	1.144	1.206	3.696
$w_k = 1/\mathrm{Var}[d_k]$	6.977	8.316	7.818	7.762	8.595	8.461	0.739
d_k	4.674						
$w_k = 1/\mathrm{Var}[d_k]$	0.536						

(a) Find a combined estimator of effect size, $\bar{d}_+$, based on these data. Also calculate $se[\bar{d}_+]$ and construct an approximate 95% confidence interval for the true effect size.

(b) Use the Q_W statistic to assess if the 43 studies appear homogeneous. Set $\alpha = 0.05$. If your result is significant, return to the inferences in part (a) and employ instead a random effects model using (7.15) and (7.17).

(c) Construct a funnel plot for the effect sizes using d_k on the horizontal axis and $\sqrt{w_k} = 1/\sqrt{\mathrm{Var}[d_k]}$ on the vertical axis. Does the pattern suggest any publication bias?

7.18. Show that under the model in (7.20), $E[Y_{1j}] - E[Y_{0j}] = \beta$.

7.19. Verify the indication in §7.4.2 that the method of moments estimator of φ is given by (7.23). [*Hint*: writing $\mu = Y_{h+}/N_{h+}$, you need to solve $N_{h+}^{-1}\mu(1 - \mu)$ $(1 + \varphi N_{h+})/(1 + \varphi) = S^2$.]

7.20. In an environmental carcinogenesis study of the chemical solvent stabilizer 1,2-epoxybutane (Dunnick *et al.*, 1988), $N_1 = 50$ male rats were exposed to 200 ppm of the chemical via inhalation. Of these, $Y_1 = 31$ (62%) exhibited leukemias by the end of the study. The concurrent control leukemia rate was $Y_0/N_0 = 25/50$ (50%).

(a) Use Fisher's exact test to determine if a significant increase in leukemia induction is present in the exposed animals, relative to the concurrent controls. Set $\alpha = 0.05$.

(b) Update your analysis in part (a) to include historical information on these animals. The historical control data show an overall incidence of 583/1977 (29.49%) leukemias in male rats, with an observed range of 0.5966. Combine the historical data with the current experimental information to test for increased leukemia induction due to 1,2-epoxybutane exposure. Set $\alpha = 0.05$. Does your inference from part (a) change?

7.21. Similar to the study described in Example 7.9, the US National Toxicology Program considered murine carcinogenesis after gavage exposure to the aromatic hydrocarbon benzene over a 105-week chronic exposure period (Huff, 1989). $N_1 = 50$ female mice were exposed to 50 mg/kg benzene over the test period, after which time $Y_1 = 5$ mice exhibited hepatocellular tumors. The concurrent control group exhibited only $Y_0/N_0 = 1/49$ tumors.

(a) Use Fisher's exact test to determine if a significant increase in hepatocellular carcinogenesis is present in the exposed animals, relative to the concurrent controls. Set $\alpha = 0.10$.

(b) Update your analysis in part (a) to include historical information on these animals. The historical control data show an overall incidence of 47/1176 hepatocellular cancers in female mice, with an observed standard deviation of $S = 0.0255$. Combine the historical data with the current experimental information to test for increased liver carcinogenesis due to benzene exposure. Set $\alpha = 0.10$. Does your inference from part (a) change?

Fundamentals of environmental sampling

Throughout the previous chapters our goal has been to present descriptive and inferential methods for analyzing environmental data. The inferences are made to some population of elements, and the data are taken from this population via some form of random sample. In §A.3, we define a random sample as a set of values observed from a population such that the elements are equally likely to be sampled and the outcomes are statistically independent of each other. The requirement that the sample be taken randomly underlies the logic of statistical inference, since it avoids introduction of any systematic biases in the consequent data. Haphazard sampling can influence the outcomes in ways that limit the scope of the statistical inferences, hindering or even invalidating study results.

Randomization is the mechanism we use to balance out uncontrolled systematic effects in an experiment. The term refers to the random assignment of experimental units to different treatment or study conditions. The concept is fundamental to proper study design: before imposition of some treatment or intervention, the experimental units are allocated to the different treatment levels such that any unit has the same chance of being assigned to any level.

Randomization is particularly useful in designed experiments, such as laboratory studies of environmental toxins (Piegorsch and Bailer, 1997, §3.3). Indeed, the larger concept of experimental design is an important one in environmental analysis. Properly designed studies provide the investigator with data as precise and powerful as possible for determining the environmental outcome or inference of interest. By contrast, no amount of sophisticated statistical analysis can salvage a poorly designed and/or poorly conducted study. Thus the investigator must plan as carefully as possible the identification of the experimental units, determination of extraneous variables and pertinent comparison or control groups, randomization of experimental

Analyzing Environmental Data   W. W. Piegorsch and A. J. Bailer
© 2005 John Wiley & Sons, Ltd   ISBN: 0-470-84836-7 (HB)

units to treatment conditions (mentioned above), determination of appropriate sample sizes, and proper sampling of the observations. It is this latter issue to which we direct attention in this chapter. For more on general issues of experimental design, see, for example, Clarke and Kempson (1996) or Cox and Reid (2000).

At the simplest level, our definition of a random sample assumes that the population is so large as to be essentially infinite. In some environmental sampling situations, however, populations are relatively small and this finite population size must taken into account. Since the associated inferences on the parameters will then depend on the various strategies used to draw the sample, we refer to them as *design-based inferences* (Levy and Lemeshow, 1999, Ch. 16). In design-based inference, the sampling design and nature of the randomization are viewed as the only sources of uncertainty. The statistical properties of the estimators – bias or lack thereof, variance, etc. – are all derived from these design components. By contrast, if the responses are assumed to follow some probability distribution, perhaps in association with some auxiliary variables, we can base inferences on the population parameters of this model. We say then that the inferences are *model-based* (Chambers, 2002; Thompson, 2002, §2.7). Of the two approaches, design-based inference is more common in environmental sampling, and we emphasize its use here. For more on the distinctions (and the debate) between design-based inference and model-based inference in survey sampling, see, for example, Gregoire (1998).

In this chapter, we focus on production of data through the selection of elements from a population using a variety of sample survey designs. We introduce some basic terminology for constructing random samples for environmental applications, and include important considerations that must be made when the population size is finite. More complete expositions on survey sampling techniques exist in dedicated texts such as Scheaffer *et al.* (1996), Levy and Lemeshow (1999), Thompson (2002), Gregoire and Valentine (2004), or the review by Stehman and Overton (2002). For the statistical theory underlying much of the material we present below, see also the classic text by Cochran (1977).

8.1 Sampling populations – simple random sampling

We define a *sampling strategy* as the combination of (i) a proper sampling plan or design, (ii) an approach for measuring pertinent attributes from the sampled elements, and (iii) a framework for using the measurements to estimate associated population descriptive parameters. All three components require the investigator's attention when designing an environmental study. At its most basic level, the strategy must be formulated such that every sample adequately represents the larger population. For example, if a study of pollutant runoff in watersheds is conducted in the southeastern United States, is the sample representative of watersheds throughout North America? Perhaps not. Indeed, data from watersheds in the US Southeast may differ drastically from those, say, on the Pacific coast. Thus when constructing a sampling strategy one must always keep in mind that the population to which we wish to draw inferences must also be the population from which the data are randomly sampled.

Ideally, proper identification of a finite target population will allow the investigator to list in some fashion every element of the population. Such a list is called a *list frame*, or sometimes simply a *frame*. For example, if the investigator is studying heights of trees in a forest stand, the population is all the trees in that particular stand, the elements are the individual trees, and the frame is the comprehensive list of all trees. (In practice, one would probably be measuring only those trees that are, say, at least 2 meters tall.) Frames need not be simple lists, however. In environmental applications they take many forms; for instance, in the pollutant runoff example mentioned above, the frame might be the longitude and latitude coordinates of the watershed centers. A sampling frame based on the spatial or geographic features of the population elements under study is often called a *map frame*. For any form of frame, we denote the number of elements contained therein (and hence, in the finite population) by N.

Notice that the population's elements are assumed discrete, or at least can be discretized in an unambiguous fashion to produce N finite response values. If this is not the case – as might occur, for example, in a study of some continuous outcome over a spatial map frame – then more advanced methods for continuously distributed populations are in order. See, for example, Overton and Stehman (1993) or Stevens (1997).

When collecting an environmental sample, elements are typically selected from the frame without replacement, that is, without returning an element to the population after measuring its attribute(s). In some cases, the sampling is performed with replacement, but this is rare. Throughout this chapter we assume that sampling is without replacement unless indicated otherwise.

To begin, we define an *inclusion probability* as the probability that a population element is selected for a sample. With this, a *probability sample* is any sampling design that randomly selects elements from the population in a planned fashion. We require that each sampled element's inclusion probability be known and that each population element have a non-zero inclusion probability. By contrast, a *non-probability sample* is any sample where no planned random selection is employed.

The simplest form of probability sample is the simple random sample (SRS) mentioned above. Formally, a random sample is 'simple' if every sample of size n from the larger population has an equal probability of selection. To draw an SRS from a fixed population, we (i) assign a unique identifier to each element in the frame; (ii) draw a series of random numbers from a table or via computer using a variable seed or start point (always recording the seed as part of proper record-keeping); and (iii) select an element when a random number matches its identifier. We discard repeats (unless sampling is performed with replacement) and any random numbers that do not correspond to a population identifier. Using simple counting arguments, it can be shown that the probability of selecting any particular SRS of size n without replacement from a population of size N is $1/\binom{N}{n}$, where $\binom{N}{n}$ is the binomial coefficient from §A.2.1.

For acquiring the random numbers, random digit tables are available in a variety of sources, including Scheaffer *et al.* (1996, Table A1.2), Levy and Lemeshow (1999, Table A.1), Samuels and Witmer (2003, Table 1), or Hogg and Tanis (2001, Table IX). They may also be generated by computer, either via a standard pseudo-random number algorithm (Press *et al.*, 1992, Ch. 7; Gentle, 2003, §1.7), or in various statistical packages that can build the SRS directly. For instance, in SAS ver. 8 the procedure PROC SURVEYSELECT can draw an SRS from any provided list frame, as shown in the next example.

Example 8.1 (Soil contamination)   Environmental contamination of soil by heavy metals is a concern in many industrialized nations. Suppose a set of investigators plans a study of soil contamination in a series of Austrian cities where mining and heavy industry are prevalent. Table 8.1 summarizes the site information.

From this population of $N = 96$ sites, an SRS is to be taken of size $n = 16$. Figure 8.1 gives sample SAS code for producing such a sample. In the code, notice the use of unique site identifiers (siteID) for each site within each city; for example, each siteID for the 16 Untertiefenbach sites is distinct from those for the 12 Reisenberg sites, etc. The call to PROC SURVEYSELECT contains a number of options for obtaining the sample. First, the method=srs option calls for a simple random sample. Next, the size of the SRS is set via the sampsize= option. Here $n = 16$. The seed= option gives a user-defined, integer-valued seed from which to start the pseudo-random number generations. Lastly, the out= option specifies the name of

Table 8.1   City and site information for Austrian soil contamination study

City	Available sites
1. Untertiefenbach	16
2. Reisenberg	12
3. Feistritz	13
4. Mitterberghütten	16
5. Ramingstein	14
6. Bleiberg	13
7. Arnoldstein	12

```
data soilsrs;
attrib city length=$17;
input city $ siteID $ @@;
datalines;
Untertiefenbach UT01    Untertiefenbach UT02
Untertiefenbach UT03    Untertiefenbach UT04
    :                        :
Untertiefenbach UT15    Untertiefenbach UT16
Reisenberg RB01         Reisenberg RB02         Reisenberg RB03
    :                        :                        :
Arnoldstein AN10        Arnoldstein AN11        Arnoldstein AN12
;

proc surveyselect data=soilsrs   method=srs   sampsize=16
                  seed=62656      out=sample81;

proc print data=sample81;
```

Figure 8.1   Sample SAS (ver. 8) program to build an SRS of size $n = 16$

an SAS data set into which the SRS information is placed. To view the resulting SRS, we employ a call to PROC PRINT in Fig. 8.1.

The (edited) output in Fig. 8.2 gives a list of the $n = 16$ elements making up the SRS, along with some summary information. With this, the investigators can now proceed to each selected site and perform their soil samples.

Before continuing, we should note that five of the 16 sites selected were from Bleiberg, a city with only 13 possible sites. By contrast, only one of the 16 possible sites from Mitterberghütten was selected. Indeed, another SRS from this same frame could fail to include any sites from Mitterberghütten (or any other city in Table 8.1), since by definition an SRS requires any sample of size $n = 16$ from the $N = 96$ sites to be as likely as any other. In some settings, however, the investigator may wish to ensure that certain cities are represented with at least one observation in the sample. This violates the basic definition of a SRS, but nonetheless may be appealing for practical purposes. We discuss alternative sampling designs that can achieve this sort of stratification in §8.2.2, below. ✪

Once an SRS has been collected, population descriptive parameters of interest are estimated. Common quantities include the population mean response, μ (e.g., the mean contaminant concentration in soils at hazardous waste sites), or the proportion of elements, π, that exhibit a critical environmental characteristic (e.g., the proportion of trees in a forest affected by a detrimental insect or disease). As mentioned above, it is important to adjust any parameter estimates for the finite population size, since if $N < \infty$ the standard formulas for point and interval estimates will not apply. Consider estimation of μ. Given measurements y_i, $i = 1, \ldots, n$, from an SRS, the sample mean $\bar{y} = \sum_{i=1}^{n} y_i/n$ is an unbiased – technically, a *design-unbiased* – estimator of μ: $E[\bar{y}] = \mu$. To find the standard error of $\bar{y}$, however, the finite population size influences the computations. When sampling without replacement from a finite population, knowledge that y_i has been selected is informative, since we now know that any other y_j $(i \neq j)$ was not selected (and that y_i cannot be selected again). This induces, in effect, a negative correlation among the y_is: written in terms of the covariance between y_i and y_j, it can be shown that $\mathrm{Cov}[y_i, y_j] = -\sigma^2/(N-1)$, where

```
                          The SAS System
                       The SURVEYSELECT Procedure

               Selection Method      Simple Random Sampling
               Input Data Set               SOILSRS
               Random Number Seed            62656
               Sample Size                      16
               Output Data Set              SAMPLE81

                            site                              site
      Obs  city             ID        Obs  city               ID
        1  Untertiefenbach  UT02       9   Ramingstein        RN03
        2  Untertiefenbach  UT04      10   Bleiberg           BB01
        3  Reisenberg       RB05      11   Bleiberg           BB02
        4  Reisenberg       RB10      12   Bleiberg           BB04
        5  Reisenberg       RB12      13   Bleiberg           BB08
        6  Feistritz        FW05      14   Bleiberg           BB09
        7  Mitterberghuetten MB08     15   Arnoldstein        AN05
        8  Ramingstein      RN02      16   Arnoldstein        AN11
```

Figure 8.2 SAS output (edited) of SRS for soil contamination study

σ^2 is the population variance of any y_i (Scheaffer *et al.*, 1996, Appx. 1). This is true even though the sample is taken in a fully random fashion. Consequently, the theoretical variance of any linear combination of the y_is – including $\bar{y}$ – will include contributions from these covariances; cf. equation (A.6). Here, the result is

$$\text{Var}[\bar{y}] = \left(\frac{N-n}{N-1}\right)\frac{\sigma^2}{n}. \tag{8.1}$$

Notice that as $N \to \infty$, (8.1) approaches σ^2/n, which is the familiar variance of a sample mean from an infinite population. To find the standard error of $\bar{y}$, we assume σ^2 is unknown and substitute an unbiased estimator for it in (8.1). The square root of the resulting quantity is $se[\bar{y}]$.

The usual, infinite-population unbiased estimator of σ^2 is the sample variance $s^2 = \sum_{i=1}^{n}(y_i - \bar{y})^2/(n-1)$. Here again, however, standard methods from infinite-population sampling require adjustment if the population size is finite. A correlation effect similar to that seen with $\bar{y}$ creeps into s^2, causing a bias: $E[s^2] = N\sigma^2/(N-1)$. Luckily, it is straightforward to compensate for the effect: simply take $(N-1)s^2/N$ as an unbiased estimator of σ^2. Substitute this into (8.1) and take the standard error of $\bar{y}$ as

$$se[\bar{y}] = \sqrt{\left(\frac{N-n}{N}\right)\frac{s^2}{n}}. \tag{8.2}$$

Notice, again, that as $N \to \infty$, the bias dissipates and (8.2) collapses to the usual standard error for $\bar{y}$. The quantity $1 - n/N$ multiplying s^2/n in (8.2) is called the *finite-population correction (FPC)* since it is the correction factor for any finite N.

In large samples, the design-based approach provides that $\bar{y}$ is approximately normal with mean μ and variance given by (8.1) (Thompson, 2002, §3.2). Thus, for example, an approximate $1 - \alpha$ design-based confidence interval for μ is

$$\bar{y} \pm z_{\alpha/2}se[\bar{y}],$$

where $se[\bar{y}]$ is given by (8.2). One can also include an adjustment for small sample sizes by employing a t-distribution critical point: $\bar{y} \pm t_{\alpha/2}(n-1)se[\bar{y}]$. This often provides reasonable coverage characteristics, even with decidedly non-normal data (Gregoire and Schabenberger, 1999). Beyond about $n > 30$, the z and t critical points produce roughly similar results.

The term $z_{\alpha/2}se[\bar{y}]$ or $t_{\alpha/2}(n-1)se[\bar{y}]$ being added to and subtracted from $\bar{y}$ in the confidence interval is often called the *bound on the error of estimation*, since it quantifies with approximate $1 - \alpha$ confidence by how much $\bar{y}$ may be in error when estimating μ. Experienced readers will recognize this also as the *margin of error* of the confidence interval. We will use both terms interchangeably.

Example 8.2 (Soil contamination, cont'd) Continuing with the soil contamination study from Example 8.1, suppose now that an SRS has been taken of sites through-out Austria. For example, Majer *et al.* (2002) give data on total acid soluble chromium (Cr) in soil from $n = 16$ sites in Table 8.1. The data, as $y = \log(\text{Cr})$, appear in

Table 8.2. (Note that the authors selected a different random sample from what we generated in Fig. 8.2.)

We find for the data in Table 8.2 that $\bar{y} = 3.4396$ and $s^2 = 0.3013$. For $N = 96$ and $n = 16$, the FPC is $1 - \frac{16}{96} = 0.8333$, so that (8.2) gives $se[\bar{y}] = 0.1253$. With the relatively small sample size we apply a $t(n-1)$ critical point in the approximate 95% confidence interval for μ: $\bar{y} \pm t_{0.025}(15)se[\bar{y}] = 3.4396 \pm (2.131 \times 0.1253) = 3.4396 \pm 0.2670$, or $3.173 < \mu < 3.707$.

We can also perform these calculations in SAS. The ver. 8 procedure PROC SURVEYMEANS can calculate point estimates and confidence limits for a variety of sampling designs, including an SRS. The sample code in Fig. 8.3 illustrates its use. In the procedure call, notice use of the `total=` option to specify the finite population size, N, and the `mean` and `clm` options to request point and 95% interval estimates, respectively, of μ. For the confidence limits, PROC SURVEYMEANS also employs the t critical point $t_{0.025}(df)$, where df are the degrees of freedom for the particular sampling design; the `df` option is used to list this value. The `var` statement specifies which variable in the data set to take as the response. Edited output appears in Fig. 8.4, where we recover values similar to those seen above. ✪

If the parameter of interest is a population proportion, π, the outcome variable y_i will typically be dichotomous. Without loss of generality, we code this as binary: $y_i = 1$ or 0 according to whether the ith element does or does not exhibit the endpoint

Table 8.2 Log-chromium (Cr) concentrations in soil from $n = 16$ Austrian sites. Sites taken from population in Table 8.1

Site	UT7	UT14	RB5	RB11	FW2	FW9	MB9	MB10
log(Cr)	3.761	3.807	4.317	2.833	3.258	4.205	3.611	2.303
Site	RN1	RN7	BB8	BB11	BB12	AN2	AN5	AN12
log(Cr)	3.434	3.367	3.970	2.565	3.091	3.611	3.466	3.434

Source: Majer *et al.* (2002).

```
data soil96;
attrib city length=$17;
input city $ siteID $ YlogCr @@;
datalines;
Untertiefenbach     UT07 3.761    Untertiefenbach     UT14 3.807
Reisenberg          RB05 4.317    Reisenberg          RB11 2.833
Feistritz           FW02 3.258    Feistritz           FW09 4.205
Mitterberghuetten MB09 3.611    Mitterberghuetten MB10 2.303
Ramingstein         RN01 3.434    Ramingstein         RN07 3.367
Bleiberg            BB08 3.970    Bleiberg            BB11 2.565
Bleiberg            BB12 3.091    Arnoldstein         AN02 3.611
Arnoldstein         AN05 3.466    Arnoldstein         AN12 3.434
;

proc surveymeans total=96  mean  clm  df;
var YlogCr;
```

Figure 8.3 Sample SAS (ver. 8) code to estimate population mean μ with soil contamination data from finite population of size $N = 96$

```
                      The SAS System
                  The SURVEYMEANS Procedure

                       Data Summary
                  Number of Observations    16

                            Std Error    Lower 95%   Upper   95%
   Variable    DF    Mean     of Mean        CL for Mean

   YlogCr      15  3.439563   0.125268    3.172560     3.706565
```

Figure 8.4 SAS output (edited) for estimating μ with soil contamination data

of interest, respectively. Estimation of π proceeds in similar fashion to that of μ; indeed, the equations are essentially the same, although some simplification is available by recognizing that y_i is binary. In particular, an unbiased estimator of π is the sample proportion

$$p = \frac{\sum\limits_{i=1}^{n} y_i}{n},$$

with finite-population-adjusted standard error

$$se[p] = \sqrt{\left(\frac{N-n}{N}\right)\frac{p(1-p)}{n-1}}; \tag{8.3}$$

see Exercise 8.5. With this, an approximate $1 - \alpha$ confidence interval for π is simply $p \pm z_{\alpha/2}se[p]$.

When the population size is finite, it is also of interest to study the population total, $\tau = N\mu$. (Notice that when N is finite, so is τ, making it a parameter of potential value for further study.) For instance, a forest ranger may wish to estimate the total number of visitors to the forest in a month, a regional planner may wish to estimate the total water requirements for a drought-stricken community, or a geologist may wish to estimate the total number of precious mineral veins in a mine.

Estimation of τ is performed by manipulating the fundamental relation between it and μ. That is, since $\tau = N\mu$ and if for an SRS we know $E[\bar{y}] = \mu$, then $E[N\bar{y}] = NE[\bar{y}] = N\mu = \tau$. Thus $\hat{\tau} = N\bar{y}$ is an unbiased estimator of τ. Also, given $se[\bar{y}]$ from (8.2), we know $se[\hat{\tau}] = se[N\bar{y}] = Nse[\bar{y}]$, so an approximate $1 - \alpha$ confidence interval on τ is $N\bar{y} \pm z_{\alpha/2}Nse[\bar{y}]$ or, to adjust for small sample sizes, $N\bar{y} \pm t_{\alpha/2}(n-1)Nse[\bar{y}]$.

Equation (8.2) can also be used to help the investigator select a sample size prior to conducting the study. For instance, one might ask how large a sample is needed so that the margin of error from an approximate 95% confidence interval for μ is no larger than, say, ε units. Notice that we are working with the population margin of error, not its sample estimate. To do so, we set $\varepsilon = 1.96\sqrt{\text{Var}[\bar{y}]}$ using (8.1), and then solve for n. For simplicity, we approximate $1.96\sqrt{\text{Var}[\bar{y}]}$ as $2\sqrt{\text{Var}[\bar{y}]}$. The result (Exercise 8.6) is

$$n \geq \frac{4N\sigma^2}{4\sigma^2 + (N-1)\varepsilon^2}. \tag{8.4}$$

In (8.4), σ^2 is the population variance, which is typically unknown in practice. To use the equation, some preliminary estimate of σ^2 must be posited. This can be an unbiased estimate from some previous study or experiment, or a rough estimate based on other prior knowledge. An example of the latter was used in §7.4.2: $\hat{\sigma}^2 \approx (\text{Range})^2/26.5$, where 'Range' is the range of possible values over which the outcome variable is thought to extend.

Results similar to (8.4) are available when estimating π or τ; see Exercise 8.6.

Example 8.3 (Soil contamination, cont'd) Suppose that the soil contamination study from Example 8.2 was only a preliminary investigation into heavy metal soil pollution in Austria. The study researchers wish to conduct a larger study and estimate mean chromium content, now with a margin of error no larger than $\varepsilon = 0.1$. Given this objective, how many sites should be sampled?

To achieve this goal, we apply equation (8.4). From the earlier study, we have $s^2 = 0.3013$ and we know that an unbiased estimate of σ^2 based on this value is $(N-1)s^2/N = 95 \times 0.3013/96 = 0.298$. Using this in (8.4) with $\varepsilon = 0.1$ yields

$$n \geq \frac{4 \times 96 \times 0.298}{(4 \times 0.298) + (95 \times 0.1^2)} = 53.4.$$

Since we require n to be a positive integer, and since we are selecting a sample size that produces a margin of error *no larger* than $\varepsilon = 0.1$, we round this value up to $n = 54$.

In some finite populations the data may be heavily skewed, invalidating the normal approximation for the margins of error and confidence intervals. Examples include concentration data spanning a wide range of magnitudes, ratio responses, or small counts. If standard transformations such as the natural logarithm or square root are unable to correct the problem, more complex operations may be required. These include adjustments to the t approximation that make it more resilient to the skewness distortion (Johnson, 1978), or use of hierarchical model constructions (Meeden, 1999). To give guidance on when the approximation *is* valid, Cochran (1977, §2.15) gave a rough rule of thumb for minimal sample sizes when extreme population skewness is evident: use $n > 25\gamma_1^2$, where γ_1 is a measure of skewness estimated via

$$\hat{\gamma}_1 = \frac{\sum_{i=1}^{n}(y_i - \bar{y})^3}{(n-1)s^3}.$$

When the population size, N, is itself small, one can adjust $\hat{\gamma}_1$ by including a finite population correction such as $\hat{\gamma}_1\sqrt{[N/(N-1)]}$. Alternatively, by appealing to a multi-term approximation for the distribution of $\bar{y}$, Sugden *et al.* (2000) recommended including an additive penalty term: $n > 28 + 25\gamma_1^2$.

8.2 Designs to extend simple random sampling

8.2.1 Systematic sampling

In many instances it is possible to improve the precision (i.e., reduce the variances/ standard errors) of our design-based estimates and confidence intervals for μ, π, or τ by extending the SRS design into a more complex sampling strategy. Perhaps the simplest extension of simple random sampling is known as *systematic sampling*. A systematic sample of size n takes a list or map frame and selects the first sample element in a random fashion from among the first k elements in the frame. Then, it selects the remaining $n - 1$ sample elements by moving through the list or map in a systematic fashion. The most common approach for doing so is to take every kth element from the frame, where $k \leq N/n$. This is known as a *1-in-k systematic sample*, where k is the *sampling interval*. A 1-in-k systematic sample can be considered a form of probability sample, since the initial sample element is chosen via a random, probability-based mechanism.

Although systematic samples are easy to implement, they can be detrimentally affected by hidden periodicities in the data (or, more technically, in the frame). That is, if some unrecognized periodic effect exists among the population elements, and if the choice of k happens to align in phase with this periodicity, then the sample measurements may not adequately represent the larger population. This can induce bias in the point estimator. Clearly, the investigator must be careful to check for and adjust against any possible periodicities in the frame before undertaking a systematic sample. A simple way to do so is to order the frame as randomly as possible prior to undertaking any form of systematic sample. Or, if an auxiliary variate is available that is (i) known for all elements of the population, (ii) highly correlated with the outcome of interest, and (iii) easy/inexpensive to acquire, gains in precision are possible by performing the systematic sample on the frame after deliberately ordering it by the auxiliary variate.

In SAS ver. 8, PROC SURVEYSELECT can be used to draw a systematic sample by using the `method=sys` option in the procedure call. If `method=sys` is specified, use of the option `samprate=`f will specify the sampling interval as $k = 1/f$. One can also use the `sampsize=` option to select n, forcing PROC SURVEYSELECT to set k equal to N/n or, equivalently, $f = n/N$.

Estimation of population quantities such as μ, τ, or π proceeds in a manner similar to that for simple random sampling. If the ratio N/k is an integer, then $\bar{y}$ will be unbiased for μ, leading also to $N\bar{y}$ being unbiased for τ and to the sample proportion being unbiased for π. If N/k is not an integer the estimators can be biased, although as N grows large these biases become negligible. For smaller populations, some modification to the systematic sampling scheme may be necessary; see Levy and Lemeshow (1999, §4.5) for one such approach.

Calculation of standard errors and confidence intervals is more problematic, however, since it is difficult to find an unbiased estimator of σ^2 under a systematic sample. Luckily, if the population and frame are arranged randomly so that no underlying periodicity or other form of ordered structure is present, then the variance estimators and standard errors from the SRS case will also be approximately valid for systemic samples. Thus, for example, with an aperiodic, randomly ordered population, we still take $se[\bar{y}]$ as in (8.2) when constructing approximate confidence intervals for μ. An alternative approach which can be more accurate in some cases was discussed by Meyer (1958). He suggested using the *order statistics* of the sample (§A.4), $y_{(1)} = \min_{i \in \{1,\dots,n\}}\{y_i\} \leq y_{(2)} \leq y_{(3)} \cdots \leq y_{(n-1)} \leq y_{(n)} = \max_{i \in \{1,\dots,n\}}\{y_i\}$, to calculate the successive difference mean square

$$s_D^2 = \frac{1}{2(n-1)} \sum_{i=1}^{n-1}\{y_{(i+1)} - y_{(i)}\}^2$$

and use this in place of s^2 in (8.2). That is, take

$$se_D[\bar{y}] = \sqrt{\left(\frac{N-n}{N}\right)\frac{s_D^2}{n}}. \tag{8.5}$$

With (8.5), an approximate $1 - \alpha$ confidence interval for μ is $\bar{y} \pm z_{\alpha/2}se_D[\bar{y}]$; for τ, use $N\bar{y} \pm z_{\alpha/2}Nse_D[\bar{y}]$. In small samples, one can replace the $z_{\alpha/2}$ critical point with $t_{\alpha/2}(n-1)$.

In the general case, equations for variances such as $\text{Var}[\bar{y}]$, $\text{Var}[\hat{\tau}]$, or $\text{Var}[p]$ become more complex, and will depend on additional quantities such as the intra-class correlation between the y_is; see Levy and Lemeshow (1999, §4.4).

A way to extend the 1-in-k systematic sample that can protect against the problem of periodicity is to break up the single systematic sample into a set of L smaller samples. Each sample is of size n/L, with its own random start point. Typical values of L include 2, 3, 4, up to perhaps 10 or so for large n. Scheaffer *et al.* (1996, §7.6) and Levy and Lemeshow (1999, §4.7) give more detail on this form of *repeated systematic sampling*.

8.2.2 Stratified random sampling

It is common in many environmental scenarios for different characteristics to exist between elements of a sampling frame. For example, ocean depth may differ across sampling locations in a trawl survey; a power plant's energy demands may vary by season; adjacent municipalities may differentially fund recycling programs; etc. To account for these known sources of differential response we can stratify the sampling design, such that each stratum represents a different level of the variable character. In such a *stratified sampling* scenario, it is assumed that the variation in measured responses is more homogeneous within a stratum than it is between strata. Thus, for example, a trawl survey of ocean-floor species abundance may encounter more species at higher depths, leading to large variations in abundance. At a fixed depth, however, the species abundance may be fairly homogeneous, and neither systematic nor simple random sampling can guarantee *a priori* that all the differing depth levels will be included in the final sample. By stratifying on depth and then requiring a sample from each depth stratum to be included, more representative inferences may be achieved. Indeed, when within-stratum variation is sufficiently small relative to between-stratum variation, stratification leads to substantive variance reduction in point estimates of μ, τ, and π (Levy and Lemeshow, 1999, §5.3). Increased study precision can result. A similar concern enters into the soil contamination study from Examples 8.1–8.2. As mentioned there, it may be desirable to stratify by cities to ensure geographic diversity and as a result achieve increased precision; see Exercise 8.13.

In its simplest form, stratified random sampling is conducted by partitioning the sampling frame into an exhaustive and mutually exclusive set of $H > 1$ strata and then drawing an SRS of size n_h from each stratum, $h = 1, \ldots, H$. We assume each stratum contain N_h elements, each of which is listed in a suitable stratum-specific frame. The total population size is $N = \sum_{h=1}^{H} N_h$, while the total sample size is $n = \sum_{h=1}^{H} n_h$. Assuming sampling in any stratum is statistically independent of that in any other stratum, the basic SRS methods from §8.1 can be applied in a straightforward manner to each stratum in order to conduct the stratified random sample.

Example 8.4 (Insect abundance) See *et al.* (2000) discuss a sampling scenario where the abundance of a certain insect is studied in $H = 4$ different forest habitats. The four regions have essentially equal rectangular areas, and are each partitioned into 100 equal-sized plots indexed as 00, 01, ..., 99. We view each of the four regions as a separate stratum.

Insect abundance is recorded by laying out nets below the canopy in each plot, fogging the trees with an insecticide, and then counting the number of insects that drop into the nets. The effort is labor-intensive, so it is of interest to sample from each 100-plot stratum rather than conduct a complete enumeration. Resources are available to sample a total of $n = 25$ plots, so we construct a stratified sampling plan that identifies an equal allocation of $n_h = 6$ plots in the first three strata ($h = 1, 2, 3$) and add an additional plot in stratum $h = 4$ (so $n_4 = 7$).

To plan the stratified sample, we can again apply PROC SURVEYSELECT. Figure 8.5 gives sample SAS code to perform the operations. Note the construction of a special SAS data set, `allocate`, used to specify the unbalanced allocation scheme via the `sampsize=` option. SAS requires that the variable _NSIZE_ be used to contain the n_hs. The stratified sample is generated via use of the `strata` statement in PROC SURVEYSELECT, which defines the stratification variable. (Both the source data set and the allocation data set must be sorted prior to use of `strata`. Towards this end, Fig. 8.5 employs PROC SORT.) Edited output containing the stratified sample appears in Fig. 8.6.

The output in Fig. 8.6 lists all 25 individual stratum elements to be sampled, ordered by strata, along with some summary information from the PROC SURVEYSELECT operation. Given this information, the investigators can now proceed to each selected site and perform their stratified sample. (See Exercise 8.12.) ✪

Estimation of population descriptive parameters from a stratified random sample takes advantage of independence among strata to construct pooled sample estimators. To establish the notation, denote by τ_h the individual per-stratum totals such that the population total is $\tau = \sum_{h=1}^{H} \tau_h$. Also denote the per-stratum

```
data insect;
input forest siteID @@;
datalines;
1 00    1 01    1 02    1 03    1 04    1 05    1 06    1 07
1 08    1 09    1 10    1 11    1 12    1 13    1 14    1 15
  :       :       :       :       :       :       :       :
4 88    4 89    4 90    4 91    4 92    4 93    4 94    4 95
4 96    4 97    4 98    4 99

proc sort data=insect;
  by forest;

data allocate;
input forest _NSIZE_ @@;
datalines;
      1  6       2  6       3  6       4  7

proc sort data=allocate;
  by forest;

proc surveyselect data=insect  method=srs  sampsize=allocate
                  seed=100956 out=sample25;
  strata forest;

proc print data=sample25;
```

Figure 8.5 Sample SAS (ver. 8) program to build a stratified random sample of size $n = 25$

```
                        The SAS System
                  The SURVEYSELECT Procedure

          Selection Method      Simple Random Sampling
          Strata Variable       forest

            Input Data Set                  EX84
            Random Number Seed            100956
            Sample Size Data Set         ALLOCATE
            Number of Strata                   4
            Total Sample Size                 25
            Output Data Set              SAMPLE25
```

Obs	forest	site ID		Obs	forest	site ID
1	1	07		13	3	26
2	1	26		14	3	29
3	1	30		15	3	50
4	1	36		16	3	53
5	1	64		17	3	65
6	1	83		18	3	78
7	2	00		19	4	14
8	2	12		20	4	25
9	2	28		21	4	27
10	2	45		22	4	38
11	2	87		23	4	58
12	2	96		24	4	78
				25	4	95

Figure 8.6 SAS output (edited) of stratified random sample for insect abundance study

means as $\mu_h = \tau_h/N_h, h = 1, \ldots, H$. The overall population mean is then $\mu = \tau/N = \sum_{h=1}^{H} \tau_h/N = \sum_{h=1}^{H} N_h\mu_h/N$. For $W_h = N_h/N$ this can be written as $\mu = \sum_{h=1}^{H} W_h\mu_h$, a weighted average of the per-stratum means where the weights, W_h, are the proportions of the population belonging to each stratum. If sampling from a map frame, one might instead use weights based on areas or other spatial measures; for example, $W_h = A_h/A$, where A_h is the area of the hth stratum and $A = \sum_{h=1}^{H} A_h$.

To estimate μ, start with the unbiased estimators of μ_h,

$$\bar{y}_h = \frac{\sum_{i=1}^{n_h} y_{hi}}{n_h}.$$

Since $\mu = \sum_{h=1}^{H} W_h\mu_h$, an unbiased estimator of μ is then

$$\hat{\mu} = \sum_{h=1}^{H} W_h\bar{y}_h.$$

To find the standard error of $\hat{\mu}$, recall from §A.1 that the variance of a sum of independent random variables is the sum of the individual variances, weighted by squares of any multiplying coefficients: $\text{Var}[\hat{\mu}] = \sum_{h=1}^{H} \text{Var}[W_h\bar{y}_h] = \sum_{h=1}^{H} W_h^2\text{Var}[\bar{y}_h]$. From an SRS, however, we know that the variance of a sample mean is given by (8.1): $\text{Var}[\bar{y}_h] = (N_h - n_h)\sigma_h^2/\{n_h(N_h - 1)\}$. We can estimate this in each stratum by simply applying (8.2) and squaring the result. The square root of the corresponding sum is the standard error of $\hat{\mu}$:

$$se[\hat{\mu}] = \sqrt{\sum_{h=1}^{H} \left(\frac{N_h - n_h}{N_h}\right) \frac{W_h^2 s_h^2}{n_h}}, \tag{8.6}$$

where s_h^2 is the sample variance of the n_h observations in the hth stratum. An approximate $1 - \alpha$ confidence interval on μ based on the stratified sample is $\hat{\mu} \pm z_{\alpha/2}se[\hat{\mu}]$; in small samples use $\hat{\mu} \pm t_{\alpha/2}(v_H)se[\hat{\mu}]$, where $v_H = \sum_{h=1}^{H} (n_h - 1)$. Similar constructions are available for estimating the population proportion, π (Scheaffer *et al.*, 1996, §5.6; Levy and Lemeshow, 1999, §5.6). To estimate the population total, τ, use $\hat{\tau} = N\hat{\mu} = N\sum_{h=1}^{H} W_h\bar{y}_h = \sum_{h=1}^{H} N_h\bar{y}_h$. Given $se[\hat{\mu}]$ from (8.6), we have $se[\hat{\tau}] = Nse[\hat{\mu}]$.

Example 8.5 (Phosphorus concentrations) Gilbert (1987, §5.3) gives data on phosphorus (P) concentrations in water of a large pond. The concentrations were thought to vary by pond depth, so the data were stratified into $H = 3$ depth layers: surface ($h = 1$), intermediate ($h = 2$), and bottom ($h = 3$). Within each stratum, observation were taken as P concentrations (in µg/100 ml) from 100 ml aliquots chosen randomly from within the stratum. The allocation pattern was $n_1 = 12$, $n_2 = 10$, and $n_3 = 8$. The data appear in Table 8.3, along with selected stratum-specific summary statistics.

From the summary values in Table 8.3, we find $\hat{\mu} = \sum_{h=1}^{H} W_h\bar{y}_h = 3.450$ µg/100 ml, with $se[\hat{\mu}] = \sqrt{0.0060 + 0.0046 + 0.0109} = 0.146$ from (8.6). From this, an approximate 95% confidence interval on the mean P concentration in this pond is $\hat{\mu} \pm t_{0.025}(27)se[\hat{\mu}] = 3.450 \pm (2.052 \times 0.146) = 3.450 \pm 0.300$ µg/100 ml.

Table 8.3 Pond phosphorus concentrations (µg/100 ml) and stratum-specific summary statistics

	Stratum		
	1. Pond surface	2. Intermediate depth	3. Pond bottom
Data, y_{hi}	1.1, 1.2, 1.4	2.9, 3.0, 3.3	3.4, 3.9, 5.1
	1.5, 1.7, 1.9	3.6, 3.7, 4.0	5.3, 5.7, 5.8
	2.1, 2.1, 2.3	4.1, 4.4, 4.5	6.0, 6.9
	2.5, 3.0, 3.1	4.8	
n_h	12	10	8
$\bar{y}_h$	1.9917	3.8300	5.2625
s_h^2	0.4299	0.4134	1.2941
N_h	3940000	3200000	2500000
$W_h = N_h/N$	0.4087	0.3320	0.2593

Source: Gilbert (1987, §5.3).

We can also perform these calculations in SAS, using the ver. 8 procedure PROC SURVEYMEANS. Sample code appears in Fig. 8.7. In the code, notice the specification and use of heterogeneous sampling weights to adjust for the different sampling rates within strata: for each observation, the weight is calculated as the inverse of the inclusion probability: $\theta_h \propto N_h/n_h$. Without this specification, PROC SURVEYMEANS assigns an equal weight to each observation, which can lead to a biased estimate of μ under a stratified sampling design.

```
data phosphor;
input depth $ Yconc  @@;
    if depth='surf' then theta=3940000/12;  *sampling weights;
    if depth='inte' then theta=3200000/10;
    if depth='bot'  then theta=2500000/8;
datalines;
    surf 1.1   surf 1.2   surf 1.4   surf 1.5   surf 1.7   surf 1.9
    surf 2.1   surf 2.1   surf 2.3   surf 2.5   surf 3.0   surf 3.1
    inte 2.9   inte 3.0   inte 3.3   inte 3.6   inte 3.7   inte 4.0
    inte 4.1   inte 4.4   inte 4.5   inte 4.8
    bot  3.4   bot  3.9   bot  5.1   bot  5.3   bot  5.7   bot  5.8
    bot  6.0   bot  6.9              =
    
data StrTotal;
input depth $  _total_  @@;
datalines;
    surf 3940000    inte 3200000    bot 2500000
    ;

proc surveymeans data=phosphor total=StrTotal   mean   df   clm;
stratum depth ;
var      Yconc;
weight   theta;
```

Figure 8.7 Sample SAS (ver. 8) code to estimate population mean μ via stratified random sample with phosphorus concentration data

In the procedure call, the `total=` option is used to specify a previously defined SAS data set that identifies the stratum totals, N_h, while the `mean` and `clm` options request point and 95% interval estimates, respectively, of μ. (In the data set containing the stratum totals, the SAS variable with the N_h values must be called `_total_`.) The `weight` statement specifies the sampling weights θ_h. As in Fig. 8.4, PROC SURVEYMEANS employs the t critical point $t_{0.025}(df)$ for the confidence limits, where df are the degrees of freedom for the particular sampling design. Edited output appears in Fig. 8.8, where we find results similar to those achieved above.

We can compare our results for these data with those attained if we had not performed a stratified sample, but rather had only conducted an SRS of $n = 30$ observations. That is, suppose we ignore the fact in Table 8.3 that sampling was stratified. The point estimate of μ becomes $\bar{y} = (1/n)\sum_{i=1}^{n} y_i = 104.3/30 = 3.477$, which is little changed from $\hat{\mu} = 3.450$. But $se[\bar{y}] = 0.285$ from (8.2), while $se[\hat{\mu}] = 0.146$: almost a 50% reduction in the standard error has been achieved by incorporating the stratification. ☉

An important consideration in stratified sampling is how to select the total sample size, $n = \sum_{h=1}^{H} n_h$, and also how to allocate the n_hs among the strata. In the former case, we can apply methods similar to those used to find equation (8.4). Suppose we wish to estimate μ to within a margin of error no larger than $\varepsilon > 0$. The optimum choice for n becomes

$$n \geq \frac{\sum_{h=1}^{H} N_h^2 \sigma_h^2 / \omega_h}{\frac{1}{4} N^2 \varepsilon^2 + \sum_{h=1}^{H} N_h \sigma_h^2}, \tag{8.7}$$

where the σ_h^2 terms are per-stratum variances (approximated, as in §8.1, from previous data or preliminary estimates) and $\omega_h = n_h/n$ are predetermined per-stratum allocation weights. An obvious choice for the latter terms is known as a *proportional allocation*: set the sampling allocation in each stratum equal to N_h/N. That is, set $\omega_h = N_h/N = W_h$, leading to

$$n \geq \frac{\sum_{h=1}^{H} N_h \sigma_h^2}{\frac{1}{4} N \varepsilon^2 + \frac{1}{N}\sum_{h=1}^{H} N_h \sigma_h^2}.$$

```
                            The SAS System
                       The SURVEYMEANS Procedure

                             Data Summary
                Number of Strata                       3
                Number of Observations               30
                Sum of Weights                  9640000

                                 Std Error    Lower 95%    Upper 95%
     Variable    DF      Mean     of Mean        CL for Mean

      Yconc      27   3.450147    0.146356    3.149850     3.750444
```

Figure 8.8 SAS output (edited) for estimating μ via stratified random sample with phosphorus concentration data

Notice that by selecting the ω_h values for use in (8.7), the investigator automatically sets an allocation scheme for the individual strata. A broad literature exists on selecting optimum allocations for stratified random sampling, including optimal designs that minimize costs of sampling when such costs can be specified in advance. The resulting allocations typically increase n_h in strata where variability is high or costs are low, and such guidance can be useful. For more on methods of optimal allocation, see Scheaffer *et al.* (1996, §5.4), Levy and Lemeshow (1999, §6.5), or Thompson (2002, §11.5).

Stratified sampling is closely related to the concept of *blocking* in experimental design. In general, blocking is advocated as a means for decreasing variation: one first subdivides the experimental units into relatively homogeneous subgroups based on some *blocking factor*, and next randomly assigns units within blocks to the experimental/treatment levels. The blocking factor is a known source of variability that affects the population in a specific, deterministic manner. When left unrecognized, it will contribute to experimental error in the observations, but by recognizing and incorporating blocking factors into the design, we reduce error variance and improve our ability to detect true differences among the treatment effects (Neter *et al.*, 1996, Ch. 27). From a sampling perspective, strata can be viewed as naturally occurring 'blocks.' By taking advantage of the homogeneity of observations within strata, a stratified sample may lead to increased precision over an SRS in the same way that a block design increases precision over simpler experimental designs.

8.2.3 Cluster sampling

Another extension of simple random sampling occurs when the population elements cluster into natural groupings which are themselves conducive to sampling. This is different from the stratified sampling scenario in §8.2.2, since there is no requirement that the clusters represent a known source of variation among the observations. Formally, a *cluster* is any arrangement or collection of population elements that are grouped together under some common theme. In environmental applications, clusters are often based on natural delineations. Examples include: a grove clustered into trees, a population of animals clustered into herds, a watershed clustered into lakes, a block of residences, etc. Clusters can, however, be artificially defined, such as daily measurements clustered into weeks, a hazardous waste site clustered into non-overlapping quadrats, damage from a hurricane clustered by county, or contributors to a wildlife fund clustered by state.

A *cluster sample* is any sample in which the *primary sampling unit (PSU)* is a cluster of population elements. We say then that the *secondary sampling unit (SSU)* is the actual element within a cluster whose observation is recorded for analysis. One could, of course, go farther into the hierarchy and sample clusters within clusters, creating tertiary sampling units, etc. Here, we restrict attention to the simple case of a population of elements studied with a single level of clustering.

Cluster sampling is most common when construction of a complete frame of population elements is difficult, but construction of a frame of clusters is straightforward. For instance, it may be difficult or costly to list all the landfills in all the counties of a state or province, but simpler to first list just the counties. Viewing the counties as PSUs and the landfills as SSUs, one takes a probability sample of counties and then measures responses from the landfills within each county. If every landfill within each sampled county is measured or enumerated, we have a *one-stage cluster sample*. (If, however, a second probability sample of landfills is taken within each county we have a *two-stage cluster sample*; see §8.2.4.)

The notation for cluster sampling is more complex than that for the other forms of sampling described above. Since the clusters are the PSUs, we begin by defining N as the number of clusters in the population. From these we sample $n < N$ of the clusters. In the ith cluster we denote the number of possible elements as K_i, $i = 1, \ldots, n$, where each element's observed outcome is y_{ij}, $j = 1, \ldots, K_i$. From these, other quantities and population descriptive parameters of interest may be derived; Table 8.4 summarizes the notation for one-stage cluster sampling.

To conduct a one-stage cluster sample, construct the full frame of PSUs (clusters) and select n of them randomly. In the simplest case, this can be an SRS of clusters determined, for example, using the methods in §8.1. Once the sample has been identified, the one-stage cluster sample is completed by adding together all the observations in each ith sampled cluster; in Table 8.4 we denote these values as $y_{i+} = \sum_{j=1}^{K_i} y_{ij}, i = 1, \ldots, n$.

Under such a sampling strategy, we estimate the population total, τ, by recognizing that the sample mean of the SRS-based cluster totals,

$$\bar{y}_+ = \frac{\sum_{i=1}^{n} y_{i+}}{n},$$

Table 8.4 Notation for one-stage cluster sampling

Quantity	Explanation
Cluster characteristics	
N	Number of clusters in population
n	Number of clusters in probability sample
K_m	Number of population elements in mth cluster, $m = 1, \ldots, N$
K_{total}	Total number of population elements: $K_{\text{total}} = \sum_{m=1}^{N} K_m$
$\bar{K}$	Average number of population elements per cluster: $\bar{K} = K_{\text{total}}/N$
Observations	
y_{ij}	Observation from jth element in ith cluster, $i = 1, \ldots, n; j = 1, \ldots, K_i$
y_{i+}	Sum of all observations in ith cluster, $i = 1, \ldots, n$: $y_{i+} = \sum_{j=1}^{K_i} y_{ij}$
Population descriptive parameters	
μ	Mean response in population: $\mu = E[y_{ij}]$
τ	Total response in population: $\tau = \mu K_{\text{total}}$

estimates the average cluster total. Thus multiplying by N will estimate the overall total $\tau = \sum_{m=1}^{N} \sum_{j=1}^{K_i} y_{mj} = \sum_{m=1}^{N} y_{m+}$. That is, an unbiased estimator of τ is the number of clusters multiplied by the average total per cluster, or

$$\hat{\tau} = \frac{N}{n} \sum_{i=1}^{n} y_{i+} = N\bar{y}_+. \tag{8.8}$$

Notice that in (8.8) we do not use the number of elements in the population, $K_{\text{total}} = \sum_{m=1}^{N} K_m$, to obtain an estimate of the population total.

To estimate $se[\hat{\tau}]$, apply the natural analog from §8.1:

$$se[\hat{\tau}] = Nse[\bar{y}_+] = N\sqrt{\frac{(N-n)s_+^2}{Nn}}$$

for $s_+^2 = \sum_{i=1}^{n}(y_{i+} - \bar{y}_+)^2/(n-1)$. Estimation of μ follows, using

$$\hat{\mu} = \frac{\hat{\tau}}{K_{\text{total}}} \tag{8.9}$$

and $se[\hat{\mu}] = se[\hat{\tau}]/K_{\text{total}}$.

If the data are binary, we define $y_{ij} = 1$ if the jth element of the ith cluster possesses the characteristic of interest and $y_{ij} = 0$ otherwise. Interest then centers on estimating the population proportion $\pi = P[y_{ij} = 1]$. In this case (8.8) and (8.9) remain valid, now with the y_{ij}s as the corresponding binary observations.

If the cluster total y_{i+} and cluster size K_i are highly correlated, it is possible to improve efficiency in estimating τ. To measure the correlation, we use the usual product-moment estimator

$$r_{YK} = \frac{\sum_{i=1}^{n}(y_{i+} - \bar{y}_+)(K_i - \bar{K})}{(n-1)s_+s_K}, \tag{8.10}$$

where $s_K^2 = \sum_{i=1}^{n}(K_i - \bar{K})^2/(n-1)$. If r_{YK} is near ± 1, we consider an alternative estimator for τ, found by multiplying the population size by the average response observed in the sampled clusters. This produces the ratio estimator

$$\hat{\tau}_R = K_{\text{total}} \frac{\sum_{i=1}^{n} y_{i+}}{\sum_{i=1}^{n} K_i}. \tag{8.11}$$

When r_{YK} is near ± 1, (8.11) will often exhibit smaller variation than that of (8.8).

Unfortunately, complications arise when employing (8.11) if the K_is are random (which is common) or if the number of population elements, K_{total}, is unknown. Of perhaps greatest concern then is that (8.11) is not unbiased, that is, $E[\hat{\tau}_R] \neq \tau$. As a function of sample size, however, the bias diminishes faster than the standard error. Thus, if the precision in the sample is felt to be reasonable, the lack of bias may not be a problem (Thompson, 2002, §12.1).

The standard error of (8.11) must be approximated; one possibility is

$$
se[\hat{\tau}_R] \approx \sqrt{\frac{N(N-n)}{n}\frac{1}{n-1}\sum_{i=1}^{n}\left(y_{i+} - \hat{\tau}_R\frac{K_i}{K_{total}}\right)^2}
\tag{8.12}
$$

$$
= \sqrt{\frac{N(N-n)}{n}\left\{s_+^2 + \left(\frac{\hat{\tau}_R}{K_{total}}\right)^2 s_K^2 - \frac{2\hat{\tau}_R r_{YK}S_+S_K}{K_{total}}\right\}}
$$

(Scheaffer *et al.*, 1996, §8.3; Thompson, 2002, §12.1). To estimate μ when r_{YK} is near ±1, simply apply (8.11) in (8.9), and take $se[\hat{\mu}_R] = se[\hat{\tau}_R]/K_{total}$. Notice that since $\hat{\mu}_R = \hat{\tau}_R/K_{total}$, some simplification is possible in (8.12); cf. Exercise 8.16.

If large disparities exist between the cluster sizes K_i, it is common to sample the PSUs such that larger clusters are favored, in order to give greater weight to these larger assemblages of population elements. Such *probability proportional to size (PPS)* sampling is performed with replacement, where the probability of selecting the *i*th cluster is taken to be the fraction of the population in the *i*th cluster, $q_i = K_i/K_{total}$. PROC SURVEYSELECT can perform PPS sampling via a variety of possible schemes in its `method=` option.

Under PPS sampling, an unbiased estimator for τ is known as the *Hansen–Hurwitz estimator*,

$$
\hat{\tau}_{HH} = \frac{1}{n}\sum_{i=1}^{n}\frac{y_{i+}}{q_i}
\tag{8.13}
$$

(Hansen and Hurwitz, 1943). This has standard error

$$
se[\hat{\tau}_{HH}] = \sqrt{\frac{1}{n(n-1)}\sum_{i=1}^{n}\left(\frac{y_{i+}}{q_i} - \hat{\tau}_{HH}\right)^2}.
$$

To estimate μ, use $\hat{\mu}_{HH} = \hat{\tau}_{HH}/K_{total}$ with $se[\hat{\mu}_{HH}] = se[\hat{\tau}_{HH}]/K_{total}$.

8.2.4 Two-stage cluster sampling

In many environmental applications, variation between SSUs within a PSU is often much smaller than the variation between PSUs. In such cases, it is inefficient to enumerate or even sample a large number of SSUs within each PSU, and hence it is unusual to see an entire PSU enumerated after being selected for the cluster sample. More common is use of a second probability sample of SSUs, producing a *two-stage cluster sample*. For instance, after taking a simple random sample of *n* clusters, one next takes a simple random sample of population elements from within each of the sampled clusters. Table 8.5 summarizes the two-stage notation.

Table 8.5 Notation for two-stage cluster sampling

Quantity	Explanation
N	Number of clusters in population
n	Number of clusters in first-stage probability sample
K_i	Number of population elements in ith cluster, $i = 1, \ldots, n$
K_{total}	Total number of population elements: $K_{total} = \sum_{m=1}^{N} K_m$
k_i	Number of elements sampled from ith cluster at second stage
y_{ij}	Observation from jth sampled element in ith cluster, $i = 1, \ldots, n; j = 1, \ldots, k_i$.
$\bar{y}_i$	Average of sampled observations in ith cluster: $\bar{y}_i = \sum_{j=1}^{k_i} y_{ij}/k_i$
μ	Mean response in population: $\mu = E[y_{ij}]$
τ	Total response in population: $\tau = \mu K_{total}$

To estimate τ from a two-stage cluster sample we imitate the unbiased estimator from (8.8): first, estimate y_{i+} – which is unknown when individual clusters are not fully enumerated – with the unbiased quantity $K_i\bar{y}_i$. Then, use this in (8.8) to yield the unbiased two-stage estimator

$$\hat{\tau}_{TS} = \frac{N}{n} \sum_{i=1}^{n} K_i\bar{y}_i. \tag{8.14}$$

Contrast the two-stage estimator in equation (8.14) with the one-stage estimator from equation (8.8): the two-stage estimator is the product of the number of clusters and the *estimated* average total per cluster, while the one-stage estimator is the product of the number of clusters and the *observed* average total per cluster.

Standard errors become more complex under two-stage sampling: for (8.14) the standard error becomes

$$se[\hat{\tau}_{TS}] = \sqrt{\frac{N(N-n)}{n} s_b^2 + \frac{N}{n} \sum_{i=1}^{n} \frac{K_i(K_i - k_i)}{k_i} s_i^2}, \tag{8.15}$$

where $s_b^2 = \sum_{i=1}^{n} (K_i\bar{y}_i - (1/N)\hat{\tau}_{TS})^2/(n-1)$ is a between-cluster variance estimator and $s_i^2 = \sum_{j=1}^{k_i} (y_{ij} - \bar{y}_i)^2/(k_i - 1)$ are the per-cluster (or 'within-cluster') sample variances, $i = 1, \ldots, n$. To estimate μ, use $\hat{\mu}_{TS} = \hat{\tau}_{TS}/K_{total}$ with $se[\hat{\mu}_{TS}] = se[\hat{\tau}_{TS}]/K_{total}$. Alternative methods for estimating τ or μ with more complex settings such as PPS sampling or if the population size K_{total} is unknown are also possible; see, for example, Scheaffer *et al.* (1996, §9.4) or Thompson (2002, Ch. 13).

Estimation of parameters such as μ or τ may be performed by computer. For example, in SAS cluster-sampled data may be analyzed using PROC SURVEYMEANS via the `cluster` statement.

Example 8.6 (Groundwater conductivity) Gilbert (1987, Ch. 6) gives data on specific conductance (in μmho/cm) of groundwater from a set of $N = 5$ wells near a

Table 8.6 Two-stage cluster sample from study of water conductivity

Cluster (well), i	Conductivity data, y_{ij}			
1	3150	3150	3145	3150
2	1400	1380	1390	1380
3	3720	3720	3719	3710

Source: Gilbert (1987, Ch. 6).

contaminated well site. We view each well as a cluster. A two-stage cluster sample is conducted, where in the first stage $n = 3$ wells are sampled randomly at the site. At the second stage, a further random sample of $k = 4$ conductance measurements is taken within each well. The data appear in Table 8.6.

If each well is of roughly equal volume such that it would produce a collection of $K = 8$ samples when sampled fully, we have $K_{total} = 40$. Consider estimation of the mean conductance, μ, throughout the site. From Table 8.6 the per-cluster sample means are $\bar{y}_1 = 3148.75, \bar{y}_2 = 1387.50$, and $\bar{y}_3 = 3717.25$. Using these, (8.14) produces $\hat{\tau}_{TS} = 110\,046.67$, and so $\hat{\mu}_{TS} = \hat{\tau}_{TS}/K_{total} = 110\,046.67/40 = 2751.167\,\mu$mho/cm. We also find the per-cluster sample variances to be $s_1^2 = 6.250, s_2^2 = 91.667$, and $s_3^2 = 23.583$. Lastly, $s_b^2 = 94\,431\,241.33$, so application of (8.15) yields $se[\hat{\tau}_{TS}] = \sqrt{314\,772\,424.44} = 17\,741.83$. Thus $se[\hat{\mu}_{TS}] = se[\hat{\tau}_{TS}]/K_{total} = 17\,741.83/40 = 443.546$. An approximate 95% confidence interval on μ is then $\hat{\mu}_{TS} \pm z_{0.25}se[\hat{\mu}_{TS}] = 2751.167 \pm (1.96 \times 443.546) = 2751.167 \pm 869.350\,\mu$mho/cm. ☉

Two-stage cluster samples may also be taken via a PPS format, as mentioned in §8.2.3. In this case, a form of Hansen–Hurwitz estimator is available for estimating τ; details can be found in Levy and Lemeshow (1999, §11.3.1) or Christman (2002).

When more than two levels of clustering exist, two-stage samples can be extended hierarchically into *multi-stage samples*. Readers should be careful not to confuse this with a much different sampling technique known as *multi-phase sampling*. Multi-phase sampling is used to incorporate auxiliary information in the population elements. One first samples to acquire the auxiliary information and then uses this information to sample for the outcome of interest. (When using only two sample phases, this is also known as *double sampling*.) For more on two-phase and multi-phase sampling, see Lesser (2002) or Thompson (2002, Ch. 14).

8.3 Specialized techniques for environmental sampling

8.3.1 Capture–recapture sampling

When an environmental sampling frame is based on spatial or geographic locations – called a *map frame* above – it is often necessary to employ techniques that are more specialized than the methods described in §§8.1–8.2. A common example occurs when estimation of the size or abundance of a population is of interest, for example, the number of animals in a herd or the number of fish in a lake. Consider the simplest case of a single population of size $\psi > 0$ whose elements are mobile or otherwise dynamic and hence cannot be easily enumerated. In this case, we use a two-sample design to estimate ψ. In the first sampling sweep, we take an SRS of $t > 0$ elements from the population and *tag* or *mark* them in some unambiguous fashion. We then release these tagged elements back into the population. A short time later, we return to the population and take another SRS of size n (where each population element has an equal probability of capture in either sampling sweep) and count the number of elements, M, that exhibit the previous tag. This is known as a *capture–recapture sampling design*, sometimes also called a *tag–recapture* or *mark–recapture* design. Note that one typically selects $t \geq n$, and constructs the design such that no element captured and tagged in the first sample can be overlooked in the second sample.

In the simplest case the population is *closed*, so that no migration or other demographic changes occur during the period of sampling. This ensures that ψ remains constant over the course of the study. (Seber, 1970, gives a more formal description of the closed-population assumptions; see also Borchers *et al.*, 2002.) To estimate ψ, notice that the observed proportion of tagged elements, M/n, estimates the closed-population proportion t/ψ under an SRS, so we expect $M/n \approx t/\psi$. Solving for ψ yields what is known as the *Lincoln–Petersen estimator* of ψ, $\hat{\psi}_{LP} = nt/M$ (Petersen, 1896; Lincoln, 1930). Since $\hat{\psi}_{LP}$ need not be a whole number, in practice it is rounded up to the nearest integer.

If sampling in the second SRS is with replacement, M has a binomial distribution with parameters n and $p = t/\psi$. In practice, however, sampling without replacement is more likely. If so, M has a hypergeometric distribution (§A.2.3) with parameters t, ψ, and n. Unfortunately, for either sampling distribution the value of $P[M = 0]$ is greater than 0, which causes $E[\hat{\psi}_{LP}]$ to diverge. This brings into question the validity of $\hat{\psi}_{LP}$. Indeed, even if we modify the model to force $P[M = 0] = 0$, $\hat{\psi}_{LP}$ will still be biased: as $\psi \to \infty$, the relative bias of $\hat{\psi}_{LP}$ is $\{E[\hat{\psi}_{LP}] - \psi\}/\psi \approx (\psi - t)/nt$ (Scheaffer *et al.*, 1996, §10.2). Thus unless n is chosen to be very large, use of $\hat{\psi}_{LP}$ has drawbacks.

Chapman (1951) suggested an alternative to $\hat{\psi}_{LP}$ based on modifying the maximum likelihood estimator of ψ under a hypergeometric model for M. He gave

$$\hat{\psi}_C = \frac{(n + 1)(t + 1)}{M + 1} - 1. \tag{8.16}$$

This estimator is also slightly biased: Robson and Regier (1964) found $E[\hat{\psi}_C] \approx \psi - \psi e^{-nt/\psi}$, thus the relative bias is approximately $e^{-nt/\psi}$. For $M \geq 7$, Robson and Regier showed that this bias is negligible.

Chapman's estimator has standard error

$$se[\hat{\psi}_C] = \sqrt{\frac{(t + 1)(n + 1)(t - M)(n - M)}{(M + 1)^2(M + 2)}} \tag{8.17}$$

(Seber, 1970). An approximate $1 - \alpha$ confidence interval for ψ is then $\hat{\psi}_C \pm z_{\alpha/2}se[\hat{\psi}_C]$. In small samples, the approximation can perform poorly; an alternative suggested by

Chapman (1948) solves iteratively for the lower limit ψ_L and the upper limit ψ_U in the following set of simultaneous equations:

$$\frac{\alpha}{2} = \sum_{i=0}^{M} \frac{\binom{t}{i}\binom{\psi_U - t}{n - i}}{\binom{\psi_U}{n}} = \sum_{i=M}^{\min\{t,n\}} \frac{\binom{t}{i}\binom{\psi_L - t}{n - i}}{\binom{\psi_L}{n}}. \qquad (8.18)$$

Due to the discrete nature of the hypergeometric distribution, it may not be possible to satisfy the equalities in (8.18) exactly. In such cases, we choose values of ψ_L and ψ_U that are closest to preserving a confidence level of at least $1 - \alpha$.

When sample sizes are small and if sufficient computing resources exist, the $1 - \alpha$ confidence limits in (8.18) may be preferred over use of the Wald interval $\hat{\psi}_C \pm z_{\alpha/2} se[\hat{\psi}_C]$. With very large samples, however, the two intervals will perform similarly. Note that for a closed population we expect $\psi \geq t$ and we typically set $t \geq n$. Thus, one never reports a lower endpoint below t. In fact, since from §A.2.3 the hypergeometric p.m.f. is undefined when $\psi < t - M + n$, a more pertinent lower bound on ψ_L is $t - M + n$.

Example 8.7 (Bobwhite quail abundance) Pollock *et al.* (1990) give data on the size of a population of bobwhite quail (*Colinus virginianus*) in a Florida field research station, estimated using closed-population capture–recapture methods. Initially, $t = 148$ quail were trapped and banded (the 'capture' sweep) over a 20-day period. After release, $n = 82$ birds were recaptured from which a total of $M = 39$ bands were recovered. Thus from (8.16) the Chapman estimator of the population size is $\hat{\psi}_C = \frac{83 \times 149}{40} - 1 = 308.175$, or 309 birds. (The Lincoln–Petersen estimate is $\hat{\psi}_{LP} = \frac{82 \times 148}{39} = 311.18$, or 312 birds.) Applying the hypergeometric-based confidence limits from (8.18), iterative calculations show that at $\alpha = 0.05$ the limits are $\psi_L = 259$ and $\psi_U = 394$ birds.

Sample sizes here are arguably too low for use of the Wald confidence interval. For completeness, however, we illustrate its calculation at $\alpha = 0.05$: using (8.17), we find $se[\hat{\psi}_C] = 29.7254$, and the interval becomes $\hat{\psi}_C \pm z_{\alpha/2} se[\hat{\psi}_C] = 308.175 \pm (1.96 \times 29.7254) = 308.175 \pm 58.2618$, or $249.9132 < \psi < 366.4368$. Since ψ must be an integer, one would report these limits as $249 < \psi < 367$ birds. ☺

Useful generalizations of capture–recapture methods are available for studying open populations where birth, death, and migration allow ψ to change while sampling takes place. The archetype for such a setting is known as the Cormack–Jolly–Seber model (Cormack, 1964; Jolly, 1965; Seber, 1965), in which multiple recaptures are allowed and where birth and survival rates can vary dynamically across sampling sweeps. An extension for modeling survival rates that incorporates use of log-linear models such as those seen in §3.3.4 is due to Cormack (1981); see also Fienberg (1972).

Advanced model formulations such as these extend beyond the scope of this section; interested readers may find details on open-population capture–recapture methods in sources such as McDonald and Amstrup (2001), Gould and Pollock (2002), and the references therein. More complex sampling designs for capture–recapture studies with both open and closed populations are reviewed by Lindberg and Rexstad (2002); their Fig. 1 is a useful template for choosing a design strategy when estimating animal abundance.

8.3.2 Quadrat sampling

For many environmental monitoring or sampling problems, the sampling elements may occupy an area or shape – such as a square, hexagon, circle, or perhaps a complex irregular form – or the frame may be reasonably tessellated into such shapes. Collectively known as *quadrats*, these areal units often become the basic sampling units/PSUs under a map frame. We saw an illustration of the use of square quadrats with the redwood seedlings data in Example 6.1, and give a further example in Exercise 8.18. Indeed, quadrat sampling is a simple way to incorporate the shape or area of the region under study into the sampling process.

An interesting application of quadrat sampling occurs when estimating an ecological population's density (number of elements per unit area) or size. Suppose an ecosystem of overall area A is partitioned into N equally sized quadrats, each of area B. Clearly, $A = NB$. Note that the quadrats here are analogous to the clusters from §8.2.3, and so we use N to denote the total number of quadrats. To estimate the size and/or density, we take an SRS of n quadrats and count the number, y_i, of elements observed in the ith quadrat, $i = 1, \ldots, n$. Denote the unknown population size as $\psi = \sum_{i=1}^{N} y_i$ and the corresponding population density as $\lambda = \psi/A$.

To estimate λ we start with the average quadrat count, $\bar{y} = \sum_{i=1}^{n} y_i/n$, such that $E[\bar{y}] = \lambda B$. With this, an unbiased estimate of the density is simply

$$\hat{\lambda} = \frac{\bar{y}}{B}.$$

(Notice that this does not require the actual value of A.) To find the standard error of $\hat{\lambda}$, we can apply basic design-based principles: find $s_y^2 = \sum_{i=1}^{n} (y_i - \bar{y})^2/(n-1)$ and take $se[\bar{y}] = s_y/\sqrt{n}$ (Scheaffer *et al.*, 1996, §10.5). This gives

$$se[\hat{\lambda}] = \frac{s_y}{\sqrt{B^2 n}}.$$

An approximate $1 - \alpha$ confidence interval for λ is then $\hat{\lambda} \pm z_{\alpha/2} se[\hat{\lambda}]$.

To estimate the population size ψ, simply apply the relationship $\psi = \lambda A$: $\hat{\psi} = \hat{\lambda} A$ with

$$se[\hat{\psi}] = A se[\hat{\lambda}] = \frac{A s_y}{\sqrt{B^2 n}}.$$

An approximate $1 - \alpha$ confidence interval for ψ is $\hat{\lambda} A \pm z_{\alpha/2} A se[\hat{\lambda}]$.

Example 8.8 (Fire ant density) Scheaffer *et al.* (1996, §10.5) present data on a quadrat sample of fire ant (*Solenopsis invicta*) mounds in a northern Florida community. To estimate the ant mound density, an SRS of $n = 50$ quadrats was taken,

Table 8.7 Frequencies of fire ant mounds in a northern Florida community

Number of mounds	0	1	2	3	4	5	≥ 6
Frequency	13	8	12	10	5	2	0

Source: Scheaffer *et al.* (1996, §10.5).

where each quadrat had area $B = 16\,\text{m}^2$. The data, as frequencies of observed mounds, appear in Table 8.7.

From the frequency table we can calculate the average quadrat count as $\bar{y} = \frac{92}{50} = 1.84$ mounds, thus an estimate of the mound density is $\hat{\lambda} = \bar{y}/B = 1.84/16 = 0.115$ mounds/m². We also find $s_y^2 = 2.178$, so that the corresponding standard error is $se[\hat{\lambda}] = \frac{1}{16}\sqrt{2.178/50} = 0.013$. Thus an approximate 95% confidence interval on the mound density is $0.115 \pm (1.96 \times 0.013) = 0.115 \pm 0.026$, or between 0.09 and 0.14 mounds/m². ✪

In practice, the *locations* of the quadrats – say, their centers or corners – often drive their selection for the sample, possibly from some structural or systematic grid. In effect, the sampling is based on these point locations rather than on the whole areal units. When this occurs, an explicit protocol or set of rules must be in place that defines how and when elements are included within a sampled quadrat. For irregularly shaped quadrats, however, this may be difficult to formalize. (Indeed, a study region of area A partitioned into N equally sized quadrats is a bit of an idealization.) If a viable, practicable protocol for quadrat definition and element inclusion cannot be identified *a priori*, then use of quadrat sampling may contraindicated (Stehman and Overton, 2002).

8.3.3 Line-intercept sampling

A variant of quadrat sampling that uses only linear transects to identify population elements is known as *line-intercept sampling*. The method in its simplest form selects n random starting points from a baseline of length B in a region of area A. From each starting point, a straight-line transect is taken across the region and each time a population element is found to cross the transect ('intercepted'), a measure of interest, x_j, is recorded. For example, a forestry researcher studying coarse woody debris may wish to measure the number, length, or volume of downed logs in an old-growth forest. From a line-intercept sample, measurements might be taken of every downed log the surveyor 'intercepts' along a series of linear transects within the forest.

Origin points for the linear transects are determined via an SRS or a systematic sample. Transiting of the linear transects can be performed directly by foot or moving vehicle if terrestrial, by boat or diver if aquatic, or via remote sensing methods such as fly-overs or satellite observation. Typical population descriptive parameters of interest from line-intercept samples are the population total $\tau = \sum_{j=1}^{N} x_j$, where N is the total number of elements in the population, or the population density $\lambda = \tau/A$.

To estimate τ, notice that along any transect the probability of encountering the jth element in the population is proportional to that element's one-dimensional cross-section or 'width,' say, w_j. (The element's length is irrelevant since the first interception along the transect counts as a detection. In addition, we assume the height of the element has no impact on its probability of detection. Indeed, height could be a measure of interest for use in x_j.) By projecting this width back on to the linear baseline, we can define the formal detection probability for the jth element as the ratio of the cross-sectional width to the baseline length: $p_j = w_j/B$. This is an observable measure: we require the surveyor to record w_j along with x_j whenever an element is encountered. Notice here that the value of w_j can be larger than the distance between transects, so an element may be encountered and recorded more than once. One views this essentially as a form of population sampling with replacement. Figure 8.9 gives a simplified schematic of the approach.

Now, along the ith transect suppose $\varphi_i \geq 0$ elements are encountered. Define from this subgroup of elements the variable

$$y_i = \sum_{j=1}^{\varphi_i} \frac{x_j}{p_j} = B \sum_{j=1}^{\varphi_i} \frac{x_j}{w_j},$$

$i = 1, \ldots, n$. Notice that use of x_j/p_j acts to weight the measurement of interest inversely to its detection probability. An unbiased estimator of τ is simply the sample mean of the y_is:

$$\hat{\tau} = \bar{y} = \frac{1}{n} \sum_{i=1}^{n} y_i.$$

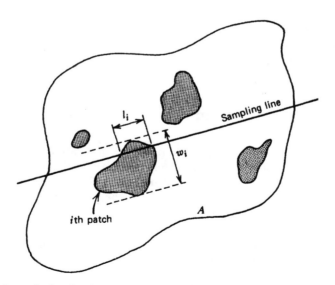

Figure 8.9 Schematic for line-intercept sampling where the ith intercept has width w_i and intercept length l_i, from Pielou (1985), in S. Kotz, N. L. Johnson and C. B. Read (eds), Encyclopedia of Statistical Sciences, Vol. 5. © 1985 John Wiley & Sons, Inc. This material is used by permission of Wiley-Liss, Inc., a subsidiary of John Wiley & Sons, Inc.

This resembles a form of Hansen–Hurwitz estimator from (8.13), since it incorporates an auxiliary measure, p_j, that provides information on the probability of sample selection.

Because $\hat{\tau}$ is a sample mean, its standard error is simple to identify: $se[\hat{\tau}] = s_y/\sqrt{n}$, where $s_y^2 = \sum_{i=1}^{n}(y_i - \hat{\tau})^2/(n-1)$. Notice that no FPC is employed in the standard error, since sampling is viewed as with replacement (Thompson, 2002, §19.1). With these, an approximate $1 - \alpha$ confidence interval for τ is $\hat{\tau} \pm z_{\alpha/2}se[\hat{\tau}]$. To estimate the density λ use $\hat{\lambda} = \hat{\tau}/A$, with $se[\hat{\lambda}] = se[\hat{\tau}]/A$.

In certain settings, the landscape/topography under study makes it difficult for the transects to remain straight, or for the sample to adequately represent the population within the region of interest. It may then be useful to employ segmented transects that orient as L-, Δ-, T-, or Y-shaped forms (or even some combination thereof). These require modifications to the point estimators and standard errors, however, and so should be applied judiciously. See Gregoire and Valentine (2003) for more on sampling with segmented transects.

Estimation of τ can also be performed by incorporating unequal sampling probabilities, using what is known as the Horvitz–Thompson method (Horvitz and Thompson, 1952). The approach is similar to Hansen–Hurwitz estimation as seen in §8.2.3, but involves a more complex weighting scheme for $\hat{\tau}$. The added complexity is counterbalanced by potential increases in estimation efficiency. For more details, see Thompson (2002, §19.1). See also Kaiser (1983) for an excellent review on the history of the method, along with some of its more technical aspects.

Readers should not confuse line-intercept sampling with the slightly different method known as *line-transect sampling*. The latter is an extension/modification of quadrat sampling that uses data acquired not only when elements intercept a transect, but also when they are detected at a known distance from the transect. Design and estimation for line-transect and other sorts of *distance sampling* approaches are by necessity more complex than for line-intercept sampling; details are available from Thompson (2002, Ch. 17), Thomas *et al.* (2002), and the references therein.

8.3.4 Ranked set sampling

When the cost of measuring an element from the population under study greatly exceeds the cost of sampling it, interest often exists in selecting many more elements than actually can be measured in order to make the observations as representative *and* as cost-effective as possible. Of course, the consequent statistical analysis must take this selection bias into account. In this subsection (and in §8.3.5), we discuss methods for performing such selective sampling and analysis. The first, known as *ranked set sampling (RSS)*, can actually improve the efficiency of estimating the population mean, μ, after incorporating the selective effect.

The RSS method was originally developed by McIntyre (1952) to estimate herbage biomass and agricultural yields. He suggested a two-stage approach: start with a sample of size $n = m^2$ and break it down randomly into m lots, each of size m. In the second stage, take a selectively ranked observation from each lot. Thus, for example, select first the smallest observation from lot 1, next the next smallest from lot 2, then the third smallest from lot 3, etc., finishing with the largest observation from lot m. In this way, the final, selected sample of size m will be less susceptible to the effects of extreme observations than if an SRS of size m had been taken. The method will be cost-effective if it is simpler to rank the observations in each lot than to actually measure them. For example, a forester could easily rank tree heights by eye in each lot, then measure only those m trees chosen for the RSS. One could also use a concomitant variable to perform the rankings; for example, use tree height to rank in the first stage, but measure tree biomass in the second stage. In either case, we assume implicitly that there are no errors of judgment in the rankings (but see below).

Formally, let y_{ij} be the jth observation in the ith lot, $i = 1, \ldots, m$; $j = 1, \ldots, m$. Define by $y_{i(j)}$ the jth ordered observation in the ith lot. Under the RSS scheme it is the $y_{i(i)}$ values that make up the sample. Table 8.8 displays the sampling schematic. To estimate $\mu = E[y_{ij}]$, McIntyre used the sample mean of the RSS:

$$\hat{\mu} = \frac{1}{m} \sum_{i=1}^{m} y_{i(i)}. \tag{8.19}$$

Here, $\hat{\mu}$ is unbiased for μ, and its variance never exceeds that of the simpler sample mean from an SRS, that is, $\mathrm{Var}[\hat{\mu}] \leq \mathrm{Var}[\bar{y}_{\mathrm{SRS}}]$ (Takahasi and Wakimoto, 1968). This embodies the sampling efficiency of the RSS approach: the RSS mean of m observations is often more precise than the SRS mean of m observations. (The variance reduction is related to the fact that RSS is essentially a form of stratification, albeit adaptive in nature.) Of course, the SRS mean from the larger sample of m^2 observations will typically be more precise than the RSS mean from only m observations, but when the cost of measuring each observation is high the RSS approach can be more effective overall.

If in practice it is not possible to ensure that the rankings are accurate – i.e., that some *judgment error* may enter into the ranking stage – some of the rank-sampled elements may be assigned ranks that differ from their true values. Dell and Clutter (1972) recognized this concern, but were nonetheless able to show that $\hat{\mu}$ from (8.19) remained both unbiased for μ and more efficient than the SRS estimator.

Calculation of the standard error of $\hat{\mu}$ is somewhat more problematic. The variance of $\hat{\mu}$ is

$$\mathrm{Var}[\hat{\mu}] = \frac{\sigma^2}{m} - \sum_{i=1}^{m} \frac{(\mu_{(i)} - \mu)^2}{m^2},$$

Table 8.8 Ranked set sampling: m ordered observations (boxed) are taken from a larger sample of m^2 observations

Lot	Observations ordered within lots						
1	$\boxed{y_{1(1)}}$	$\leq$	$y_{1(2)}$	$\leq$	$\cdots$	$\leq$	$y_{1(m)}$
2	$y_{2(1)}$	$\leq$	$\boxed{y_{2(2)}}$	$\leq$	$\cdots$	$\leq$	$y_{2(m)}$
$\vdots$	$\vdots$			$\vdots$		$\ddots$	$\vdots$
m	$y_{m(1)}$	$\leq$	$y_{m(2)}$	$\leq$	$\cdots$	$\leq$	$\boxed{y_{m(m)}}$

where σ^2 is the population variance of the y_{ij}s and $\mu_{(i)}$ is the expected value of the ith ordered observation in any lot: $\mu_{(i)} = E[y_{i(i)}]$ (Dell and Clutter, 1972). These terms will depend upon the underlying distribution of the y_{ij}s, and they require specific determination in every case. For example, suppose the y_{ij}s come from an exponential distribution with mean μ. Assuming no judgment error, results from the theory of order statistics can be used to find $\mathrm{Var}[\hat{\mu}] = 11\mu^2/54$ (Kvam and Samaniego, 1993). This distributional specificity hinders application of RSS, since it is not always clear which family of distributions should be specified for the original observations.

In practice, values of m larger than 2, 3, or perhaps 4 become unmanageable and even counterproductive: the ranking of a large number of elements per lot can be time-consuming and hence cost- or resource-intensive. As a result, sample sizes are typically restricted under simple RSS to measurements on only 4, 9, or perhaps 16 observations. When resources permit a larger number of elements to be measured, we can extend the RSS scheme: group the data are into r *cycles* of m lots, where each lot continues to contain m observations. That is, from an original data set of size $n = m^2 r$ take measurements on mr elements. (This may appear to require many more elements, but by reducing the lot size to only $m = 2$ or 3, the consequent value of $m^2 r$ remains manageable.) Then, apply RSS to each kth cycle ($k = 1, \ldots, r$), producing the observed values $y_{i(i)k}$. An unbiased estimator for μ is

$$\hat{\mu}_r = \frac{1}{mr} \sum_{k=1}^{r} \sum_{i=1}^{m} y_{i(i)k}. \tag{8.20}$$

This retains the sampling efficiency acquired in (8.19) (Patil *et al.*, 1994b). An estimator for $\mathrm{Var}[\hat{\mu}_r]$ is s_r^2/r, where

$$s_r^2 = \frac{1}{m^2(r-1)} \sum_{k=1}^{r} \sum_{i=1}^{m} (y_{i(i)k} - \bar{y}_{(i)+})^2$$

and $\bar{y}_{(i)+} = \sum_{k=1}^{r} y_{i(i)k}/r$ (Nussbaum and Sinha, 1997). From this, take $se[\hat{\mu}_r] = s_r/\sqrt{r}$ in the approximate $1 - \alpha$ confidence interval $\hat{\mu}_r \pm z_{\alpha/2} se[\hat{\mu}_r] = \hat{\mu}_r \pm z_{\alpha/2} s_r/\sqrt{r}$. The approximation improves as r grows large.

Example 8.9 (Gasoline volatility) Nussbaum and Sinha (1997) give data from a US Environmental Protection Agency study of automobile gasoline emissions. Of interest was the volatility of the gasoline after reformulation to reduce pollutant emissions during automobile use. It is usually quite simple to acquire and analyze gasoline samples at the pump; however, the cost of shipping the volatile samples to the laboratory can be expensive. RSS methods seem natural in this case.

Suppose that $n = 36$ gasoline samplings are available, from which an RSS is to be drawn. With $n = 36$ an obvious assignment is $r = 4$ and $m = 3$, so that the data are grouped into four cycles of $3^2 = 9$ observations. In each cycle, the $m = 3$ RSS data points are taken as the smallest observation from the first lot of gasoline samplings, the median from the second lot, and the largest observation from the third lot. This produces a RSS of $mr = 12$ measurements.

Table 8.9 gives data on the gasoline samplings as vapor pressure measurements (in psi). Lower pressures suggest lower pollutant emissions. The boxed values are the actual laboratory measurements taken for the RSS; the other values are not used here. (In practice, to perform the ordering and to produce each cycle's RSS one would require some form of judgment ranking. Nussbaum and Sinha suggest use of a concomitant variable such as a less reliable but faster field measurement taken directly at the pump.)

Using (8.20), we find for the RSS data (boxes) in Table 8.9 that $\hat{\mu}_4 = 100.45/12 = 8.3708$ psi. For the standard error, we first calculate the rank means as $\bar{y}_{(1)+} = \frac{1}{4}(7.59 + 8.56 + 7.83 + 7.73) = 7.9275$, $\bar{y}_{(2)+} = \frac{1}{4}(9.01 + 7.99 + 7.83 + 7.98) = 8.2025$, and $\bar{y}_{(3)+} = \frac{1}{4}(9.28 + 10.67 + 7.83 + 8.15) = 8.9825$. With these, s_4^2 is $\frac{1}{27}\{(7.59 - 7.9275)^2 + \ldots + (8.15 - 8.9825)^2\} = 6.4054/27 = 0.2372$, leading to $se[\hat{\mu}_4] = \sqrt{0.2372/4} = 0.2435$. Thus an approximate 95% confidence interval for μ is $8.3708 \pm (1.96 \times 0.2435) = 8.3708 \pm 0.4773$ psi. By way of comparison, if we had treated these 12 observations as an SRS, the estimate of μ assuming an infinite population size would still have been the sample mean 8.3708. Its standard error would have increased, however, to 0.2582. ✪

We should note that in order to estimate μ via (8.19) or (8.20) and construct the associated standard errors we require the implicit assumption that the population from which the first-stage sample is drawn is infinite. If the population is only of finite size, as seen throughout the discussion in §8.2, then some correction is required in the estimation equations. Patil *et al.* (1995) give details for this setting. An important consequence of their results is that even when corrected for a finite population size, the RSS estimator of μ continues to exhibit smaller variance than the corresponding SRS-type estimator (Takahasi and Futatsuya, 1998).

Table 8.9 Gasoline vapor pressures (psi) taken as $r = 4$ cycles of $m = 3$ lots each. Ranked set sampled data are given in boxes

Cycle	Lot	Observations ordered within lots		
1	1	7.59	$y_{1(2)1}$	$y_{1(3)1}$
	2	$y_{2(1)1}$	9.01	$y_{2(3)1}$
	3	$y_{3(1)1}$	$y_{3(2)1}$	9.28
2	1	8.56	$y_{1(2)2}$	$y_{1(3)2}$
	2	$y_{2(1)2}$	7.99	$y_{2(3)2}$
	3	$y_{3(1)2}$	$y_{3(2)2}$	10.67
3	1	7.83	$y_{1(2)3}$	$y_{1(3)3}$
	2	$y_{2(1)3}$	7.83	$y_{2(3)3}$
	3	$y_{3(1)3}$	$y_{3(2)3}$	7.83
4	1	7.73	$y_{1(2)4}$	$y_{1(3)4}$
	2	$y_{2(1)4}$	7.98	$y_{2(3)4}$
	3	$y_{3(1)4}$	$y_{3(2)4}$	8.15

Source: Nussbaum and Sinha (1997).

The literature on RSS continues to grow, as more environmental (and other) studies are planned that employ its cost-effective features. Useful reviews on the RSS approach are given by Stokes (1986), Patil *et al.* (1994b), and Patil (2002). See also the special issue of *Environmental and Ecological Statistics* (Ross and Stokes, 1999) devoted to RSS.

Before moving on we should note that there are many other ways to incorporate cost into the construction of a random sample; a subtlety such as rank-set ordering is only one example. Mentioned also in §8.2.2, the larger concept of sampling for cost optimization dates back to the seminal work of Neyman (1934) and is given in more detail by, for example, Scheaffer *et al.* (1996, §5.5) and Thompson (2002, §11.5). A caveat, however: Stehman and Overton (2002) note that environmental surveys often have multiple goals, and optimizing for a single attribute such as minimum cost may overlook other important factors that affect the study outcome(s). Designers should always keep in mind the larger goals of their study.

8.3.5   Composite sampling

Another approach used to reduce costs with complex environmental samples involves compositing of the sample material into groups or batches. As with ranked set sampling, when the cost of measuring or analyzing the sample material greatly exceeds the cost of acquiring the sample, compositing can become cost-efficient. This concern is especially prevalent in chemometrics and environmental/biomedical monitoring, where chemical analytes or biochemical/pharmaceutical metabolites are assessed for their levels of occurrence and whether they have exceeded or dropped below some critical threshold. Patil *et al.* (1994a) describe a number of similar applications. The approach is also known as or closely associated with *group testing* (where the data are often binary indicators of an environmental event), *group screening*, or *bulk sampling*.

In the simplest case, suppose independent observations, X_{ij} ($i = 1, \ldots, n$; $j = 1, \ldots, m_i$) are available from a population whose unknown mean response is $\mu = E[X_{ij}]$, and whose variance is $\sigma^2 = E[(X_{ij} - \mu)^2]$. Rather than record the observations directly, the sampling procedure pools a group of individual samples into a single measurement that is then analyzed. In this case, the X_{ij}s are not separately observed; all that is available are the composite observations $Y_i = \sum_{j=1}^{m_i} X_{ij}/m_i$. For example, suppose a brownfield site (a location where industrial or other activities have led to environmental contamination of the surrounding earth) is to be assayed for levels of polychlorinated biphenyls (PCBs). PCBs can exist in soils at minute concentrations, so when costs of chemical determination are high the investigator may group multiple soil samples together to both increase detection and decrease the cost of the sampling effort.

Note that we use the term 'sampling' here in a rather different manner than throughout the rest of this chapter. Composite 'samples' are not so much a form of SRS as they are a way to collect sampled material together in order to reduce resource expenditures. Thus they share certain features with the RSS approach from §8.3.4, but less so than with the basic design-based sampling methods described earlier. Nonetheless, the term 'composite sampling' is standard throughout the environmetric literature, leading us to include the method here.

Under composite sampling $E[Y_i] = \sum_{j=1}^{m_i} E[X_{ij}]/m_i = m_i\mu/m_i = \mu$, so Y_i is unbiased for μ. Using the independence assumption among the original outcomes, we also find $\mathrm{Var}[Y_i] = \sum_{j=1}^{m_i} \mathrm{Var}[X_{ij}]/m_i^2 = m_i\sigma^2/m_i^2 = \sigma^2/m_i$. Thus as with any operation where a sample mean is used to summarize the central value of a set of data, the compositing has led to a decrease in variation. The effect improves as m_i increases, and as a result it is natural to make m_i as large as possible. If, due to resource constraints, this leads to a decrease in n, however, a negative consequence ensues: estimation of the variation in X may be compromised, leading to a loss in precision (see Example 8.10, below). That is, n must be at least 2, and preferably much larger, for estimates of σ^2, standard errors, and confidence intervals on μ to be useful.

For $n \geq 2$ we use as an estimator of μ the weighted average

$$\hat{\mu}_n = \frac{\sum_{i=1}^{n} m_i Y_i}{\sum_{j=1}^{n} m_j} = \sum_{i=1}^{n} w_i Y_i, \tag{8.21}$$

where $w_i = m_i / \sum_{j=1}^{n} m_j$. This is, in fact, a weighted least squares estimator (§A.4.1) of μ, and is also unbiased (Edland and van Belle, 1994). An unbiased estimator for σ^2 is

$$\hat{\sigma}^2 = \frac{1}{n-1} \sum_{i=1}^{n} m_i (Y_i - \hat{\mu}_n)^2, \tag{8.22}$$

which is used to calculate the standard error for $\hat{\mu}_n$ as

$$se[\hat{\mu}_n] = \frac{\hat{\sigma}}{\sqrt{\sum_{i=1}^{n} m_i}}.$$

If the original data are normally distributed, a $1 - \alpha$ confidence interval for μ is $\hat{\mu}_n \pm t_{\alpha/2}(n-1)se[\hat{\mu}_n]$. If the normality assumption is in doubt, and if no suitable transformation can be found to bring it closer to viability, then in large samples an approximate $1 - \alpha$ interval is $\hat{\mu}_n \pm z_{\alpha/2}se[\hat{\mu}_n]$. The approximation improves as n grows large.

Example 8.10 (Soil compositing) Correll (2001) gives data on heavy metal contamination in a landfill, where lead concentrations in soil cores were sampled randomly between 0 and 500 mm from the landfill surface. A total of 108 concentrations were measured, the natural logarithms of which appear in Table 8.10. (The log transform was used to correct for excessive right skew in the concentration data.) For all 108 observations, the sample mean is 5.1736 with standard error 0.1406. Thus a 95% confidence interval on μ is $5.1736 \pm \{t_{0.025}(107) \times 0.1406\} = 5.1736 \pm (1.9824 \times 0.1406) = 5.1736 \pm 0.2787$.

Since soil core extraction is much less expensive than separation and laboratory analysis of the heavy metals in the soil, compositing here is a natural alternative to full random sampling. To study the effect of compositing with these data, we will

Table 8.10   Log-transformed Pb concentrations in surface soil near a landfill

Original log-concentrations, X_{ij}	Composite means (over 9 successive observations), Y_i
3.5835, 2.6391, 4.0073, 4.4427, 7.0901, 4.3820, 6.3279, 10.0605, 10.5187	5.8946
5.9915, 6.2916, 5.7366, 6.2538, 3.7377, 4.7875, 5.8289, 6.0868, 4.4998	5.4682
5.2470, 6.1527, 5.8861, 4.7875, 7.0901, 5.5607, 7.2079, 5.0106, 7.3460	6.0321
6.2538, 4.4998, 4.0073, 6.9565, 5.4806, 5.5215, 4.4998, 3.0910, 3.8286	4.9043
6.4615, 5.3471, 4.4998, 5.9661, 4.5539, 4.4998, 5.1930, 5.6699, 9.1160	5.7008
3.0910, 4.7005, 4.4427, 2.6391, 6.7334, 6.2538, 5.2470, 5.3471, 4.3175	4.7525
5.0106, 5.2983, 6.5221, 6.4615, 3.4012, 7.6497, 7.0901, 4.0073, 5.9915	5.7147
3.5835, 2.3026, 4.0943, 3.8286, 4.7875, 5.3936, 4.3175, 7.0901, 5.1930	4.5101
4.0943, 6.2538, 2.6391, 5.2470, 6.1527, 4.7005, 2.8904, 3.4012, 2.9957	4.2639
3.6376, 4.3820, 3.8712, 4.0943, 4.1744, 3.5264, 6.0162, 4.1744, 4.8675	4.3049
4.4427, 4.4998, 5.1358, 5.5607, 5.3471, 5.1930, 6.5793, 5.4381, 5.9661	5.3514
5.0752, 8.0392, 4.4067, 3.6376, 5.2204, 4.6540, 3.8712, 6.2916, 5.4806	5.1863

Source: Correll (2001).

take the original set of 108 observations and manipulate them into a composite sample. Start by supposing that a composite sample of size $n = 12$ is desired, requiring us to composite every $m = 9$ soil cores. Apply this to the data in Table 8.10 by averaging each group of nine consecutive observations: the first composite measurement is $Y_1 = (3.5835 + 2.6391 + 4.0073 + \cdots + 10.5187)/9 = 5.8946$. Proceeding in a similar fashion with $Y_2, \ldots, Y_{12}$ produces the final 'composite' column in Table 8.10.

Now apply (8.21): $\hat{\mu}_{12} = (5.8946 + 5.4682 + \cdots + 5.1863)/12 = 5.1736$, which as expected is also the sample mean of the original observations. To estimate σ^2, use (8.22): $\hat{\sigma}^2 = 37.9671/(12 - 1) = 3.4516$. From this, the standard error is $se[\hat{\mu}_{12}] = \hat{\sigma}/\sqrt{\sum_{i=1}^{n} m_i} = \sqrt{3.4516/108} = 0.1788$ and a 95% confidence interval on μ is $\hat{\mu}_{12} \pm t_{\alpha/2}(11)se[\hat{\mu}_{12}] = 5.1736 \pm (2.201 \times 0.1788) = 5.1736 \pm 0.3935$. Compare this with the 95% interval calculated above: the raw-data standard error is 21% smaller, leading to a 29% narrower interval. Thus as anticipated, compositing has brought some loss in precision. But it also requires processing of only one-ninth as much material as the full SRS, and if costs for processing and analyzing the Pb concentrations markedly overwhelm the costs of sampling, compositing may be more effective.

Similar outcomes are seen with different levels of compositing: for example, if we increase to 18 composites per observation, we set $n = 6$ and find $\hat{\sigma}^2 = 20.7187/(6 - 1) = 4.1437$. ($\hat{\mu}_6$ remains 5.1736.) The 95% confidence interval for μ widens farther, to $5.1736 \pm \{t_{0.025}(5) \times 0.1959\} = 5.1736 \pm (2.571 \times 0.1959) = 5.1736 \pm 0.5037$. Clearly, composite sampling must be applied wisely: cost/resource reductions often lead to decreases in precision, and the two effects must be balanced carefully against each other.   🌎

As with RSS in §8.3.4, the underlying population from which the original data are drawn is assumed infinite. This is a natural assumption for most environmental or chemometric composites taken from soil, water, vegetation, blood, tissue, etc. If only a finite number of possible composites can be drawn, however, then as in §8.2 some correction is required in the estimation equations. Gilbert (1987, §7.2.3) and Brus *et al.* (2002) give appropriate formulas for this case.

A further manipulation sometimes seen with composited environmental data involves *subsampling*, where a series of, say, b subsamples is drawn from each individual composite. Thus, for example, if the original $\sum_{i=1}^{n} m_i$ data points lead to n composites, the b subsamples produce a total of nb data values, say, U_{ik} ($i = 1, \ldots, n; k = 1, \ldots, b$). The motivation for this strategy appears to be one of increasing the number of data points in the sample, so as to avoid losses in precision such as those seen in Example 8.10. This is, of course, a fallacy, since there is no new source of independent information. Indeed, the subsamples actually *add* a source of variation to the overall analysis. For normally distributed data, this is essentially a *nested design* (Neter *et al.*, 1996, Ch. 28), seen, for example, in certain agricultural experiments. Although estimation of μ remains fairly straightforward under subsampling, estimation of σ^2 – indeed, proper identification and extraction of the various *components* of variation in the data – becomes more difficult. One might think of it as having an unknown proportion, π_{ijk}, of the original sample quantified by X_{ij} in each subsampled composite value U_{ik}.

Subsampling is not without value, however, since information in the composites' subsamples may be useful to the investigator. Unfortunately, the statistical models become much more complex than what we describe here, and we refer readers to Gilbert (1987, Ch. 7) or Lovison *et al.* (1994) for more detail.

Compositing is also very useful when the trait under study is binary, such as whether or not a contaminant has entered a water supply or if an individual has been infected by a rare virus. This leads naturally to use of proportion data and the binomial distribution. Lovison *et al.* (1994) emphasize use of composite sampling for this setting, among other issues. More detail on these and other advanced uses of composite sampling may be found in reviews by Patil *et al.* (1994a), Lancaster and Keller-McNulty (1998), and the references therein.

Exercises

8.1. For the following environmental studies, give the population, the elements, and the sample frame.

(a) A study of recreational parks to identify usage rates by the public.

(b) A study of automobile use in miles per annum at a specific locality.

(c) A study to determine if clean-up efforts at US EPA Superfund hazardous waste sites have achieved desired level(s) of remediation.

(d) A soil survey to determine soil pH levels in a county or region.

(e) A study of atomic bomb survivors to determine their rates of thyroid cancer.

(f) A study of the levels of carbon monoxide emissions among dwellings in a rural county.

(g) A study of costs associated with decreased pesticide use in farms within an agricultural growing region.

(h) A study to estimate proportions of trees damaged by exposure to low-level ozone in a mountain range.

8.2. Suppose you are given a list frame of 265 elements from a population. Code these simply as E1, E2, ..., E265. Use a random number device (tabular or computer) to draw an SRS of $n = 53$ elements from this population, without replacement.

8.3. Use a random number device (tabular or computer) to draw another SRS, without replacement, from the population of 96 sites summarized in Table 8.1. Set your sample size to $n = 54$.

8.4. In Example 8.2, Majer *et al.* (2002) also reported on soil pH and on carbon biomass (as mg CO_2/100 g DM/h) at the 16 sites from their SRS. The data are:

Site	UT7	UT14	RB5	RB12	FW2	FW9	MB9	MB10
pH	6.27	6.04	6.72	6.50	3.57	3.74	4.62	5.41
Biomass	3.7	1.8	2.8	0.9	1.6	0.8	1.8	1.4
Site	RN1	RN7	BB8	BB11	BB12	AN2	AN5	AN13
pH	4.71	5.41	7.15	7.20	7.32	6.60	7.33	6.84
Biomass	1.2	1.6	4.0	1.8	4.1	4.6	5.2	2.9

For each outcome variable estimate the mean, μ, and construct an approximate 95% confidence interval for μ.

8.5. Show that equation (8.3) can be derived from equation (8.2) by applying (8.2) to data y_i that are strictly binary, $y_i = 0$ or 1. (*Hint:* if y_i can only be 0 or 1, then $y_i^2 = y_i$.)

8.6. Given a maximum desirable margin of error, ε, set the margin of error equal to ε and show how to derive the following sample size equations under simple random sampling.

(a) Derive equation (8.4), the sample size for estimating the population mean, μ.

(b) Show that for estimating the population total, $\tau = N\mu$,

$$n \geq \frac{4N^3\sigma^2}{4N^2\sigma^2 + (N-1)\varepsilon^2}.$$

[*Hint:* to find the margin of error, recall that $\text{Var}[\hat{\tau}] = \text{Var}[N\bar{y}] = N^2\text{Var}[\bar{y}]$.]

(c) Show that for estimating the population proportion, π,

$$n \geq \frac{4Np(1-p)}{4p(1-p) + (N-1)\varepsilon^2}.$$

You will need to know that the margin of error for the unbiased estimator $p = \sum_{i=1}^{n} y_i/n$ is based on $\text{Var}[p] = p(1-p)(N-n)/\{n(N-1)\}$.

(d) In part (c), note that the result for n is a function of p. Show that this is maximized at $p = \frac{1}{2}$.

8.7. For estimating the mean soil pH in Exercise 8.4, the variation in this sample was thought to be high. A larger sample with greater accuracy was desired. How large a sample should be taken to estimate μ with a margin of error no larger than 0.20?

8.8. Return to the population of 96 soil sampling sites summarized in Table 8.1 and draw a systematic sample of size 16. What is the largest possible period that you could use?

8.9. Levy and Lemeshow (1999, p.115) give data on concentrations of the pesticide dieldrin along an 11.5 m stretch of river. The concentrations (in μg/L) were recorded a few centimeters below water level at 36 different 'zones' along the river's route. The data are:

Zone code	1L	2R	3L	4R	5L	6R	7L	8R	9L
Mile marker	0.35	0.35	0.9	0.9	1.4	1.4	2.0	2.0	2.5
Concentration	0	0	0	1	1	0	2	1	2
Zone code	10R	11L	12R	13L	14R	15L	16R	17L	18R
Mile marker	2.5	3.3	3.3	4.0	4.0	4.7	4.7	5.5	5.5
Concentration	2	5	6	6	5	6	5	5	5
Zone code	19L	20R	21L	22R	23L	24R	25L	26R	27L
Mile marker	6.1	6.1	7.0	7.0	7.6	7.6	8.3	8.3	9.0
Concentration	4	3	4	2	2	2	2	2	1
Zone code	28R	29L	30R	31L	32R	33L	34R	35L	36R
Mile marker	9.0	9.7	9.7	10.3	10.3	10.8	10.8	11.3	11.3
Concentration	2	1	1	1	1	1	1	0	0

Since dieldrin is a suspected carcinogen (Exercise 4.11), it is of interest to estimate the average concentration, μ, of the pesticide along this stretch of river.

(a) Draw an SRS of size $n = 9$ from the 'population' of $N = 36$ concentrations given above. Estimate μ from the resulting data. Also compute an approximate 95% confidence interval for μ.

(b) Since the sample is to be taken along a river, it is operationally simpler to take a 1-in-k systematic sample than to go back and forth along the river taking an SRS. Set $k = 36/9 = 4$ and draw such a sample from the $N = 36$ concentrations given above. Estimate μ from the resulting data. (Notice that here $N/k = 36/9 = 4$ is an integer.) Also compute an approximate 95% confidence interval for μ, using (8.2) to approximate $se[\bar{y}]$. How does this compare to the results in part (a)?

(c) In your sample from part (b), use instead (8.5) for $se[\bar{y}]$. How does this compare to the results in part (b)? In part (a)?

(d) Notice that the 'true' mean concentration is $\mu = \frac{82}{36} = 2.2778 \, \mu g/L$. Did any of the intervals in parts (a), (b), or (c) cover this value? To study this further, access a computer and for each of the sampling strategies used above draw 2000 Monte Carlo samples and use each to construct an approximate 95% confidence interval for μ. Determine how often each of these 2000 intervals covers the true value of μ. Is this approximately 95%?

(e) Repeat parts (a)–(c), but now for the total concentration $\tau = N\mu$.

8.10. When conducting trawl surveys along the northeast coast of North America, the US National Oceanic and Atmospheric Administration draws a stratified random sample based on $H = 73$ different-depth strata (Nordahl and Clark, 2002). Suppose for simplicity that in each stratum there are a constant number of sampling locations, say $N_h = 15$ for $h = 1, \ldots, 73$. Plan a stratified random sample from this frame of 1095 locations (elements), with equal stratum sample sizes $n_h = 5$ for all $h = 1, \ldots, 73$.

8.11. Show that equation (8.6) can be written in the alternative form

$$se[\hat{\mu}] = \frac{1}{N}\sqrt{\sum_{h=1}^{H}(N_h - n_h)\frac{N_h s_h^2}{n_h}}.$$

8.12. Applying the stratified selection scheme in Fig. 8.6 to the population of insect abundances given by See et al. (2000) produces the following stratified random sample:

Forest 1	Forest 2	Forest 3	Forest 4
21.23	101.14	30.14	28.39
28.75	67.44	31.37	25.83
21.90	15.97	30.36	21.82
19.64	23.71	30.96	26.19
27.82	13.09	31.69	36.03
21.82	14.95	30.36	40.71
			33.12

(a) Use these data to estimate the mean insect abundance across all four forests. Find the standard error of your point estimator and use it to construct an approximate 95% confidence interval on this mean abundance.

(b) Estimate the total abundance, τ, across all four forests. Also construct an approximate 95% confidence interval for τ.

(c) Ignore the stratification across forests and estimate the mean insect abundance using methods appropriate for an SRS. Also find the standard error of your point estimator. How do these values compare with those found in part (a)?

8.13. Although not constructed as stratified random sample, the soil contamination data from Example 8.2 do share similar features with such a design. Indeed, it would be natural to consider using the $H = 7$ cities as strata in some future study. For now, return to these data and reanalyze them as if they had been drawn as a stratified random sample. The allocation pattern is $n_1 = 2, n_2 = 2, n_3 = 2, n_4 = 2, n_5 = 2, n_6 = 3, n_7 = 3$, while the stratum totals here would be $N_1 = 16, N_2 = 12, N_3 = 13, N_4 = 16, N_5 = 14, N_6 = 13$, and $N_7 = 12$. Find the per-stratum means $\bar{y}_h$ and variances s_h^2, and use these to compute the stratified sampling estimate, $\hat{\mu}$, of the mean log-chromium concentration. Also find $se[\hat{\mu}]$ and place an approximate 95% confidence on μ. How do these values compare with their SRS-based counterparts from Example 8.2? If there are any anomalies with this comparison, can you see why?

8.14. Lewis (2002) presents a study on sediment disposition and delivery as it was transported through the Van Duzen river watershed in the US state of California. For example, data were taken on non-earthflow sediment delivery (in $m^3/km^2/yr$) throughout the river basin. Since sediment delivery is thought to vary by terrain, sampling is stratified over five different terrain types. The data are summarized as follows:

Stratum	N_h	n_h	$\bar{y}_h$	$se^2[\bar{y}_h]$
1. Stable sandstone	3180	8	4.00	3.0976
2. Potentially unstable sandstone	675	25	18.88	27.9460
3. Stable melange	690	4	6.00	25.4016
4. Older slump-earthflow melange	1860	25	6.04	3.5059
5. Active slump-earthflow melange	160	18	19.89	54.1532
Total	6565	80		

(The standard errors were found using equation (8.2).) From this summary information, calculate the stratum weights $W_h = N_h/N$ and estimate the total sediment, τ, delivered through the river basin. Also construct an approximate 95% confidence interval for τ.

8.15. (a) Show that equation (8.11) reduces to (8.8) when all the cluster sizes are equal.

(b) Suppose in a cluster sampling scenario that the number of elements in the population, K_{total}, is unknown. An estimate for it can be constructed as the number of clusters in the population times the average cluster size in sampled clusters: $\hat{K}_{total} = N \sum_{i=1}^{n} K_i/n$. Substitute this into (8.11) and compare your result to (8.8).

8.16. Show that $(n-1)^{-1}\sum_{i=1}^{n}(y_{i+} - \hat{\mu}K_i)^2 = s_+^2 + \hat{\mu}^2 s_K^2 - 2\hat{\mu}r_{YK}s_+s_K$, where $\hat{\mu} = \sum_{i=1}^{n} y_{i+} / \sum_{i=1}^{n} K_i$, $s_+^2 = (n-1)^{-1}\sum_{i=1}^{n}(y_{i+} - \bar{y}_+)^2$, $s_K^2 = (n-1)^{-1}\sum_{i=1}^{n}(K_i - \bar{K})^2$, and r_{YK} is given in equation (8.10).

8.17. To study average height of trees affected by the invasive vine known as kudzu (*Pueraria lobata*) along a southeastern greenway, the region was divided into $N = 386$ one-acre clusters. A random sample of $n = 10$ clusters gave the following measurements on all the affected trees in those clusters:

Cluster code	C1	C2	C3	C4	C5	C6	C7	C8	C9	C10
No. of trees	1	2	3	4	2	6	1	2	5	2
Total height	6.1	12.4	17.4	19.6	12.2	36.6	7.0	10.6	24.5	13.4

Summary statistics for these data include: $s_K = 1.687$, $s_+ = 9.150$, and $r_{YK} = 0.976$. From previous data, it is known that the total number of affected trees in the greenway is $K_{total} = 1081$. Use (8.9) to estimate the average height of affected trees throughout the greenway, find the associated standard error, and give an approximate 99% confidence interval.

8.18. Greenwood (1996) describes a species abundance study of skylarks (*Alauda arvensis*) in a rural region where $N = 16$ clusters were identified, each of size $1\,\text{km}^2$. An SRS of $n = 4$ clusters was taken for study; within each cluster the sampling elements were square quadrats of size $200\,\text{m}^2$, so that each cluster had a constant number of 25 elements. Data were taken as the number of skylarks observed in the jth sampled $200\,\text{m}^2$ quadrat from the ith sampled $1\,\text{km}^2$ cluster. Summary values from this two-stage cluster sample were:

Sampled cluster	K_i	k_i	$\bar{y}_i$	s_i^2	$K_i(K_i - k_i)s_i^2/k_i$
$i = 1$	25	10	1.6	2.044	76.667
$i = 2$	25	10	2.3	0.678	25.417
$i = 3$	25	10	1.4	1.378	51.667
$i = 4$	25	10	2.4	2.044	76.667

From these data, estimate the total number of skylarks in the region. Also find the standard error of your estimate, and give an approximate 95% confidence interval for τ.

8.19. Buckland (1984) gives data on the size of a colony of yellow ants (*Lasius flavus*). In the colony, the total number of worker ants was to be estimated using closed-population capture–recapture methods. Initially, $t = 500$ workers were captured and marked with a radioactive phosphorus (^{32}P) tracer. After release and reintegration into the colony, $n = 189$ workers were recaptured, from which a total of $M = 17$ ^{32}P-marked ants were recovered.

(a) From these data, estimate the total number of worker ants in the colony and include a 95% confidence interval. If resources permit, apply both the approximate Wald interval and the hypergeometric interval from (8.18).

(b) If radioactive marking causes some workers to be rejected from the colony, this would violate the closed-population assumptions. In such a case would you expect the resulting point estimate of the total to be too small or too large? Why?

8.20. To estimate the number of adults who contracted a respiratory disease spread by an airborne pathogen, an environmental epidemiologist collected data over $n = 25$ equally sized quadrats in a small city. Each quadrat measured 205 acres. She found the following:

Number of diseased adults	0	1	2	3	4	5	6	7
Frequency	9	4	6	3	1	1	0	1

The city's metropolitan area covers 5125 acres. Use these data to estimate the total number of diseased adults in this city. Also give an approximate 95% confidence interval for the total.

8.21. In some line-intercept studies, the cross-sectional width is constant by design. For example, to study the abundance of tarnished plant bugs (*Lygus lineolaris*) in cotton fields, a drop cloth is employed to collect insects along rows of cotton, and the number of insects shaken on to the cloth is counted. Thus w_j is the fixed, known width of the cloth. Willers *et al.* (1999) describe such a study, where x_j is the number of tarnished plant bugs per drop cloth observed over $n = 6$ linear transects in zones of heavy cotton fruit. Samplings occur whenever a transect crosses a row of cotton; here, eight times per transect. Since the drop cloths' widths are smaller than the distances between transects, each drop-cloth sampling is viewed as a separate encounter; hence $\varphi_i = 8$ for $i = 1, \ldots, 6$. The data are:

Transect	Number of insects							
1	2	2	2	0	2	0	2	3
2	0	2	0	1	0	0	0	0
3	0	1	0	0	0	0	0	0
4	0	0	0	0	0	0	1	0
5	0	0	0	0	0	0	1	0
6	0	0	0	0	0	2	0	0

Assuming $w = 2.5$ ft and that the baseline length is $B = 450$ ft, calculate the values of $y_i = B \sum_{j=1}^{\varphi_i} (x_j/w_j)$, $i = 1, \ldots, 6$, and from these estimate the total number of insects in the region. Also include an approximate 90% confidence interval.

8.22. Thompson (2002, §19.1) gives data on a wildlife survey to estimate abundance of Alaskan wolverines (*Gulo gulo*). Over a rectangular region of area $A = 120\,\mathrm{m}^2, n = 4$ aerial transects were taken to identify wolverine tracks in the snow. Starting points for the transects were systematically sampled from a 12 m baseline, thus $B = 12$. Once a transect crossed a set of wolverine tracks, the cross-sectional track width, w_j, and number of wolverines, x_j, were recorded. The data involved four distinct encounters: $w_1 = 5.25\,\mathrm{m}, x_1 = 1$; $w_2 = 7.50\,\mathrm{m}, x_2 = 2$; $w_3 = 2.40\,\mathrm{m}, x_3 = 2$; and $w_4 = 7.05\,\mathrm{m}, x_4 = 1$. Over the first transect, the 1st, 2nd, and 4th sets of tracks intercepted, so $\varphi_1 = 3$. Over the second transect, again the 1st, 2nd, and 4th sets of tracks intercepted, so $\varphi_2 = 3$. Over the third transect, the 3rd and 4th sets of tracks intercepted, so $\varphi_3 = 2$. And, over the last transect, again the 3rd and 4th sets of tracks intercepted, so $\varphi_4 = 2$. Find the values of $y_i, i = 1, \ldots, 4$, and estimate the total number of wolverines in the $120\,\mathrm{m}^2$ region. Also find an approximate 95% confidence interval for the total.

8.23. Recall the vapor pressure data of Nussbaum and Sinha (1997) from Example 8.9. Those authors actually provided the complete data set, so with $n = 36$ another possible RSS configuration may be constructed at $m = 2$ and $r = 9$:

Cycle	Lot	Observations ordered within lots	
1	1	[7.59]	$y_{1(2)1}$
	2	$y_{2(1)1}$	[9.03]
2	1	[8.92]	$y_{1(2)2}$
	2	$y_{2(1)2}$	[9.28]
3	1	[8.56]	$y_{1(2)3}$
	2	$y_{2(1)3}$	[8.70]
4	1	[7.83]	$y_{1(2)4}$
	2	$y_{2(1)4}$	[8.29]
5	1	[8.21]	$y_{1(2)5}$
	2	$y_{2(1)5}$	[7.86]
6	1	[7.83]	$y_{1(2)6}$
	2	$y_{2(1)6}$	[7.85]
7	1	[7.80]	$y_{1(2)7}$
	2	$y_{2(1)7}$	[7.77]
8	1	[7.73]	$y_{1(2)8}$
	2	$y_{2(1)8}$	[7.98]
9	1	[7.76]	$y_{1(2)9}$
	2	$y_{2(1)9}$	[8.15]

Use this new RSS to estimate μ via (8.20). Also find the corresponding standard error and use this to construct an approximate 95% confidence interval for μ. Compare your results from those in Example 8.9.

8.24. Barnett (1999) gives $n = 40$ soil concentrations of potassium (K) in a contaminated waste site as the following values:

24	32	28	28	28	28	28	41	24	41
19	24	36	20	17	15	55	18	20	26
24	20	25	19	28	23	19	20	14	25
18	16	22	27	18	20	31	24	24	36

These data are highly skewed, so Barnett suggests use of a natural logarithmic transform. Apply this transformation to the data. Construct from the resulting values an RSS with $r = 10$ cycles comprised of 4 values in $m = 2$ lots; for example, the first cycle would be the first 4 observations $\{24, 32, 28, 28\}$, with lot no. $1 = \{24, 32\}$ and lot no. $2 = \{28, 28\}$. Estimate the mean log-K concentration from the resulting RSS and find its standard error. Use these to calculate an approximate 90% confidence interval for the mean log-concentration.

8.25. Edland and van Belle (1994) give data on biomass (in g) of pink salmon fry (*Oncorhynchus gorbuscha*), composited to reduce costs of sampling the tiny individual fry. The composited data comprised $n = 4$ points: $Y_1 = 0.292$, $Y_2 = 0.299$, $Y_3 = 0.280$, and $Y_4 = 0.316$, with corresponding composite sizes $m_1 = 92, m_2 = 105, m_3 = 110$, and $m_4 = 102$. Find a 99% confidence interval for the underlying mean biomass.

8.26. Return to the soil compositing data in Table 8.10 and construct a composite sample using the following unequal composite sizes. Comment on how your results change, if at all, from those in Example 8.10.

(a) $m_1 = 9$, $m_2 = 18$, $m_3 = 9$, $m_4 = 18$, $m_5 = 9$, $m_6 = 18$, $m_7 = 9$, $m_8 = 18$.

(b) $m_1 = 9$, $m_2 = 18$, $m_3 = 27$, $m_4 = 36$, $m_5 = 18$.

(c) Select any unequal-m design of your own, such that $\sum_{i=1}^{n} m_i = 108$.

A

Review of probability and statistical inference

A.1 Probability functions

This appendix provides a review of basic concepts in probability and statistical inference along with their underlying vocabulary and notation. At the core of these concepts is the notion of a *probability function*: a unifying mathematical description of how a random outcome varies. Probability functions can describe two basic types of random variables, discrete and continuous. A discrete random variable takes only discrete values; examples include simple dichotomies (say, 'diseased' $= 1$ vs. 'healthy' $= 0$ in an environmental health study), counts of occurrences (numbers of animals recovered in a tag–recapture study, numbers of days without rain during a drought), or even studies that result in an infinite, yet countable, number of outcomes (i.e., counts without a clear upper bound, such as the number of different plant species in a tropical forest, or the number of fissures in a rock stratum near a fault line). A continuous random variable takes on values over a continuum, and is encountered when some measured level of response is under study (ground concentrations of an ore or metal, temperature change, etc.). A discrete random variable can often be thought of as arising from a counting or classification process, while a continuous random variable can be thought of as arising from some measurement process.

In both cases, the probability functions will depend on the nature of the random outcome. Suppose the random variable X is discrete and consider the 'event' that X takes some specific value, say $X = k$. Then the values of $P[X = k]$ over all possible

Analyzing Environmental Data   W. W. Piegorsch and A. J. Bailer
© 2005 John Wiley & Sons, Ltd   ISBN: 0-470-84836-7 (HB)

values of k describe the *probability distribution* of X. We write $f_X(k) = P[X = k]$, and call this the *probability mass function* (p.m.f.) of X. (Notice the convention that an upper-case letter, X, is used to denote the random variable, while lower-case letters, k or x, are used to represent realized values of the random variable).

Since $f_X(k)$ is a probability, it must satisfy a series of basic axioms. These are: (i) $0 \le f_X(k) \le 1$ for all arguments k, and (ii) $\sum f_X(k) = 1$, where the sum is taken over all possible values of k that X can take. (This set of all such possible values is called the *sample space* or *support space*, $\mathbb{S}$.) Summing the p.m.f. over increasing k produces the *cumulative distribution function* (c.d.f.) of X:

$$F_X(k) = P[X \le k] = \sum_{i \le k} f_X(i). \tag{A.1}$$

Important properties of any c.d.f. are (i) $0 \le F_X(k) \le 1$ (since it is itself a probability), and (ii) it must be a non-decreasing function (since it gives cumulative probabilities).

In some instances, it is of interest to find the complementary probability $P[X > k]$, which can be calculated as $P[X > k] = 1 - P[X \le k] = 1 - F_X(k)$. We often call this an application of the *complement rule* from probability theory.

If X is a continuous random variable its probability function, $f_X(x)$, must describe the continuous density of probability. This is expressed as probabilities over interval subsets of the real numbers using definite integrals, for example,

$$P[a \le X \le b] = \int_a^b f_X(x)\,dx.$$

Similarly, the c.d.f. of a continuous random variable is the area under $f(x)$ integrated from $-\infty$ to the argument, x, of the function:

$$F_X(x) = P[X \le x] = \int_{-\infty}^x f(y)\,dy. \tag{A.2}$$

From (A.2), we see $P[a \le X \le b] = F_X(b) - F_X(a)$ and so $P[X = a] = P[a \le X \le a] = F_X(a) - F_X(a) = 0$ for any a. Thus, while non-zero probability can be assigned to events that correspond to particular values of a discrete random variable, non-zero probability can be assigned only to events that correspond to intervals of values for a continuous random variable. As a result, if X is continuous, $P[X \le a] = P[X < a] + P[X = a] = P[X < a] + 0 = P[X < a]$. (This is not necessarily true for discrete random variables, where $P[X = a]$ can be non-zero.)

We call $f_X(x)$ the *probability density function* (p.d.f.) of a continuous random variable X. As with discrete random variables and their p.m.f.s, the p.d.f. from a continuous random variable must satisfy two basic axioms: (i) $f_X(x) \ge 0$ for all arguments x, and (ii) $\int_{-\infty}^{\infty} f_X(x)dx = 1$. Notice also that if $F_X(x)$ is a differentiable function, then its derivative at the point x is the p.d.f.: $dF_X(x)/dx = f_X(x)$.

For more than one random variable, multivariate extensions of the probability functions may be developed. For instance, if X and Y are discrete random variables, the *joint bivariate p.m.f.* is $f_{X,Y}(j, k) = P[X = j$ and $Y = k]$, and the *joint bivariate c.d.f.* is $F_{X,Y}(j, k) = P[X \leq j$ and $Y \leq k]$. Individually, X is itself a random variable; its *marginal p.m.f.* is derived from the joint p.m.f. by summing over all possible values of $Y = k$: $f_X(j) = \sum_k f_{X,Y}(j, k)$. For continuous random variables, the results are similar:

$$\text{joint p.d.f.:} \quad f_{X,Y}(x, y);$$
$$\text{joint c.d.f.:} \quad F_{X,Y}(x, y) = P[X \leq x \text{ and } Y \leq y];$$

$$\text{marginal p.d.f.:} \quad f_X(x) = \int_{-\infty}^{\infty} f_{X,Y}(x, y)\, dy.$$

For greater detail, see texts on probability and statistics such as Hogg and Tanis (2001).

If two (or more) random variables occur in such a way that one has absolutely no impact on the other's occurrence, we say the two variables are *independent*. One important characteristic of independent random variables is that their joint probability functions factor into the marginal components: $f_{X,Y}(j, k) = f_X(j)f_Y(k)$. Extended to multiple variables, say $X_1, \ldots, X_n$, this is

$$f_{X_1,\ldots,X_n}(k_1, \ldots, k_n) = \prod_{i=1}^{n} f_{X_i}(k_i). \tag{A.3}$$

The theoretical *population mean* of a random variable X represents a measure of the central tendency of X. This is called the *expected value* of X, and is denoted as $E[X]$. In the discrete case, $E[X]$ is defined as the probability-weighted sum over all possible outcomes in the support space,

$$E[X] = \sum_{\mathbb{S}} x f_X(x), \tag{A.4}$$

while in the continuous case it is an integral,

$$E[X] = \int_{-\infty}^{\infty} x f_X(x) dx. \tag{A.5}$$

A common symbol for the mean is $\mu = E[X]$.

The expectation operator can also be applied to any function of X, say, $E[g(X)] = \sum g(x)f_X(x)$ in the discrete case, or $E[g(X)] = \int g(x)f_X(x)dx$ in the continuous case. An important example of this is the *population variance*, $\text{Var}[X] = E[(X - \mu)^2]$, which can be shown to simplify to $\text{Var}[X] = E[X^2] - \mu^2$. A common symbol for the variance is $\sigma^2 = \text{Var}[X]$. The *standard deviation* of X is the square root of the variance: $\sigma = \sqrt{\text{Var}[X]}$. More generally, we say that the *jth population moment* of X is $E[X^j]$.

Bivariate and multivariate expectations are also possible. For instance, with two discrete random variables, X and Y, the expected value of some bivariate function $g(X, Y)$ is

$$E[g(X, Y)] = \sum_{S_X} \sum_{S_Y} g(x, y) f_{X,Y}(x, y).$$

(The continuous case is similar; simply replace the sums with integrals.) In particular, a summary measure of the joint variability between X and Y is known as the *covariance*: $\text{Cov}[X, Y] = E[(X - \mu_X)(Y - \mu_Y)]$, which can be shown to simplify to $\text{Cov}[X, Y] = E[XY] - \mu_X \mu_Y$. Common notation for the covariance is $\sigma_{XY} = \text{Cov}[X, Y]$. If X and Y are independent, then their covariance is zero.

Related to the covariance is the *correlation coefficient* ρ, which is calculated by dividing the covariance by the standard deviations of X and Y: $\rho = \sigma_{XY}/\sigma_X \sigma_Y$. ρ is always bounded between -1 and 1.

The expectation operator is linear: $E[a + bX + cY] = a + bE[X] + cE[Y]$ for any constants a, b, c and random variables X and Y. The same is not true of the variance: $\text{Var}[a + bX + cY]$ takes the more complex form $b^2\text{Var}[X] + c^2\text{Var}[Y] + 2bc\text{Cov}[X, Y]$. If the covariance is zero, however – e.g., if the individual random variables are independent – the variance of the sum will equal the sum of the variances, weighted by the squares of the multiplying coefficients: $\text{Var}[bX + cY] = b^2\text{Var}[X] + c^2\text{Var}[Y] + 2bc(0) = b^2\text{Var}[X] + c^2\text{Var}[Y]$.

These results can be extended to any linear combination of random variables: $E[\sum_{i=1}^{n} \omega_i Y_i] = \sum_{i=1}^{n} \omega_i E[Y_i]$, while

$$\text{Var}\left[\sum_{i=1}^{n} \omega_i Y_i\right] = \sum_{i=1}^{n} \omega_i^2 \text{Var}[Y_i] + 2 \sum_{i=1}^{n-1} \sum_{j=i+1}^{n} \omega_i \omega_j \text{Cov}[Y_i, Y_j]. \qquad (A.6)$$

A.2 Families of distributions

When the variable nature of a random outcome occurs with a regular structure, we often ascribe to it a specific distributional pattern. If this pattern can be formulated as a mathematical function, we say the distribution belongs to a family of such functions. These families typically possess one or more unknown *parameters* that describe the distribution's characteristics. (It is often the goal of an environmental study to estimate the unknown parameters of a random variable, and possibly to compare them with parameters from other, associated variables.) Thus it is common to refer to a parametric family of distributions when specifying the p.m.f. or p.d.f. of a random variable. We summarize below some important families of distributions, both discrete and continuous. More extensive descriptions can be found in dedicated texts on statistical distributions, such as Evans *et al.* (2000) or the series by Johnson, Kotz, and colleagues (Johnson *et al.*, 1993, 1994, 1995, 1997; Kotz *et al.*, 2000).

A.2.1 Binomial distribution

The *binomial distribution* is the basic form used to describe data in the form of proportions. Given a random variable X as the number of positive outcomes (or 'successes') from N statistically independent 'trials,' we let π be the probability that a success is observed on any individual trial. We say X has a binomial distribution if its p.m.f. takes the form

$$f_X(x) = \binom{N}{x} \pi^x (1 - \pi)^{N-x} I_{\{0,1,\ldots,N\}}(x), \tag{A.7}$$

where

$$\binom{N}{x} = \frac{N!}{x!(N-x)!}$$

is the *binomial coefficient* (the number of ways of selecting x items from a collection of N elements), $x!$ is the *factorial operator* $x! = x(x-1)(x-2)\ldots 2 \times 1$, and the notation $I_{\mathbb{A}}(x)$ represents the indicator function over the set $\mathbb{A}$: $I_{\mathbb{A}}(x) = 1$ if x is a member of the set $\mathbb{A}$, and 0 otherwise. Note that for convenience in the factorial operator we define $0! = 1$.

The notation to indicate this p.m.f. is $X \sim \text{Bin}(N, \pi)$. (The tilde symbol, $\sim$, is read as 'is distributed as.') If $X \sim \text{Bin}(N, \pi)$, its population mean is $\text{E}[X] = N\pi$ and its population variance is $\text{Var}[X] = N\pi(1 - \pi)$. In the special case of $N = 1$, we say that the single, dichotomous outcome X has a *Bernoulli distribution*.

The binomial distribution exhibits an important feature: it is closed under addition. That is, if $X_1, X_2, \ldots, X_k$ are all independent binomials with parameters N_i and (common) π — or, simply, $X_i \sim$ indep. $\text{Bin}(N_i, \pi), i = 1, \ldots, k$ — then their sum is also binomial: $\sum_{i=1}^{k} X_i \sim \text{Bin}(\sum_{i=1}^{k} N_i, \pi)$.

A.2.2 Beta-binomial distribution

A crucial feature of the binomial distribution is that its variance is a strict function of the success probability, π: $\text{Var}[X] = N\pi(1 - \pi)$. If this is thought to constrain the variability of the observations too severely, one can extend the binomial structure. A common extension takes the form $\text{Var}[X] = N\pi(1 - \pi)\{1 + [\varphi(N - 1)/(1 + \varphi)]\}$ or, equivalently,

$$\text{Var}[X] = N\pi(1 - \pi)\frac{1 + \varphi N}{1 + \varphi}, \tag{A.8}$$

where φ is an additional parameter that allows for extra-binomial variability. Notice that at $\varphi = 0$ we recover the binomial variance, while for $\varphi > 0$, variability exceeds that predicted under the binomial model. The latter effect is referred to as *overdispersion*.

A specific statistical distribution that achieves this form of extra-binomial variability is known as the *beta-binomial distribution*, with p.m.f.

$$f_X(x) = \binom{N}{x} \frac{\Gamma(x + \{\pi/\varphi\}) \, \Gamma(N - x + \{[1 - \pi]/\varphi\}) \, \Gamma(1/\varphi)}{\Gamma(\pi/\varphi) \, \Gamma([1 - \pi]/\varphi) \, \Gamma(\{1/\varphi\} + N)} I_{\{0,1,\dots,N\}}(x). \quad (A.9)$$

In (A.9), $\Gamma(a)$ is the *gamma function*

$$\Gamma(a) = \int_0^\infty x^{a-1} e^{-x} \, dx, \quad (A.10)$$

defined here for any $a > 0$ (Wolpert, 2002). The function satisfies the recursive relationship that $\Gamma(a + 1) = a\Gamma(a)$, hence if a is a positive integer, $\Gamma(a) = (a - 1)!$.

The beta-binomial has mean $E[X] = N\pi$ and variance (A.8). The limiting form of the distribution as $\varphi \to 0$ recovers the binomial p.m.f. in (A.7).

A.2.3 Hypergeometric distribution

Suppose a finite population of size N contains $T > 0$ elements that are labeled as 'successes,' such that the remaining $N - T$ elements are labeled as 'failures.' If sampling proceeds independently and with replacement, the probability of drawing a success at random from the population is $p = T/N$, and the number of successes, X, out of, say, $n > 0$ trials would have a binomial distribution as described above. If sampling proceeds *without* replacement, however, then the probability of drawing a success changes from trial to trial, and the binomial no longer applies. Instead, the p.m.f. of X must be modified to account for the sampling without replacement, and doing so leads to the form

$$f_X(x) = \frac{\binom{T}{x} \binom{N - T}{n - x}}{\binom{N}{n}} I_{\{0,1,\dots,n\}}(x), \quad (A.11)$$

In this case we say X has a *hypergeometric distribution* with parameters T, N, and n, and we denote it as $X \sim \mathrm{HG}(T, N, n)$. An additional requirement for (A.11) to be a true p.m.f. is that $T - (N - n) \leq x \leq T$. Since T is often unknown, however, this is more of a mathematical nicety than a practical property.

The mean of a hypergeometric distribution is $E[X] = nT/N$, the number of trials multiplied by the initial proportion of successes in the population. In this sense, the hypergeometric mimics the binomial. Also, the variance is $\mathrm{Var}[X] = n\{T(N - T) (N - n)\}/\{N^2(N - 1)\}$. As N grows larger than n, the hypergeometric variance approximates a corresponding binomial variance, and (A.11) draws closer to the corresponding binomial form; see Piegorsch and Bailer (1997, Fig. 1.2).

A.2.4 Poisson distribution

The *Poisson distribution* is the basic family of distributions used to describe data in the form of unbounded counts. Given a random variable X as the number of outcomes of some random process, we say X has a Poisson distribution if its p.m.f. takes the form

$$f_X(x) = \frac{\lambda^x e^{-\lambda}}{x!} I_{\{0,1,...\}}(x), \tag{A.12}$$

where $\lambda > 0$ is the rate parameter of the distribution. The mean and variance of a Poisson distribution are $E[X] = \text{Var}[X] = \lambda$. We denote this as $X \sim \text{Poisson}(\lambda)$. As with the binomial distribution, the Poisson family is closed under addition: if $X_i \sim$ indep. Poisson(λ_i), $i = 1, \ldots, k$, then $\sum_{i=1}^{k} X_i \sim \text{Poisson}(\sum_{i=1}^{k} \lambda_i)$.

A.2.5 Negative binomial distribution

As with the binomial p.m.f. for proportions, the structure of the Poisson p.m.f. for count data constrains the variance of the distribution to be a simple function of the mean (in fact, the Poisson mean and variance are identical). When this is felt to be too restrictive, one can extend the distribution to account for extra-Poisson variability. A specific statistical distribution that achieves this goal is the *negative binomial distribution*, with p.m.f.

$$f_X(x) = \frac{\Gamma(x + [1/\delta])}{x! \, \Gamma(1/\delta)} \left(\frac{\delta\mu}{1 + \delta\mu}\right)^x \frac{1}{(1 + \delta\mu)^{1/\delta}} I_{\{0,1,...\}}(x). \tag{A.13}$$

In (A.13), $\mu > 0$ is the population mean of X, while $\delta > 0$ is a form of dispersion parameter. We write $X \sim \text{NB}(\mu, \delta)$, where $E[X] = \mu$ and $\text{Var}[X] = \mu + \delta\mu^2$. As $\delta \to 0$, the negative binomial converges to the simpler Poisson p.m.f. in (A.12).

A special case of the negative binomial distribution occurs when $\delta = 1$: the *geometric distribution* has p.m.f.

$$f_X(x) = \left(\frac{\mu}{1 + \mu}\right)^x \frac{1}{(1 + \mu)} I_{\{0,1,...\}}(x).$$

Since $\mu > 0$, we can take $\pi = (1 + \mu)^{-1}$ and $1 - \pi = \mu/(1 + \mu)$ so that $0 < \pi < 1$. With this, most authors write the geometric p.m.f. in the simpler form $f_X(x) = \pi(1 - \pi)^x I_{\{0,1,...\}}(x)$. In some applications, the geometric distribution is used for representing an unbounded set of dichotomous trials, where X is the number of trials conducted before the first 'success' is observed. In similar fashion, a variant of the negative binomial can be motivated as the number of trials conducted before the rth 'success' is observed (Hogg and Tanis, 2001, §3.4).

A.2.6   Discrete uniform distribution

One last discrete distribution of interest is also the simplest: the *discrete uniform distribution* over a set of integers $\{1, 2, \ldots, M\}$ has p.m.f. $f_X(x) = M^{-1} I_{\{1,2,\ldots,M\}}(x)$, such that each of the integers $1, 2, \ldots, M$ has uniform probability of occurrence. The mean and variance of this distribution are $E[X] = (M+1)/2$ and $\mathrm{Var}[X] = (M^2 - 1)/12$. We denote this as $X \sim U\{1, 2, \ldots, M\}$.

A.2.7   Continuous uniform distribution

A continuous analog to the discrete uniform distribution is the (continuous) *uniform distribution* over the interval $A < Y < B$, with p.d.f. $f_Y(y) = (B - A)^{-1} I_{(A,B)}(y)$. (Some authors use the term *rectangular distribution* to refer to this p.d.f.) We denote this as $Y \sim U(A, B)$, with $E[Y] = (A + B)/2$ and $\mathrm{Var}[Y] = (B - A)^2/12$. A special case is the uniform distribution over the unit interval, $f_U(u) = I_{(0,1)}(u)$. The $U(0,1)$ distribution is often used in generating pseudo-random numbers by computer (Gentle, 2003).

A.2.8   Exponential, gamma, and chi-square distributions

After the uniform, perhaps the simplest continuous p.d.f. is that of the *exponential distribution*,

$$f_Y(y) = \frac{1}{\beta} e^{-y/\beta} I_{(0,\infty)}(y), \tag{A.14}$$

which is common for describing waiting times between discrete events (Hogg and Tanis, 2001, §4.2). The rate parameter β is assumed positive. This distribution has mean $E[Y] = \beta$ and variance $\mathrm{Var}[Y] = \beta^2$. We write $Y \sim \mathrm{Exp}(\beta)$.

An interesting property associated with the exponential distribution is that its ratio of standard deviation to mean is constant. (This ratio is known as the *coefficient of variation*: $CV = \sqrt{\mathrm{Var}[Y]}/E[Y]$.) Here, $\sqrt{\mathrm{Var}[Y]}/E[Y] = \sqrt{\beta^2}/\beta = 1$.

An extension of the p.d.f. in (A.14) is the *gamma distribution*,

$$f_Y(y) = \frac{1}{\Gamma(\lambda)\beta^\lambda} y^{\lambda-1} e^{-y/\beta} I_{(0,\infty)}(y), \tag{A.15}$$

where both β and λ assume values strictly greater than zero. The mean is $E[Y] = \lambda\beta$ and the variance is $\mathrm{Var}[Y] = \lambda\beta^2$. We write $Y \sim \mathrm{Gamma}(\lambda, \beta)$. As with the exponential distribution, the gamma distribution's coefficient of variation does not depend upon its mean. Indeed, when $\lambda = 1$ (A.15) reduces to (A.14).

The gamma family is closed under addition: if $Y_i \sim$ indep. $\mathrm{Gamma}(\lambda_i, \beta)$, $i = 1, \ldots, k$, then $\sum_{i=1}^{k} Y_i \sim \mathrm{Gamma}(\sum_{i=1}^{k} \lambda_i, \beta)$. Notice in particular that when $Y_i \sim$ i.i.d. $\mathrm{Exp}(\beta)$, $i = 1, \ldots, k$, $\sum_{i=1}^{k} Y_i \sim \mathrm{Gamma}(k, \beta)$. ('i.i.d.' is a shorthand notation for *independent* and *identically distributed.*) It is also interesting to note that since (A.15) must satisfy $\int_0^\infty f_Y(y)dy = 1$, we can write

$$\int_0^\infty y^{a-1}e^{-y/\beta}dy = \beta^a\Gamma(a),$$

which can be viewed as a generalization of the gamma function in (A.10). .

If $Y \sim \text{Gamma}(v/2, 2)$ we say Y has a χ^2 (or *chi-square*) *distribution* and write $Y \sim \chi^2(v)$, where the parameter v is referred to as the *degrees of freedom* for this distribution. Notice that then $E[Y] = v$ and $\text{Var}[Y] = 2v$. Also, if $Y_i \sim$ indep. $\chi^2(v_i), i = 1, \ldots, k$, then $\sum_{i=1}^k Y_i \sim \chi^2(\sum_{i=1}^k v_i)$.

A.2.9 Weibull and extreme-value distributions

A slightly different way to extend the exponential distribution from (A.14) leads to the *Weibull distribution*, with p.d.f. $f_Y(y) = \frac{\tau}{\beta}y^{\tau-1}e^{-y^\tau\beta^{-1}}I_{(0,\infty)}(y)$, for $\beta > 0$ and $\tau > 0$. The mean and variance are $E[Y] = \beta^{1/\tau}\Gamma(\tau^{-1} + 1)$ and $\text{Var}[Y] = \beta^{2/\tau}\{\Gamma(2\tau^{-1} + 1) - \Gamma^2(\tau^{-1} + 1)\}$. We write $Y \sim \text{Weibull}(\tau, \beta)$. Notice that at $\tau = 1$ we recover $Y \sim \text{Exp}(\beta)$.

Taking the natural logarithm of a Weibull random variable produces the *extreme-value distribution*: if $Y \sim \text{Weibull}(\tau, \beta)$ then for $V = \log(Y)$ we say $V \sim \text{EV}(\theta, \delta)$, where $\theta = 1/\tau$ and $\delta = \tau^{-1}\log(\beta)$. One has $E[V] = \delta - \tilde{\gamma}$ and $\text{Var}[V] = \theta^2\pi^2/6$, where $\tilde{\gamma} = -\Gamma'(1) = 0.577216$ is Euler's constant and π is used here to denote the area of a unit circle, $\pi = 3.1415926\ldots$.

A.2.10 Normal distribution

One of the most important continuous distributions used in statistical practice is the *normal distribution*, with p.d.f.

$$f_Y(y) = \frac{1}{\sigma\sqrt{2\pi}}\exp\left\{-\frac{(y-\mu)^2}{2\sigma^2}\right\}I_{(-\infty,\infty)}(y), \tag{A.16}$$

where μ is any real number and $\sigma > 0$. These two parameters also describe the mean and variance: $E[Y] = \mu$, $\text{Var}[Y] = \sigma^2$. We write $Y \sim N(\mu, \sigma^2)$.

Normal random variables possess a number of important characteristics. They are closed under addition, so that if $Y_i \sim$ indep. $N(\mu_i, \sigma_i^2), i = 1, \ldots, k$, then $\sum_{i=1}^k Y_i \sim N(\sum_{i=1}^k \mu_i, \sum_{i=1}^k \sigma_i^2)$. This can be extended to any linear combination of (independent) normal random variables: if $Y_i \sim$ indep. $N(\mu_i, \sigma_i^2), i = 1, \ldots, k$, then $\sum_{i=1}^k a_iY_i \sim N(\sum_{i=1}^k a_i\mu_i, \sum_{i=1}^k a_i^2\sigma_i^2)$. In particular, the sample mean of k normal variates is again normal: if $Y_i \sim$ i.i.d. $N(\mu, \sigma^2), i = 1, \ldots, k$, then $\bar{Y} = (1/k)\sum_{i=1}^k Y_i \sim N(\mu, \sigma^2/k)$

Normal random variables may also be *standardized*: if $Y \sim N(\mu, \sigma^2)$, then $Z = (Y - \mu)/\sigma \sim N(0, 1)$. We say Z is a *standard normal random variable*, and we tabulate the probabilities associated with this special case. The standard normal table gives the cumulative probabilities $P[Z \leq z]$, for a given argument z. Common notation for this is $\Phi(z) = P[Z \leq z]$. (Table B.1 reverses this to give the complementary upper-tail probabilities, $P[Z > z] = 1 - \Phi(z)$.) These can be inverted to produce values z_α such that $P[Z \leq z_\alpha] = 1 - \alpha$. That is, suppose we denote by $\Phi^{-1}(\cdot)$ the inverse function of the standard normal c.d.f. Then if the point z_α satisfies $z_\alpha = \Phi^{-1}(1 - \alpha)$, the area under the standard normal p.d.f. to the left of z_α is $1 - \alpha$. Since the total area must be 1, however, the area to the right of z_α must then be $1 - (1 - \alpha) = \alpha$. In this case, we call z_α the *upper-α critical point* of the standard normal.

The statistical computing package SAS® (SAS Institute Inc., 2000) can compute cumulative probabilities from a standard normal p.d.f. For example, SAS code and (edited) output for the probability $P[Z \leq -1.322]$ appear in Fig. A.1, where the computed value is seen to be 0.093. (In some installations of SAS, the statement run; may be required at the end of the input code; however, we do not use it in the examples herein.) Standard normal critical points may be similarly obtained; Fig. A.2 inverts the operation from Fig. A.1 to find $z_{0.907} = -1.322\,51$.

If a random variable W occurs such that $Y = \log(W) \sim N(\mu, \sigma^2)$, we say W has the *lognormal distribution*. Here, $E[W] = \exp\{\mu + \frac{1}{2}\sigma^2\}$ and $Var[W] = \exp\{2(\mu + \sigma^2)\} - \exp\{2\mu + \sigma^2\}$. Notice that W is defined over only the positive real numbers.

```
* SAS code to find Std. Normal Cumulative Probab.;
data StdNorm;
   cdf = probnorm(-1.322);

proc print;
```

```
                     The SAS System

                  Obs          cdf
                   1        0.093084
```

Figure A.1   Sample SAS code and output (edited) for standard normal cumulative probability

```
* SAS code to find Std. Normal Crit. Point;
data zCritPt;
   alpha = 0.907;
   arg  = 1 - alpha;
   z    = probit(arg);

proc print;
```

```
                   The SAS System
            Obs     alpha      arg        z
             1      0.907     0.093    -1.32251
```

Figure A.2   Sample SAS program and output (edited) to find standard normal critical point

A.2.11 Distributions derived from the normal

A number of important distributions may be derived from the normal. For instance, if $Z \sim N(0, 1)$ is independent of $W \sim \chi^2(v)$, then $T = Z/\sqrt{W/v}$ is distributed as per the p.d.f.

$$f_T(t) = \frac{\Gamma(\{v+1\}/2)}{\Gamma(v/2)\sqrt{v\pi}} \left(1 + \frac{t^2}{v}\right)^{-(v+1)/2} I_{(-\infty,\infty)}(t),$$

where v are the *degrees of freedom* (df) of the p.d.f. We call this *Student's t-distribution* after the work of W.S. Gosset, who wrote under the pseudonym 'Student' (Student, 1908). For notation, we write $T \sim t(v)$; the mean and variance are $E[T] = 0$ (if $v > 1$) and $\text{Var}[T] = v/(v - 2)$ (if $v > 2$). The $t(v)$ p.d.f. graphs very similar to the standard normal p.d.f.; both are centered at zero with a symmetric, 'bell' shape (see Appendix B). The $t(v)$ has heavier tails than the standard normal, however. As $v \to \infty$, $t(v)$ converges to $N(0, 1)$.

Probabilities and/or critical points from $t(v)$ can be derived using computer software. For example, suppose $T \sim t(31)$ and we wish to find $P[T \geq 1.322] = 1 - P[T < 1.322]$. (This is an *upper-tail* probability calculation.) Figure A.3 gives sample SAS code and (edited) output to find this as $P[T \geq 1.322] = 0.0979$.

We can also invert the upper-tail operation to find critical points of the t distribution. That is, suppose we wish to find the point $t_\alpha(v)$ of $T \sim t(v)$ such that $P[T \geq t_\alpha(v)] = \alpha$. Similar to the standard normal case, this is called the $t(v)$ upper-α critical point. These points are often tabulated (see Table B.2) or may be generated via computer. For example, Fig. A.4 gives sample SAS code and (edited) output to find $t_{0.10}(31) = 1.3095$.

Another continuous distribution derived from the normal is the χ^2 distribution seen in §A.2.8: if $Z \sim N(0, 1)$, then $Z^2 \sim \chi^2(1)$. Extending this, and using the closure of χ^2 under addition, if $Z_i \sim$ i.i.d. $N(0, 1)$, $i = 1, \ldots, n$, then $\sum_{i=1}^{n} Z_i^2 \sim \chi^2(n)$. Many statistical operations lead to χ^2 p.d.f.s; for example, if $Y_i \sim$ i.i.d. $N(\mu, \sigma^2), i = 1, \ldots, n$, then the *sample variance* $S^2 = \sum_{i=1}^{n} (Y_i - \overline{Y})^2/(n - 1)$ is related to a χ^2 distribution via $(n - 1)S^2/\sigma^2 \sim \chi^2(n - 1)$.

```
*  SAS code to find t-dist'n upper-tail probab.;
*  cdf = Pr(T(31) <= 1.322);

  data tpval;
    cdf = probt(1.322 , 31);
    pval = 1 - cdf;

  proc print;
```

```
              The SAS System
         Obs        cdf           pval
          1       0.90208       0.097920
```

Figure A.3 Sample SAS code and output (edited) for *t*-distribution upper-tail probability

```
*  SAS code to find t-dist'n Crit. Point;
data tCritPt;
   alpha = 0.100;    cdf = 1 - alpha;
   df = 31;
   t = tinv(cdf , df);

proc print;
```

```
              The SAS System

        Obs    alpha    cdf    df       t
         1      0.1     0.9    31    1.30946
```

Figure A.4 Sample SAS program and output (edited) to find t-distribution critical point

To find tail probabilities from $\chi^2(n)$, we again use the computer. For example, suppose $W \sim \chi^2(18)$ and we wish to calculate $P[W \geq 29.497]$. Figure A.5 gives sample SAS code and (edited) output to find this value as $P[W \geq 29.497] = 0.0426$.

We can also invert the upper-tail operation to find critical points of the χ^2 distribution. That is, suppose we wish to find the upper-α critical point $\chi^2_\alpha(n)$ of $W \sim \chi^2(n)$, such that $P[W \geq \chi^2_\alpha(n)] = \alpha$. These critical points are tabulated (see Table B.3) or may be generated by computer. For example, Fig. A.6 gives sample SAS code and (edited) output to find $\chi^2_{0.05}(18) = 28.8693$.

One last distribution derived from the normal that is of use in environmetric applications is the *F-distribution*. The F is derived as the scaled ratio of two independent χ^2 distributions. That is, suppose $W_1 \sim \chi^2(v_1)$, independent of $W_2 \sim \chi^2(v_2)$. Then we say $F = (W_1/v_1)/(W_2/v_2) = (v_2 W_1)/(v_1 W_2)$ has an F-distribution with v_1 and v_2 df. We write $F \sim F(v_1, v_2)$. The mean is $E[F] = v_2/(v_2 - 2)$ for $v_2 > 2$ and for any $v_1 \geq 1$, while the variance is $Var[F] = 2v_2^2(v_1 + v_2 - 2)/\{v_1(v_2 - 2)^2(v_2 - 4)\}$ for $v_2 > 4$. Note that the order of the degrees of freedom is important; in fact, if $F \sim F(v_1, v_2)$, then $1/F \sim F(v_2, v_1)$.

```
*  SAS code to find chi-square upper-tail probab.;
data chisq;
   arg = 29.497;
   df = 18;
   cdf = probchi(arg , df);
   pval = 1 - cdf;
proc print;
```

```
                The SAS System
        Obs     arg      df     cdf        pval
         1     29.497    18    0.95737    0.042633
```

Figure A.5 Sample SAS code and output (edited) for χ^2 distribution upper-tail probability

```
* SAS code to find chi-square Crit. Point;
data CritPt;
  alpha = 0.05;
  cdf = 1 - alpha;
  df = 18;
  c = cinv(cdf , df);

proc print;
```

                        The SAS System

          Obs   alpha   cdf   df   c
           1   0.05   0.95   18   28.8693

Figure A.6 Sample SAS program and output (edited) to find χ^2 critical point

The F-distribution is also related to the t-distribution: if $T \sim t(v)$, then $T^2 \sim F(1, v)$. Or, as $v_2 \to \infty$, $v_1 F$ converges to $\chi^2(v_1)$; see Piegorsch and Bailer (1997, Fig. 1.5) or Leemis (1986).

To find tail probabilities from $F(v_1, v_2)$, we again use the computer. For example, suppose $F \sim F(18, 28)$ and we wish to calculate $P[F \geq 2.529]$. Figure A.7 gives sample SAS code and (edited) output to find this value as $P[F \geq 2.529] = 0.0134$. We can also invert the upper-tail operation to find critical points of the F-distribution. That is, suppose we wish to find the upper-α critical point $F_\alpha(v_1, v_2)$ of $F \sim F(v_1, v_2)$, such that $P[F \geq F_\alpha(v_1, v_2)] = \alpha$. These critical points are extensively tabulated (Neter et al., 1996, Table B.4; Hogg and Tanis, 2001, Table VII) or may be generated by computer. For example, Fig. A.8 gives sample SAS code and (edited) output to find $F_{0.01}(18, 28) = 2.6532$. Alternatively, useful Internet sites for F critical points include http://www.stat.sc.edu/~west/applets/fdemo.html or http://calculators.stat.ucla.edu/cdf/

```
* SAS code to find F-dist'n upper-tail probab.;
data fPval;
  arg = 2.529;
  df1 = 18;     df2 = 28;
  cdf = probf(arg, df1 , df2);
  pval = 1 - cdf;

proc print;
```

                       The SAS System

      Obs   arg   df1   df2   cdf   pval
       1   2.529   18   28   0.98658   0.013421

Figure A.7 Sample SAS code and output (edited) for F-distribution upper-tail probability

```
* SAS code to find F-dist'n Crit. Point;
data fCritPt;
    alpha = 0.01;
    cdf = 1 - alpha;
    df1 = 18;        df2 = 28;
    f = finv(cdf , df1 , df2);

proc print;
```

                        The SAS System

    Obs   alpha   cdf   df1   df2   f
     1   0.01   0.99   18   28   2.65322

Figure A.8 Sample SAS program and output (edited) to find F-distribution critical point

A.2.12 Bivariate normal distribution

All of the various univariate distributions reviewed above can be extended into bivariate forms. Perhaps the most useful of these is the *bivariate normal distribution*: if two continuous random variables, X_1 and X_2, are jointly distributed as bivariate normal, then their marginal (univariate) distributions are $X_i \sim N(\mu_i, \sigma_i^2)$, $i = 1, 2$, with $\text{Cov}[X_1, X_2] = \sigma_{12}$. The covariance, σ_{12}, can be any real number. (Here, if it is zero, X_1 and X_2 are statistically independent, and vice versa.) The bivariate normal distribution is fully described by these five separate parameters, $\mu_1, \mu_2, \sigma_1^2, \sigma_2^2$, and σ_{12}.

The correlation between X_1 and X_2 is $\text{Corr}[X_1, X_2] = \sigma_{12}/(\sigma_1 \sigma_2)$, typically denoted as ρ_{12}. As such, the bivariate normal distribution may also be described in terms of the five parameters $\mu_1, \mu_2, \sigma_1^2, \sigma_2^2$, and ρ_{12}.

We often write the bivariate variables together as a single *vector*, $\mathbf{X} = \begin{bmatrix} X_1 \\ X_2 \end{bmatrix}$, with mean vector $\boldsymbol{\mu} = \begin{bmatrix} \mu_1 \\ \mu_2 \end{bmatrix}$ and with variance and covariance terms collected into a *variance–covariance matrix*

$$\mathbf{V} = \begin{bmatrix} \sigma_1^2 & \sigma_{12} \\ \sigma_{12} & \sigma_2^2 \end{bmatrix}.$$

(A matrix is an array of numbers arranged in a square or rectangular fashion; a variance–covariance matrix is a special form where variances are arranged along the diagonal and covariances are placed in corresponding positions off the diagonal. A vector is a single row or column array. Where necessary, a superscript T, $^\mathrm{T}$, is used to indicate transposition of a vector or matrix, so that, for example, $[X_1 \ X_2]^\mathrm{T} = \begin{bmatrix} X_1 \\ X_2 \end{bmatrix}$.)

Notice that since $\text{Cov}[X_1, X_2] = \text{Cov}[X_2, X_1]$, the matrix $\mathbf{V}$ is *symmetric*, in that its off-diagonal elements correspond to each other.

This construction can be extended to any number of random variables, $X_1, X_2, \ldots, X_n$. In this case we say the n-dimensional random vector $\mathbf{X}$ has a *multivariate normal distribution*, whose mean vector $\boldsymbol{\mu}$ contains the values μ_i and whose (symmetric) variance–covariance matrix has diagonal elements $\sigma_i^2, i = 1, \ldots, n$, and off-diagonal elements $\text{Cov}[X_i, X_j] = \sigma_{ij}, i \neq j$.

A.3 Random sampling

A.3.1 Random samples and independence

When sampling from a population of discrete or continuous random variables, the data values generated for statistical calculations are referred to as a *random sample*. Many types of random samples can be formed, depending on the sampling strategy employed (see Chapter 8). In the simplest case, we suppose that the observed data are all sampled from the same (identical) probability distribution, $f_Y(y)$, and are statistically independent of each other. Under statistical independence, any sampled unit's observed response does not impact or influence any other sampled unit's observed response. This is a *simple random sample*, and we write $Y_i \sim \text{i.i.d. } f_Y(y)$, $i = 1, \ldots, n$.

A.3.2 The likelihood function

An important feature of a random sample is that it contains information about the original population under study. We can quantify this information into what is known as the *likelihood function*. Mathematically, this is simply the joint probability function (p.m.f. or p.d.f.) of the entire random sample viewed as a function of the parameters. In the special case of a sample of i.i.d. random variables, equation (A.3) tells us that the joint probability function and hence the likelihood will factor into n individual contributions from the same p.m.f. or p.d.f.

As seen in §A.2, most families of p.m.f.s or p.d.f.s depend on one or more unknown parameters, such as a population mean μ or a scale parameter β. Collecting these together into a generic vector of parameters, $\vartheta = [\theta_1 \ldots \theta_p]^T$, we view the likelihood function from an i.i.d. sample as a function of both this vector and of the observations:

$$L(\vartheta; y_1, \ldots, y_n) = \prod_{i=1}^{n} f_Y(y_i; \vartheta).$$

The notation emphasizes that $L(\vartheta; y_1, \ldots, y_n)$ is a function of the unknown parameters in ϑ, given the information in the data $y_1, \ldots, y_n$. In many cases, it is easier to work with the natural logarithm of the likelihood, known as the *log-likelihood function*:

$$\ell(\vartheta) = \sum_{i=1}^{n} \log\{f_Y(y_i; \vartheta)\}. \tag{A.17}$$

The first (partial) derivative of the log-likelihood with respect to θ_j is known as a *score function*; common notation for this is $U_j(\vartheta) = \partial\ell(\vartheta)/\partial\theta_j$, $j = 1, \ldots, p$.

Likelihood as a concept was developed by R.A. Fisher (1912, 1922b), who also noted that statistical information about ϑ available in the random sample was related to the log-likelihood. With one unknown parameter, θ, the *Fisher information number* is found as the expected value of the negative second derivative of ℓ: $f(\theta) = $ $E[-\ell''(\theta)] = E[-U'(\theta)]$. For $p > 1$ unknown parameters, we take the expected negative mixed partials $f_{jk}(\vartheta) = E[-\partial^2 \ell(\vartheta)/\partial\theta_j\partial\theta_k]$ and collect them into a *Fisher information matrix*:

$$\mathbf{F}(\vartheta) = \begin{bmatrix} f_{11}(\vartheta) & f_{12}(\vartheta) & \cdots & f_{1p}(\vartheta) \\ f_{21}(\vartheta) & f_{22}(\vartheta) & \cdots & f_{2p}(\vartheta) \\ \vdots & \vdots & \ddots & \vdots \\ f_{p1}(\vartheta) & f_{p2}(\vartheta) & \cdots & f_{pp}(\vartheta) \end{bmatrix}. \tag{A.18}$$

Since it is based on the expected values of the log-likelihood derivatives, (A.18) is also called the *expected information matrix*.

A.4 Parameter estimation

Given a random sample, $Y_1, Y_2, \ldots, Y_n$, statistical estimation of the p unknown parameters in ϑ can proceed via various methods. Here, we give a short review of some of the more common approaches. We use as generic notation $\hat{\vartheta}$ for any estimator of ϑ (unless a more specific or traditional notation presents itself).

In some cases we may need to order the data, that is, to rearrange the observations in increasing order from smallest to largest. If so, we denote the ordered data as $Y_{(i)}, i = 1, \ldots, n$. The '()' subscript indicates the ordering; for example, $Y_{(1)}$ is the smallest sample observation, $Y_{(1)} = \min_{i\in\{1,\ldots,n\}}\{Y_i\}$, while $Y_{(n)}$ is the largest sample observation, $Y_{(n)} = \max_{i\in\{1,\ldots,n\}}\{Y_i\}$. (The symbol '$\in$' is read as 'is an element of.') We often call the set of ordered observations the *order statistics* of the sample.

A.4.1 Least squares and weighted least squares

A traditional method for estimating an unknown parameter is known as the method of *least squares* (LS). The method does not require specification of the likelihood, and is thus quite general. For an individual mean $\theta = E[Y_i]$, the LS method estimates θ by minimizing the objective quantity

$$D = \sum_{i=1}^{n}(Y_i - \theta)^2, \tag{A.19}$$

which is the sum of squared deviations of each observation from θ. If there are $p > 1$ unknown parameters, one simply minimizes the squared difference between the Y_is and some function of the parameters, $g(\theta_1, \theta_2, \ldots, \theta_p)$, resulting in a p-dimensional vector of estimators for the θ_js. When $E[Y_i]$ depends on the θ_js we might set

$g(\theta_1, \theta_2, \ldots, \theta_p) = E[Y_i]$, although any function pertinent to the environmental application under study may be used for $g(\theta_1, \theta_2, \ldots, \theta_p)$.

An implicit assumption made when applying the LS method is that each Y_i
contributes equally to the information about θ. If this is not true, then we apply
instead *weighted least squares* (WLS) by minimizing the weighted objective quantity

$$D_w = \sum_{i=1}^{n} w_i (Y_i - \theta)^2, \tag{A.20}$$

where the weights, w_i, in (A.20) are chosen to account for the differential quality of
each Y_i when estimating θ. For example, suppose $\mathrm{Var}[Y_i]$ varies with i so that some
observations are more variable and hence less precise regarding the information they
provide about θ. In this case, we set the weights proportional to the reciprocals of the
variances: $w_i \propto 1/\mathrm{Var}[Y_i]$. With this, a more variable (i.e., less precisely measured)
quantity has lower impact on the estimation process.

A.4.2   The method of moments

By making some very limited assumptions on the nature of the Y_is, we can appeal to
another useful estimation approach. Known as the *method of moments* (MOM), the
procedure equates the first $p \geq 1$ population moments of Y_i to their corresponding
sample moments, then solves for the p unknown parameters.

Suppose that the data Y_i are from an i.i.d. sample whose (common) *j*th population
moment is $E[Y^j]$. We assume that these moments are functionally dependent on the
unknown parameters in ϑ. Define the corresponding sample moment as
$M_j = (1/n) \sum_{i=1}^{n} X_i^j, j = 1, \ldots, p$. Then, the MOM estimators are found by solving
the p-dimensional system of equations $E[Y^j] = M_j, j = 1, \ldots, p$. We call these equations the moment *estimating equations* for ϑ.

A.4.3   Maximum likelihood

When the full parametric structure of the random sample can be specified via a
likelihood function $L(\vartheta; y_1, \ldots, y_n)$, more powerful methods may be brought to bear
for estimating ϑ. Given the likelihood, ones finds the log-likelihood function $\ell(\vartheta)$ and
then maximizes it with respect to the p unknown parameters. This is the *method of
maximum likelihood* (ML). In most cases, the ML estimators (MLEs) may be found
by appeal to differential calculus: set the first partial derivatives of $\ell(\vartheta)$ equal to zero
and solve the resulting system of p estimating equations. (When dealing with likelihood functions, the resulting stationary points are generally global maxima,
although one should verify this in each case.) When the underlying likelihood is
based on a normal distribution, the MLE of ϑ will also be the LS estimate.

An important quality of MLEs is that their large-sample distributions are often
known, at least to a good approximation. Under certain large-sample regularity conditions, the ML vector ϑ will possess a p-variate normal distribution where, individually,

$\dot{\theta}_j \sim N(\theta, \text{Var}[\hat{\theta}_j]), j = 1, \ldots, p$. (The dot notation above the $\sim$ indicates that the distributional relationship is only approximate. The approximation improves as $n \rightarrow \infty$.) The variance–covariance matrix (§A.2.12) of $\hat{\vartheta}$ is found by inverting the Fisher information matrix, $\mathbf{F}(\vartheta)$, from (A.18). For a shorthand notation, we say $E[\hat{\vartheta}] = \vartheta$ and $\text{Var}[\hat{\vartheta}] = \mathbf{F}^{-1}(\vartheta)$. Often, the latter relationship is written simply as $\text{Var}[\hat{\vartheta}] = \mathbf{F}^{-1}$. Readers unfamiliar with matrix operations such as inversion may consult a text on matrix algebra, such as Searle (1982) or Harville (1997). Here, we note briefly that the *inverse* of a square matrix $\mathbf{M}$ is another square matrix $\mathbf{M}^{-1}$ such that the products $\mathbf{M}^{-1}\mathbf{M}$ and $\mathbf{M}\mathbf{M}^{-1}$ are also square, with ones along the diagonal and zeros off of it. The latter matrix is known as the *identity matrix*, and is denoted by $\mathbf{I} = \text{diag}\{1, 1, \ldots, 1\}$. More generally, a *diagonal matrix* is a square matrix with zeros in all off-diagonal locations and with other elements, typically non-zero, along the diagonal.

If $\mathbf{F}$ is itself diagonal, say $\mathbf{F} = \text{diag}\{f_{11}, f_{22}, \ldots, f_{pp}\}$, then $\mathbf{F}^{-1}$ will also be diagonal: $\mathbf{F}^{-1} = \text{diag}\{1/f_{11}, 1/f_{22}, \ldots, 1/f_{pp}\}$. From this, the individual large-sample variances become the reciprocals of the Fisher information quantities: $\text{Var}[\hat{\theta}_j] \approx 1/f_{jj} = 1/E[-\partial^2\ell(\vartheta)/\partial\theta_j^2]$. We call the square root of an estimator's variance the *standard error* of the estimator: $se[\hat{\theta}_j] = \sqrt{\text{Var}[\hat{\theta}_j]}$. The standard error measures uncertainty associated with estimating a population parameter. By contrast, the simpler standard deviation measures the variability in a population.

A.4.4 Bias

When estimating an unknown parameter θ, we define the difference between θ and its estimator's expected value as the *bias* of estimation: $B(\hat{\theta}) = E[\hat{\theta}] - \theta$. If $B(\hat{\theta}) = 0$, that is, if $E[\hat{\theta}] = \theta$, we say $\hat{\theta}$ is *unbiased*. This is a sort of 'scientific objectivity': on average, $\hat{\theta}$ will equal what it is estimating. This is a desirable property, and we try whenever possible to employ unbiased estimators. For example, the sample mean of a set of observations is unbiased for estimating the underlying population mean: $E[\bar{Y}] = \mu$ for any population where μ exists. This helps explain the popularity of the sample mean for estimating μ.

Lack of bias in a point estimator is not a guaranteed property; for none of the estimation methods described herein is the investigator assured that $E[\hat{\theta}] = \theta$. One favorable property of maximum likelihood is, however, that as $n \rightarrow \infty$, $\hat{\theta}$ approaches θ. This is a form of large-sample *consistency* of the MLE. As a result, it is often the case that $E[\hat{\theta}] \approx \theta$ in small samples.

A.5 Statistical inference

In most environmental data analyses, estimation of the unknown parameters is only a first step. Along with the point estimates, it is also important to infer from the data the parameter's interpretation/impact on the larger population. This is *statistical inference*, of which there are two general forms: interval estimation and hypothesis testing. We briefly review each of these in turn. Readers desiring greater detail should consult dedicated texts on mathematical statistics, such as Hogg and Tanis (2001), Bickel and Doksum (2001), or Casella and Berger (2002).

A.5.1 Confidence intervals

A *confidence interval* is a form of estimator for an unknown parameter θ that uses information about the variability in the data to construct an interval within which θ may lie. In its broadest sense, a confidence interval is a pair of values $L_\theta(Y_1, \ldots, Y_n)$ and $U_\theta(Y_1, \ldots, Y_n)$ that satisfy the probability statement $P[L_\theta(Y_1, \ldots, Y_n) < \theta < U_\theta(Y_1, \ldots, Y_n)] = 1 - \alpha$, where $1 - \alpha$ is called the *confidence coefficient* of the interval. Typical values are 90%, 95%, or 99%. If L_θ and U_θ are both finite, we say the confidence interval is *two-sided*. If either limit is infinite, we say the result is a *one-sided confidence bound* on θ.

It is important to recognize that confidence is *not* probability. The probability that a calculated interval actually contains the true value of θ is not $1 - \alpha$; it is either 0 or 1. That is, suppose we calculate a 95% interval for θ and find $7.72 < \theta < 12.01$. Clearly, the probability that θ is included in this interval is 0 (if it is not) or 1 (if it is). Since we ascribe no random variability to θ, neither can we place any other form of probability on the statement $7.72 < \theta < 12.01$. Thus, instead of a probabilistic interpretation, we say that the interval 'covers' θ with confidence $1 - \alpha$. This is a *frequentist interpretation* for coverage: if, over repeated sampling we were able to count the number of times an interval covers the true θ, we would find that $100(1 - \alpha)\%$ of the intervals cover it correctly. Thus confidence is a measure of the frequency of correct coverage of the unknown parameter by the interval estimator.

There are many approaches available for constructing confidence intervals. One of the simplest is known as the *Wald interval*, based on Wald's (1943) pioneering work in this area. The interval is constructed by recognizing a basic feature of the normal p.d.f.: if $X \sim N(\mu, \sigma^2)$, then $P[|X - \mu|/\sigma \leq z_{\alpha/2}] = P[\mu - z_{\alpha/2}\sigma < X < \mu + z_{\alpha/2}\sigma] = 1 - \alpha$, where $z_{\alpha/2}$ is the upper-$(\alpha/2)$ critical point from the standard normal distribution. Recall, however, that MLEs are approximately normal: $\hat{\theta} \sim N(\theta, \mathrm{Var}[\hat{\theta}])$. Thus in large samples, $P[\theta - z_{\alpha/2}se(\hat{\theta}) < \hat{\theta} < \theta + z_{\alpha/2}se(\hat{\theta})] \approx 1 - \alpha$, where $se[\hat{\theta}] = \sqrt{\mathrm{Var}[\hat{\theta}]}$ is the standard error of $\hat{\theta}$. This relationship may be manipulated into

$$P[\hat{\theta} - z_{\alpha/2}se(\hat{\theta}) < \theta < \hat{\theta} + z_{\alpha/2}se(\hat{\theta})] \approx 1 - \alpha.$$

Thus a set of (approximate) $1 - \alpha$ confidence limits for θ is $\hat{\theta} \pm z_{\alpha/2}se(\hat{\theta})$.

By recognizing that the Wald construction is based on a simple $1 - \alpha$ probability statement, $P[|\hat{\theta} - \theta|/se(\hat{\theta}) \leq z_{\alpha/2}] \approx 1 - \alpha$, we can develop other methods for building confidence intervals (by basing them on other forms of $1 - \alpha$ probability statements). For instance, recall that the first derivative of the log-likelihood function is the score function: $U(\theta) = d\ell(\theta)/d\theta$. The notation emphasizes that the scores are functions of the unknown parameter θ, although they are also functions of the data, $Y_1, \ldots, Y_n$. In large samples the score is approximately normal: $U(\theta) \sim N(0, \mathcal{I}[\theta])$, where

$\mathit{f}(\theta) = E[-\ell''(\theta)] = E[-U'(\theta)]$ is the Fisher information number (Rao, 1947). Hence, standardizing the score yields $Z(\theta) = U(\theta)/\sqrt{\mathit{f}(\theta)} \overset{.}{\sim} N(0, 1)$. Recall, however, that the square of a standard normal is $\chi^2(1)$, so we can write $Z^2(\theta) \overset{.}{\sim} \chi^2(1)$, and from this, $P[Z^2(\theta) \le \chi^2_\alpha(1)] \approx 1 - \alpha$. Depending on the structure of the score function $U(\theta)$ and of the Fisher information $\mathit{f}(\theta)$, this probability statement can be manipulated into a confidence interval for θ. We call the result a $1 - \alpha$ *score (confidence) interval* for θ. Extensions are also possible to the multi-parameter case, where in large samples the score vector $\mathbf{U}(\vartheta)$ is approximately multivariate normal: $\mathbf{U}(\vartheta) \overset{.}{\sim} N(\mathbf{0}, \mathbf{F}[\vartheta])$; see Bickel and Doksum (2001, §6.3.2).

A different construction employs a ratio of likelihood functions. Specifically, let $\Lambda(\theta) = L(\theta; y_1, \ldots, y_n)/L(\hat{\theta}; y_1, \ldots, y_n)$, where $\hat{\theta}$ is the MLE of θ. In large samples, $G^2(\theta) = -2 \log(\Lambda) \overset{.}{\sim} \chi^2(1)$, so that $P[G^2(\theta) \le \chi^2_\alpha(1)] \approx 1 - \alpha$ (Wilks, 1938). As with the score interval, this probability statement may be manipulated into a confidence interval for θ. We call this a $1 - \alpha$ *likelihood ratio* (LR) *interval* for θ.

Unfortunately, both the score and LR confidence intervals can be difficult to implement, since they do not always allow for closed-form expression for the actual confidence limits. Indeed, in the general case we cannot write the actual limits $L_\theta(Y_1, \ldots, Y_n) < \theta < U_\theta(Y_1, \ldots, Y_n)$ without more specific details about the nature of the likelihood. As a result, computer calculation is common for both methods. Nonetheless, along with the Wald interval, all of these likelihood-based approaches can serve as very useful paradigms for constructing $1 - \alpha$ confidence intervals.

One can also construct confidence intervals on a case-by-case basis, sculpting each procedure to fit the features of the distribution under study. For example, if $Y \sim \text{Exp}(\theta)$, an exact $1 - \alpha$ confidence interval for the rate parameter θ is $-Y/\log(\frac{1}{2}\alpha) < \theta < -Y/\log(1 - \frac{1}{2}\alpha)$ if $\alpha < 0.5$ (George and Elston, 1993). Similar sorts of constructions exists for many other families of distributions; however, it is beyond the scope of this appendix to detail them here. We gave an exposition on confidence intervals for selected distributions in Piegorsch and Bailer (1997, §2.6).

A.5.2 Bootstrap-based confidence intervals

When the parent distribution of the data is unknown, likelihood-based methods for building confidence intervals are unavailable, since it is impossible to construct a complete likelihood for the unknown parameter. An alternative approach in this case is to approximate the unknown distribution of the data by simulating random outcomes a large number of times via computer. This is an application of the *Monte Carlo method*, a name coined by John von Neumann and Stanislaw Ulam while both were working at the Los Alamos National Laboratory in the 1940s (see Anonymous, 1949, p. 546). The term associates with the random outcomes in games of chance seen in Monte Carlo, the capital of Monaco and a well-known center for gambling (Ulam was apparently an avid poker player). An early statistical introduction was given by Metropolis and Ulam (1949). Of course, random simulation of stochastic outcomes was common well before the term 'Monte Carlo' described the approach (Stigler, 1991). Other names include *stochastic simulation*, *Monte Carlo simulation*, and *synthetic data generation*.

One specialized form of Monte Carlo simulation useful for building confidence intervals is the method of *bootstrap resampling* (Efron, 2000). The bootstrap method is based upon an elegantly simple idea: since the sampling distribution for a statistic is based upon repeated samples with replacement – or 'resamples' – from the same population, one can use the computer to simulate repeated sampling, calculating the statistic for each simulated sample. The resulting, simulated sampling distribution for the statistic is used to approximate the true sampling distribution of the statistic of interest, leading to approximate interval estimates. The approximation improves as the number of simulated samples increases.

In the simplest case the empirical distribution of the data is used as the basis for the simulated resamples. Simulated resamples are drawn by computer from a theoretical distribution that matches the empirical distribution, and the statistic of interest is calculated for each simulated resample. The resampled values of the statistic provide an approximate distribution from which to construct confidence intervals.

We illustrate the approach with a simple bootstrap-based confidence interval. Take a random sample, $X_1, \ldots, X_n$ from some population with unknown c.d.f. $F_X(x)$. Suppose interest exists in obtaining an interval estimate of some unknown parameter θ based upon an estimator $\hat{\theta}$ calculated from the data. We can estimate $F_X(x)$ from the data via the empirical c.d.f.

$$\hat{F}_X(x) = \frac{\{\text{number of } X_i \text{s} \leq x\}}{n}.$$

From this, assuming the data are discrete, we can construct the corresponding empirical p.m.f., $\hat{f}_X(x)$, as

$$\hat{f}_X(x) = \frac{\{\text{number of } X_i \text{s} = x\}}{n}, \qquad \text{where } x = x_1, \ldots, x_n.$$

If $\hat{F}_X(x)$ is a good estimate of $F_X(x)$ and $\hat{f}_X(x)$ is a good estimate of $f_X(x)$, we can generate bootstrap samples as follows:

1. Generate a bootstrap sample, say $X_1^*, \ldots, X_n^*$, at random from the empirical p.m.f. $\hat{f}_X(x)$.

2. Calculate the statistic of interest, say $\hat{\theta}^*$, from the bootstrap sample.

3. Repeat steps 1 and 2 a large number of times, say $B \geq 1000$. (Babu and Singh, 1983, give theoretical arguments for $B = n(\log\{n\})^2$, although if $n < 60$ a common recommendation is to set at least $B = 1000$ for confidence interval construction.) Collect together all the statistics that were calculated from each of the bootstrap samples. Let $\hat{\theta}_1^*, \ldots, \hat{\theta}_B^*$ be the collection of all of these statistics.

The collection of bootstrap statistics forms an empirical estimate of the sampling distribution of $\hat{\theta}$. From this, we obtain a confidence interval for θ by selecting specified percentiles from the empirical distribution of $\hat{\theta}$ formed from the collection of all bootstrap statistics. This is known as the *percentile method*. For example, for a 95% confidence interval on θ based on the percentile method, take the 2.5th percentile of the empirical distribution of $\hat{\theta}$ for the lower bound and the 97.5th percentile for the upper bound.

More detail on these and other forms of bootstrap methodology is available, for example, in Efron and Tibshirani (1993), Dixon (2002a), and the references therein.

A.5.3 Hypothesis tests

A second form of statistical inference useful in environmetric practice is based on the concept of comparing two *hypotheses* about an unknown parameter θ. Suppose interest centers upon a particular value of θ, say θ_0. We write the *null hypothesis* as $H_0: \theta = \theta_0$. This is also called the no-effect hypothesis, since θ_0 often indicates some lack of environmental impact or effect. The *alternative hypothesis* or *research hypothesis*, H_a, is a specification for θ that represents a plausible research alternative to H_0. (Some authors denote the alternative hypothesis as H_1.) The goal is determination of whether H_0 or H_a represents a credible indication of the nature of θ. To achieve this, we structure the process to arrive at a decision regarding H_0, that is, whether to reject it or accept it in light of the information about θ observed in the data.

Two fundamental probabilities lie at the core of this *hypothesis testing*. The first is the false positive rate: $\alpha = P[\text{reject } H_0 | H_0 \text{ true}]$. (The symbol '|' is a conditioning operator. Read it as 'given that' or 'conditional on'.) This is called the *Type I* or *false positive error rate*, and it is usually fixed in advance by the investigator. The second is the false negative rate: $\beta = P[\text{accept } H_0 | H_0 \text{ false}]$. This is called the *Type II* or *false negative error rate*. Associated with β is the *power* or *sensitivity* of the test: $P[\text{reject } H_0 | H_0 \text{ false}] = 1 - \beta$. For fixed α we wish to find a test of H_0 (vs. H_a) that minimizes β and thus maximizes power to the greatest extent possible. A test whose power exceeds that of all other tests for all possible values of θ is called a *uniformly most powerful (UMP)* test. Whether or not such a test exists depends on a number of factors, including the set of hypotheses under study and the underlying likelihood assumed for the data (Casella and Berger, 2002, §8.3). When they exist, UMP tests are the best test procedures that can be derived for a given parametric testing problem. In some cases, a test may be most powerful for only selected values of θ; if so, test is called *locally most powerful*.

Given a set of data, a test statistic is calculated that addresses the effect being studied in H_0. From this a *rejection region* (also called a *critical region*) is formed where, if the test statistic falls into this region, H_0 is rejected in favor of H_a. (If not, we fail to reject H_0.) Equivalently, one can find the *P-value* of the test, defined as the probability under H_0 of observing a test statistic as extreme as or more extreme than that actually observed. (Note that 'more extreme' is defined in the context of H_a. For example, when testing $H_0: \theta = \theta_0$ vs. $H_a: \theta > \theta_0$, 'more extreme' corresponds to values of the test statistic supporting $\theta > \theta_0$.) Small P-values indicate departure from H_0, thus for a fixed significance level α, reject H_0 when $P \leq \alpha$. Since the significance of departure from H_0 is captured by whether or not α exceeds P, we often call α the *significance level* of the test. Some authors also refer to this general inferential paradigm as *significance testing*, although technically the two processes are slightly different (Kempthorne, 1976). For our purposes, however, the technical differences are not as crucial as proper use of the methodology. It is this latter concern that we emphasize here.

Hypothesis tests may be derived under a number of different criteria. These correspond to similar criteria for building confidence intervals. Indeed, the two concepts are tautologously related: suppose one is testing $H_0: \theta = \theta_0$ vs. $H_a: \theta \neq \theta_0$. It is possible to build a $1 - \alpha$ confidence interval for θ such that if the interval fails to contain θ_0, one will reject H_0 at the α level of significance and vice versa (Casella and Berger, 2002, §9.2). For example, the Wald interval described in §A.5.1 can be inverted in this manner into a hypothesis test: for testing $H_0: \theta = \theta_0$ vs. $H_a: \theta \neq \theta_0$, find the MLE $\hat{\theta}$ and its standard error $se[\hat{\theta}]$. Use these to build the test statistic $W_{\text{calc}} = (\hat{\theta} - \theta_0)/se[\hat{\theta}]$. (We use the subscript 'calc' to indicate a statistic that is wholly calculable from the data.) If $\hat{\theta} \sim N(\theta, se^2[\hat{\theta}])$, then $P[\text{reject } H_0|H_0 \text{ true}] = P[|W_{\text{calc}}| \geq z_{\alpha/2}|\theta = \theta_0] \approx \alpha$. Hence the Type I error requirement is satisfied, at least approximately. This defines the *Wald test* of $H_0: \theta = \theta_0$ vs. $H_a: \theta \neq \theta_0$, via the rejection region $|W_{\text{calc}}| \geq z_{\alpha/2}$. (Equivalently, one can use $W_{\text{calc}}^2 \geq \chi_\alpha^2(1)$, since a squared standard normal variate is χ^2 with 1 df. The corresponding approximate *P*-value is $P[\chi^2(1) \geq W_{\text{calc}}^2]$. This can be calculated in SAS, as per Fig. A.5. Alternatively, Piegorsch, 2002, gives *P*-values for the χ^2 and *t*-distributions.) For a one-sided alternative such as $H_a: \theta > \theta_0$, use the rejection region $W_{\text{calc}} \geq z_\alpha$.

In similar fashion, one can also invert score intervals or LR intervals into hypothesis tests of $H_0: \theta = \theta_0$. In the former case, form the *score statistic* $T_{\text{calc}}^2 = U^2(\theta_0)/\mathcal{I}(\theta_0)$, where $U(\theta) = \ell'(\theta)$ is the score function and $\mathcal{I}(\theta) = E[-U'(\theta)]$ is the (expected) Fisher information. In T_{calc}^2, both these are evaluated at θ_0. Under H_0, $T_{\text{calc}}^2 \overset{\cdot}{\sim} \chi^2(1)$. Hence, we reject H_0 in favor of $H_a: \theta \neq \theta_0$ when $T_{\text{calc}}^2 \geq \chi_\alpha^2(1)$; the approximate *P*-value is $P[\chi^2(1) \geq T_{\text{calc}}^2]$. Notice that the score test does not require calculation of the MLE $\hat{\theta}$.

For the LR test, form the test statistic $G_{\text{calc}}^2 = -2\log\{L(\theta_0; y_1, \ldots, y_n)/L(\hat{\theta}; y_1, \ldots, y_n)\}$. Under $H_0, G_{\text{calc}}^2 \overset{\cdot}{\sim} \chi^2(1)$. Hence, we reject H_0 in favor of $H_a: \theta \neq \theta_0$ when $G_{\text{calc}}^2 \geq \chi_\alpha^2(1)$; the approximate *P*-value is $P[\chi^2(1) \geq G_{\text{calc}}^2]$. Extensions to the multi-parameter case are also possible for all of these likelihood-based testing methods; see, for example, Cox (1988).

Alert readers will notice that all three test statistics W_{calc}^2, T_{calc}^2, and G_{calc}^2 appeal in large samples to the same reference distribution, $\chi^2(1)$. This is no coincidence: in most settings the limiting value and large-sample reference distribution for each of these three likelihood-based methods will be the same. We call this a form of *asymptotic equivalence* among the three test statistics. As $n \to \infty$, the three methods provide essentially the same inference on θ.

Perhaps the best-known hypothesis test is the so-called *t-test*, based on Student's *t* distribution from §A.2.11. Many forms exist for the *t*-test; we consider here the simplest case: $Y_i \sim$ i.i.d. $N(\mu, \sigma^2), i = 1, \ldots, n$, for unknown μ and σ^2. We are interested in testing $H_0: \mu = \mu_0$ vs. $H_a: \mu \neq \mu_0$ at significance level α. We estimate μ with its MLE, $\overline{Y} = \sum_{i=1}^n Y_i/n$, and use the sample variance, $S^2 = \sum_{i=1}^n (Y_i - \overline{Y})^2/(n - 1)$, to estimate σ^2. The standard error of $\overline{Y}$ is $se[\overline{Y}] = S/\sqrt{n}$, hence the Wald test for this setting employs the statistic $W_{\text{calc}} = (\overline{Y} - \mu_0)/(S/\sqrt{n})$. Under $H_0, W_{\text{calc}} \sim t(n - 1)$ and we reject H_0 in favor of H_a when $|W_{\text{calc}}| \geq t_{\alpha/2}(n - 1)$. The (exact) *P*-value here is $2P[t(n - 1) \geq |W_{\text{calc}}|]$.

Interestingly, if instead we use the LR approach here for testing $H_0: \mu = \mu_0$, the rejection region becomes $n(\bar{Y} - \mu_0)^2/S^2 \geq F_\alpha(1, n-1)$ (Neyman and Pearson, 1933; Hogg and Tanis, 2001, §9.4). Notice that this is simply $W^2_{\text{calc}} \geq t^2_\alpha(n-1)$, since a squared $t(n-1)$ variate is distributed as $F(1, n-1)$. Thus in this simple normal setting the Wald and LR methodologies coincide. (The score statistic is only slightly different: $T^2_{\text{calc}} = n(\bar{Y} - \mu_0)^2/\hat{\sigma}_0^2$ for $\hat{\sigma}_0^2 = \sum_{i=1}^n (Y_i - \mu_0)^2/n$.)

A.5.4   Multiple comparisons and the Bonferroni inequality

An important consideration in environmetric inference is that of *multiple comparisons*, that is, when more than a single inference is made from a single set of data. This is common when there are $p > 1$ population parameters under study and confidence intervals or hypothesis tests are desired on each parameter. For instance, suppose a single hypothesis test is conducted for each parameter using the same set of data; without correction for the multiplicity of tests being undertaken, there will be p opportunities to make a false positive error. Thus the *familywise* false positive error rate will be much larger than the *pointwise* significance level, α. (Analogous concerns occur regarding the coverage coefficient, $1 - \alpha$, when constructing multiple pointwise confidence intervals.) To account for this error inflation, some adjustment is required.

Perhaps the simplest way to adjust for multiple inferences is to build the multiplicity into the estimation or testing scenario: construct joint hypothesis tests or joint confidence regions that simultaneously contain all p parameters of interest. All of the methods we mentioned above can be extended in this fashion; for example, a simultaneous, p-dimensional, $1 - \alpha$ Wald confidence region for a parameter vector ϑ is the ellipsoid defined by the matrix inequality $(\hat{\vartheta} - \vartheta)^{\mathsf{T}} \mathbf{F}(\hat{\vartheta})(\hat{\vartheta} - \vartheta) \leq \chi^2_\alpha(p)$, where $\mathbf{F}(\hat{\vartheta})$ is the Fisher information matrix evaluated at the MLE $\hat{\vartheta}$.

Of course, there are many instances where inferences are desired on the individual parameters, θ_j $(j = 1, \ldots, p)$, of ϑ. For example, one may require individual interval statements on each θ_j. In such cases, it is still possible to adjust for the multiplicity, using one of a variety of probability inequalities (Galambos and Simonelli, 1996). One of the most general of these is due to Bonferroni (1936): the familywise probability that any set of events occurs is always at least as large as 1 minus the sum of the probabilities that each individual event did not occur. Written symbolically, let $P[\mathcal{B}_1 \mathcal{B}_2 \ldots \mathcal{B}_p]$ denote the probability that the p events $\mathcal{B}_j$ $(j = 1, \ldots, p)$ occur simultaneously. Then, if $\mathcal{B}_j^c$ denotes the complementary event that $\mathcal{B}_j$ did not occur, the *Bonferroni inequality* may be written as

$$P[\mathcal{B}_1 \mathcal{B}_2 \cdots \mathcal{B}_p] \geq 1 - \sum_{j=1}^p P[\mathcal{B}_j^c]. \tag{A.21}$$

In the context of a set of p confidence intervals, $\mathcal{B}_j$ corresponds to the event that θ_j was covered correctly by the jth interval, $j = 1, \ldots, p$. Alternatively, in the context of a set of p hypothesis tests, each $\mathcal{B}_j$ corresponds to the event that a correct statistical decision was made for the jth test. For example, if $H_{0j}: \theta_j = \theta_{0j}$, then $\mathcal{B}_j$ is the event that H_{0j} is not rejected when it is true, while $\mathcal{B}_j^c$ represents the event that H_{0j} is rejected when it is true. Thus, the complementary event, $\mathscr{B}_j^c$, corresponds to a false positive error on the jth test, and $P[\mathscr{B}_j^c]$ is the associated false positive (Type I) error rate.

To put (A.21) into practice, suppose that the individual pointwise probabilities $P[\mathscr{B}_j^c]$ are known, but that $P[\mathscr{B}_1^c\mathscr{B}_2^c \cdots \mathscr{B}_p^c]$ and $P[\mathscr{B}_1\mathscr{B}_2 \cdots \mathscr{B}_p]$ are unknown. Then equation (A.21) provides a lower bound for the simultaneous probability $P[\mathscr{B}_1\mathscr{B}_2 \cdots \mathscr{B}_p]$. The simplest way to construct such a lower bound is to set the complementary probabilities, $P[\mathscr{B}_j^c]$, equal to each other, say $P[\mathscr{B}_j^c] = \gamma$ for all j. Then, the lower bound from (A.21) becomes $1 - p\gamma$. If, for example, we require that the simultaneous confidence level among p individual confidence intervals be no smaller than $1 - \alpha$, taking $\gamma = \alpha/p$ produces the desired result. That is, if each complementary event – here, failure to cover θ_j – occurs with probability α/p, the simultaneous collection of coverage events occurs with probability at least $1 - \alpha$, using (A.21). (We call this *minimal simultaneous coverage*.) Similar operations can be performed for testing multiple hypotheses.

Notice that for both hypothesis tests and confidence intervals the multiple events $\mathscr{B}_j$ are often correlated, since they are derived from the same set of data. This will not affect the Bonferroni method; it is valid for any form of correlation structure among the $\mathscr{B}_j$s.

Beyond Bonferroni's inequality, there is a much larger theory and practice of multiple comparisons and other simultaneous inferences. A full description exceeds the scope of this Appendix; interested readers should consult the various texts on the topic, such as Hsu (1996) or Hochberg and Tamhane (1987). For executing multiple comparisons in SAS, the workbook by Westfall *et al.* (1999) is also useful.

A.6 The delta method

A.6.1 Inferences on a function of an unknown parameter

We often find when performing statistical inferences that the unknown parameter, θ, of the distribution under study is not the quantity upon which we wish to make the final inferences. Rather, a function of the unknown parameter, say $h(\theta)$, is of greater interest. If we find the MLE of θ to be $\hat{\theta}$, then it is in fact the case that the MLE of $h(\theta)$ is $h(\hat{\theta})$. This is known as the *invariance property* of maximum likelihood (Casella and Berger, 2002, §7.2), and it is a very important reason why the ML approach is so widely accepted.

Interval estimates for $h(\theta)$ are somewhat more difficult to achieve, however. For instance, a $1 - \alpha$ Wald interval for $h(\theta)$ takes the form $h(\hat{\theta}) \pm z_{\alpha/2}se[h(\hat{\theta})]$. This will be a valid interval estimator if (i) $h(\hat{\theta})$ is (approximately) normal, so that the standard normal critical point $z_{\alpha/2}$ may be employed, and (ii) we can find the standard error of $h(\hat{\theta})$, at least to a good approximation. In some cases the result may be straightforward; for example, suppose $h(\theta)$ transforms θ in a linear fashion, say, $h(\theta) = a + b\theta$, and that we already have a Wald interval on θ of the form $\hat{\theta} \pm z_{\alpha/2}se[\hat{\theta}]$. From §A.2.10 we know that a linear transformation of a normal random variable is also normal, here with variance $\mathrm{Var}[a + b\hat{\theta}] = b^2\mathrm{Var}[\hat{\theta}]$. Since the variance is the squared standard error of an estimator, this means $se[a + b\hat{\theta}] = \sqrt{b^2 se^2[\hat{\theta}]} = |b|se[\hat{\theta}]$. Thus the $1 - \alpha$ Wald interval is simply $a + b\hat{\theta} \pm z_{\alpha/2}|b|se[\hat{\theta}]$.

For more complex functions of $\hat{\theta}$, however, the effort is not so simple. Fortunately, a general result exists that allows us to construct interval estimators or hypothesis tests for any (smooth) function of $\hat{\theta}$, known as the *delta method*. The result is based on approximation of $h(\theta)$ via Taylor's theorem (Casella and Berger, 2002, §5.5). Readers versed in differential calculus will recall that a *Taylor series expansion* approximates a smooth function, $h(x)$, by a polynomial of any desired order expanded about a known point, $x = a$, using the derivatives of $h(x)$. We write

$$h(x) \approx h(a) + (x - a)h'(a) + \frac{1}{2!}(x - a)^2 h''(a) + \frac{1}{3!}(x - a)^3 h^{(3)}(a) + \cdots .$$

Thus, for example, a first-order Taylor expansion takes the form $h(x) \approx h(a) + (x - a)h'(a)$.

The delta method applies this to $h(\hat{\theta})$ by expanding the estimator about the point $E[\hat{\theta}]$. For simplicity, assume that $\hat{\theta}$ is unbiased, so that the expansion is carried out about $E[\hat{\theta}] = \theta$. Then, for $x = \hat{\theta}$ and $a = \theta$, a first-order Taylor expansion of $h(\hat{\theta})$ is $h(\hat{\theta}) \approx h(\theta) + (\hat{\theta} - \theta)h'(\theta)$. Now, take expectations: $E[h(\hat{\theta})] \approx E[h(\theta) + (\hat{\theta} - \theta)h'(\theta)] = E[h(\theta)] + E[(\hat{\theta} - \theta)h'(\theta)] = h(\theta) + h'(\theta)E[\hat{\theta} - \theta]$, which, if $\hat{\theta}$ is unbiased, simplifies to $E[h(\hat{\theta})] \approx h(\theta) + 0 = h(\theta)$. This is a powerful result: if $\hat{\theta}$ is unbiased then for any smooth function $h(\cdot)$ so is $h(\hat{\theta})$, at least to a first-order approximation.

The approximate variance of $h(\hat{\theta})$ follows in similar fashion: if $\text{Var}[\hat{\theta}]$ is sufficiently stable, $\text{Var}[h(\hat{\theta})] \approx \text{Var}[h(\theta) + (\hat{\theta} - \theta)h'(\theta)] = \text{Var}[(\hat{\theta} - \theta)h'(\theta)] = \{h'(\theta)\}^2 \text{Var}[\hat{\theta} - \theta] = \{h'(\theta)\}^2 \text{Var}[\hat{\theta}]$. By taking square roots, we find that

$$se[h(\hat{\theta})] \approx |h'(\theta)|se[\hat{\theta}] \qquad (A.22)$$

is then a first-order approximation to the standard error of $h(\hat{\theta})$. In practice, if θ remains in the expression after completing the approximation, we replace it with its unbiased estimator $\hat{\theta}$.

Example A.1 (Natural logarithm)   We often have need to apply a natural logarithm to a positive random quantity or point estimator, and having done so, to find the variance or standard error of the resulting quantity. This calls for application of the delta method. Suppose an unbiased estimator $\hat{\theta}$ has variance $\text{Var}[\hat{\theta}]$. Take $h(\hat{\theta}) = \log(\hat{\theta})$ and note that $h'(\hat{\theta}) = 1/\hat{\theta}$. Then, from (A.22), the standard error of $h(\hat{\theta})$ is approximately $se[h(\hat{\theta})] \approx se[\hat{\theta}]/|\hat{\theta}|$.   ✪

The delta method may also be used to approximate the full distribution of $h(\hat{\theta})$. The particulars are beyond the scope here, but we can state the general result: under appropriate regularity conditions, $h(\hat{\theta}) \sim N(h[\theta], \{h'(\theta)\}^2 \text{Var}[\hat{\theta}])$, where the approximation improves as $n \to \infty$. Notice that $\text{Var}[h(\hat{\theta})]$ coincides with the result we achieved above. Readers familiar with various forms of the central limit theorem (Hogg and Tanis, 2001, §6.4) will notice that the delta method has features similar in style and structure to this important statistical theorem; also see Lehmann and Casella (1998, §1.8).

This result on large-sample normality of $h(\hat{\theta})$ allows for construction of an approximate Wald interval: $h(\hat{\theta}) \pm z_{\alpha/2}|h'(\theta)|se[\hat{\theta}]$. With careful manipulation, likelihood-ratio and/or score-based statistics may also be constructed in some cases. Bickel and Doksum (2001, §5.3.2) give more detail on the theoretical underpinnings of the delta method's large-sample features.

A.6.2 Inferences on a function of multiple parameters

When there is more than one unknown parameter under study, a multivariate version of the delta method may be developed. Once again the particulars extend beyond the scope of detail here, but we can state the general result: suppose we have p parameters $\theta_1, \theta_2, \ldots, \theta_p$, with unbiased estimators $\hat{\theta}_1, \hat{\theta}_2, \ldots, \hat{\theta}_p$. Let $\text{Var}[\hat{\theta}_j] = \sigma_j^2$ and $\text{Cov}[\hat{\theta}_j, \hat{\theta}_k] = \sigma_{jk}$ $(j \neq k)$. Assume we are interested in some (univariate) function $h(\hat{\theta}_1, \ldots, \hat{\theta}_p)$. Then, under appropriate regularity conditions the delta method states that to first order, $\text{E}[h(\hat{\theta}_1, \ldots, \hat{\theta}_p)] \approx h(\theta_1, \ldots, \theta_p)$ and

$$\text{Var}[h(\hat{\theta}_1, \ldots, \hat{\theta}_p)] \approx \sum_{j=1}^{p} \left(\frac{\partial h}{\partial \theta_j}\right)^2 \sigma_j^2 + 2\sum_{j=1}^{p-1} \sum_{k=j+1}^{p} \frac{\partial h}{\partial \theta_j} \frac{\partial h}{\partial \theta_k} \sigma_{jk}, \qquad (A.23)$$

where $\partial h / \partial \theta_j$ is the partial derivative of $h(\theta_1, \ldots, \theta_p)$ with respect to θ_j. For the simplest case, $p = 2$, (A.23) simplifies to

$$\text{Var}[h(\hat{\theta}_1, \hat{\theta}_2)] \approx \left(\frac{\partial h}{\partial \theta_1}\right)^2 \text{Var}[\hat{\theta}_1] + 2\frac{\partial h}{\partial \theta_1} \frac{\partial h}{\partial \theta_2} \text{Cov}[\hat{\theta}_1, \hat{\theta}_2] + \left(\frac{\partial h}{\partial \theta_2}\right)^2 \text{Var}[\hat{\theta}_2].$$

In practice, if θ_j remains in the expression for the variance approximation we replace it with its unbiased estimator $\hat{\theta}_j$.

In large samples, approximate normality also holds: under appropriate regularity conditions, $h(\hat{\theta}_1, \ldots, \hat{\theta}_p) \overset{\cdot}{\sim} N(h[\theta_1, \ldots, \theta_p], \text{Var}[h(\hat{\theta}_1, \ldots, \hat{\theta}_p)])$, where the variance is given by (A.23).

Example A.2 (Variance of product) Suppose we have two unbiased estimators $\hat{\theta}_1$ and $\hat{\theta}_2$, and we wish to estimate the product $h(\theta_1, \theta_2) = \theta_1\theta_2$. From the delta method, we know that $h(\hat{\theta}_1, \hat{\theta}_2) = \hat{\theta}_1\hat{\theta}_2$ will be approximately unbiased for $\theta_1\theta_2$, with approximate first-order variance

$$\text{Var}[\hat{\theta}_1\hat{\theta}_2] \approx \left(\frac{\partial\{\theta_1\theta_2\}}{\partial \theta_1}\right)^2 \text{Var}[\hat{\theta}_1] + 2\left(\frac{\partial\{\theta_1\theta_2\}}{\partial \theta_1}\right)\left(\frac{\partial\{\theta_1\theta_2\}}{\partial \theta_2}\right) \text{Cov}[\hat{\theta}_1, \hat{\theta}_2]$$

$$+ \left(\frac{\partial\{\theta_1\theta_2\}}{\partial \theta_2}\right)^2 \text{Var}[\hat{\theta}_2],$$

$$= \theta_2^2 \text{Var}[\hat{\theta}_1] + 2\theta_2\theta_1 \text{Cov}[\hat{\theta}_1, \hat{\theta}_2] + \theta_1^2 \text{Var}[\hat{\theta}_2].$$

For practical applications, replace the θ_js with their unbiased estimators; for example, an approximate standard error is $se[\hat\theta_1\hat\theta_2] \approx \sqrt{\hat\theta_2^2 \mathrm{Var}[\hat\theta_1] + 2\hat\theta_1\hat\theta_2 \mathrm{Cov}[\hat\theta_1,\hat\theta_2] + \hat\theta_1^2 \mathrm{Var}[\hat\theta_2]}$. ☯

Example A.3 (Variance of ratio) Many operations in environmental statistics require us to take a ratio of parameters, $h(\theta_1, \theta_2) = \theta_1/\theta_2$. Ratios are notorious for their instability (especially if θ_2 is near zero), and any calculations using them must be performed with care. With this warning in mind, we can find a delta-method approximation for the variance of the ratio estimator, $h(\hat\theta_1, \hat\theta_2) = \hat\theta_1/\hat\theta_2$. We know that this will be approximately unbiased for θ_1/θ_2, with approximate first-order variance

$$\mathrm{Var}\left[\frac{\hat\theta_1}{\hat\theta_2}\right] \approx \left(\frac{\partial\{\theta_1/\theta_2\}}{\partial\theta_1}\right)^2 \mathrm{Var}[\hat\theta_1] + 2\left(\frac{\partial\{\theta_1/\theta_2\}}{\partial\theta_1}\right)\left(\frac{\partial\{\theta_1/\theta_2\}}{\partial\theta_2}\right) \mathrm{Cov}[\hat\theta_1,\hat\theta_2]$$
$$+ \left(\frac{\partial\{\theta_1/\theta_2\}}{\partial\theta_2}\right)^2 \mathrm{Var}[\hat\theta_2].$$
$$= \theta_2^{-2}\{\mathrm{Var}[\hat\theta_1] - 2\theta_1\theta_2^{-1}\mathrm{Cov}[\hat\theta_1,\hat\theta_2] + \theta_1^2\theta_2^{-2}\mathrm{Var}[\hat\theta_2]\}.$$

For use in practice, take the square root and replace the θ_js with their unbiased estimators to find an approximate value for $se[\hat\theta_1/\hat\theta_2]$. From this, a $1-\alpha$ Wald interval for the ratio is $\hat\theta_1/\hat\theta_2 \pm z_{\alpha/2}se[\hat\theta_1/\hat\theta_2]$.

One caveat: this Wald interval for θ_1/θ_2 can be unstable in small samples, depending on the values of θ_1 and θ_2 and on the underlying distribution of the data. In practice, an alternative approach known as *Fieller's theorem* often operates with much better stability and accuracy (Freedman, 2001); but see also Cox (1990) and Example A.4, below. We describe Fieller's theorem in Exercise 1.16, and employ it for building confidence intervals throughout the text. ☯

Example A.4 (Variance of ratio, cont'd) Continuing with the ratio of parameters θ_1/θ_2 from Example A.3, notice that $\theta_1/\theta_2 = \exp\{\log(\theta_1/\theta_2)\}$. To derive an alternative confidence interval on the ratio, take $h(\theta_1, \theta_2) = \log(\theta_1/\theta_2) = \log(\theta_1) - \log(\theta_2)$ and consider manipulating this to find $\mathrm{Var}[h(\theta_1, \theta_2)] = \mathrm{Var}[\log(\theta_1/\theta_2)]$ via the delta method. This is

$$\mathrm{Var}\left[\log\left\{\frac{\hat\theta_1}{\hat\theta_2}\right\}\right] \approx \left(\frac{\partial\log\{\theta_1/\theta_2\}}{\partial\theta_1}\right)^2 \mathrm{Var}[\hat\theta_1]$$
$$+ 2\left(\frac{\partial\log\{\theta_1/\theta_2\}}{\partial\theta_1}\right)\left(\frac{\partial\log\{\theta_1/\theta_2\}}{\partial\theta_2}\right) \mathrm{Cov}[\hat\theta_1,\hat\theta_2]$$
$$+ \left(\frac{\partial\log\{\theta_1/\theta_2\}}{\partial\theta_2}\right)^2 \mathrm{Var}[\hat\theta_2]$$
$$= \theta_1^{-2}\mathrm{Var}[\hat\theta_1] - 2\theta_1^{-1}\theta_2^{-1}\mathrm{Cov}[\hat\theta_1,\hat\theta_2] + \theta_2^{-2}\mathrm{Var}[\hat\theta_2].$$

One again, we replace the θ_js with appropriate estimators and take the square root to find an approximate value for $se[\log\{\hat\theta_1/\hat\theta_2\}]$. From this, a $1-\alpha$ Wald interval for the log of the ratio is $\log\{\hat\theta_1/\hat\theta_2\} \pm z_{\alpha/2} se[\log\{\hat\theta_1/\hat\theta_2\}]$. For an interval on the original ratio, we then exponentiate both limits: $\exp\{\log(\hat\theta_1/\hat\theta_2) \pm z_{\alpha/2} se[\log\{\hat\theta_1/\hat\theta_2\}]\}$. ☺

If more than one function of the p parameters is of interest, the delta method can be further extended into a multidimensional version. As above, suppose the parameters $\theta_1, \theta_2, \ldots, \theta_p$ have unbiased estimators $\hat\theta_1, \hat\theta_2, \ldots, \hat\theta_p$, with $\mathrm{Var}[\hat\theta_j] = \sigma_j^2$ and $\mathrm{Cov}[\hat\theta_j, \hat\theta_k] = \sigma_{jk}$ ($j \neq k$). Suppose we define a vector-valued function of the θ_js, $\mathbf{h}(\theta_1, \ldots, \theta_p)$ with individual elements $h_m(\theta_1, \ldots, \theta_p), m = 1, \ldots, q$. Then, under appropriate regularity conditions the distribution of the vector $\mathbf{h}(\hat\theta_1, \ldots, \hat\theta_p)$ is approximately multivariate normal, where the approximation improves as $n \to \infty$ (Bickel and Doksum, 2001, Lemma 5.3.3). In particular, the mean of each component is $\mathrm{E}[h_m(\hat\theta_1, \ldots, \hat\theta_p)] \approx h_m(\theta_1, \ldots, \theta_p)$, and the variance of each component is

$$\mathrm{Var}[h_m(\hat\theta_1, \ldots, \hat\theta_p)] \approx \sum_{t=1}^{p} \left(\frac{\partial h_m}{\partial \theta_t}\right)^2 \sigma_t^2 + 2\sum_{t=1}^{p-1}\sum_{u=t+1}^{p} \frac{\partial h_m}{\partial \theta_t}\frac{\partial h_m}{\partial \theta_u}\sigma_{tu}, \tag{A.24}$$

where $\partial h_m/\partial\theta_j$ is the partial derivative of $h_m(\theta_1, \ldots, \theta_p)$ with respect to θ_j. Also, the covariance between any two components is

$$\mathrm{Cov}[h_\ell(\hat\theta_1, \ldots, \hat\theta_p), h_m(\hat\theta_1, \ldots, \hat\theta_p)] \approx \sum_{t=1}^{p} \frac{\partial h_\ell}{\partial\theta_t}\frac{\partial h_m}{\partial\theta_t}\sigma_t^2 + \sum_{t=1}^{p-1}\sum_{u=t+1}^{p} \left\{\frac{\partial h_m}{\partial\theta_u}\frac{\partial h_\ell}{\partial\theta_t} + \frac{\partial h_m}{\partial\theta_t}\frac{\partial h_\ell}{\partial\theta_u}\right\}\sigma_{tu}$$

$$\tag{A.25}$$

($\ell \neq m$). As above, if any θ_j remains in the expression for the variance or covariance approximation, we replace it with its unbiased estimator $\hat\theta_j$.

The variances and covariances in (A.24) and (A.25) may be collected together into a variance–covariance matrix (§A.2.12), with the terms from (A.24) placed along the diagonal and the terms from (A.25) placed off the diagonal. This can be written as a matrix product using the following shorthand notation: let $\boldsymbol\vartheta = [\hat\theta_1 \ldots \hat\theta_p]^{\mathrm{T}}$ be the vector of unbiased point estimators, and let $\mathbf{V} = \mathrm{Var}[\boldsymbol\vartheta]$ be its variance–covariance matrix. As employed above, $\mathbf{V}$ has diagonal elements $\sigma_j^2 = \mathrm{Var}[\hat\theta_j]$ and off-diagonal element $\sigma_{jk} = \mathrm{Cov}[\hat\theta_j, \hat\theta_k]$ ($j \neq k$). Also, denote the matrix of partial derivatives of $\mathbf{h}(\theta_1, \ldots, \theta_p)$ as $\nabla\mathbf{h}$, with (m, j)th element $\partial h_m/\partial\theta_j$. Then, $\mathrm{Var}[\mathbf{h}]$ is approximately equal to the matrix product $(\nabla\mathbf{h})\mathbf{V}(\nabla\mathbf{h})^{\mathrm{T}}$ (Cox, 1998).

For instance, take the simplest multivariate case $p = q = 2$. Then,

$$\mathrm{Var}[h_m(\hat\theta_1, \hat\theta_2)] \approx \left(\frac{\partial h_m}{\partial\theta_1}\right)^2 \sigma_1^2 + 2\frac{\partial h_m}{\partial\theta_1}\frac{\partial h_m}{\partial\theta_2}\sigma_{12} + \left(\frac{\partial h_m}{\partial\theta_2}\right)^2 \sigma_2^2$$

($m = 1, 2$) and

$$\mathrm{Cov}[h_1(\hat\theta_1, \hat\theta_2), h_2(\hat\theta_1, \hat\theta_2)] \approx \frac{\partial h_1}{\partial\theta_1}\frac{\partial h_2}{\partial\theta_1}\sigma_1^2 + \left\{\frac{\partial h_1}{\partial\theta_1}\frac{\partial h_2}{\partial\theta_2} + \frac{\partial h_1}{\partial\theta_2}\frac{\partial h_2}{\partial\theta_1}\right\}\sigma_{12} + \frac{\partial h_1}{\partial\theta_2}\frac{\partial h_2}{\partial\theta_2}\sigma_2^2.$$

$$\tag{A.26}$$

Tables

Table B.1 Standard normal upper-tail probabilities, $P[Z > z] = 1 - \Phi(z)$

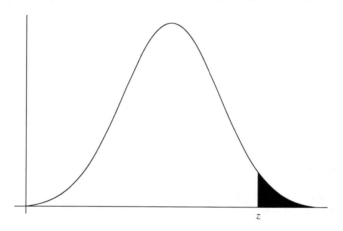

z	0.00	0.01	0.02	0.03	0.04	0.05	0.06	0.07	0.08	0.09
0.0	0.5000	0.4960	0.4920	0.4880	0.4840	0.4801	0.4761	0.4721	0.4681	0.4641
0.1	0.4602	0.4562	0.4522	0.4483	0.4443	0.4404	0.4364	0.4325	0.4286	0.4247
0.2	0.4207	0.4168	0.4129	0.4090	0.4052	0.4013	0.3974	0.3936	0.3897	0.3859
0.3	0.3821	0.3783	0.3745	0.3707	0.3669	0.3632	0.3594	0.3557	0.3520	0.3483
0.4	0.3446	0.3409	0.3372	0.3336	0.3300	0.3264	0.3228	0.3192	0.3156	0.3121
0.5	0.3085	0.3050	0.3015	0.2981	0.2946	0.2912	0.2877	0.2843	0.2810	0.2776
0.6	0.2743	0.2709	0.2676	0.2643	0.2611	0.2578	0.2546	0.2514	0.2483	0.2451
0.7	0.2420	0.2389	0.2358	0.2327	0.2296	0.2266	0.2236	0.2206	0.2177	0.2148

Continued

Analyzing Environmental Data   W. W. Piegorsch and A. J. Bailer
© 2005 John Wiley & Sons, Ltd   ISBN: 0-470-84836-7 (HB)

Table B.1   Continued

z	0.00	0.01	0.02	0.03	0.04	0.05	0.06	0.07	0.08	0.09
0.8	0.2119	0.2090	0.2061	0.2033	0.2005	0.1977	0.1949	0.1922	0.1894	0.1867
0.9	0.1841	0.1814	0.1788	0.1762	0.1736	0.1711	0.1685	0.1660	0.1635	0.1611
1.0	0.1587	0.1562	0.1539	0.1515	0.1492	0.1469	0.1446	0.1423	0.1401	0.1379
1.1	0.1357	0.1335	0.1314	0.1292	0.1271	0.1251	0.1230	0.1210	0.1190	0.1170
1.2	0.1151	0.1131	0.1112	0.1093	0.1075	0.1056	0.1038	0.1020	0.1003	0.0985
1.3	0.0968	0.0951	0.0934	0.0918	0.0901	0.0885	0.0869	0.0853	0.0838	0.0823
1.4	0.0808	0.0793	0.0778	0.0764	0.0749	0.0735	0.0721	0.0708	0.0694	0.0681
1.5	0.0668	0.0655	0.0643	0.0630	0.0618	0.0606	0.0594	0.0582	0.0571	0.0559
1.6	0.0548	0.0537	0.0526	0.0516	0.0505	0.0495	0.0485	0.0475	0.0465	0.0455
1.7	0.0446	0.0436	0.0427	0.0418	0.0409	0.0401	0.0392	0.0384	0.0375	0.0367
1.8	0.0359	0.0351	0.0344	0.0336	0.0329	0.0322	0.0314	0.0307	0.0301	0.0294
1.9	0.0287	0.0281	0.0274	0.0268	0.0262	0.0256	0.0250	0.0244	0.0239	0.0233
2.0	0.0228	0.0222	0.0217	0.0212	0.0207	0.0202	0.0197	0.0192	0.0188	0.0183
2.1	0.0179	0.0174	0.0170	0.0166	0.0162	0.0158	0.0154	0.0150	0.0146	0.0143
2.2	0.0139	0.0136	0.0132	0.0129	0.0125	0.0122	0.0119	0.0116	0.0113	0.0110
2.3	0.0107	0.0104	0.0102	0.0099	0.0096	0.0094	0.0091	0.0089	0.0087	0.0084
2.4	0.0082	0.0080	0.0078	0.0075	0.0073	0.0071	0.0069	0.0068	0.0066	0.0064
2.5	0.0062	0.0060	0.0059	0.0057	0.0055	0.0054	0.0052	0.0051	0.0049	0.0048
2.6	0.0047	0.0045	0.0044	0.0043	0.0041	0.0040	0.0039	0.0038	0.0037	0.0036
2.7	0.0035	0.0034	0.0033	0.0032	0.0031	0.0030	0.0029	0.0028	0.0027	0.0026
2.8	0.0026	0.0025	0.0024	0.0023	0.0023	0.0022	0.0021	0.0021	0.0020	0.0019
2.9	0.0019	0.0018	0.0018	0.0017	0.0016	0.0016	0.0015	0.0015	0.0014	0.0014
3.0	0.0013	0.0013	0.0013	0.0012	0.0012	0.0011	0.0011	0.0011	0.0010	0.0010
3.1	0.0010	0.0009	0.0009	0.0009	0.0008	0.0008	0.0008	0.0008	0.0007	0.0007
3.2	0.0007	0.0007	0.0006	0.0006	0.0006	0.0006	0.0006	0.0005	0.0005	0.0005
3.3	0.0005	0.0005	0.0005	0.0004	0.0004	0.0004	0.0004	0.0004	0.0004	0.0003
3.4	0.0003	0.0003	0.0003	0.0003	0.0003	0.0003	0.0003	0.0003	0.0003	0.0002
3.5	0.0002	0.0002	0.0002	0.0002	0.0002	0.0002	0.0002	0.0002	0.0002	0.0002
3.6	0.0002	0.0002	0.0001	0.0001	0.0001	0.0001	0.0001	0.0001	0.0001	0.0001

Table B.2 Student's t-distribution upper-α critical points, $t_\alpha(\text{df})$

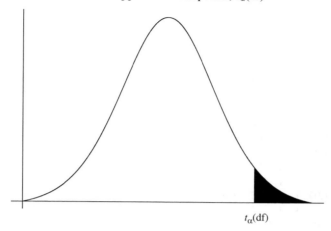

$t_\alpha(\text{df})$

α

df	0.20	0.15	0.10	0.05	0.025	0.02	0.01	0.005	0.001	0.0005
1	1.3764	1.9626	3.0777	6.3138	12.706	15.895	31.821	63.657	318.31	636.62
2	1.0607	1.3862	1.8856	2.9200	4.3027	4.8487	6.9646	9.9248	22.327	31.599
3	0.9785	1.2498	1.6377	2.3534	3.1824	3.4819	4.5407	5.8409	10.215	12.924
4	0.9410	1.1896	1.5332	2.1318	2.7764	2.9985	3.7469	4.6041	7.1732	8.6103
5	0.9195	1.1558	1.4759	2.0150	2.5706	2.7565	3.3649	4.0321	5.8934	6.8688
6	0.9057	1.1342	1.4398	1.9432	2.4469	2.6122	3.1427	3.7074	5.2076	5.9588
7	0.8960	1.1192	1.4149	1.8946	2.3646	2.5168	2.9980	3.4995	4.7853	5.4079
8	0.8889	1.1081	1.3968	1.8595	2.3060	2.4490	2.8965	3.3554	4.5008	5.0413
9	0.8834	1.0997	1.3830	1.8331	2.2622	2.3984	2.8214	3.2498	4.2968	4.7809
10	0.8791	1.0931	1.3722	1.8125	2.2281	2.3593	2.7638	3.1693	4.1437	4.5869
11	0.8755	1.0877	1.3634	1.7959	2.2010	2.3281	2.7181	3.1058	4.0247	4.4370
12	0.8726	1.0832	1.3562	1.7823	2.1788	2.3027	2.6810	3.0545	3.9296	4.3178
13	0.8702	1.0795	1.3502	1.7709	2.1604	2.2816	2.6503	3.0123	3.8520	4.2208
14	0.8681	1.0763	1.3450	1.7613	2.1448	2.2638	2.6245	2.9768	3.7874	4.1405
15	0.8662	1.0735	1.3406	1.7531	2.1314	2.2485	2.6025	2.9467	3.7328	4.0728
16	0.8647	1.0711	1.3368	1.7459	2.1199	2.2354	2.5835	2.9208	3.6862	4.0150
17	0.8633	1.0690	1.3334	1.7396	2.1098	2.2238	2.5669	2.8982	3.6458	3.9651
18	0.8620	1.0672	1.3304	1.7341	2.1009	2.2137	2.5524	2.8784	3.6105	3.9216
19	0.8610	1.0655	1.3277	1.7291	2.0930	2.2047	2.5395	2.8609	3.5794	3.8834
20	0.8600	1.0640	1.3253	1.7247	2.0860	2.1967	2.5280	2.8453	3.5518	3.8495
21	0.8591	1.0627	1.3232	1.7207	2.0796	2.1894	2.5176	2.8314	3.5272	3.8193
22	0.8583	1.0614	1.3212	1.7171	2.0739	2.1829	2.5083	2.8188	3.5050	3.7921
23	0.8575	1.0603	1.3195	1.7139	2.0687	2.1770	2.4999	2.8073	3.4850	3.7676
24	0.8569	1.0593	1.3178	1.7109	2.0639	2.1715	2.4922	2.7969	3.4668	3.7454
25	0.8562	1.0584	1.3163	1.7081	2.0595	2.1666	2.4851	2.7874	3.4502	3.7251
26	0.8557	1.0575	1.3150	1.7056	2.0555	2.1620	2.4786	2.7787	3.4350	3.7066

Continued

Table B.2   Continued

df	0.20	0.15	0.10	0.05	0.025	0.02	0.01	0.005	0.001	0.0005
					α					
27	0.8551	1.0567	1.3137	1.7033	2.0518	2.1578	2.4727	2.7707	3.4210	3.6896
28	0.8546	1.0560	1.3125	1.7011	2.0484	2.1539	2.4671	2.7633	3.4082	3.6739
29	0.8542	1.0553	1.3114	1.6991	2.0452	2.1503	2.4620	2.7564	3.3962	3.6594
30	0.8538	1.0547	1.3104	1.6973	2.0423	2.1470	2.4573	2.7500	3.3852	3.6460
31	0.8534	1.0541	1.3095	1.6955	2.0395	2.1438	2.4528	2.7440	3.3749	3.6335
32	0.8530	1.0535	1.3086	1.6939	2.0369	2.1409	2.4487	2.7385	3.3653	3.6218
33	0.8526	1.0530	1.3077	1.6924	2.0345	2.1382	2.4448	2.7333	3.3563	3.6109
34	0.8523	1.0525	1.3070	1.6909	2.0322	2.1356	2.4411	2.7284	3.3479	3.6007
35	0.8520	1.0520	1.3062	1.6896	2.0301	2.1332	2.4377	2.7238	3.3400	3.5911
40	0.8507	1.0500	1.3031	1.6839	2.0211	2.1229	2.4233	2.7045	3.3069	3.5510
50	0.8489	1.0473	1.2987	1.6759	2.0086	2.1087	2.4033	2.6778	3.2614	3.4960
60	0.8477	1.0455	1.2958	1.6706	2.0003	2.0994	2.3901	2.6603	3.2317	3.4602
70	0.8468	1.0442	1.2938	1.6669	1.9944	2.0927	2.3808	2.6479	3.2108	3.4350
80	0.8461	1.0432	1.2922	1.6641	1.9901	2.0878	2.3739	2.6387	3.1953	3.4163
90	0.8456	1.0424	1.2910	1.6620	1.9867	2.0839	2.3685	2.6316	3.1833	3.4019
100	0.8452	1.0418	1.2901	1.6602	1.9840	2.0809	2.3642	2.6259	3.1737	3.3905
150	0.8440	1.0400	1.2872	1.6551	1.9759	2.0718	2.3515	2.6090	3.1455	3.3566
200	0.8434	1.0391	1.2858	1.6525	1.9719	2.0672	2.3451	2.6006	3.1315	3.3398
∞	0.8416	1.0364	1.2816	1.6449	1.9600	2.0537	2.3263	2.5758	3.0902	3.2905

Table B.3 χ^2 distribution upper-α critical points, $\chi_\alpha^2(df)$

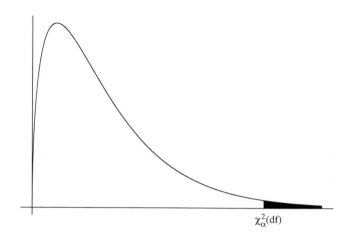

$\chi_\alpha^2(df)$

					α					
df	0.20	0.15	0.10	0.05	0.025	0.02	0.01	0.005	0.001	0.0005
1	1.642	2.072	2.706	3.841	5.024	5.412	6.635	7.879	10.828	12.116
2	3.219	3.794	4.605	5.991	7.378	7.824	9.210	10.597	13.816	15.202
3	4.642	5.317	6.251	7.815	9.348	9.837	11.345	12.838	16.266	17.730
4	5.989	6.745	7.779	9.488	11.143	11.668	13.277	14.860	18.467	19.997
5	7.289	8.115	9.236	11.070	12.833	13.388	15.086	16.750	20.515	22.105
6	8.558	9.446	10.645	12.592	14.449	15.033	16.812	18.548	22.458	24.103
7	9.803	10.748	12.017	14.067	16.013	16.622	18.475	20.278	24.322	26.018
8	11.030	12.027	13.362	15.507	17.535	18.168	20.090	21.955	26.124	27.868
9	12.242	13.288	14.684	16.919	19.023	19.679	21.666	23.589	27.877	29.666
10	13.442	14.534	15.987	18.307	20.483	21.161	23.209	25.188	29.588	31.420
11	14.631	15.767	17.275	19.675	21.920	22.618	24.725	26.757	31.264	33.137
12	15.812	16.989	18.549	21.026	23.337	24.054	26.217	28.300	32.909	34.821
13	16.985	18.202	19.812	22.362	24.736	25.472	27.688	29.819	34.528	36.478
14	18.151	19.406	21.064	23.685	26.119	26.873	29.141	31.319	36.123	38.109
15	19.311	20.603	22.307	24.996	27.488	28.259	30.578	32.801	37.697	39.719
16	20.465	21.793	23.542	26.296	28.845	29.633	32.000	34.267	39.252	41.308
17	21.615	22.977	24.769	27.587	30.191	30.995	33.409	35.718	40.790	42.879
18	22.760	24.155	25.989	28.869	31.526	32.346	34.805	37.156	42.312	44.434
19	23.900	25.329	27.204	30.144	32.852	33.687	36.191	38.582	43.820	45.973
20	25.038	26.498	28.412	31.410	34.170	35.020	37.566	39.997	45.315	47.498
21	26.171	27.662	29.615	32.671	35.479	36.343	38.932	41.401	46.797	49.011
22	27.301	28.822	30.813	33.924	36.781	37.659	40.289	42.796	48.268	50.511
23	28.429	29.979	32.007	35.172	38.076	38.968	41.638	44.181	49.728	52.000
24	29.553	31.132	33.196	36.415	39.364	40.270	42.980	45.559	51.179	53.479
25	30.675	32.282	34.382	37.652	40.646	41.566	44.314	46.928	52.620	54.947

Continued

Table B.3 Continued

					α					
df	0.20	0.15	0.10	0.05	0.025	0.02	0.01	0.005	0.001	0.0005
26	31.795	33.429	35.563	38.885	41.923	42.856	45.642	48.290	54.052	56.407
27	32.912	34.574	36.741	40.113	43.195	44.140	46.963	49.645	55.476	57.858
28	34.027	35.715	37.916	41.337	44.461	45.419	48.278	50.993	56.892	59.300
29	35.139	36.854	39.087	42.557	45.722	46.693	49.588	52.336	58.301	60.735
30	36.250	37.990	40.256	43.773	46.979	47.962	50.892	53.672	59.703	62.162
31	37.359	39.124	41.422	44.985	48.232	49.226	52.191	55.003	61.098	63.582
32	38.466	40.256	42.585	46.194	49.480	50.487	53.486	56.328	62.487	64.995
33	39.572	41.386	43.745	47.400	50.725	51.743	54.776	57.648	63.870	66.403
34	40.676	42.514	44.903	48.602	51.966	52.995	56.061	58.964	65.247	67.803
35	41.778	43.640	46.059	49.802	53.203	54.244	57.342	60.275	66.619	69.199
40	47.269	49.244	51.805	55.758	59.342	60.436	63.691	66.766	73.402	76.095
50	58.164	60.346	63.167	67.505	71.420	72.613	76.154	79.490	86.661	89.561
60	68.972	71.341	74.397	79.082	83.298	84.580	88.379	91.952	99.607	102.70
70	79.715	82.255	85.527	90.531	95.023	96.388	100.43	104.22	112.32	115.58
80	90.405	93.106	96.578	101.88	106.63	108.07	112.33	116.32	124.84	128.26
90	101.05	103.90	107.57	113.15	118.14	119.65	124.12	128.30	137.21	140.78
100	111.67	114.66	118.50	124.34	129.56	131.14	135.81	140.17	149.45	153.17
120	132.81	136.06	140.23	146.57	152.21	153.92	158.95	163.65	173.62	177.60
150	164.35	167.96	172.58	179.58	185.80	187.68	193.21	198.36	209.27	213.61
200	216.61	220.74	226.02	233.99	241.06	243.19	249.45	255.26	267.54	272.42

References

Abbott, W. S. (1925). A method of computing the effectiveness of an insecticide. *Journal of Economic Entomology* **18**, 265–267.

Adler, R. J. (2002). Random fields. In A. H. El-Shaarawi and W. W. Piegorsch (eds), *Encyclopedia of Environmetrics*, Vol. 3, pp. 1677–1678. Chichester: John Wiley & Sons.

Agresti, A. (1990). *Categorical Data Analysis*. New York: John Wiley & Sons.

Agresti, A. (1996). *An Introduction to Categorical Data Analysis*. New York: John Wiley & Sons.

Ahn, H., and Kodell, R. L. (1998). Analysis of long-term carcinogenicity studies. In S.-C. Chow and J.-P. Liu (eds), *Design and Analysis of Animal Studies in Pharmaceutical Development*, pp. 259–289. New York: Marcel Dekker.

Akaike, H. (1973). Information theory and an extension of the maximum likelihood principle. In B. N. Petrov and B. Csáki (eds), *Proceedings of the Second International Symposium on Information Theory*, pp. 267–281. Budapest: Akadémiai Kiadó.

Al-Saidy, O. M., Piegorsch, W. W., West, R. W., and Nitcheva, D. K. (2003). Confidence bands for low-dose risk estimation with quantal response data. *Biometrics* **59**, 1056–1062.

Andrews, D. F., and Hertzberg, A. M. (1985). *Data: A Collection of Problems from Many Fields for the Student and Research Worker*. New York: Springer-Verlag.

Anonymous (1949). News. *Mathematical Tables and Other Aids to Computation* **3**, 544–547.

Armitage, P. (1955). Tests for linear trends in proportions and frequencies. *Biometrics* **11**, 375–386.

Armitage, P., and Doll, R. (1954). The age distribution of cancer and a multi-stage theory of carcinogenesis. *British Journal of Cancer* **8**, 1–12.

Ashraf, W., and Jaffar, M. (1997). Concentrations of selected metals in scalp hair of an occupationally exposed population segment of Pakistan. *International Journal of Environmental Studies, Section A* **51**, 313–321.

Astuti, E. T., and Yanagawa, T. (2002). Trend test for count data with extra-Poisson variability. *Biometrics* **58**, 398–402.

Babu, G. J., and Singh, K. (1983). Inference on means using the bootstrap. *Annals of Statistics* **11**, 999–1003.

Bacastow, R. B., Keeling, C. D., and Whorf, T. P. (1985). Seasonal amplitude increase in atmospheric CO_2 concentration at Mauna Loa, Hawaii, 1959–82. *Journal of Geophysical Research, Series D – Atmospheres* **90**, 10 529–10 540.

Analyzing Environmental Data   W. W. Piegorsch and A. J. Bailer
© 2005 John Wiley & Sons, Ltd   ISBN: 0-470-84836-7 (HB)

Bailer, A. J., and Dankovic, D. A. (1997). An introduction to the use of physiologically based pharmacokinetic models in risk assessment. *Statistical Methods in Medical Research* **6**, 341–358.

Bailer, A. J., and Oris, J. T. (1997). Estimating inhibition concentrations for different response scales using generalized linear models. *Environmental Toxicology and Chemistry* **16**, 1554–1560.

Bailer, A. J., and Oris, J. T. (2000). Defining the baseline for inhibition concentration calculations for hormetic hazards. *Journal of Applied Toxicology* **20**, 121–125.

Bailer, A. J., and Piegorsch, W. W. (2000). Quantitative potency estimation to measure risk with bio-environmental hazards. In P. K. Sen and C. R. Rao (eds), *Handbook of Statistics Volume 18: Bioenvironmental and Public Health Statistics*, pp. 441–463. Amsterdam: North-Holland/Elsevier.

Bailer, A. J., and Portier, C. J. (1988). Effects of treatment-induced mortality and tumor-induced mortality on tests for carcinogenicity in small samples. *Biometrics* **44**, 417–431.

Bailer, A. J., and Portier, C. J. (1990). A note on fitting one-compartment models: Non-linear least squares versus linear least squares using transformed data. *Journal of Applied Toxicology* **10**, 303–306.

Bailer, A. J., and See, K. (1998). Individual-based risk estimation for count responses. *Environmental Toxicology and Chemistry* **17**, 530–533.

Bailer, A. J., Walker, S., and Venis, K. J. (2000). Estimating and testing bioconcentration factors. *Environmental Toxicology and Chemistry* **19**, 2338–2340.

Barnett, V. (1999). Ranked set sample design for environmental investigations. *Environmental and Ecological Statistics* **6**, 59–74.

Barnett, V. (2004). *Environmental Statistics: Methods and Applications*. Chichester: John Wiley & Sons.

Bartlett, M. S. (1946). On the theoretical specification and sampling properties of autocorrelated time-series. *Journal of the Royal Statistical Society* (Supplement) **8**, 27–41.

Bartlett, M. S. (1955). *Stochastic Processes*. Cambridge: Cambridge University Press.

Bartlett, M. S. (1964). The spectral analysis of two-dimensional point processes. *Biometrika* **51**, 299–311.

Bates, D. M., and Watts, D. G. (1988). *Nonlinear Regression Analysis and Its Applications*. New York: John Wiley & Sons.

Becka, M. (2002). Compartmental analysis. In A. H. El-Shaarawi and W. W. Piegorsch (eds), *Encyclopedia of Environmetrics*, Vol. 1, pp. 371–381. Chichester: John Wiley & Sons.

Becka, M., Bolt, H. M., and Urfer, W. (1993). Statistical evaluation of toxicokinetic data. *Environmetrics* **4**, 311–322.

Becker, R., Nikolova, T., Wolff, I., Lovell, D., Huttner, E., and Foth, H. (2001). Frequency of HPRT mutants in humans exposed to vinyl chloride via an environmental accident. *Mutation Research* **494**, 87–96.

Begg, C. B., and Mazumdar, M. (1994). Operating characteristics of a rank correlation test for publication bias. *Biometrics* **50**, 1088–1101.

Beirlant, J., and Matthys, G. (2002). Generalized extreme value distribution. In A. H. El-Shaarawi and W. W. Piegorsch (eds), *Encyclopedia of Environmetrics*, Vol. 2, pp. 863–869. Chichester: John Wiley & Sons.

Bellio, R., Jensen, J. E., and Seiden, P. (2000). Applications of likelihood asymptotics for nonlinear regression in herbicide bioassays. *Biometrics* **56**, 1204–1212.

Belsley, D. A., Kuh, E., and Welsch, R. E. (1980). *Regression Diagnostics: Identifying Influential Data and Sources of Collinearity*. New York: John Wiley & Sons.

Berke, O. (1999). Estimation and prediction in the spatial linear model. *Water, Air, and Soil Pollution* **110**, 215–237.

Besag, J. E., and Diggle, P. J. (1977). Simple Monte Carlo tests for spatial pattern. *Applied Statistics* **26**, 327–333.

Bhattacharya, R. N. (2002). Stochastic process. In A. H. El-Shaarawi and W. W. Piegorsch (eds), *Encyclopedia of Environmetrics*, Vol. 4, pp. 2137–2142. Chichester: John Wiley & Sons.

Bickel, P. J., and Doksum, K. A. (2001). *Mathematical Statistics: Basic Ideas and Selected Topics, Volume I*, 2nd edn. Upper Saddle River, NJ: Prentice Hall.

Birge, R. T. (1932). The calculation of errors by the method of least squares. *Physical Review* **16**, 1–32.

Bliss, C. I. (1935). The calculation of dosage-mortality curves. *Annals of Applied Biology* **22**, 134–167.

Blood, E. R., Phillips, J. S., Calhoun, D., and Edwards, D. (1997). The role of the Floridan aquifer in depressional wetlands hydrodynamics and hydroperiod. In K. J. Hatcher (ed.), *Proceedings of the Georgia Water Resources Conference*, pp. 273–279. Athens, GA: Institute of Natural Resources, University of Georgia.

Bloomfield, P. (2000). *Fourier Analysis of Time Series: An Introduction*, 2nd edn. New York: John Wiley & Sons.

Bonferroni, C. E. (1936). Teoria statistica delle classi e calcolo delle probabilità. *Pubblicazioni del R. Istituto Superiore di Scienze Economiche e Commerciali di Firenze* **8**, 3–62.

Borchers, D. L., Buckland, S. T., and Zucchini, W. (2002). *Estimating Animal Abundance: Closed Populations*. New York: Springer-Verlag.

Borwein, P., and Erdélyi, T. (1995). *Polynomials and Polynomial Inequalities*. New York: Springer-Verlag.

Box, G. E. P., and Cox, D. R. (1964). An analysis of transformations (with discussion). *Journal of the Royal Statistical Society, Series B* **26**, 211–252.

Box, G. E. P., and Tiao, G. C. (1975). Intervention analysis with applications to economic and environmental problems. *Journal of the American Statistical Association* **70**, 70–79.

Box, G. E. P., Jenkins, G. M., and Reinsel, G. C. (1994). *Time Series Analysis: Forecasting and Control*, 3rd edn. Englewood Cliffs, NJ: Prentice Hall.

Bradley, R. A., and Srivastava, S. S. (1979). Correlation in polynomial regression. *American Statistician* **33**, 11–14.

Braña, F., and Ji, X. (2000). Influence of incubation temperature on morphology, locomotor performance, and early growth of hatchling wall lizards (*Podarcis muralis*). *Journal of Experimental Zoology* **286**, 422–433.

Brillinger, D. R. (1994). Trend analysis: Time series and point process problems. *Environmetrics* **5**, 1–19.

Brockwell, P. J., and Davis, R. A. (1991). *Time Series: Theory and Methods*, 2nd edn. New York: Springer-Verlag.

Brown, C., and Koziol, J. (1983). Statistical aspects of the estimation of human risk from suspected environmental carcinogens. *SIAM Review* **25**, 151–181.

Brus, D. J., Jansen, M. J. W., and De Gruijter, J. J. (2002). Optimizing two- and three-stage designs for spatial inventories of natural resources by simulated annealing. *Environmental and Ecological Statistics* **9**, 71–88.

Buckland, S. T. (1984). Monte Carlo confidence intervals. *Biometrics* **40**, 811–817.

Buhlman, P. (2002). Time series. In A. H. El-Shaarawi and W. W. Piegorsch (eds), *Encyclopedia of Environmetrics*, Vol. 4, pp. 2187–2202. Chichester: John Wiley & Sons.

Buonaccorsi, J. P. (2002). Fieller's theorem. In A. H. El-Shaarawi and W. W. Piegorsch (eds), *Encyclopedia of Environmetrics*, Vol. 2, pp. 773–775. Chichester: John Wiley & Sons.

Burkhart, J. G., Helgen, J. C., Fort, D. J., Gallagher, K., Bowers, D., Propst, T. L., Gernes, M., Magner, J., Shelby, M. D., and Lucier, G. (1998). Induction of mortality and malformation in *Xenopus laevis* embryos by water sources associated with field frog deformities. *Environmental Health Perspectives* **106**, 841–848.

Calabrese, E. J., and Baldwin, L. A. (1994). A toxicological basis to derive a generic inter-species uncertainty factor. *Environmental Health Perspectives* **102**, 14–17.

Campbell, M. J., and Walker, A. M. (1977). A survey of statistical work on the Mackenzie River series of annual Canadian lynx trappings for the years 1821–1934 and a new analysis. *Journal of the Royal Statistical Society, Series A* **140**, 411–431.

Campolongo, F., Saltelli, A., Sørensen, T., and Tarantola, S. (2000). Hitchhiker's guide to sensitivity analysis. In A. Saltelli, K. Chan, and E. M. Scott (eds), *Sensitivity Analysis*, pp. 15–47. New York: John Wiley & Sons.

Carlin, B. P., and Louis, T. A. (2000). *Bayes and Empirical Bayes Methods for Data Analysis*, 2nd edn. Boca Raton, FL: Chapman & Hall/CRC Press.

Carlstein, E., Müller, H.-G., and Siegmund, D. (eds) (1995). *Change-Point Problems*. Hayward, CA: Institute of Mathematical Statistics.

Carr, G. J., and Gorelick, N. J. (1995). Statistical design and analysis of mutation studies in transgenic mice. *Environmental and Molecular Mutagenesis* **25**, 246–255.

Carroll, R. J., and Ruppert, D. (1988). *Transformation and Weighting in Regression*. New York: Chapman & Hall.

Carroll, S. S. (1998). Modelling abiotic indicators when obtaining spatial predictions of species richness. *Environmental and Ecological Statistics* **5**, 257–276.

Carrothers, T. J., Thompson, K. M., and Burmaster, D. E. (2002). Risk assessment, probabilistic. In A. H. El-Shaarawi and W. W. Piegorsch (eds), *Encyclopedia of Environmetrics*, Vol. 3, pp. 1833–1837. Chichester: John Wiley & Sons.

Casella, G., and Berger, R. L. (2002). *Statistical Inference*, 2nd edn. Pacific Grove, CA: Duxbury.

Chambers, R. (2002). Model-based inference. In A. H. El-Shaarawi and W. W. Piegorsch (eds), *Encyclopedia of Environmetrics*, Vol. 3, pp. 1284–1286. Chichester: John Wiley & Sons.

Chan, K., Tarantola, S., and Saltelli, A. (2000). Variance-based methods. In A. Saltelli, K. Chan, and E. M. Scott (eds), *Sensitivity Analysis*, pp. 167–197. New York: John Wiley & Sons.

Chand, N., and Hoel, D. G. (1974). A comparison of models for determining safe levels of environmental agents. In F. Proschan and R. Serfling (eds), *Reliability and Biometry: Statistical Analysis of Lifelength*, pp. 681–700. Philadelphia: Society for Industrial and Applied Mathematics.

Chapman, D. G. (1948). A mathematical study of confidence limits of salmon populations calculated from sample tag ratios. *International Pacific Salmon Fisheries Commission Bulletin* **II**, 69–85.

Chapman, D. G. (1951). Some properties of the hypergeometric distribution with application to zoological sample censuses. *University of California Publications in Statistics* **1**, 131–160.

Chapman, P. M., Caldwell, R. S., and Chapman, P. F. (1996). A warning: NOECs are inappropriate for regulatory use. *Environmental Toxicology and Chemistry* **15**, 77–79.

Chen, D. G., and Pounds, J. G. (1998). A nonlinear isobologram model with Box–Cox transformation to both sides for chemical mixtures. *Environmental Health Perspectives* **106**, Supplement 6, 1367–1371.

Chen, J. J. (2002). Dose-response modeling for clustered data. In A. H. El-Shaarawi and W. W. Piegorsch (eds), *Encyclopedia of Environmetrics*, Vol. 1, pp. 569–574. Chichester: John Wiley & Sons.

Christensen, K. M., and Whitham, T. G. (1993). Impact of insect herbivores on competition between birds and mammals for pinyon pine seeds. *Ecology* **74**, 2270–2278.

Christensen, R. (1996). *Analysis of Variance, Design and Regression*. London: Chapman & Hall.

Christman, M. C. (2002). Hansen–Hurwitz estimator. In A. H. El-Shaarawi and W. W. Piegorsch (eds), *Encyclopedia of Environmetrics*, Vol. 2, pp. 981–982. Chichester: John Wiley & Sons.

Clarke, G., and Kempson, R. (1996). *Introduction to the Design and Analysis of Experiments.* London: Hodder & Stoughton.

Cliff, A. D., and Ord, J. K. (1981). *Spatial Processes: Models & Applications.* London: Pion.

Cocchi, D., and Trivisano, C. (2002). Ozone. In A. H. El-Shaarawi and W. W. Piegorsch (eds), *Encyclopedia of Environmetrics*, Vol. 3, pp. 1518–1523. Chichester: John Wiley & Sons.

Cochran, W. G. (1937). Problems arising in the analysis of a series of similar experiments. *Journal of the Royal Statistical Society* (Supplement) **4**, 102–118.

Cochran, W. G. (1954). Some methods for strengthening the common χ^2 tests. *Biometrics* **10**, 417–451.

Cochran, W. G. (1977). *Sampling Techniques*, 3rd edn. New York: John Wiley & Sons.

Cohen, J. (1969). *Statistical Power Analysis for the Behavioral Sciences.* New York: Academic Press.

Coles, S. (2001). *An Introduction to Statistical Modelling of Extreme Values.* New York: Springer-Verlag.

Collings, B. J., and Margolin, B. H. (1985). Testing goodness of fit for the Poisson assumption when observations are not identically distributed. *Journal of the American Statistical Association* **80**, 411–418.

Cooke, R. M., and van Noortwijk, J. M. (2000). Graphical methods. In A. Saltelli, K. Chan, and E. M. Scott (eds), *Sensitivity Analysis*, pp. 245–264. New York: John Wiley & Sons.

Cooley, J. W., and Tukey, J. W. (1965). An algorithm for the machine calculation of complex Fourier series. *Mathematics of Computation* **19**, 297–301.

Cordeiro, G. M. (2004). On Pearson's residuals in generalized linear models. *Statistics and Probability Letters* **66**, 213–219.

Cordeiro, G. M., and McCullagh, P. (1991). Bias correction in generalized linear models. *Journal of the Royal Statistical Society, Series B* **53**, 629–643.

Cormack, R. M. (1964). Estimates of survival from the sighting of marked animals. *Biometrika* **51**, 429–438.

Cormack, R. M. (1981). Loglinear models for capture-recapture experiments on open populations. In R. W. Hiorns and D. Cooke (eds), *The Mathematical Theory of the Dynamics of Biological Populations II*, pp. 217–235. London: Academic Press.

Correll, R. L. (2001). The use of composite sampling in contaminated sites – a case study. *Environmental and Ecological Statistics* **8**, 185–200.

Coull, B. A., and Ryan, L. M. (2002). Biological assay. In A. H. El-Shaarawi and W. W. Piegorsch (eds), *Encyclopedia of Environmetrics*, Vol. 1, pp. 189–192. Chichester: John Wiley & Sons.

Cox, C. (1990). Fieller's theorem, the likelihood and the delta method. *Biometrics* **46**, 709–718.

Cox, C. (1998). Delta method. In P. Armitage and T. Colton (eds), *Encyclopedia of Biostatistics*, Vol. 2, pp. 1125–1127. Chichester: John Wiley & Sons.

Cox, C., and Ma, G. (1995). Asymptotic confidence bands for generalized nonlinear regression models. *Biometrics* **51**, 142–150.

Cox, D. R. (1958). The regression analysis of binary sequences (with discussion). *Journal of the Royal Statistical Society, Series B* **20**, 215–242.

Cox, D. R. (1961). Tests of separate families of hypotheses. In J. Neyman (ed.), *Proceedings of the Fourth Berkeley Symposium on Mathematical Statistics and Probability*, pp. 105–123. Berkeley: University of California Press.

Cox, D. R. (1962). Further results on testing separate families of hypotheses. *Journal of the Royal Statistical Society, Series B* **24**, 406–424.

Cox, D. R. (1988). Some aspects of conditional and asymptotic inference: A review. *Sankhyā, Series A* **50**, 314–337.

Cox, D. R., and Reid, N. (2000). *The Theory of the Design of Experiments.* Boca Raton, FL: Chapman & Hall/CRC.

Cox, D. R., and Snell, E. J. (1989). *Analysis of Binary Data*, 2nd edn. London: Chapman & Hall.

Cressie, N. A. (1990). The origins of kriging. *Mathematical Geology* **22**, 239–252.

Cressie, N. A. (1993). *Statistics for Spatial Data*. New York: John Wiley & Sons.

Cressie, N. A. (2002a). Variogram. In A. H. El-Shaarawi and W. W. Piegorsch (eds), *Encyclopedia of Environmetrics*, Vol. 4, pp. 2313–2316. Chichester: John Wiley & Sons.

Cressie, N. A. (2002b). Variogram estimation. In A. H. El-Shaarawi and W. W. Piegorsch (eds), *Encyclopedia of Environmetrics*, Vol. 4, pp. 2316–2321. Chichester: John Wiley & Sons.

Cressie, N. A., and Hawkins, D. M. (1980). Robust estimation of the variogram, I. *Mathematical Geology* **1**, 115–125.

Cressie, N. A., and Lahiri, S. N. (1996). Asymptotics for REML estimation of spatial covariance parameters. *Journal of Statistical Planning and Inference* **50**, 327–341.

Cressie, N. A., and Pardo, L. (2002). Phi-divergence statistic. In A. H. El-Shaarawi and W. W. Piegorsch (eds), *Encyclopedia of Environmetrics*, Vol. 3, pp. 1551–1555. Chichester: John Wiley & Sons.

Crump, K. S. (1984). A new method for determining allowable daily intake. *Fundamental and Applied Toxicology* **4**, 854–871.

Crump, K. S. (1995). Calculation of benchmark doses from continuous data. *Risk Analysis* **15**, 79–89.

Crump, K. S. (2002). Benchmark analysis. In A. H. El-Shaarawi and W. W. Piegorsch (eds), *Encyclopedia of Environmetrics*, Vol. 1, pp. 163–170. Chichester: John Wiley & Sons.

Crump, K. S., and Howe, R. (1985). A review of methods for calculating confidence limits in low dose extrapolation. In D. B. Clayson, D. Krewski, and I. Munro (eds), *Toxicological Risk Assessment, Volume I: Biological and Statistical Criteria*, pp. 187–203. Boca Raton, FL: CRC Press.

Crump, K. S., Guess, H. A., and Deal, K. L. (1977). Confidence intervals and tests of hypotheses concerning dose response relations inferred from animal carcinogenicity data. *Biometrics* **33**, 437–451.

Crump, K., Allen, B., Faustman, E., Donison, M., Kimmel, C., and Zenich, H. (1995). The use of the benchmark dose approach in health risk assessment. Technical Report no. EPA/630/R-94/007. U.S. Environmental Protection Agency, Washington, DC.

Daae, E. B., and Ison, A. P. (1998). A simple structured model describing the growth of *Streptomyces lividans*. *Biotechnology and Bioengineering* **58**, 263–266.

Dalgård, C., Grandjean, P., Jørgensen, P. J., and Weihe, P. (1994). Mercury in the umbilical cord: Implications for risk assessment for Minamata disease. *Environmental Health Perspectives* **102**, 548–550.

Davenport, A. G., and Kopp, G. A. (2002). Natural disasters. In A. H. El-Shaarawi and W. W. Piegorsch (eds), *Encyclopedia of Environmetrics*, Vol. 3, pp. 1356–1362. Chichester: John Wiley & Sons.

Davidian, M., and Carroll, R. J. (1987). Variance function estimation. *Journal of the American Statistical Association* **82**, 1079–1091.

Davidian, M., and Giltinan, D. M. (1995). *Nonlinear Models for Repeated Measurement Data*. New York: Chapman & Hall.

Davison, A. C., and Tsai, C. L. (1992). Regression model diagnostics. *International Statistical Review* **60**, 337–353.

de Jong, P., and Penzer, J. (1998). Diagnosing shocks in time series. *Journal of the American Statistical Association* **93**, 796–806.

De Silva, H. N., Lai, C. D., and Ball, R. D. (1997). Fitting S_B distributions to fruit sizes with implications for prediction models. *Journal of Agricultural, Biological, and Environmental Statistics* **2**, 333–346.

Dean, C. B. (1992). Testing for overdispersion in Poisson and binomial regression models. *Journal of the American Statistical Association* **87**, 451–457.

DeLeve, L. D. (1996). Dinitrochlorobenzene is genotoxic by sister chromatid exchange in human skin fibroblasts. *Mutation Research* **371**, 105–108.

Dell, T. R., and Clutter, J. L. (1972). Ranked set sampling theory with order statistics background. *Biometrics* **28**, 545–553.

Demidenko, E. (2004). *Mixed Models: Theory and Applications*. Chichester: John Wiley & Sons.

Derr, R. E. (2000). Performing exact logistic regression with the SAS® System. Technical Report no. P254–25. SAS Institute Inc., Cary, NC.

DerSimonian, R., and Laird, N. (1986). Meta-analysis in clinical trials. *Controlled Clinical Trials* **7**, 177–188.

Diggle, P. J. (2003). *Statistical Analysis of Spatial Point Patterns*, 2nd edn. London: Arnold.

Diggle, P. J., Liang, K.-Y., and Zeger, S. L. (1994). *Analysis of Longitudinal Data*. Oxford: Oxford University Press.

Dillon, M. (1992). Search for efficient local kriging algorithm. *The Statistician* **41**, 383.

Dixon, P. M. (2002a). Bootstrap resampling. In A. H. El-Shaarawi and W. W. Piegorsch (eds), *Encyclopedia of Environmetrics*, Vol. 1, pp. 212–220. Chichester: John Wiley & Sons.

Dixon, P. M. (2002b). Nearest neighbor methods. In A. H. El-Shaarawi and W. W. Piegorsch (eds), *Encyclopedia of Environmetrics*, Vol. 3, pp. 1370–1383. Chichester: John Wiley & Sons.

Dixon, P. M. (2002c). Ripley's K function. In A. H. El-Shaarawi and W. W. Piegorsch (eds), *Encyclopedia of Environmetrics*, Vol. 3, pp. 1796–1803. Chichester: John Wiley & Sons.

Donnelly, K. P. (1978). Simulations to determine the variance and edge effect of total nearest-neighbour distances. In I. Hodder (ed.), *Simulations Studies in Archaeology*, pp. 91–95. Cambridge: Cambridge University Press.

Dourson, M. L., and Derosa, C. T. (1991). The use of uncertainty factors in establishing safe levels of exposure. In D. Krewski and C. Franklin (eds), *Statistics in Toxicology*, pp. 613–627. New York: Gordon and Breach.

Dourson, M. L., and Stara, J. F. (1983). Regulatory history and experimental support of uncertainty (safety) factors. *Regulatory Toxicology and Pharmacology* **3**, 224–238.

Drum, M. (2002). Mixed effects. In A. H. El-Shaarawi and W. W. Piegorsch (eds), *Encyclopedia of Environmetrics*, Vol. 3, pp. 1268–1270. Chichester: John Wiley & Sons.

Dunnick, J. K., Eustis, S. L., Piegorsch, W. W., and Miller, R. A. (1988). Respiratory tract lesions in F344/N rats and B6C3F$_1$ mice after exposure to 1,2-epoxybutane. *Toxicology* **50**, 69–82.

Dunson, D. B. (2001). Modeling of changes in tumor burden. *Journal of Agricultural, Biological, and Environmental Statistics* **6**, 38–48.

Durbin, J., and Watson, G. S. (1950). Testing for serial correlation in least squares regression. I. *Biometrika* **37**, 409–428 (corr. **38**, 177–178).

Durbin, J., and Watson, G. S. (1951). Testing for serial correlation in least squares regression. II. *Biometrika* **38**, 159–177.

Duval, S., and Tweedie, R. (2000). A nonparametric 'trim and fill' method of accounting for publication bias in meta-analysis. *Journal of the American Statistical Association* **95**, 89–98.

Ecker, M. D., and Heltshe, J. F. (1994). Geostatistical estimates of scallop abundance. In N. Lange, L. Ryan, L. Billard, D. Brillinger, L. Conquest, and J. Greenhouse (eds), *Case Studies in Biometry*, pp. 107–124. New York: John Wiley & Sons.

Eckhardt, D. A., Flipse, W. J., and Oaksford, E. T. (1989). Relation between land use and ground-water quality in the upper glacial aquifer in Nassau and Suffolk Counties, Long Island. Water-Resources Investigations Report no. 86–4142. US Geological Survey, Syosset, NY.

Edenharder, R., Ortseifen, M., Koch, M., and Wesp, H. F. (2000). Soil mutagens are airborne mutagens: variation of mutagenic activities induced in *Salmonella typhimurium* TA98 and TA100 by organic extracts of agricultural and forest soils in dependence on location and season. *Mutation Research* **472**, 23–36.

Edland, S. D., and van Belle, G. (1994). Decreased sampling costs and improved accuracy with composite sampling. In C. R. Cothern and N. P. Ross (eds), *Environmental Statistics, Assessment, and Forecasting*, pp. 29–55. Boca Raton, FL: Lewis Publishers.

Efron, B. (2000). The boostrap and modern statistics. *Journal of the American Statistical Association* **95**, 1293–1296.

Efron, B., and Hinkley, D. V. (1978). Assessing the accuracy of the maximum likelihood estimator: Observed versus expected Fisher information. *Biometrika* **65**, 457–487.

Efron, B., and Tibshirani, R. (1993). *Introduction to the Bootstrap*. New York: Chapman & Hall.

El-Shaarawi, A. H., and Hunter, J. S. (2002). Environmetrics, overview. In A. H. El-Shaarawi and W. W. Piegorsch (eds), *Encyclopedia of Environmetrics*, Vol. 2, pp. 698–702. Chichester: John Wiley & Sons.

El-Shaarawi, A. H., and Piegorsch, W. W. (eds) (2002). *Encyclopedia of Environmetrics*. Chichester: John Wiley & Sons.

Elton, C., and Nicholson, M. (1942). The ten-year cycle in numbers of lynx in Canada. *Journal of Animal Ecology* **11**, 215–244.

Esterby, S. R. (2002). Change, detecting. In A. H. El-Shaarawi and W. W. Piegorsch (eds), *Encyclopedia of Environmetrics*, Vol. 1, pp. 318–322. Chichester: John Wiley & Sons.

Evans, M., Hastings, N. A. J., and Peacock, J. B. (2000). *Statistical Distributions*, 3rd edn. New York: John Wiley & Sons.

Fassó, A., and Perri, P. F. (2002). Sensitivity analysis. In A. H. El-Shaarawi and W. W. Piegorsch (eds), *Encyclopedia of Environmetrics*, Vol. 4, pp. 1968–1982. Chichester: John Wiley & Sons.

Fieller, E. C. (1940). The biological standardization of insulin. *Journal of the Royal Statistical Society, Series B* **7**, 1–53.

Fienberg, S. E. (1972). The multiple recapture census for closed populations and incomplete 2^k contingency tables. *Biometrika* **59**, 591–603.

Finette, B. A., Sullivan, L. M., O'Neill, J. P., Nicklas, J. A., Vacek, P. M., and Albertini, R. J. (1994). Determination of *hprt* mutant frequencies in T-lymphocytes from a healthy pediatric population: statistical comparison between newborn, children and adult mutant frequencies, cloning efficiency and age. *Mutation Research* **308**, 223–232.

Finney, D. J. (1952). *Statistical Method in Biological Assay*. London: Chas. Griffin & Co.

Firth, D. (1993). Bias reduction of maximum likelihood estimates. *Biometrika* **80**, 27–38 (corr. **82**, 667).

Fisher, R. A. (1912). On an absolute criterion for fitting frequency curves. *Messenger of Mathematics* **41**, 155–160.

Fisher, R. A. (1921). On the 'probable error' of a coefficient of correlation deduced from a small sample. *Metron* **1**, 3–32.

Fisher, R. A. (1922a). On the interpretation of χ^2 from contingency tables, and the calculation of *P. Journal of the Royal Statistical Society* **85**, 87–94.

Fisher, R. A. (1922b). On the mathematical foundations of theoretical statistics. *Philosophical Transactions of the Royal Society* **222**, 309–368.

Fisher, R. A. (1929). Tests of significance in harmonic analysis. *Proceedings of the Royal Society of London. Series A* **125**, 54–59.

Fisher, R. A. (1935). The logic of inductive inference (with discussion). *Journal of the Royal Statistical Society, Series A* **98**, 39–82.

Fisher, R. A. (1948). Combining independent tests of significance. *American Statistician* **2**, 30.

Fisher, R. A., Thornton, H. G., and MacKenzie, W. A. (1922). The accuracy of the plating method of estimating the density of bacterial populations. *Journal of Applied Biology* **9**, 325–359.

Fortin, M.-J., Dale, M. R. T., and Ver Hoef, J. (2002). Spatial analysis in ecology. In A. H. El-Shaarawi and W. W. Piegorsch (eds), *Encyclopedia of Environmetrics*, Vol. 4, pp. 2051–2058. Chichester: John Wiley & Sons.

Fourier, J. B. J. (1822). *Théorie Analytique de la Chaleur*. Paris: Chez Firmin Didot.

Freedman, L. S. (2001). Confidence intervals and statistical power of the 'validation' ratio for surrogate or intermediate endpoints. *Journal of Statistical Planning and Inference* **96**, 143–153.

Freeman, M. F., and Tukey, J. W. (1950). Transformations related to the angular and the square root. *Annals of Mathematical Statistics* **21**, 607–611.

Galambos, J., and Simonelli, I. (1996). *Bonferroni-type Inequalities with Applications*. New York: Springer-Verlag.

Gandin, L. S. (1963). *Objective Analysis of Meteorological Fields*. Leningrad: GIMIZ [in Russian].

García-Soidán, P. H., Febrero-Bande, M., and González-Manteiga, W. (2004). Nonparametric kernel estimation of an isotropic variogram. *Journal of Statistical Planning and Inference* **121**, 65–92.

Gardner, R. H., O'Neill, R. V., Mankin, J. B., and Carney, J. H. (1981). A comparison of sensitivity analysis and error analysis based on a stream ecosystem model. *Ecological Modelling* **12**, 173–190.

Gaudard, M., Karson, M., Linder, E., and Sinha, D. (1999). Bayesian spatial prediction (with discussion). *Environmental and Ecological Statistics* **6**, 147–182.

Gaver, D. P., Draper, D., Goel, P. K., Greenhouse, J. B., Hedges, L. V., Morris, C. N., and Waternaux, C. (1992). *Combining Information: Statistical Issues and Opportunities for Research*. Washington, DC: National Academies Press.

Gaylor, D. W., and Kodell, R. L. (2000). Percentiles of the product of uncertainty factors for establishing probabilistic reference doses. *Risk Analysis* **20**, 245–250.

Gaylor, D. W., Kodell, R. L., Chen, J. J., and Krewski, D. (1999). A unified approach to risk assessment for cancer and noncancer endpoints based on benchmark doses and uncertainty/ safety factors. *Regulatory Toxicology and Pharmacology* **29**, 151–157.

Geary, R. C. (1954). The contiguity ratio and statistical mapping. *Incorporated Statistician* **5**, 115–127, 129–145.

Gentle, J. E. (2003). *Random Number Generation and Monte Carlo Methods*, 2nd edition. New York: Springer-Verlag.

Genton, M. G. (1998). Highly robust variogram estimation. *Mathematical Geology* **30**, 213–221.

George, V. T., and Elston, R. C. (1993). Confidence limits based on the first occurrence of an event. *Statistics in Medicine* **12**, 685–690.

Gilbert, R. O. (1987). *Statistical Methods for Environmental Pollution Monitoring*. New York: Van Nostrand Reinhold.

Givens, G. H., Smith, D. D., and Tweedie, R. L. (1997). Publication bias in meta-analysis: A Bayesian data-augmentation approach to account for issues exemplified in the passive smoking debate (with discussion). *Statistical Science* **12**, 221–250.

Glass, G. V. (1976). Primary, secondary, and meta-analysis of research. *Educational Researcher* **5**, 3–8.

Goddard, M. J., and Krewski, D. (1995). The future of mechanistic research in risk assessment: Where are we going and can we get there from here? *Toxicology* **102**, 53–70.

Gompertz, B. (1825). On the nature of the function expressive of the law of human mortality, and on a new mode of determining the value of life contingencies. *Philosophical Transactions of the Royal Society* **155**, 513–593.

Gonzalez, L., and Manly, B. F. J. (1998). Analysis of variance by randomization with small data sets. *Environmetrics* **9**, 53–65.

Gould, W. R., and Pollock, K. H. (2002). Capture-recapture methods. In A. H. El-Shaarawi and W. W. Piegorsch (eds), *Encyclopedia of Environmetrics*, Vol. 1, pp. 243–251. Chichester: John Wiley & Sons.

Gouriéroux, C., Monfort, A., and Trognon, A. (1984). Pseudo-maximum likelihood methods: theory. *Econometrica* **52**, 681–700.

Gray, H. N., and Bergbreiter, D. E. (1997). Applications of polymeric smart materials to environmental problems. *Environmental Health Perspectives* **105**, Supplement 1, 55–63.

Greenwood, J. J. D. (1996). Basic techniques. In W. J. Sutherland (ed.), *Ecological Census Techniques: A Handbook*, pp. 11–110. Cambridge: Cambridge University Press.

Gregoire, T. G. (1998). Design-based and model-based inference in survey sampling: appreciating the difference. *Canadian Journal of Forest Research* **28**, 2346–2351.

Gregoire, T. G., and Schabenberger, O. (1999). Sampling-skewed biological populations: Behavior of confidence intervals for the population total. *Ecology* **80**, 1056–1065.

Gregoire, T. G., and Valentine, H. T. (2003). Line intersect sampling: Ell-shaped transects and multiple intersections. *Environmental and Ecological Statistics* **10**, 263–279.

Gregoire, T. G., and Valentine, H. T. (2004). *Sampling Techniques for Natural and Environmental Resources*. Boca Raton, FL: Chapman & Hall/CRC Press.

Gregorczyk, A. (1998). Richards' plant growth model. *Journal of Agronomy and Crop Science* **181**, 243–247.

Griffiths, R. (2002). Risk assessment, management and uncertainties. In A. H. El-Shaarawi and W. W. Piegorsch (eds), *Encyclopedia of Environmetrics*, Vol. 3, pp. 1812–1833. Chichester: John Wiley & Sons.

Grimston, M. (2002). Nuclear risk. In A. H. El-Shaarawi and W. W. Piegorsch (eds), *Encyclopedia of Environmetrics*, Vol. 3, pp. 1425–1432. Chichester: John Wiley & Sons.

Guan, Y., Sherman, M., and Calvin, J. A. (2004). A nonparametric test for spatial isotropy using subsampling. *Journal of the American Statistical Association* **99**, 810–821.

Gurevitch, J., and Hedges, L. V. (1993). Meta-analysis: Combining the results of independent experiments. In S. M. Scheiner and J. Gurevitch (eds), *The Design and Analysis of Ecological Experiments*, pp. 378–398. New York: Chapman & Hall.

Guttorp, P. (2003). Environmental statistics – A personal view. *International Statistical Review* **71**, 169–179.

Guttorp, P., and Sampson, P. D. (1994). Methods for estimating heterogeneous spatial covariance functions with environmental applications. In G. P. Patil and C. R. Rao (eds), *Handbook of Statistics Volume 12: Environmental Statistics*, pp. 661–690. New York: North-Holland/Elsevier.

Hamilton, M. A. (1991). Estimation of the typical lethal dose in acute toxicity studies. In D. Krewski and C. Franklin (eds), *Statistics in Toxicology*, pp. 61–88. New York: Gordon and Breach.

Hampel, F. (2002). Robust inference. In A. H. El-Shaarawi and W. W. Piegorsch (eds), *Encyclopedia of Environmetrics*, Vol. 3, pp. 1865–1885. Chichester: John Wiley & Sons.

Hansen, M. M., and Hurwitz, W. N. (1943). On the theory of sampling from finite populations. *Annals of Mathematical Statistics* **14**, 333–362.

Hardy, R. J., and Thompson, S. G. (1998). Detecting and describing heterogeneity in meta-analysis. *Statistics in Medicine* **17**, 841–856.

Härkönen, K., Viitanen, T., Larsen, S. B., Bonde, J. P., ASCLEPIOS Project, and Lähdetie, J. (1999). Aneuploidy in sperm and exposure to fungicides and lifestyle factors. *Environmental and Molecular Mutagenesis* **34**, 39–46.

Hartung, J. (1999). A note on combining dependent tests of significance. *Biometrical Journal* **41**, 849–855.

Harvey, A. (2002). Trend analysis. In A. H. El-Shaarawi and W. W. Piegorsch (eds), *Encyclopedia of Environmetrics*, Vol. 4, pp. 2243–2257. Chichester: John Wiley & Sons.

Harville, D. A. (1997). *Matrix Algebra From a Statistician's Perspective*. New York: Springer-Verlag.

Haseman, J. K. (1990). Use of statistical decision rules for evaluating laboratory animal carcinogenicity studies. *Fundamental and Applied Toxicology* **14**, 637–648.

Hauck, W. W., and Donner, A. (1977). Wald's test as applied to hypotheses in logit analysis. *Journal of the American Statistical Association* **72**, 851–853.

Hedges, L. V. (1992). Modeling publication selection effects in meta-analysis. *Statistical Science* **7**, 246–255.

Hedges, L. V., and Olkin, I. (1985). *Statistical Methods for Meta-analysis*. Orlando, FL: Academic Press.

Heidel, J., and Maloney, J. (1999). When can sigmoidal data be fit to a Hill curve? *Journal of the Australian Mathematical Society, Series B – Applied Mathematics* **41**, 83–92.

Helgesen, R., and Pribble, R. (eds) (2001). *Air Quality in Minnesota: Problems and Approaches*. St. Paul, MN: Minnesota Pollution Control Agency.

Helton, J. C., and Davis, F. J. (2000). Sampling-based methods. In A. Saltelli, K. Chan, and E. M. Scott (eds), *Sensitivity Analysis*, pp. 101–153. New York: John Wiley & Sons.

Hertzberg, R. C., Rice, G., and Teuschler, L. K. (1999). Methods for health risk assessment of combustion mixtures. In S. Roberts, C. Teaf and J. Bean (eds), *Hazardous Waste Incineration: Evaluating the Human Health and Environmental Risks*, pp. 105–148. Boca Raton, FL: Lewis.

Heyde, C. C. (1994). A quasi-likelihood approach to the REML estimating equations. *Statistics & Probability Letters* **21**, 381–384.

Hilbe, J. M. (1994). Log negative binomial regression using the GENMOD procedure SAS/STAT software. *Proceedings of the SAS Users Group International Conference*, **19**, 1199–1204.

Hill, A. V. (1910). The possible effects of the aggregation the molecules of haemoglobin on its dissociation curves. *Proceedings of the Physiological Society* **40**, iv–vii.

Hinde, J. (1992). Choosing between non-nested models: a simulation approach. In L. Fahrmeir, B. Francis, R. Gilchrist, and G. Tutz (eds), *Advances in GLIM and Statistical Modelling, Proceedings of the GLIM92 Conference and the 7th International Workshop on Statistical Modelling*, pp. 119–124. New York: Springer-Verlag.

Hinde, J., and Demetrio, C. G. B. (1998). Overdispersion: Models and estimation. *Computational Statistics & Data Analysis* **27**, 151–170.

Hirji, K. F., Mehta, C. R., and Patel, N. R. (1987). Computing distributions for exact logistic regression. *Journal of the American Statistical Association* **82**, 1110–1117.

Hochberg, Y., and Tamhane, A. C. (1987). *Multiple Comparison Procedures*. New York: John Wiley & Sons.

Hocking, R. R. (1996). *Methods and Applications of Linear Models*. New York: John Wiley & Sons.

Hoel, D. G. (1983). Conditional two sample tests with historical controls. In P. K. Sen (ed.), *Contributions to Statistics: Essays in Honor of Norman L. Johnson*, pp. 229–236. Amsterdam: North-Holland.

Hoel, D. G., and Yanagawa, T. (1986). Incorporating historical controls in testing for a trend in proportions. *Journal of the American Statistical Association* **81**, 1095–1099.

Hogg, R. V., and Tanis, E. A. (2001). *Probability and Statistical Inference*, 6th edn. Upper Saddle River, NJ: Prentice Hall.

Horvitz, D. G., and Thompson, D. J. (1952). A generalization of sampling without replacement from a finite universe. *Journal of the American Statistical Association* **47**, 663–685.

Hsu, J. C. (1996). *Multiple Comparisons*. New York: Chapman & Hall.

Huang, L.-S., and Smith, R. L. (1999). Meteorologically-dependent trends in urban ozone. *Environmetrics* **10**, 103–118.

Huber, P. (1967). The behaviour of maximum likelihood estimators under nonstandard conditions. In L. LeCam and J. Neyman (eds), *Proceedings of the Fifth Berkeley Symposium on Mathematical Statistics and Probability*, Vol. 1, pp. 221–233. Berkeley: University of California Press.

Huff, J. E. (1989). Multiple-site carcinogenicity of benzene in Fischer 344 rats and B6C3F$_1$ mice. *Environmental Health Perspectives* **82**, 125–163.

Hunter, J. E., and Schmidt, F. L. (2004). *Methods of Meta-analysis: Correcting Error and Bias in Research Findings*, 2nd edn. Newbury Park, CA: Sage Publications.

Hussell, D. J. T. (1972). Factors affecting clutch size in arctic passerines. *Ecological Monographs* **42**, 317–364.

Iman, R. L., and Helton, J. C. (1988). An investigation of uncertainty and sensitivity analysis techniques for computer models. *Risk Analysis* **8**, 71–90.

Iman, R. L., and Shortencarier, M. J. (1984). A FORTRAN program and user's guide for the generation of Latin hypercube and random samples for use with computer models. Technical Report no. NUREG/CR-3624 (SAND83-2365), Sandia National Laboratories, Albuquerque, NM.

Imrey, P. B., and Simpson, D. G. (2002). Categorical data. In A. H. El-Shaarawi and W. W. Piegorsch (eds), *Encyclopedia of Environmetrics*, Vol. 1, pp. 290–312. Chichester: John Wiley & Sons.

Irwin, J. O. (1937). Statistical method applied to biological assays (with discussion). *Journal of the Royal Statistical Society* (supplement), **4**, 1–60.

Isaacson, J. D., and Zimmerman, D. L. (2000). Combining temporally correlated environmental data from two measurement systems. *Journal of Agricultural, Biological, and Environmental Statistics* **5**, 398–416.

Jandhyala, V. K., Zacks, S., and El-Shaarawi, A. H. (1999). Change-point problems and their applications: Contributions of Ian MacNeill. *Environmetrics* **10**, 657–676.

Jandhyala, V. K., Fotopoulos, S. B., and El-Shaarawi, A. H. (2002). Change-point methods. In A. H. El-Shaarawi and W. W. Piegorsch (eds), *Encyclopedia of Environmetrics*, Vol. 1, pp. 324–332. Chichester: John Wiley & Sons.

Johnson, N. L. (1978). Modified *t* tests and confidence intervals for asymmetrical populations. *Journal of the American Statistical Association* **73**, 536–544.

Johnson, N. L., Kotz, S., and Kemp, A. W. (1993). *Univariate Discrete Distributions*, 2nd edn. New York: John Wiley & Sons.

Johnson, N. L., Kotz, S., and Balakrishnan, N. (1994). *Continuous Univariate Distributions, Volume 1*, 2nd edn. New York: John Wiley & Sons.

Johnson, N. L., Kotz, S., and Balakrishnan, N. (1995). *Continuous Univariate Distributions, Volume 2*, 2nd edn. New York: John Wiley & Sons.

Johnson, N. L., Kotz, S., and Balakrishnan, N. (1997). *Discrete Multivariate Distributions*. New York: John Wiley & Sons.

Jolicoeur, P., and Heusner, A. (1986). Log-normal variation belts for growth curves. *Biometrics* **42**, 785–794.

Jolly, G. M. (1965). Explicit estimates for capture-recapture data with both death and immigration-stochastic model. *Biometrika* **52**, 225–247.

Jones, R. H., and Boadi-Boateng, F. (1991). Unequally spaced longitudinal data with AR(1) serial correlation. *Biometrics* **47**, 161–175.

Jones, R. H., Sharitz, R. R., James, S. M., and Dixon, P. M. (1994). Tree population dynamics in 7 South Carolina mixed-species forests. *Bulletin of the Torrey Botanical Club* **121**, 360–368.

Jorgensen, M. (2002). Iteratively reweighted lest squares. In A. H. El-Shaarawi and W. W. Piegorsch (eds), *Encyclopedia of Environmetrics*, Vol. 2, pp. 1084–1088. Chichester: John Wiley & Sons.

Jørgensen, B. (2002). Generalized linear models. In A. H. El-Shaarawi and W. W. Piegorsch (eds), *Encyclopedia of Environmetrics*, Vol. 2, 873–880. Chichester: John Wiley & Sons.

Journel, A. G., and Huijbregts, C. J. (1978). *Mining Geostatistics*. London: Academic Press.

Juliano, S. A. (1993). Nonlinear curve fitting: Predation and functional response curves. In S. M. Scheiner and J. Gurevitch (eds), *The Design and Analysis of Ecological Experiments*, pp. 159–182. New York: Chapman & Hall.

Kaiser, L. (1983). Unbiased estimation in line-intercept sampling. *Biometrics* **39**, 965–976.

Kalbfleisch, J. D., and Prentice, R. L. (2002). *The Statistical Analysis of Failure Time Data*, 2nd edn. New York: John Wiley & Sons.

Kalf, G. F., Renz, J. F., and Niculescu, R. (1996). *p*-benzoquinone, a reactive metabolite of benzene, prevents the processing of pre-interleukins-1α and -1β to active cytokines by inhibition of the processing enzymes, calpain, and interleukin-1β coverting enzyme. *Environmental Health Perspectives* **104**, Supplement 6, 1251–1256.

Kamakura, T., and Takizawa, T. (1994). Multiple comparisons among groups of growth curves. *Environmental Health Perspectives* **102**, Supplement 1, 39–42.

Kathirgamatamby, N. (1953). A note on the Poisson index of dispersion. *Biometrika* **40**, 225–228.

Keeling, C. D., and Whorf, T. P. (2000). Atmospheric CO_2 records from sites in the SIO air sampling network. In *Trends: A Compendium of Data on Global Change*. Oak Ridge, TN: US Department of Energy, Oak Ridge National Laboratory, Carbon Dioxide Information Analysis Center.

Kelly, R. J., and Mathew, T. (1994). Improved nonnegative estimation of variance components in some mixed models with unbalanced data. *Technometrics* **36**, 171–181.

Kempthorne, O. (1976). Of what use are tests of significance and tests of hypothesis? *Communications in Statistics — Theory and Methods* **A5**, 763–777.

Kendall, M. G., and Stuart, A. (1967). *The Advanced Theory of Statistics, Volume 2. Inference and Relationship*, 2nd edn. New York: Hafner.

Kenward, M. G. (2002). Repeated measures. In A. H. El-Shaarawi and W. W. Piegorsch (eds), *Encyclopedia of Environmetrics*, Vol. 3, pp. 1755–1757. Chichester: John Wiley & Sons.

Khorasani, F., and Milliken, G. A. (1982). Simultaneous confidence bands for nonlinear regression functions. *Communications in Statistics — Theory and Methods* **22**, 1241–1253.

Khuri, A. I. (1993). *Advanced Calculus with Applications in Statistics*. New York: John Wiley & Sons.

Khuri, A. I., and Cornell, J. A. (1996). *Response Surfaces: Designs and Analyses*, 2nd edn. New York: Marcel Dekker.

Kitanidis, P. K. (1997). *Introduction to Geostatistics: Applications in Hydrogeology*. New York: Cambridge University Press.

Kitchin, K. T., Brown, J. L., and Setzer, R. W. (1994). Dose–response relationship in multistage carcinogenesis: Promoters. *Environmental Health Perspectives* **102**, Supplement 1, 255–264.

Kodell, R. L., and Pounds, J. G. (1991). Assessing the toxicity of mixtures of chemicals. In D. Krewski and C. Franklin (eds), *Statistics in Toxicology*, pp. 559–591. New York: Gordon and Breach.

Kodell, R. L., and West, R. W. (1993). Upper confidence intervals on excess risk for quantitative responses. *Risk Analysis* **13**, 177–182.

Kolmogorov, A. N. (1941). The local structure of turbulence in incompressible viscous fluid for very large Reynolds numbers. *Doklady Akademii Nauk SSSR* **30**, 301–305.

Kotz, S., Johnson, N. L., and Read, C. B. (1983). Method of Lagrange multipliers. In S. Kotz, N. L. Johnson, and C. B. Read (eds), *Encyclopedia of Statistical Sciences*, Vol. 4, p. 456. New York: John Wiley & Sons.

Kotz, S., Balakrishnan, N., and Johnson, N. L. (2000). *Continuous Multivariate Distributions, Volume 1: Models and Applications*, 2nd edn. New York: John Wiley & Sons.

Krewski, D., and van Ryzin, J. (1981). Dose response models for quantal response toxicity data. In M. Csörgő, D. A. Dawson, J. N. K. Rao, and A. K. M. E. Saleh (eds), *Statistics and Related Topics*, pp. 201–231. Amsterdam: North-Holland.

Krewski, D., Smythe, R. T., Dewanji, A., and Colin, D. (1988). Statistical tests with historical controls. In H. C. Grice and J. L. Ciminera (eds), *Carcinogenicity: The Design, Analysis, and Interpretation of Long-Term Animal Studies*, pp. 23–38. New York: Springer-Verlag.

Krige, D. G. (1951). A statistical approach to some basic mine valuation problems on the Witwatersrand. *Journal of the Chemical, Metallurgical, and Mining Society of South Africa* **5**, 119–139.

Kshirsagar, A. M., and Smith, W. B. (1995). *Growth Curves*. New York: Marcel Dekker.

Kupper, L. L. (2002). Litter effect. In A. H. El-Shaarawi and W. W. Piegorsch (eds), *Encyclopedia of Environmetrics*, Vol. 2, pp. 1169–1172. Chichester: John Wiley & Sons.

Kvam, P. H., and Samaniego, F. J. (1993). On the inadmissibility of empirical averages as estimators in ranked set sampling. *Journal of Statistical Planning and Inference* **36**, 39–55.

Lancaster, V. A., and Keller-McNulty, S. (1998). A review of composite sampling methods. *Journal of the American Statistical Association* **93**, 1216–1230.

Larralde-Corona, C. P., López-Isunza, F., and Viniegra-González, G. (1997). Morphometric evaluation of the specific growth rate of *Aspergillus niger* grown in agar plates at high glucose levels. *Biotechnology and Bioengineering* **56**, 288–294.

Lawless, J. F. (1982). *Statistical Models and Methods for Lifetime Data*. New York: John Wiley & Sons.

Lawless, J. F. (1987). Negative binomial and mixed Poisson regression. *Canadian Journal of Statistics* **15**, 209–225.

Lee, T. C. M. (2001). A stabilized bandwidth selection method for kernel smoothing of the periodogram. *Signal Processing* **81**, 419–430.

Leemis, L. M. (1986). Relationships among common univariate distributions. *American Statistician* **40**, 143–146.

Lefkopoulou, M., Rotnitzky, A., and Ryan, L. (1996). Trend tests for clustered data. In B. J. T. Morgan (ed.), *Statistics in Toxicology*, pp. 179–197. Oxford: Clarendon Press.

Lehmann, E. L., and Casella, G. (1998). *Theory of Point Estimation*, 2nd edn. New York: Springer-Verlag.

Leroux, B. G., and Krewski, D. (1993). AMESFIT: A microcomputer program for fitting linear-exponential dose-response models in the Ames *Salmonella* assay. *Environmental and Molecular Mutagenesis* **22**, 78–84.

Lesser, V. M. (2002). Multiphase sampling. In A. H. El-Shaarawi and W. W. Piegorsch (eds), *Encyclopedia of Environmetrics*, Vol. 3, pp. 1312–1314. Chichester: John Wiley & Sons.

Levenberg, K. (1944). A method for the solution of certain non-linear problems in least squares. *Quarterly of Applied Mathematics* **2**, 164–168.

Levy, P. S., and Lemeshow, S. (1999). *Sampling of Populations*, 3rd edn. New York: John Wiley & Sons.

Lewis, J. (2002). Quantifying recent erosion and sediment delivery using probability sampling: a case study. *Earth Surface Processes and Landforms* **27**, 559–572.

Liang, K.-Y., and Zeger, S. L. (1986). Longitudinal data analysis using generalized linear models. *Biometrika* **73**, 13–22.

Liang, K.-Y., Zeger, S. L., and Qaqish, B. (1992). Multivariate regression analysis for categorical data. *Journal of the Royal Statistical Society, series B* **54**, 3–40.

Light, R. J., and Pillemer, D. B. (1984). *Summing Up: The Science of Reviewing Research*. Cambridge, MA: Harvard University Press.

Lincoln, F. C. (1930). Calculating waterfowl abundance on the basis of banding returns. Circular number 118, US Department of Agriculture, Washington, DC.

Lindberg, M., and Rexstad, E. (2002). Capture-recapture sampling designs. In A. H. El-Shaarawi and W. W. Piegorsch (eds), *Encyclopedia of Environmetrics*, Vol. 1, pp. 251–262. Chichester: John Wiley & Sons.

Lindsey, J. K. (1997). *Applying Generalized Linear Models*. New York: Springer-Verlag.

Lineweaver, H., and Burk, D. (1934). The determination of enzyme dissociation constants. *Journal of the American Chemical Society* **56**, 658–666.

Link, W. A., and Sauer, J. R. (1996). Extremes in ecology: avoiding the misleading effects of sampling variation in summary analyses. *Ecology* **77**, 1633–1640.

Liptak, T. (1958). On the combination of independent tests. *Magyar Tudományos Akadémia Matematikai Kutató Intézetenek Közleményei* **3**, 171–197.

Littell, R. C., Milliken, G. A., Stroup, W. W., and Wolfinger, R. D. (1996). *SAS® System for Mixed Models*. Cary, NC: SAS Institute Inc.

Littell, R. C., Stroup, W. W., and Freund, R. J. (2002). *SAS® for Linear Models*, 4th edn. Cary, NC: SAS Institute Inc.

Liu, C., Bennett, D. H., Kastenberg, W. E., McKone, T. E., and Browne, D. (1999). A multimedia, multiple pathway exposure assessment of atrazine: fate, transport and uncertainty analysis. *Reliability Engineering & System Safety* **63**, 169–184.

Longley, P. A., Goodchild, M. F., Maguire, D. J., and Rhind, D. W. (2001). *Geographic Information Systems and Science*. Chichester: John Wiley & Sons.

Lovison, G., Gore, S. D., and Patil, G. P. (1994). Design and analysis of composite sampling procedures: A review. In G. P. Patil and C. R. Rao (eds), *Handbook of Statistics Volume 12: Environmental Statistics*, pp. 103–166. New York: North-Holland/Elsevier.

Lu, J., Liu, S., Yin, M., and Hughes-Oliver, J. M. (1999). Modeling restricted bivariate censored lowflow data. *Environmetrics* **10**, 125–136.

Lund, R., and Reeves, J. (2002). Detection of undocumented changepoints: a revision of the two-phase regression model. *Journal of Climate* **15**, 2547–2554.

Majer, B. J., Tscherko, D., Paschke, A., Wennrich, R., Kundi, M., Kandeler, E., and Knasmüller, S. (2002). Effects of heavy metal contamination of soils on micronucleus induction in *Tradescantia* and on microbial enzyme activities: a comparative investigation. *Mutation Research* **515**, 111–124.

Malling, H. V., and Delongchamp, R. R. (2001). Direct separation of in vivo and in vitro *am3* revertants in transgenic mice carrying the phiX174 *am3, cs70* vector. *Environmental and Molecular Mutagenesis* **37**, 345–355.

Manly, B. F. J. (1994). *Multivariate Statistical Methods: A Primer*, 2nd edn. New York: Chapman & Hall.

Manly, B. F. J. (2001). *Statistics for Environmental Science and Management*. Boca Raton, FL: Chapman & Hall/CRC.

Mardia, K. V., and Marshall, R. J. (1984). Maximum likelihood estimation of models for residual covariance in spatial regression. *Biometrika* **71**, 135–146.

Mardia, K. V., and Watkins, A. J. (1989). On multimodality of the likelihood in the spatial linear model. *Biometrika* **76**, 289–295.

Margolin, B. H., and Risko, K. J. (1984). The use of historical control data in laboratory studies. *Proceedings of the International Biometric Conference* **12**, 21–30.

Margolin, B. H., Resnick, M. A., Rimpo, J. Y., Archer, P., Galloway, S. M., Bloom, A. D., and Zeiger, E. (1986). Statistical analyses for in vitro cytogenetic assays using Chinese hamster ovary cells. *Environmental Mutagenesis* **8**, 183–204.

Marquardt, D. W. (1963). An algorithm for least-squares estimation of nonlinear parameters. *Journal of the Society for Industrial and Applied Mathematics* **11**, 431–441.

Mason, R. L., Gunst, R. F., and Hess, J. L. (2003). *Statistical Design and Analysis of Experiments: With Applications to Engineering and Science*, 2nd edn. New York: John Wiley & Sons.

Matheron, G. (1962). *Traité de Géostatistique Appliquée, Tome 1* (Mémoires du Bureau de Recherches Géologiques et Minières, No. 14). Paris: Editions Technip.

Matheron, G. (1963). Principles of geostatistics. *Economic Geology* **58**, 1246–1266.

Matheron, G. (1971). *The Theory of Regionalized Variables and its Applications* (Les Cahiers du Centre de Morphologie Mathématique de Fountainebleau, Fascicule 5). Fountainebleau: École Nationale Supérieure des Mines de Paris.

Matis, J. H., and Wehrly, T. E. (1994). Compartmental models of ecological and environmental systems. In G. P. Patil and C. R. Rao (eds), *Handbook of Statistics Volume 12: Environmental Statistics*, pp. 583–614. New York: North-Holland/Elsevier.

McCullagh, P. (1983). Quasi-likelihood functions. *Annals of Statistics* **11**, 59–67.

McCullagh, P., and Nelder, J. A. (1989). *Generalized Linear Models*, 2nd edn. London: Chapman & Hall.

McDonald, T. L., and Amstrup, S. C. (2001). Estimation of population size using open capture-recapture models. *Journal of Agricultural, Biological, and Environmental Statistics* **6**, 206–220.

McGarigal, K., Cushman, S., and Stafford, S. (2000). *Multivariate Statistics for Wildlife and Ecology Research*. New York: Springer-Verlag.

McIntyre, G. A. (1952). A method for unbiased selective sampling, using ranked sets. *Australian Journal of Agricultural Research* **3**, 385–390.

McKay, M. D., Beckman, R. J., and Conover, W. J. (1979). A comparison of three methods for selecting values of input variables in the analysis of output from a computer code. *Technometrics* **21**, 239–245.

McLeod, A. I., and Hipel, K. W. (1995). Exploratory spectral analysis of hydrological time series. *Stochastic Hydrology and Hydraulics* **9**, 171–205.

Meeden, G. (1999). Interval estimators for the population mean for skewed distributions with a small sample size. *Journal of Applied Statistics* **26**, 81–96.

Mehta, C. R., and Patel, N. R. (1995). Exact logistic regression: Theory and examples. *Statistics in Medicine* **14**, 2143–2160.

Menke, M., Angelis, K. J., and Schubert, I. (2000). Detection of specific DNA lesions by a combination of Comet assay and FISH in plants. *Environmental and Molecular Mutagenesis* **35**, 132–138.

Meredith, M. P., and Stehman, S. V. (1991). Repeated measures experiments in forestry: Focus on analysis of response curves. *Canadian Journal of Forest Research* **21**, 957–965.

Metropolis, N., and Ulam, S. (1949). The Monte Carlo method. *Journal of the American Statistical Association* **44**, 335–341.

Meyer, H. A. (1958). The calculation of the sampling error of a cruise from the mean square successive difference. *Journal of Forestry* **54**, 182–184.

Michaelis, L., and Menten, M. L. (1913). Die Kinetik der Invertionwirkung. *Biochemische Zeitschrift* **49**, 333–369.

Millard, S. P., and Neerchal, N. K. (2001). *Environmental Statistics with S-PLUS*. Boca Raton, FL: Chapman & Hall/CRC.

Moolgavkar, S. (1999). Stochastic models for estimation and prediction of cancer risk. In V. Barnett, A. Stein, and K. F. Turkman (eds), *Statistics for the Environment 4: Statistical Aspects of Health and the Environment*, pp. 237–257. Chichester: John Wiley & Sons.

Moore, D. F. (1986). Asymptotic properties of moment estimators for overdispersed counts and proportions. *Biometrika* **73**, 583–588.

Moran, P. A. P. (1948). Some theorems on time series: II. The significance of the serial correlation coefficient. *Biometrika* **35**, 255–260.

Moran, P. A. P. (1950). Notes on continuous stochastic phenomena. *Biometrika* **37**, 17–23.

Morgan, P. H., Mercer, L. P., and Flodin, N. W. (1975). General model for nutritional responses of higher organisms. *Proceedings of the National Academy of Sciences, USA* **72**, 4327–4331.

Moser, E. B. (1987). The analysis of mapped spatial point patterns. In *Proceedings of the 12th Annual SAS Users Group International Conference*, pp. 1141–1145.

Mugglestone, M. A., and Renshaw, E. (2001). Spectral tests of randomness for spatial point patterns. *Environmental and Ecological Statistics* **8**, 237–251.

Nair, R. S., Sherman, J. H., Stevens, M. W., and Johannsen, F. R. (1995). Selecting a more realistic uncertainty factor: reducing compounding effects of multiple uncertainties. *Human and Ecological Risk Assessment* **1**, 576–589.

Narula, S. C. (1979). Orthogonal polynomial regression. *International Statistical Review* **47**, 31–36.

Nash, J. C., and Quon, T. (1996). Software for modeling kinetic phenomena. *American Statistician* **50**, 368–378.

National Toxicology Program (1989). Toxicology and carcinogenesis studies of pentachlorophenol in B6C3F1 mice. Technical Report no. 349, US Department of Health and Human Services, Public Health Service, Research Triangle Park, NC.

National Toxicology Program (2004). Toxicology and carcinogenesis studies of urethane, ethanol, and urethane/ethanol in $B6C3F_1$ mice. Technical Report no. 510, US Department of Health and Human Services, Public Health Service, Research Triangle Park, NC.

Nelder, J. A., and Pregibon, D. (1987). An extended quasi-likelihood function. *Biometrika* **74**, 221–231.

Nelder, J. A., and Wedderburn, R. W. M. (1972). Generalized linear models. *Journal of the Royal Statistical Society, Series A* **135**, 370–384.

Neter, J., Kutner, M. H., Nachtsheim, C. J., and Wasserman, W. (1996). *Applied Linear Statistical Models*, 4th edn. Chicago: R.D. Irwin.

Neyman, J. (1934). On the two different aspects of the representative method: the method of stratified sampling and the method of purposive sampling (with discussion). *Journal of the Royal Statistical Society* **97**, 558–625.

Neyman, J., and Pearson, E. S. (1933). On the problem of most efficient tests of statistical hypotheses. *Philosophical Transactions of the Royal Society, Series A* **231**, 289–337.

Nilson, T., Anniste, J., Lang, M., and Praks, J. (1999). Determination of needle area indices of coniferous forest canopies in the NOPEX region by ground-based optical measurements and satellite images. *Agricultural and Forest Meteorology* **98–99**, 449–462.

Nitcheva, D. K., Piegorsch, W. W., West, R. W., and Kodell, R. L. (2005). Multiplicity-adjusted inferences in risk assessment: Benchmark analysis with quantal response data. *Biometrics* **61** (in press).

Nordahl, V., and Clark, S. (2002). Trawl surveys. In A. H. El-Shaarawi and W. W. Piegorsch (eds), *Encyclopedia of Environmetrics*, Vol. 4, pp. 2234–2237. Chichester: John Wiley & Sons.

Normand, S.-L. T. (1999). Meta-analysis: formulating, evaluating, combining, and reporting. *Statistics in Medicine* **18**, 321–359.

North, E. R., Shen, S. S., and Basist, A. S. (2002). Global warming. In A. H. El-Shaarawi and W. W. Piegorsch (eds), *Encyclopedia of Environmetrics*, Vol. 2, pp. 929–933. Chichester: John Wiley & Sons.

Nussbaum, B. D., and Sinha, B. K. (1997). Cost effective gasoline sampling using ranked set sampling. *Proceedings of the American Statistical Association, Section on Statistics and the Environment*, 83–87.

Olden, J. D., and Jackson, D. A. (2000). Torturing data for the sake of generality: How valid are our regression models? *Ecoscience* **7**, 501–510.

Orcutt, G. H., and Irwin, J. O. (1948). A study of the autoregressive nature of the time series used for Tinbergen's model of the economic system of the United States, 1919–1932. *Journal of the Royal Statistical Society, Series B* **10**, 1–53.

Oris, J. T., and Bailer, A. J. (2003). Quantitative models in ecological toxicology: Application in ecological risk assessment. In J. Cole and C. Canham (eds), *The Role of Models in*

Ecosystem Science, Cary Conference IX, pp. 346–364. Princeton, NJ: Princeton University Press.

Overton, W. S., and Stehman, S. V. (1993). Properties of designs for sampling continuous spatial resources from a triangular grid. *Communications in Statistics – Theory and Methods* **22**, 2641–2660.

Patil, G. P. (2002). Ranked set sampling. In A. H. El-Shaarawi and W. W. Piegorsch (eds), *Encyclopedia of Environmetrics*, Vol. 3, pp. 1684–1690. Chichester: John Wiley & Sons.

Patil, G. P., Gore, S. D., and Sinha, A. K. (1994a). Environmental chemistry, statistical modeling, and observational economy. In C. R. Cothern and N. P. Ross (eds), *Environmental Statistics, Assessment, and Forecasting*, pp. 57–97. Boca Raton, FL: Lewis.

Patil, G. P., Sinha, A. K., and Tallie, C. (1994b). Ranked set sampling. In G. P. Patil and C. R. Rao (eds), *Handbook of Statistics Volume 12: Environmental Statistics*, pp. 167–200. Amsterdam: North-Holland/Elsevier.

Patil, G. P., Sinha, A. K., and Tallie, C. (1995). Finite population corrections for ranked set sampling. *Annals of the Institute of Statistical Mathematics* **47**, 621–636.

Patterson, H. D., and Thompson, R. (1971). Recovery of interblock information when block sizes are unequal. *Biometrika* **58**, 545–554.

Pearson, K. (1900). On the criterion that a given system of deviations from the probable in the case of a correlated system of variables is such that it can be reasonably supposed to have arisen from random sampling. *The London, Edinburgh and Dublin Philosophical Magazine and Journal of Science, 5th Series* **50**, 157–175.

Pearson, K. (1903). Mathematical contributions to the theory of evolution. – On homotyposis in homologous but differentiated organs. *Proceedings of the Royal Society of London* **71**, 288–313.

Pendergast, J. F., Gange, S. J., Newton, M. A., Lindstrom, M. J., Palta, M., and Fisher, M. R. (1996). A survey of methods for analyzing clustered binary response data. *International Statistical Review* **64**, 89–118.

Petersen, C. G. J. (1896). The yearly immigration of young plaice into the Limfjord from the German Sea. *Report of the Danish Biological Station* **6**, 5–48.

Pickles, A. (2002). Generalized estimating equations. In A. H. El-Shaarawi and W. W. Piegorsch (eds), *Encyclopedia of Environmetrics*, Vol. 2, pp. 853–863. Chichester: John Wiley & Sons.

Piegorsch, W. W. (1987). Discretizing a normal prior for change point estimation in switching regressions. *Biometrical Journal* **29**, 777–782.

Piegorsch, W. W. (1993). Biometrical methods for testing dose effects of environmental stimuli in laboratory studies. *Environmetrics* **4**, 483–505.

Piegorsch, W. W. (2002). Tables of *P*-values for *t*- and chi-square reference distributions. Technical Report no. 194. Department of Statistics, University of South Carolina, Columbia, SC. Available at http://www.stat.sc.edu/~piegorsc/TR194.pdf

Piegorsch, W. W., and Bailer, A. J. (1997). *Statistics for Environmental Biology and Toxicology*. London: Chapman & Hall.

Piegorsch, W. W., and Cox, L. H. (1996). Combining environmental information II: Environmental epidemiology and toxicology. *Environmetrics* **7**, 309–324.

Piegorsch, W. W., Simmons, S. J., Margolin, B. H., Zeiger, E., Gidrol, X. M., and Gee, P. (2000). Statistical modeling and analysis of a base-specific *Salmonella* mutagenicity assay. *Mutation Research* **467**, 11–19.

Piegorsch, W. W., West, R. W., Pan, W., and Kodell, R. L. (2005). Low-dose risk estimation via simultaneous statistical inferences. *Applied Statistics* **54**, 245–258.

Pielou, E. C. (1985). Line intercept sampling. In S. Kotz, N. L. Johnson, and C. B. Read (eds), *Encyclopedia of Statistical Sciences*, Vol. 5, pp. 70–71. New York: John Wiley & Sons.

Pierce, D. A., and Schafer, D. W. (1986). Residuals in generalized linear models. *Journal of the American Statistical Association* **81**, 977–986.

Pires, A. M., Branco, J. A., Picado, A., and Mendonca, E. (2002). Models for the estimation of a 'no effect concentration'. *Environmetrics* **13**, 15–27.

Pocock, S. J. (1976). The combination of randomized and historical controls in clinical trials. *Journal of Chronic Diseases* **29**, 175–188.

Pollock, K. H., Nichols, J. D., Brownie, C., and Hines, J. E. (1990). Statistical inferences for capture-recapture experiments. *Wildlife Monograph* **107**, 1–97.

Porter, D. E., Edwards, D., Scott, G., Jones, B., and Street, W. S. (1997). Assessing the impacts of anthropogenic and physiographic influences on grass shrimp in localized salt-marsh estuaries. *Aquatic Botany* **58**, 289–306.

Potscher, B. M., and Novak, A. J. (1998). The distribution of estimators after model selection: Large and small sample results. *Journal of Statistical Computation and Simulation* **60**, 19–56.

Potvin, C., Lechowicz, M. J., and Tardif, S. (1990). The statistical analysis of ecophysiological response curves obtained from experiments involving repeated measures. *Ecology* **71**, 1389–1400.

Prentice, R. L., and Zhao, L. P. (1991). Estimating equations for parameters in means and covariances of multivariate discrete and continuous responses. *Biometrics* **47**, 825–839.

Press, W. H., Teukolsky, S. A., Vettering, W. T., and Flannery, B. P. (1992). *Numerical Recipes in FORTRAN: The Art of Scientific Computing*, 2nd edn. New York: Cambridge University Press.

Priyadarshi, A., Khuder, S. A., Schaub, E. A., and Priyadarshi, S. S. (2001). Environmental risk factors and Parkinson's disease: A meta-analysis. *Environmental Research* **86**, 122–127.

Ralph, S., and Petras, M. (1997). Genotoxicity monitoring of small bodies of water using two species of tadpoles and the alkaline single cell gel (Comet) assay. *Environmental and Molecular Mutagenesis* **29**, 418–430.

Rao, C. R. (1947). Large sample tests of statistical hypotheses concerning several parameters with applications to problems of estimation. *Proceedings of the Cambridge Philosophical Society* **44**, 50–57.

Rasmussen, P. W., Heisey, D. M., Nordheim, E. V., and Frost, T. M. (1993). Time-series intervention analysis: Unreplicated large-scale experiments. In S. M. Scheiner and J. Gurevitch (eds), *The Design and Analysis of Ecological Experiments*, pp. 138–158. New York: Chapman & Hall.

Ratkowsky, D. A. (1990). *Handbook of Nonlinear Regression Models*. New York: Marcel Dekker.

Redfern, A., Bunyan, M., and Lawrence, T. (eds) (2003). *The Environment in Your Pocket*, 7th edn. London: UK Department for Environment, Food and Rural Affairs.

Regan, M. M. (2002). Developmental toxicity study. In A. H. El-Shaarawi and W. W. Piegorsch (eds), *Encyclopedia of Environmetrics*, Vol. 1, pp. 515–518. Chichester: John Wiley & Sons.

Rhees, B. K., and Atchley, W. R. (2000). Body weight and tail length divergence in mice selected for rate of development. *Journal of Experimental Zoology* **288**, 151–164.

Ribakov, Y., Gluck, J., and Reinhorn, A. M. (2001). Active viscous damping system for control of MDOF structures. *Earthquake Engineering and Structural Dynamics* **30**, 195–212.

Richards, F. J. (1959). A flexible growth function for empirical use. *Journal of Experimental Biology* **10**, 290–300.

Ripley, B. D. (1976). The second-order analysis of stationary point processes. *Journal of Applied Probability* **13**, 255–266.

Ripley, B. D. (1977). Modelling spatial patterns (with discussion). *Journal of the Royal Statistical Society, Series B* **39**, 172–212.

Ripley, B. D. (1979). Tests of 'randomness' for spatial point patterns. *Journal of the Royal Statistical Society, Series B* **41**, 368–374.

Robinson, G. K. (1991). That BLUP is a good thing: The estimation of random effects (with discussion). *Statistical Science* **6**, 15–51.

Robson, D. S., and Regier, H. A. (1964). Sample size in Petersen mark-recapture experiments. *Transactions of the American Fisheries Society* **93**, 215–226.

Ross, N. P., and Stokes, S. L. (1999). Editorial: Special issue on statistical design and analysis with ranked set samples. *Environmental and Ecological Statistics* **6**, 5–10.

Ross, S. M. (1995). *Stochastic Processes*, 2nd edn. New York: John Wiley & Sons.

Rowlingson, B. S., and Diggle, P. J. (1993). Splancs: spatial point pattern analysis code in S-Plus. *Computers & Geosciences* **19**, 627–655.

Ruppert, D., Cressie, N., and Carroll, R. J. (1989). A transformation/weighting model for Michaelis–Menten parameters. *Biometrics* **45**, 637–656.

Ryan, L. M., Catalano, P. J., Kimmel, C. A., and Kimmel, G. L. (1991). Relationship between fetal weights and malformation in developmental toxicology studies. *Teratology* **44**, 215–223.

S-Plus (1997). *S-Plus User's Guide, Version 4.0*. Seattle, WA: MathSoft, Inc., Data Analysis Products Division.

Sakamoto, H., Ohe, T., Hayatsu, T., and Hayatsu, H. (1996). Evaluation of blue-chitin column, blue-rayon hanging, and XAD-resin column techniques for concentrating mutagens from two Japanese rivers. *Mutation Research* **371**, 79–85.

Saltelli, A. (2002). Sensitivity analysis for importance assessment. *Risk Analysis* **22**, 579–590.

Saltelli, A., Chan, K., and Scott, E. M. (eds) (2000). *Sensitivity Analysis*. New York: John Wiley & Sons.

Samuels, M. L., and Witmer, J. A. (2003). *Statistics for the Life Sciences*, 3rd edn. Upper Saddle River, NJ: Prentice Hall.

SAS Institute Inc. (2000). *SAS/STAT® User's Guide, Version 8, Volumes 1, 2, and 3*. Cary, NC: SAS Institute Inc.

Satterthwaite, F. E. (1946). An approximate distribution of estimates of variance components. *Biometrics* **2**, 110–114.

Sawada, M. C. (1999). ROOKCASE: An Excel 97/2000 Visual Basic (VB) add-in for exploring global and local spatial autocorrelation. *Bulletin of the Ecological Society of America* **80**, 231–234.

Schabenberger, O., and Pierce, F. J. (2002). *Contemporary Statistical Models for the Plant and Soil Sciences*. Boca Raton, FL: Chapman & Hall/CRC.

Schaber, G. G., Kirk, R. L., and Strom, R. G. (1998). Data base of impact craters on Venus based on analysis of Magellan radar images and altimetry data. Open-File Report no. 98-104, US Department of the Interior, US Geological Survey Information Service, Denver, CO.

Scheaffer, R. L., Mendenhall, W., and Ott, L. (1996). *Elementary Survey Sampling*, 5th edn. Belmont, CA: Duxbury Press.

Scheffé, H. (1953). A method for judging all contrasts in the analysis of variance. *Biometrika* **40**, 87–104.

Schipper, M., and Meelis, E. (1997). Sequential analysis of environmental monitoring data: Refined SPRT's for testing against a minimal relevant trend. *Journal of Agricultural, Biological, and Environmental Statistics* **2**, 467–489.

Schlekat, C. E., Scott, K. J., Swartz, R. C., Albrecht, B., Antrim, L., Doe, K., Douglas, S., Ferretti, J. A., Hansen, D. J., Moore, D. M., Mueller, C., and Tang, A. (1995). Intralaboratory comparison of a 10-day sediment toxicity test method using *Ampelisa abdita*, *Eohaustorius estuarius* and *Leptocherirus plumulosus*. *Environmental Toxicology and Chemistry* **14**, 2163–2174.

Schork, N. (1993). Combining Monte Carlo and Cox tests of nonnested hypotheses. *Communications in Statistics – Simulation and Computation* **22**, 939–954.

Schuster, A. (1898). On the investigation of hidden periodicities with application to a supposed 26 day period of meteorological phenomena. *Terrestrial Magnetism* **3**, 13–41.

Searle, S. R. (1982). *Matrix Algebra Useful for Statistics*. New York: John Wiley & Sons.

Searle, S. R., Casella, G., and McCulloch, C. E. (1992). *Variance Components*. New York: John Wiley & Sons.

Seber, G. A. F. (1965). A note on the multiple-recapture census. *Biometrika* **52**, 249–259.

Seber, G. A. F. (1970). The effects of trap response on tag-recapture estimates. *Biometrics* **26**, 13–22.

Seber, G. A. F., and Wild, C. J. (1989). *Nonlinear Regression*. New York: John Wiley & Sons.

See, K., and Bailer, A. J. (1998). Added risk and inverse estimation for count responses in reproductive aquatic toxicology studies. *Biometrics* **54**, 67–73.

See, K., Stufken, J., Song, S. Y., and Bailer, A. J. (2000). Relative efficiencies of sampling plans for selecting a small number of units from a rectangular region. *Journal of Statistical Computation and Simulation* **66**, 273–294.

Seshu, D. V., and Cady, F. B. (1984). Response of rice to solar radiation and temperature estimated from international yield trials. *Crop Science* **24**, 649–654.

Sherman, C., and Portier, C. J. (1998). Multistage carcinogenesis models. In P. Armitage and T. Colton (eds), *Encyclopedia of Biostatistics*, Vol. 4, pp. 2808–2814. Chichestes: John Wiley & Sons.

Sielken, R. L., and Stevenson, D. E. (1997). Opportunities to improve quantitative risk assessment. *Human and Ecological Risk Assessment* **3**, 479–489.

Simmons, S. J., Piegorsch, W. W., Nitcheva, D. K., and Zeiger, E. (2003). Combining environmental information via hierarchical modeling: An example using mutagenic potencies. *Environmetrics* **14**, 159–168.

Slob, W. (1999). Deriving safe exposure levels for chemicals from animal studies using statistical methods: Recent developments. In V. Barnett, A. Stein, and K. F. Turkman (eds), *Statistics for the Environment 4: Statistical Aspects of Health and the Environment*, pp. 153–174. Chichester: John Wiley & Sons.

Smith, E. P. (2002a). BACI design. In A. H. El-Shaarawi and W. W. Piegorsch (eds), *Encyclopedia of Environmetrics*, Vol. 1, pp. 141–148. Chichester: John Wiley & Sons.

Smith, E. P. (2002b). Uncertainty analysis. In A. H. El-Shaarawi and W. W. Piegorsch (eds), *Encyclopedia of Environmetrics*, Vol. 4, pp. 2283–2297. Chichester: John Wiley & Sons.

Smythe, G. K. (2002). Optimization. In A. H. El-Shaarawi and W. W. Piegorsch (eds), *Encyclopedia of Environmetrics*, Vol. 3, pp. 1481–1487. Chichester: John Wiley & Sons.

Solow, A. R. (1993). Spectral estimation by variable span log periodogram smoothing: An application to annual lynx numbers. *Biometrical Journal* **35**, 627–633.

Soper, K. A. (1998). Interval estimation with small samples for median lethal dose or median effective dose. In S.-C. Chow and J.-P. Liu (eds) *Design and Analysis of Animal Studies in Pharmaceutical Development*, pp. 43–78. New York: Marcel Dekker.

Sorokine-Durm, I., Durand, V., Le Roy, A., Paillole, N., Roy, L., and Voison, P. (1997). Is FISH painting an appropriate biological marker for dose estimates of suspected accidental radiation overexposure? A review of cases investigated in France from 1995 to 1996. *Environmental Health Perspectives* **105**, Supplement 6, 1427–1432.

Spearman, C. (1908). The method of 'right and wrong cases' ('constant stimuli') without Gauss's formulae. *Journal of Psychology* **2**, 227–242.

Stehman, S. V., and Overton, W. S. (2002). Sampling, environmental. In A. H. El-Shaarawi and W. W. Piegorsch (eds), *Encyclopedia of Environmetrics*, Vol. 4, pp. 1914–1937. Chichester: John Wiley & Sons.

Stein, M. (1999). *Interpolation of Spatial Data: Some Theory for Kriging*. New York: Springer-Verlag.

Stelling, M., and Sjerps, M. (1999). A sampling and hypothesis testing method for the legal control of solid hazardous waste. *Environmetrics* **10**, 247–259.

Stern, A. (2002). Risk assessment, quantitative. In A. H. El-Shaarawi and W. W. Piegorsch (eds), *Encyclopedia of Environmetrics*, Vol. 3, pp. 1837–1843. Chichester: John Wiley & Sons.

Stevens, D. L. (1997). Variable density grid-based sampling designs for continuous spatial populations. *Environmetrics* **8**, 167–195.

Stewart-Oaten, A. (2002). Impact assessment. In A. H. El-Shaarawi and W. W. Piegorsch (eds), *Encyclopedia of Environmetrics*, Vol. 2, pp. 1025–1035. Chichester: John Wiley & Sons.

Stewart-Oaten, A., Murdoch, W. M., and Parker, K. R. (1986). Environmental impact assessment: 'pseudoreplication' in time? *Ecology* **67**, 929–940.

Stigler, S. M. (1991). Stochastic simulation in the nineteenth century. *Statistical Science* **6**, 89–97.

Stokes, S. L. (1986). Ranked set sampling. In S. Kotz, N. L. Johnson and C. B. Read (eds), *Encyclopedia of Statistical Sciences*, Vol. 7, pp. 585–588. New York: John Wiley & Sons.

Stouffer, S. A., Suchman, E. A., DeVinney, L. C., Star, S. A., and Williams, R. M., Jr. (1949). *The American Soldier, Volume I. Adjustment during Army Life*. Princeton, NJ: Princeton University Press.

Strauss, D. J. (1975). A model for clustering. *Biometrika* **62**, 467–475.

Strodl Andersen, J., Holst, H., Spliid, H., Andersen, H., Baun, A., and Nyholm, N. (1998). Continuous ecotoxicological data evaluated relative to a control response. *Journal of Agricultural, Biological, and Environmental Statistics* **3**, 405–420.

Student (1908). The probable error of a mean. *Biometrika* **6**, 1–25.

Sugden, R. A., Smith, T. M. F., and Jones, R. P. (2000). Cochran's rule for simple random sampling. *Journal of the Royal Statistical Society, Series B* **62**, 787–793.

Sun, J. Y., Raz, J., and Faraway, J. J. (1999). Confidence bands for growth and response curves. *Statistica Sinica* **9**, 679–698.

Sutton, A. J., Abrams, K. R., Sheldon, T. A., and Song, F. (2000). *Methods for Meta-analysis in Medical Research*. New York: John Wiley & Sons.

Switzer, P. (2002). Kriging. In A. H. El-Shaarawi and W. W. Piegorsch (eds), *Encyclopedia of Environmetrics*, Vol. 2, pp. 1109–1118. Chichester: John Wiley & Sons.

Takahasi, K., and Futatsuya, M. (1998). Dependence between order statistics in samples from finite population and its application to ranked set sampling. *Annals of the Institute of Statistical Mathematics* **50**, 49–70.

Takahasi, K., and Wakimoto, K. (1968). On unbiased estimates of the population mean based on the sample stratified by means of ordering. *Annals of the Institute of Statistical Mathematics* **20**, 1–31.

Tamhane, A. C. (1986). A survey of literature on quantal response curves with a view towards application of the problem of selecting the curve with the smallest q-quantile (ED100q). *Communications in Statistics – Theory and Methods* **15**, 2679–2718.

Tarone, R. E. (1979). Testing the goodness of fit of the binomial distribution. *Biometrika* **66**, 585–590.

Tarone, R. E. (1982). The use of historical control information in testing for a trend in Poisson means. *Biometrics* **38**, 457–462.

Tarone, R. E. (1986). Correcting tests for trend in proportions for skewness. *Communications in Statistics – Theory and Methods* **15**, 317–328.

Tarone, R. E., and Gart, J. J. (1980). On the robustness of combined tests for trends in proportions. *Journal of the American Statistical Association* **75**, 110–116.

Tez, M. (1991). Confidence bands for the Michaelis–Menten kinetic model. *Statistical Papers* **32**, 253–260.

Thiébaux, H. J. (1994). *Statistical Data Analysis for Ocean and Atmospheric Sciences*. New York: Academic Press.

Thomas, L., Buckland, S. T., Burnham, K. P., Anderson, D. R., Laake, J. L., Borchers, D. L., and Strindberg, S. (2002). Distance sampling. In A. H. El-Shaarawi and W. W. Piegorsch (eds), *Encyclopedia of Environmetrics*, Vol. 1, pp. 544–552. Chichester: John Wiley & Sons.

Thompson, S. K. (2002). *Sampling*, 2nd edn. New York: John Wiley & Sons.

Thornton, A., and Lee, P. (2000). Publication bias in meta-analysis: its causes and consequences. *Journal of Clinical Epidemiology* **53**, 207–216.

Tinwell, H., Paton, D., Guttenplan, J. B., and Ashby, J. (1996). Unexpected genetic toxicity to rodents of the N′, N′-dimethyl analogues of MNU and ENU. *Environmental and Molecular Mutagenesis* **27**, 202–210.

Tippett, L. H. C. (1931). *The Methods of Statistics*. London: Williams & Norgate.

Todd, A. C., Parsons, P. J., Tang, S., and Moshier, E. L. (2001). Individual variability in human tibia lead concentration. *Environmental Health Perspectives* **109**, 1139–1143.

Trevan, J. W. (1927). The error of determination of toxicity. *Proceedings of the Royal Society, Series B* **101**, 483–514.

Tweedie, R. L. (2002). Meta-analysis. In A. H. El-Shaarawi and W. W. Piegorsch (eds), *Encyclopedia of Environmetrics*, Vol. 3, pp. 1245–1251. Chichester: John Wiley & Sons.

US EPA (1992). Respiratory health effects of passive smoking: Lung cancer and other disorders. Technical Report no. EPA/600/6-90/006F, US Environmental Protection Agency, Washington, DC.

US EPA (1994). Methods for derivation of inhalation reference concentrations and application of inhalation dosimetry. Technical Report no. EPA/600/8-90/066F, US Environmental Protection Agency, Washington, DC.

US EPA (1997). Integrated Risk Information System (IRIS) www.epa.gov/iris/. National Center for Environmental Assessment, US Environmental Protection Agency, Cincinnati, OH.

Umbach, D. M. (2002). Effect size. In A. H. El-Shaarawi and W. W. Piegorsch (eds), *Encyclopedia of Environmetrics*, Vol. 2, pp. 629–631. Chichester: John Wiley & Sons.

Væth, M. (1985). On the use of Wald's test in exponential families. *International Statistical Review* **53**, 199–214.

Van Ewijk, P. H., and Hoekstra, J. A. (1993). Calculation of the EC_{50} and its confidence interval when subtoxic stimulus is present. *Ecotoxicology and Environmental Safety* **25**, 25–32.

Van Groenendaal, W. J. H., and Kleijnen, J. P. C. (2002). Deterministic versus stochastic sensitivity analysis in investment problems: An environmental case study. *European Journal of Operational Research* **141**, 8–20.

Venables, W. N., and Ripley, B. D. (2002). *Modern Applied Statistics with S*, 4th edn. New York: Springer-Verlag.

Verhulst, P.-F. (1845). Recherches mathématiques sur la loi d'accroissement de la population. *Nouveaux Mémoires de l'Académie Royale des Sciences et Belles-lettres de Bruxelles* **18**, 1–41.

von Ende, C. N. (1993). Repeated-measures analysis: Growth and other time-dependent measures. In S. M. Scheiner and J. Gurevitch (eds), *The Design and Analysis of Ecological Experiments*, pp. 113–137. New York: Chapman & Hall.

von Neumann, J. (1941). Distribution of the ratio of the mean square successive difference to the variance. *Annals of Mathematical Statistics* **14**, 367–395.

Vonesh, E. F., and Chinchilli, V. M. (1996). *Linear and Nonlinear Models for the Analysis of Repeated Measurements*. New York: Marcel Dekker.

Vroblesky, D. A., Robertson, J. F., Petkewich, M. D., Chapelle, F. H., Bradley, P. M., and Landmeyer, J. E. (1997). Remediation of petrolium hydrocarbon-contaminated ground water in the vicinity of a jet-fuel tank farm, Hanahan, South Carolina. Water-Resources Investigations Report no. 96-4251, US Geological Survey, Columbia, SC.

Wahba, G. (1980). Automatic smoothing of the log periodogram. *Journal of the American Statistical Association* **75**, 122–132.

Wald, A. (1943). Tests of statistical hypotheses concerning several parameters when the number of observations is large. *Transactions of the American Mathematical Society* **54**, 426–482.

Waldmeier, M. (1961). *The Sunspot Activity in the Years 1610–1960*. Zurich: Swiss Federal Observatory.

Walter, S. D. (1992). The analysis of regional patterns in health data. 2. The power to detect environmental effects. *American Journal of Epidemiology* **136**, 742–759.

Wang, F. J., and Wall, M. A. (2003). Incorporating parameter uncertainty into prediction intervals for spatial data modeled via a parametric variogram. *Journal of Agricultural, Biological, and Environmental Statistics* **8**, 296–309.

Wang, S. C. D., and Smith, E. P. (2000). Adjusting for mortality effects in chronic toxicity testing: mixture model approach. *Environmental Toxicology and Chemistry* **19**, 204–209.

Wang, Y. Y. (1971). Probabilities of the Type I errors of the Welch tests for the Behrens–Fisher problem. *Journal of the American Statistical Association* **66**, 605–608.

Weber, J. E., and Monarchi, D. E. (1982). Performance of the Durbin–Watson test and WLS estimation when the disturbance term includes serial dependence in addition to first-order autocorrelation. *Journal of the American Statistical Association* **77**, 117–128.

Webster, R., and Oliver, M. (2001). *Geostatistics for Environmental Scientists*. Chichester: John Wiley & Sons.

Wedderburn, R. W. M. (1974). Quasi-likelihood functions, generalized linear models, and the Gauss–Newton method. *Biometrika* **61**, 439–447.

Welch, B. L. (1938). The significance of the difference between two means when the population variances are unequal. *Biometrika* **29**, 350–362.

Westfall, P. H., Tobias, R. D., Rom, D., Wolfinger, R. D., and Hochberg, Y. (1999). *Multiple Comparisons and Multiple Tests Using the SAS® System*. Cary, NC: SAS Institute Inc.

Whittemore, A. S., and Keller, J. B. (1978). Quantitative theories of carcinogenesis. *SIAM Review* **20**, 1–30.

Wilhelm, M. S., Carter, E. M., and Hubert, J. J. (1998). Multivariate iteratively re-weighted least squares, with applications to dose-response data. *Environmetrics* **9**, 303–315.

Wilkinson, B. (1951). A statistical consideration in psychological research. *Psychological Bulletin* **48**, 156–158.

Wilks, S. S. (1938). The large-sample distribution of the likelihood ratio for testing composite hypotheses. *Annals of Mathematical Statistics* **9**, 60–62.

Willers, J. L., Seal, M. R., and Luttrell, R. G. (1999). Remote sensing, line-intercept sampling for tarnished plant bugs (Heteroptera: Miridae) in mid-south cotton. *Journal of Cotton Science* **3**, 160–170.

Williams, P. L. (1998). Trend test for counts and proportions. In P. Armitage and T. Colton (eds), *Encyclopedia of Biostatistics*, Vol. 6, pp. 4573–4584. Chichester: John Wiley & Sons.

Winsor, C. P. (1932). The Gompertz curve as a growth curve. *Proceedings of the National Academy of Sciences, USA* **18**, 1–8.

Wold, H. (1938). *A Study in the Analysis of Stationary Time Series*. Uppsala: Almqvist and Wiksells.

Wolf, J. R. (1856). Mittheilungen über die Sonnenflecken. *Vierteljahrsschrift der Naturforschenden Gesellschaft in Zürich* **1**, 151–161.

Wolfinger, R. D. (1993). Covariance structure selection in general mixed models. *Communications in Statistics – Simulation and Computation* **22**, 1079–1106.

Wolpert, R. L. (2002). Gamma function. In A. H. El-Shaarawi and W. W. Piegorsch (eds), *Encyclopedia of Environmetrics*, Vol. 2, pp. 837–839. Chichester: John Wiley & Sons.

Woolf, B. (1955). On estimating the relation between blood group and disease. *Annals of Human Genetics* **19**, 251–253.

Yates, F. (1948). The analysis of contingency tables based on quantitative characters. *Biometrika* **35**, 176–181.

Young, L. J., Campbell, N. L., and Capuano, G. (1999). Analysis of overdispersed count data from single-factor experiments: A comparative study. *Journal of Agricultural, Biological, and Environmental Statistics* **4**, 258–275.

Young, P. C., and Pedregal, D. J. (2002). Forecasting, environmental. In A. H. El-Shaarawi and W. W. Piegorsch (eds), *Encyclopedia of Environmetrics*, Vol. 2, pp. 786–796. Chichester: John Wiley & Sons.

Yule, G. U. (1927). On a method of investigating periodicities in disturbed series, with special reference to Wolfer's sunspot numbers. *Philosophical Transactions of the Royal Society of London. Series A, Containing Papers of a Mathematical or Physical Character* **226**, 267–298.

Zeger, S. L., and Liang, K.-Y. (1992). An overview of methods for the analysis of longitudinal data. *Statistics in Medicine* **11**, 1825–1839.

Zeise, L., Wilson, R., and Crouch, E. A. C. (1987). Dose-response relationships for carcinogens: A review. *Environmental Health Perspectives* **73**, 259–308.

Zhang, P. (1992). Inference after variable selection in linear regression models. *Biometrika* **79**, 741–746.

Zhu, Y., and Fung, K. (1996). Statistical methods in developmental toxicity risk assessment. In A. M. Fan and L. W. Chang (eds), *Toxicology and Risk Assessment. Principles, Methods, and Applications*, pp. 413–446. New York: Marcel Dekker.

Zimmerman, D. L. (1989). Computationally efficient restricted maximum likelihood estimation of generalized covariance functions. *Mathematical Geology* **21**, 655–672.

Zimmerman, D. L. (1993). Another look at anisotropy in geostatistics. *Mathematical Geology* **25**, 453–470.

Author index

Analyzing Environmental Data   W. W. Piegorsch and A. J. Bailer
© 2005 John Wiley & Sons, Ltd   ISBN: 0-470-84836-7 (HB)

Subject index

Analyzing Environmental Data   W. W. Piegorsch and A. J. Bailer
© 2005 John Wiley & Sons, Ltd   ISBN: 0-470-84836-7 (HB)